Bionanotechnology

Bionanotechnology

Editor: Chloe Clark

CALLISTO REFERENCE

www.callistoreference.com

Callisto Reference,
118-35 Queens Blvd., Suite 400,
Forest Hills, NY 11375, USA

Visit us on the World Wide Web at:
www.callistoreference.com

ISBN: 978-1-63239-861-1 (Hardback)

The publisher's policy is to use permanent paper from mills that operate a sustainable forestry policy. Furthermore, the publisher ensures that the text paper and cover boards used have met acceptable environmental accreditation standards.

Trademark Notice: Registered trademark of products or corporate names are used only for explanation and identification without intent to infringe.

Printed in the United States of America.

Cataloging-in-publication Data

Bionanotechnology / edited by Chloe Clark.
   p. cm.
Includes bibliographical references and index.
ISBN 978-1-63239-861-1
1. Nanobiotechnology. 2. Nanotechnology. 3. Biotechnology. I. Clark, Chloe.
TP248.25.N35 B56 2017
660.6--dc23

Table of Contents

Permissions

List of Contributors

Index

Preface

Since its emergence, bionanotechnology has been rapidly expanding. It is an interdisciplinary field, which unifies the principles of biology and nanotechnology. It primarily focuses on developing and applying Nano-tools to solve diverse biological problems. This book is a valuable compilation of topics, ranging from the basic to the most complex advancements in the field of bionanotehnology. There has been rapid progress in this field and its applications are finding their way across multiple industries. From theories to research to practical applications, case studies related to all contemporary topics of relevance to this field have been included in this book. Scientists and students actively engaged in this field will find this book full of crucial and unexplored concepts.

The world is advancing at a fast pace like never before. Therefore, the need is to keep up with the latest developments. This book was an idea that came to fruition when the specialists in the area realized the need to coordinate together and document essential themes in the subject. That's when I was requested to be the editor. Editing this book has been an honour as it brings together diverse authors researching on different streams of the field. The book collates essential materials contributed by veterans in the area, which can be utilized by students and researchers alike.

Each chapter is a sole-standing publication that reflects each author's interpretation. Thus, the book displays a multi-facetted picture of our current understanding of applications and diverse aspects of the field. I would like to thank the contributors of this book and my family for their endless support.

Editor

Effects of Internalized Gold Nanoparticles with Respect to Cytotoxicity and Invasion Activity in Lung Cancer Cells

Zhengxia Liu[ꝺ], Yucheng Wu[ꝺ], Zhirui Guo, Ying Liu, Yujie Shen, Ping Zhou, Xiang Lu*

Department of Geriatrics, the Second Affiliated Hospital, Nanjing Medical University, Jiangsu, China

Abstract

The effect of gold nanoparticles on lung cancer cells is not yet clear. In this study, we investigated the cytotoxicity and cell invasion activity of lung cancer cells after treatment with gold nanoparticles and showed that small gold nanoparticles can be endocytosed by lung cancer cells and that they facilitate cell invasion. The growth of A549 cells was inhibited after treatment with 5-nm gold nanoparticles, but cell invasion increased. Endocytosed gold nanoparticles (size, 10 nm) notably promoted the invasion activity of 95D cells. All these effects of gold nanoparticles were not seen after treatment with larger particles (20 and 40 nm). The enhanced invasion activity may be associated with the increased expression of matrix metalloproteinase 9 and intercellular adhesion molecule-1. In this study, we obtained evidence for the effect of gold nanoparticles on lung cancer cell invasion activity in vitro. Moreover, matrix metalloproteinase 9 and intercellular adhesion molecule-1, key modulators of cell invasion, were found to be regulated by gold nanoparticles. These data also demonstrate that the responses of the A549 and 95D cells to gold nanoparticles have a remarkable relationship with their unique size-dependent physiochemical properties. Therefore, this study provides a new perspective for cell biology research in nanomedicine.

Editor: Hidayatullah G. Munshi, Northwestern University, United States of America

Funding: This work is supported by grants from the National Science Foundation of China(No.81270428 and 81300999). The funders had no role in study design, data collection and analysis, decision to publish, or preparation of the manuscript.

Competing Interests: The authors have declared that no competing interests exist.

* E-mail: luxiang66@njmu.edu.cn

ꝺ These authors contributed equally to this work.

Introduction

Previous studies identified that gold nanoparticles (Au-NPs) show little cytotoxicity despite their efficient uptake into human cells by endocytosis [1,2], making them suitable candidates for nanomedicine. Besides their biocompatibility, the fact that they are easy to synthesize, characterize, and surface modify contributed to attract much attention in various biomedical applications. Au-NPs have been investigated as drug delivery vehicles and photothermal therapy and molecular imaging tools for potential biodiagnosis [3,4]. Nanoparticle-based therapeutic strategies for cancer treatment are mainly based on the delivery of chemotherapeutic agents to induce apoptosis [5]. The primary reasons for using nanoparticles as carriers for therapeutic delivery are to enable multimodal functionalities, such as imaging or specific targeting, to increase tissue permeability and site-specific drug accumulation, and to reduce side effects to healthy tissues [6]. Currently, Au-NPs are used in different biomedical applications: not only can they be used as scaffolds for increasingly potent cancer drug delivery but they can also serve as transfection agents for selective gene therapy and as intrinsic antineoplastic agents[7–9]. Dreaden et al. [8] have shown that targeted Au-NPs are capable of altering the cell cycle, including cell division, signaling, and proliferation.

Despite the widespread application of Au-NPs, a clear understanding of how biological systems respond to the nanoparticles is vital, and characterization of the unique size-dependent physicochemical properties of the Au-NPs is a critical component. A previous study proved that the surface size of Au-NPs plays a large role in their therapeutic effect [10]. Au-NPs of very small diameter (<2 nm) can penetrate cells and cellular compartments such as the nucleus and be extremely toxic [11]. For example, it was found that spherical Au-NPs with a diameter of 1.4 nm induce necrosis and mitochondrial damage in various cell lines via oxidative stress mechanisms, which may be associated with their well-known catalytic activity at that size [12]. A recent study by Connor et al. [1] reported that significant amounts of larger Au-NPs (e.g., 18 nm in diameter) penetrate into cells, but that these Au-NPs are not inherently toxic to human cells. Chithrani et al. [13] studied the relationship between Au-NPs and HeLa cells and suggested that Au-NPs entered the cells via receptor-mediated endocytosis at a threshold size of approximately 50 nm. Since there are no safety regulations yet, the effect of Au-NPs on cells still requires further study.

Invasion and metastasis are important pathologic features of cancer cells. Invasive capacity is the single most important trait that distinguishes benign from malignant lesions [14]. Indeed, invasive tumor cells can escape surgical resection and be responsible for tumor recurrence. Despite advances in surgery, chemotherapy, and radiotherapy, relapse is almost inevitable in the presence of an aggressive metastatic spread [15,16]. The process of invasion and metastasis includes cell proliferation, dissociation from the primary lesions, degradation, and permeation into the extracellular matrix (ECM), migration in the blood or lymph stream, adhesion and growth in a secondary organ [17]. Previous reports have described that intercellular adhesion molecule-1 (ICAM-1) and matrix metalloproteinase 9 (MMP-9) are involved in cancer cell adhesion, invasion, and migration, which contribute to cancer metastasis [18]. These factors have been considered as prognostic biomarkers for lung cancer progression [19]. Further studies on the effects of Au-NPs on the expression of these proteins are needed.

Since cytotoxicity may not be the only effect that nanoparticles can induce, in this study, we focused on cell proliferation, invasion activity, and protein expression, all of which may be affected by the presence of Au-NPs. To investigate the importance of particle size in nanomedicine, we chose four different Au-NPs sizes (i.e., 5 nm, 10 nm, 20 nm, 40 nm) for an in-depth analysis of this system.

Finally, we chose to study these effects on two human lung cancer cell lines (i.e., A549 and 95D) because lung cancer is a major malignant tumor in China and has increasing incidence in most cities. In 2010, there were approximately 600,000 new cases of lung cancer in China [20]. The rates of morbidity continue to rise rapidly because of the serious air pollution [21]; however, knowledge of the biological interactions and responses of Au-NPs with human lung cancer cells is very limited. Moreover, to gain a fundamental understanding of the processes involved, we felt that it was important to initially focus on the effects of nanoparticles on individual cancer cells, rather than on entire organs, where other factors can complicate the interpretation of the results. Therefore, the goal of this research was to provide an important basis for applying Au-NPs in lung cancer therapy.

Methods

Preparation and Characterization of Citrate-Capped Au-NPs

Au-NPs (5 nm and 10 nm) were synthesized using a tannic acid/citrate solution, in which tannic acid plays the role of a reducing agent and citrate acts as a stabilizing agent [22]. For each synthesis, two initial solutions were required: (a) 1 mL of 1% (w/v) $HAuCl_4$ solution, which was added to 79 mL of water; and (b) a mixture consisting of 4 mL of 1% (w/v) citrate solution, 0.7 mL or 0.1 mL of 1% tannic acid, and water (to achieve a final volume of 20 mL). Both (a) and (b) solutions were heated to 60°C in water bath and then solution (b) was added to (a) under constant stirring. The resulting Au-NPs were cooled to room temperature (RT) for subsequent experiments.

Syntheses of 20-nm and 40-nm Au-NPs stabilized by citrate ions were conducted using the classic citrate reduction method [23]. For each synthesis, 100 mL of 0.01% $HAuCl_4$ solution was heated to boiling. Selected volumes (4.5 mL or 1.0 mL) of 1% citrate solution were added and boiling was continued until the color of the solution turned into ruby red. The citrate-capped Au-NPs solution was cooled naturally to RT.

The morphology of the Au-NPs was observed by using transmission electron microscopy (TEM; JEM-2100EX, JEOL, Tokyo, Japan) at accelerating voltage of 200 kV. Ultraviolet-visible (UV-Vis) spectra were acquired with a Shimadzu UV-3600 spectrophotometer in the range of 300–1100 nm.

Cell Culture

A549 and 95D cells (Shanghai Institutes for Biological Sciences, Chinese Academy of Sciences) were cultured in Dulbecco's Modified Eagle's Medium (DMEM) supplemented with 10% (v/v) fetal bovine serum (FBS), 100 units/mL penicillin, and 100 µg/mL streptomycin (all from GIBCO, Invitrogen) at 37°C in a humidified atmosphere containing 5% CO_2. Cells were pre-cultured in the medium overnight and then treated either with or without 50 µg/mL Au-NPs solution for an additional 48 h. Cells were harvested by trypsin-ethylenediamintetraacetic acid (EDTA; GIBCO, Invitrogen) detachment at the logarithmic growth phase and centrifuged. The cells were then resuspended in DMEM with 10% FBS for subculture or other uses.

Uptake and TEM Studies

The uptake of Au-NPs was also examined using TEM. Before exposure to the Au-NPs, the A549 cells and 95D cells were plated at a concentration of 1×10^6 cells per dish on a 100-mm culture dish (corning, USA) containing the growth medium and incubated at 37°C with 5% CO_2 for 24 h. The Au-NPs were then added. After exposure to Au-NPs for 48 h, the cells were fixed in 3.7% (v/v) paraformaldehyde in phosphate-buffered saline (PBS) for 20 min at RT. Subsequently, cells were prepared for TEM analysis as follows: cells were fixed in 1% (w/v) osmium tetroxide for 2 h, dehydrated in a graded series of 30%, 50%, 70%, 80%, and 90% ethanol, and treated three times with 100% ethanol for 15 min each. The samples were then embedded in a mixture of resin in propylene oxide polymerized at 80°C. Ultrathin sections for TEM were prepared using a diamond knife and the samples were analyzed using a transmission electron microscope.

Cell Viability Assay

A549 and 95D cells were seeded in 96-well plates at a low confluence (2000 cells/well), allowed to attach overnight, and treated with Au-NPs (control, 5 nm, 10 nm, 20 nm, and 40 nm). After 24 h, 48 h, and 72 h, a cell viability assay reagent (Cell counting kit-8, Kaiji, Nanjing) was added to each well and incubated for 1, 2, 3, and 4 h, respectively. Absorbance values at 450 nm were recorded using a microculture plate reader (BioRad) and the cell viability expressed as a percentage of the untreated control (100% cell viability). The effect of the Au-NPs with different sizes on the viability of the cells was measured in triplicate, and the experiments were repeated at least thrice.

Apoptosis Detection by Annexin V/Propidium Iodide (PI) Staining

Cells in the log phase were seeded onto a 6-well culture plate at a density of 1×10^5 cells per well, incubated at 37°C in a CO_2 incubator, and allowed to attach overnight. After treatment with Au-NPs (control, 5 nm, 10 nm, 20 nm, and 40 nm) for 48 h, apoptosis and necrosis were analyzed with the Annexin V-PI (BD Biosciences) apoptosis detection kit following the manufacturer's instructions. The samples were analyzed using a BD FACS CantoII instrument (BD Biosciences).

Flow Cytometry Analysis of the Cell Cycle

The cells were harvested using 0.25% trypsin with 1 mM EDTA solution and fixed for 12 h in 70% ethanol at 4°C. The fixed cells were then centrifuged at 3,000 rpm for 15 min to remove the ethanol thoroughly. The cells were then washed twice with 3 mL of PBS, resuspended in 1 mL of PI staining solution, and incubated for 15 min at RT. The staining solution consisted of 20 µg/mL PI and 0.2 mg/mL RNase A in PBS. The samples were subsequently analyzed using a BD FACS CantoII instrument (BD Biosciences). Twenty thousand events were collected from each sample. The percentages of cells in the G0/G1, S, and G2/M phases of the cell cycle were determined using the ModFit software (BD Biosciences).

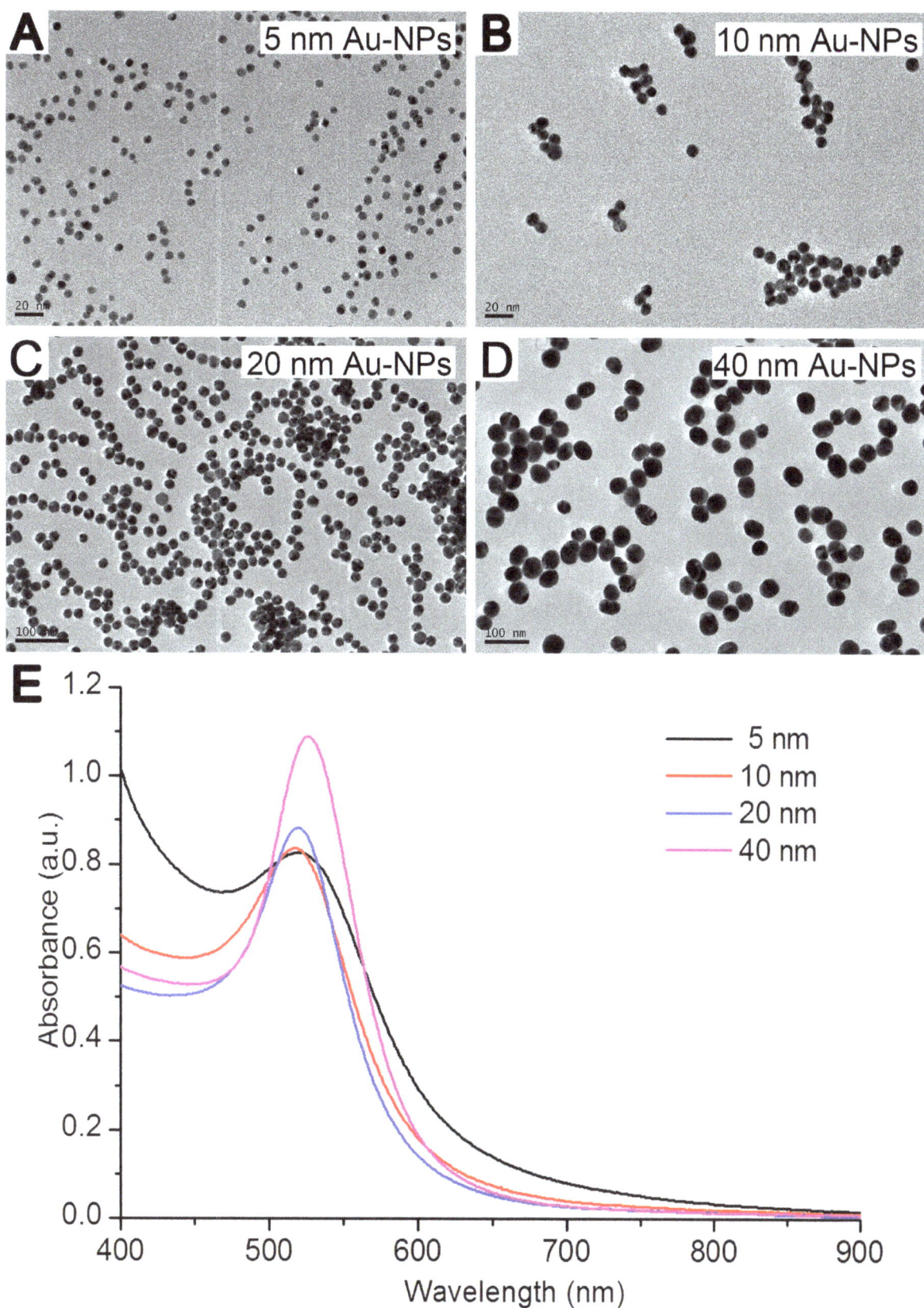

Figure 1. Characterization of the differently sized Au-NPs. Transmission electron microscopy (TEM) images of Au-NPs with diameters of (A) 5 nm, (B) 10 nm, (C) 20 nm, or (D) 40 nm. The insets show high-resolution images. Scale bars, 20 nm and 100 nm (as marked). E: UV-vis spectra of the Au-NPs with 5-nm, 10-nm, 20-nm, and 40-nm diameters.

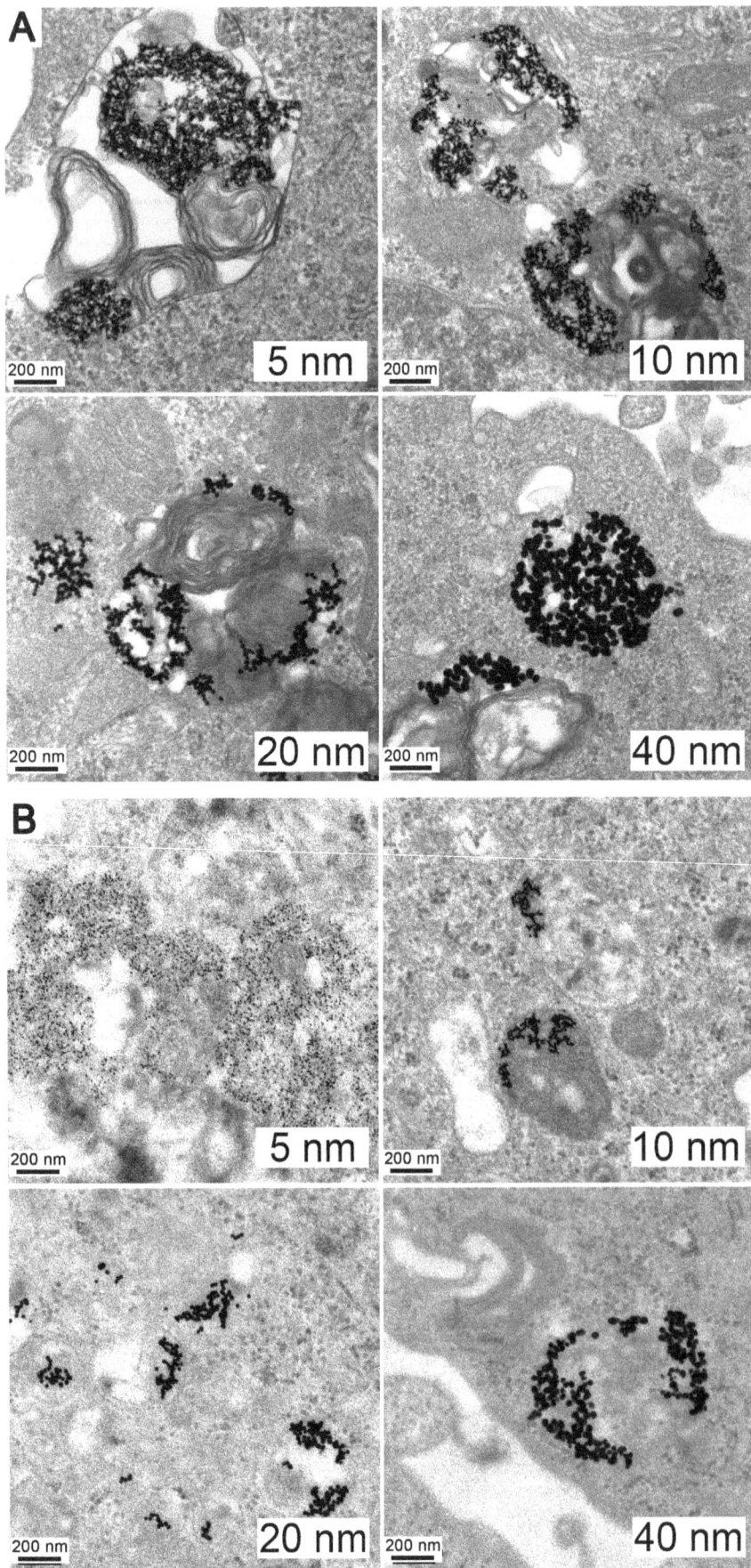

A

200 nm | **5 nm**

200 nm | **10 nm**

200 nm | **20 nm**

200 nm | **40 nm**

B

200 nm | **5 nm**

200 nm | **10 nm**

200 nm | **20 nm**

200 nm | **40 nm**

Figure 2. Transmission electron microscopy (TEM) images of Au-NPs trapped inside the cells. TEM images at 200-nm magnification showing the internalization of 25 μg/mL Au-NPs with various sizes (5 nm, 10 nm, 20 nm, and 40 nm) into (A) A549 and (B) 95D cells after 48 h of treatment.

Invasion Assay

For the invasion assay, hanging cell culture inserts (8.0-μm pore size) were pre-coated with 50 μg/mL Matrigel (BD Biosciences) on the upper surface. Cells were pre-cultured in the medium overnight and then treated either with or without 50 μg/mL Au-NPs solution for additional 48 h, then cells were harvested and 2.5×10^5 cells were plated in 200 μL DMEM supplemented with 0.2% bovine serum albumin (BSA) in the top of the chamber. The bottom of the well was added with 750 μL of DMEM containing 5% FBS. The invasion assay was carried out for 48 h in the cell culture incubator.

The cells were fixed by replacing the culture medium in the bottom and top of the chamber with 4% formaldehyde dissolved in PBS. After fixing for 15 min at RT, the chambers were rinsed in PBS and stained with 0.2% crystal violet for 10 min. After washing the chambers five times by dipping them in a large beaker filled with dH$_2$O, the cells (now blue in color) at the top of the Matrigel membrane were removed by using several Q-tips. Cells were removed until no more blue dye could be removed with the Q-tips. Then, the cells that remained were those that had invaded and had reached the bottom of the membrane.

We also quantified the invasion cells using the QCM^{TM} 24-well Cell Invasion Fluorometric Assay (Millipore) with ECMatrix-coated inserts, according to the manufacturer's instructions. This assay provides an efficient system for quantitative evaluation of the invasion of tumor cells through a basement membrane model. A549 and 95D cells were grown for 2 d in complete medium, either in the absence (control) or in the presence of Au-NPs (5 nm, 10 nm, 20 nm, 40 nm). At the end of the treatment, cells were suspended in serum-free medium and seeded (2.5×10^5 cells/ 250 μL) in each insert of a multiwall plate chamber. Serum (10%) was added to the serum-free medium in the lower chamber as chemoattractant. After a 48-h incubation period at 37°C, non-invading cells were removed from the top of the inserts. Next, the inserts were placed into a clean well containing a pre-warmed cell detachment solution and incubated at 37°C for 30 min. The inserts were removed from the wells, and lysis buffer/dye solution was added to the medium containing the detached cells for 15 min at RT. The mixtures were then read with a fluorescence plate reader (Bio-Tek, Synergy HT) using a 480/520-nm filter set. Fluorescence measurements were reported as relative fluorescence unit (RFU) values.

Quantitative Reverse Transcription Polymerase Chain Reaction (qRT-PCR) Assays

Total RNA was isolated from untreated and Au-NPs–treated A549 and 95D cells using Trizol Reagent (Invitrogen Life Technology). Reverse transcription (RT) reaction was performed using 1 μg of total RNA, which was reverse transcribed into cDNA using oligo dT primer, and then qRT-PCR was carried out using the SYBR Green Mix (Applied Bio-systems). The primer sequences used for PCR were as follows: ICAM-1, forward ACACTAGGCCACGCATCTGAT, reverse AGCATACCCAA-TAGGCAGCAA; MMP-9, forward GGCTACGTGACCTAT-GACATCCT, reverse TCCTCCCTTTCCTCCAGAACA; glyceraldehyde 3-phosphate dehydrogenase (GAPDH), forward GGAGCCAAACGGGTCATCATCTC, reverse GAGGGGC-CATCCACAGTCTTCT. PCR was carried out at 95°C for 30 s, at 60°C for 30 s, and 1 min at 70°C for 35 cycles. The comparative Ct method was used to calculate the relative abundance of the mRNA and the results for target gene expression were compared with those for GAPDH expression. The results were obtained from three independent experiments.

Protein Quantitative Assays

MMP9 protein expression in the supernatant and cell lysate was measured using an MMP Panel 2 magnetic bead kit based on the Luminex technology. Briefly, MMP9 capture antibodies (Millipore) were conjugated to Luminex beads (beads region 33). MMP9 detection antibodies (Millipore) were conjugated to biotin through custom service provided by Millipore. Cells were lysed in MILLIPLEX MAP lysis buffer (Millipore) and diluted with equal volume of MILLIPLEX MAP cell assay buffer (Millipore). MMP9 capture antibody beads were diluted in 25 μL of MILLIPLEX MAP cell assay buffer and added to a magnetic plate (Millipore). Then, 25 μL of the diluted cell lysate or cell culture supernatant was transferred to each well of the solid plate and incubated for 2 h at RT with shaking. After the incubation, beads were washed twice with wash buffer, and 25 μL of detection antibodies was added into each well and incubated for 1 h at RT with shaking. After that, 25 μL of MILLIPLEX MAP streptavidin-phycoery-thrin (Millipore) was added and incubated for 30 min at RT with shaking. Finally, sheath fluid was added after washing, and the signal was read using a Luminex FLEXMAP 3D^{TM}.

Western Blot Analysis

For western blot analysis, A549 and 95D cells were plated on culture flasks and treated with Au-NPs (control, 5 nm, 10 nm, 20 nm, and 40 nm). After 48 h, $5–10 \times 10^6$ cells were harvested and lysed with ice-cold lysis buffer. Equal amounts of protein were separated by sodium dodecyl sulfate-polyacrylamide gel electrophoresis (SDS-PAGE) and electrophoretically transferred onto polyvinylidene fluoride membranes. Membranes were blocked with 5% non-fat dry milk for 2 h and incubated overnight at 4°C with rabbit anti-MMP-9 antibody (1:1000, Abcam), rabbit anti-ICAM-1 antibody (1:500, Cell Signaling), or mouse anti-GAPDH antibody (1:10000, Abmart). GAPDH was used as the housekeeping gene control and the expression levels of the MMP9 and ICAM-1 were normalized with respect to GAPDH. The proteins were detected with horseradish peroxidase-conjugated anti-rabbit or anti-mouse secondary antibodies and visualized with chemiluminescence reagents provided with the ECL kit (BioRad, USA). Immunoreactive bands were detected by enhanced chemiluminescence and quantified using a ChemiDoc XRS molecular imager (BioRad).

Statistical Analysis

GraphPad Prism 5 statistical analyses software was used in all statistical analyses performed in this study (GraphPad Prism version 5.00 for Windows, GraphPad Software, San Diego California, USA). All results have been presented as mean ± standard deviation (SD). Statistical comparisons were conducted using one-way analysis of variance (ANOVA), followed by the Dunnett's t-test for comparison with the control group. Differences were considered significant at $P < 0.05$.

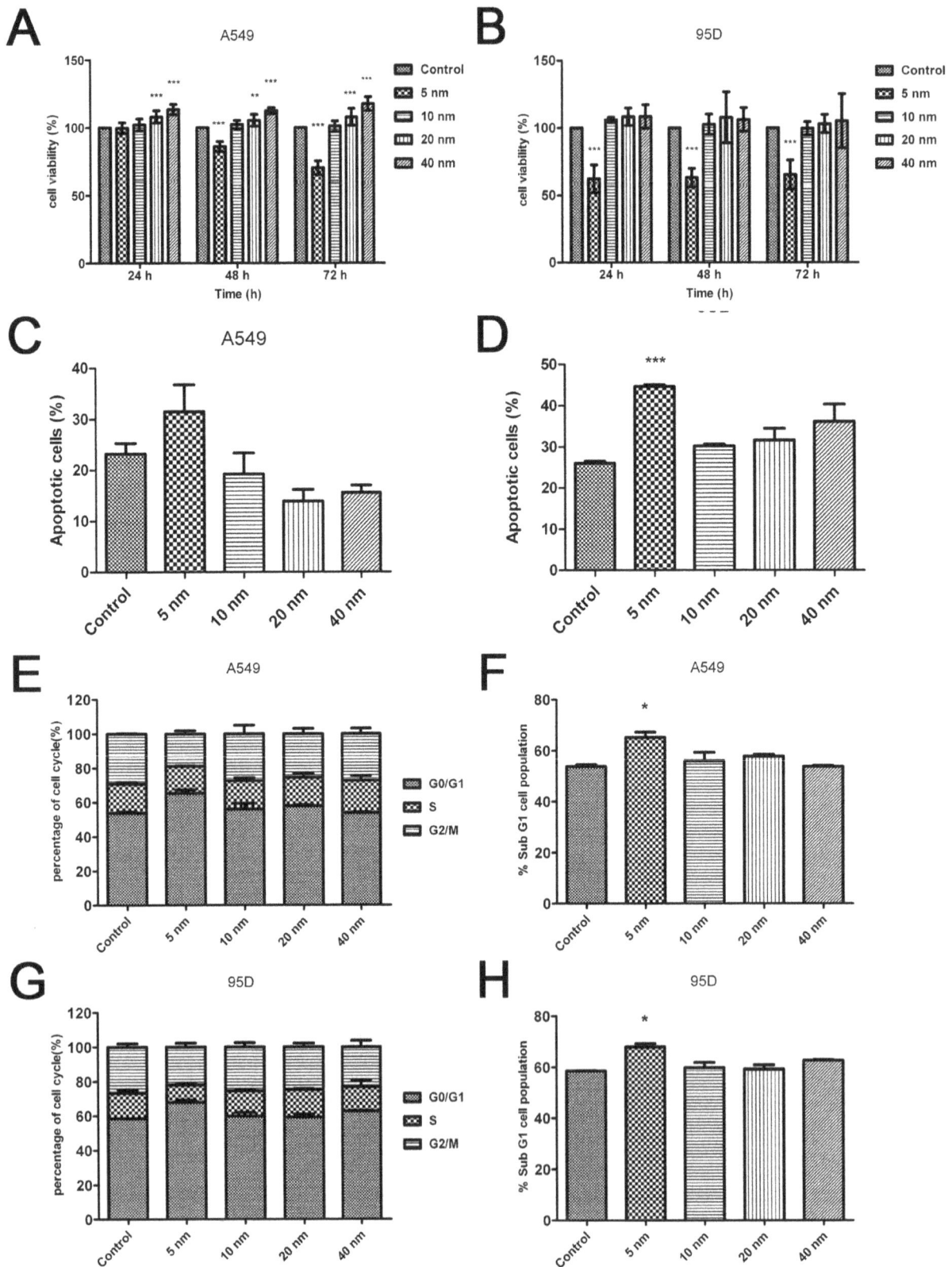

Figure 3. Comparison of the cytotoxicity of the A549 and 95D cell lines exposed to the Au-NPs. The cell lines A549 (A) and 95D (B) in the logarithmic growth phase were exposed to Au-NPs for 24 h, 48 h, and 72 h. Cell viability was calculated as the percentage of the viable cells compared to the untreated controls. Each result represents the mean viability ± standard deviation (SD) of three independent experiments. Apoptosis (C–D) and cell cycle (E–H) analyses of A549 and 95D cells treated with differently sized Au-NPs. Error bars indicate SD values from four independent experiments. *P<0.05, **P<0.01, ***P<0.001.

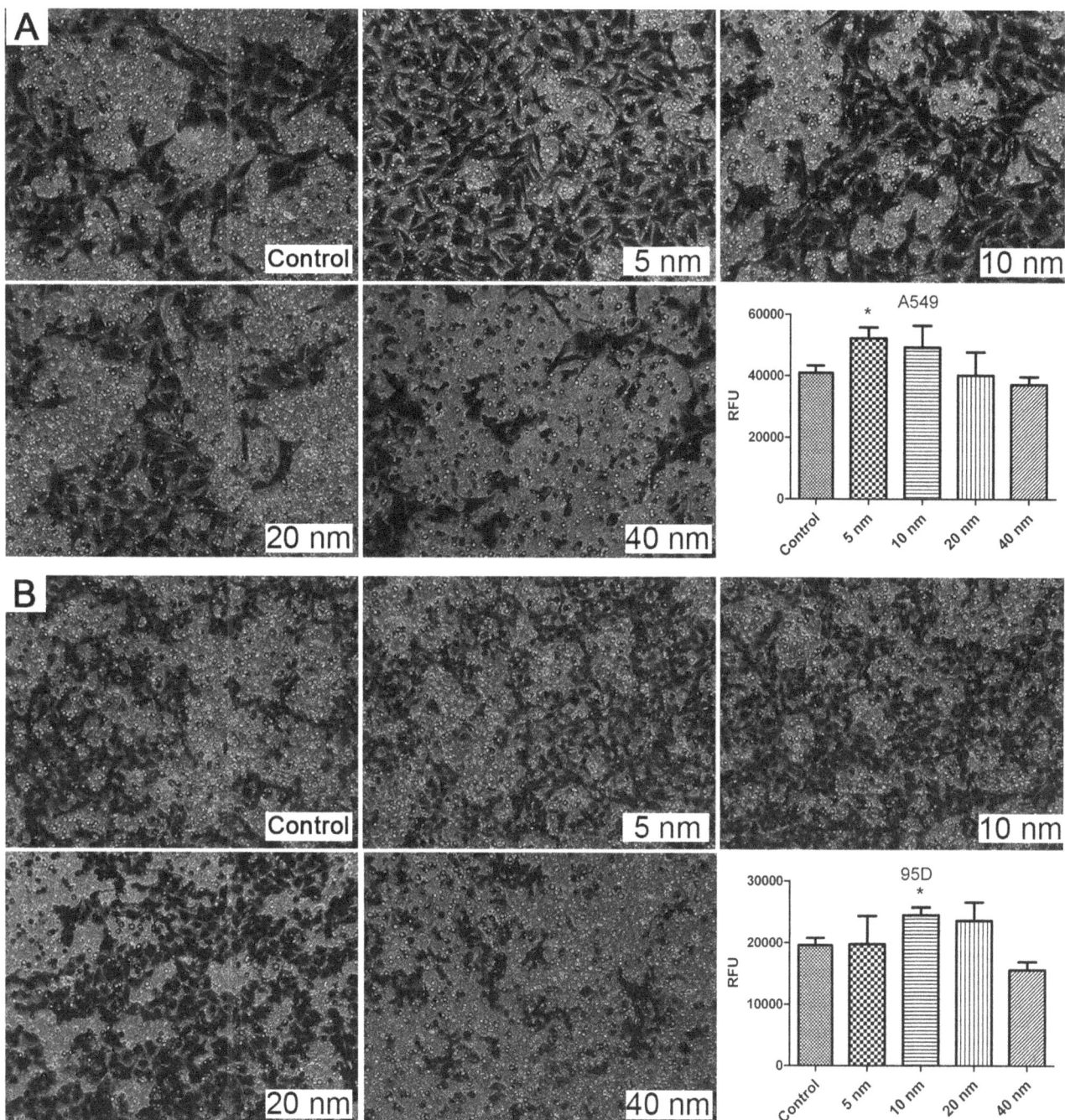

Figure 4. Optical images and fluorescence measurements of cell invasion on Au-NP treatment. The images represent A549 cells (A) and 95D cells (B) that have crossed the pores of the Matrigel invasion chamber and corresponding fluorescence intensity results. *P<0.05. Error bars indicate the SD values from three independent experiments.

Results

Synthesis and Characterization of Au-NPs

Au-NPs synthesized with the citrate reduction method without any further modification were used for all the studies and are here referred to as unmodified nanoparticles. To determine the relationship between the size and the biological function of the nanoparticles, we used Au-NPs of four different sizes (5, 10, 20, or 40 nm) and characterized them by TEM and UV-Vis spectra (Figure 1). The particles were shown to be all spheres with narrow

size distributions. The high electron densities of Au-NPs as well as the homogeneity of their shape and size made them highly evident under the transmission electron microscope.

Internalization of Au-NPs

To investigate whether Au-NPs with various sizes crossed the cell membrane and where they located, A549 and 95D cells were incubated for 48 h in the presence of Au-NPs (5, 10, 20, and 40 nm) in complete cell culture medium. Figure 2 shows the internalization of the differently sized Au-NPs. Most of the

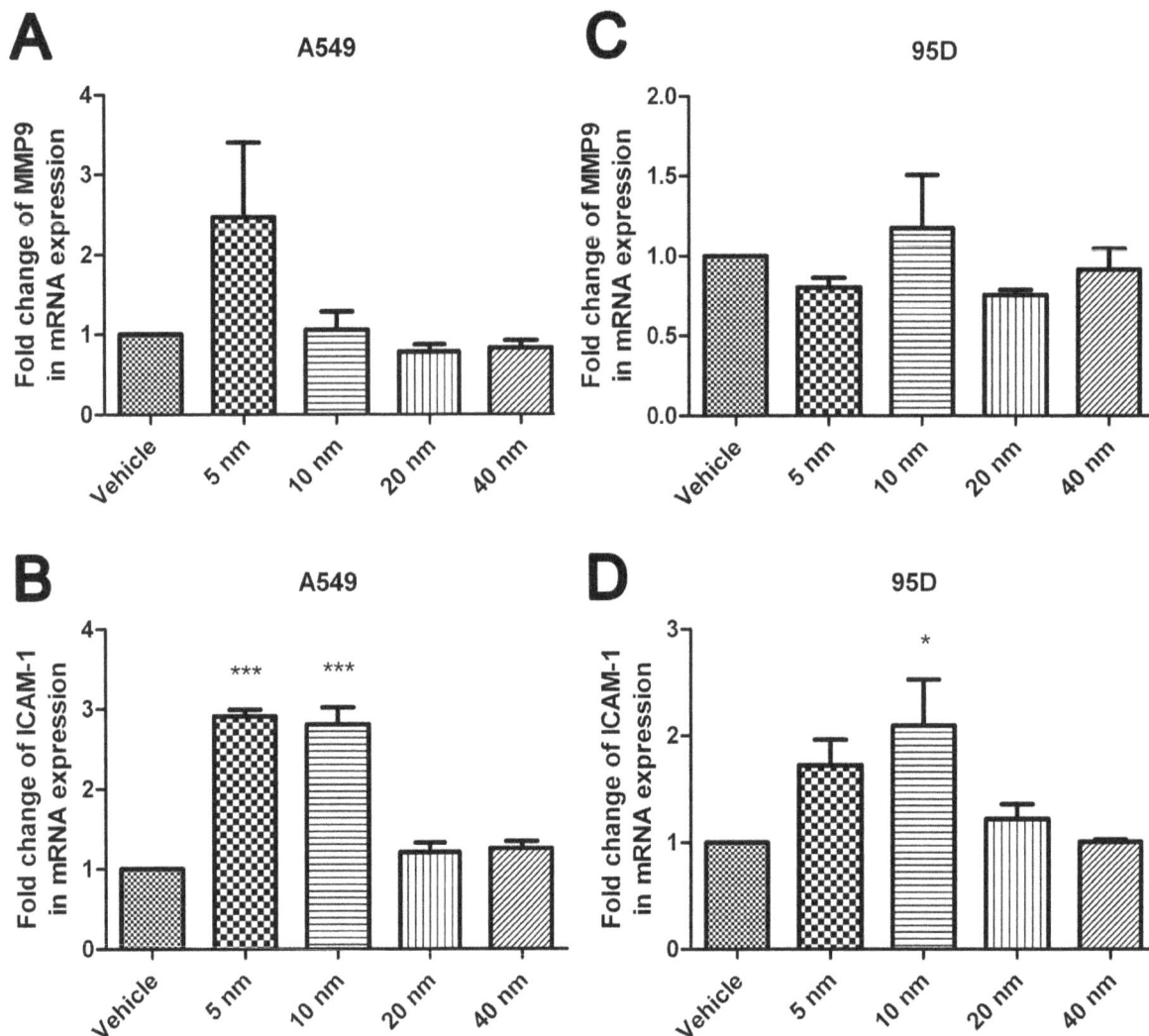

Figure 5. Fold changes in mRNA expression tested by qRT-PCR in A549 and 95D cells treated with differently sized nanoparticles. Results of qRT-PCR for (A–B) A549 and (C–D) 95D cells after 48-h incubation with 5-nm, 10-nm, 20-nm, or 40-nm Au-NPs. *P<0.05, ***P<0.001 vs. control. Error bars indicate the SD values from three independent experiments.

particles were found in membrane-bound vesicles or in the perinuclear region within the cells.

Measurement of the Cytotoxicity of Au-NPs

Next, we sought to determine the effect of these Au-NPs on the proliferation of two different lung cancer cell lines, A549 and 95D cells. The cytotoxic effect on the four Au-NPs with different diameters was tested at the same concentration. Our findings demonstrated that 5-nm Au-NPs play a pivotal role in inhibiting the proliferation of both A549 and 95D cells at 48 h and 72 h (Figures 3A and B). Au-NPs with diameters of 20 nm and 40 nm exhibited notable efficacy from 24 h onwards in promoting the proliferation of A549 cells, whereas no promotion was observed in 95D cells (Figures 3A and B). Au-NPs with a diameter of 10 nm had no effect on the proliferation of both A549 and 95D cells. Inhibiting the cell proliferation, increasing cell apoptosis, and arresting cells in the G0/G1 cycle manifested the cytotoxicity of 5-nm Au-NPs. There was no significant difference (P>0.05) in apoptosis and cell cycle distribution for the 10-nm, 20-nm, and 40-nm AuNP-treated groups (Figures 3C–H). These results indicated that small Au-NPs may be cytotoxic and that the diameter is not

the only factor influencing cell proliferation and the interaction between Au-NPs, which is a very complex process that warrants further investigation.

Invasion Assay

A basement membrane model was used for evaluating the invasion activity of cells. The results are presented in Figure 4. Our results indicated that cell invasion was promoted significantly after treatment of A549 with 5-nm Au-NPs and of 95D cells with 10-nm Au-NPs (P<0.05). In contrast, the cells internalized with 20-nm or 40-nm Au-NPs were not significantly affected by invasion activity (Figure 4). These results clearly suggested that the invasion effects were particle size- and cell type-dependent.

Effects of Au-NPs on the Expression of MMP9 and ICAM-1 in Lung Cancer Cells

To gain a deeper understanding of this phenomenon, qRT-PCR was performed to measure the mRNA levels of MMP9 and ICAM-1 after treatment with Au-NPs. The results showed that 5-nm and 10-nm Au-NPs notably facilitate the mRNA expression of

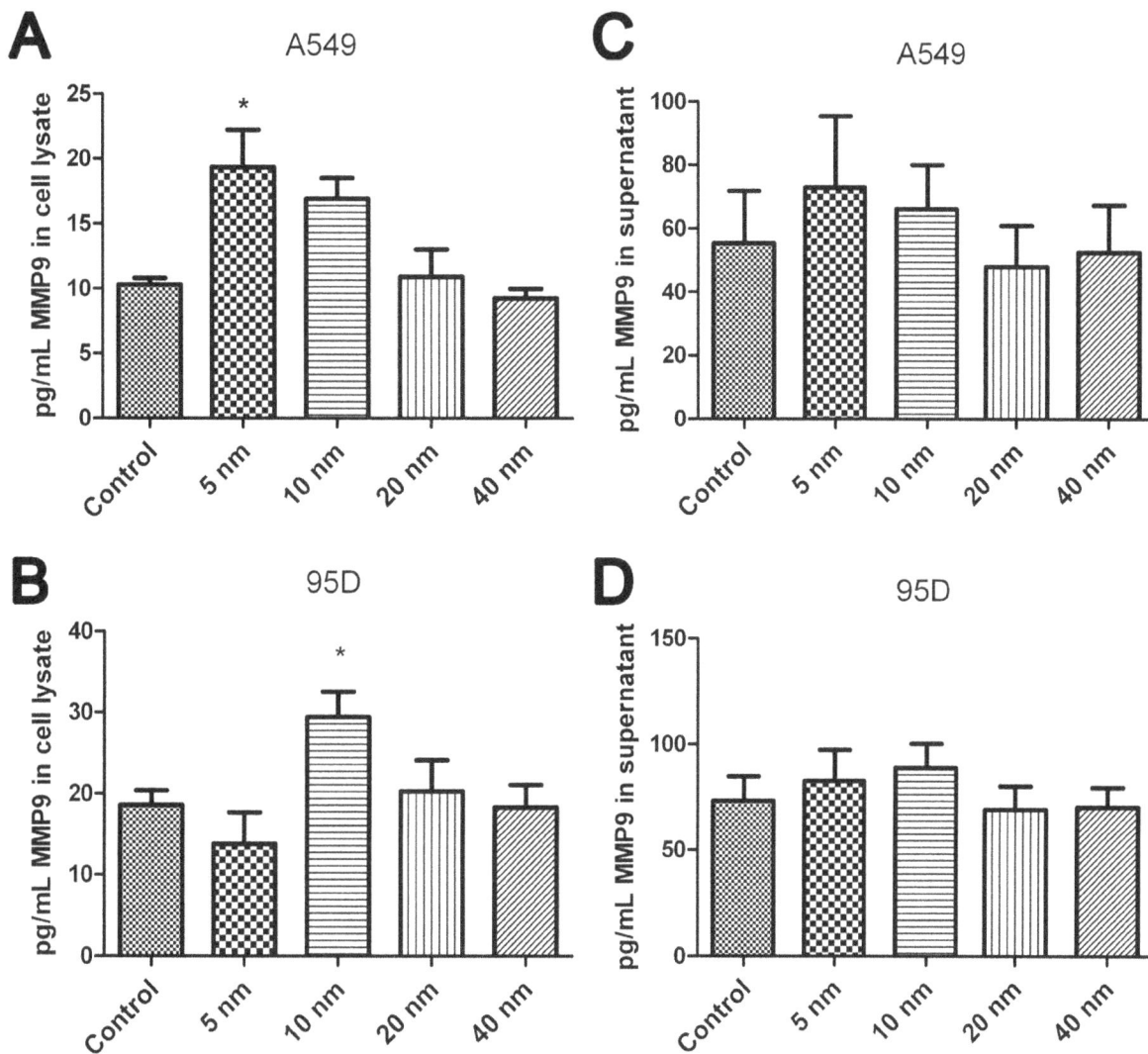

Figure 6. Expression levels of MMP9 in A549 and 95D cells after Au-NP treatment for 48 h. Quantification of the expression of MMP9 in the cell lysate (A–B) and the supernatant (C–D) of A549 and 95D cells, respectively, was conducted by using the Luminex technology. *$P<0.05$. Error bars indicate the SD values from three independent experiments.

ICAM-1 in both cell lines (Figures 5B and D, $P<0.05$). The mRNA expression of MMP9 increased in 5-nm Au-NPs–treated A549 cells and 10-nm Au-NPs–treated 95D cells but the differences were not statistically significant (Figure 5A and C). We then performed a Luminex-based experiment in the presence of MMP9 magnetic beads to quantify the protein expression of MMP9 both in the cell lysate and culture supernatant. Our results illustrated that treatment with 5-nm Au-NPs markedly increased the levels of MMP9 in the lysate of A549 cells, while MMP9 was notably upregulated in the lysate of 95D cells by treatment with 10-nm Au-NPs (Figures 6 A–B). The comparisons with the level of MMP9 in the supernatant revealed that there was no significant difference ($P>0.05$) between the Au-NP–treated and Au-NP–untreated A549 and 95D (Figures 6 C–D).

Furthermore, we investigated the protein expression using western blotting. The results showed that MMP9 and ICAM-1 were upregulated in the lysates of A549 and 95D cells by treatment with 5-nm and 10-nm Au-NPs, respectively (Figure 7); in contrast, MMP9 and ICAM-1 were downregulated by treatment with 40-nm Au-NPs, suggesting that Au-NPs particle size

plays an important role in the regulation of these proteins. These results provided the evidence that MMP9 and ICAM-1, key modulators of cell invasion, could be regulated by Au-NPs.

Discussion

The low production cost and relative ease to synthesize Au-NPs make them feasible for future biomedical applications. However, with the transition of Au-NPs from the benchtop to the clinic, much more knowledge is required about the fundamental interactions between nanoscale materials and biological systems. The biosafety and biocompatibility of Au-NPs are vital concerns that should be demonstrated before such materials are applied to biological systems. Similar to many other reports [24,25], in this study, Au-NPs were easily taken up by A549 and 95D cells via nonspecific endocytosis and localized within cytoplasmic vesicles. In this study, we demonstrated that small Au-NPs with a diameter of 5 nm exhibit high efficacy in inhibiting proliferation, promoting apoptosis, and arresting the cell cycle at the G0/G1 phase in two lung cancer cell lines. In contrast, no obvious cytotoxicity was

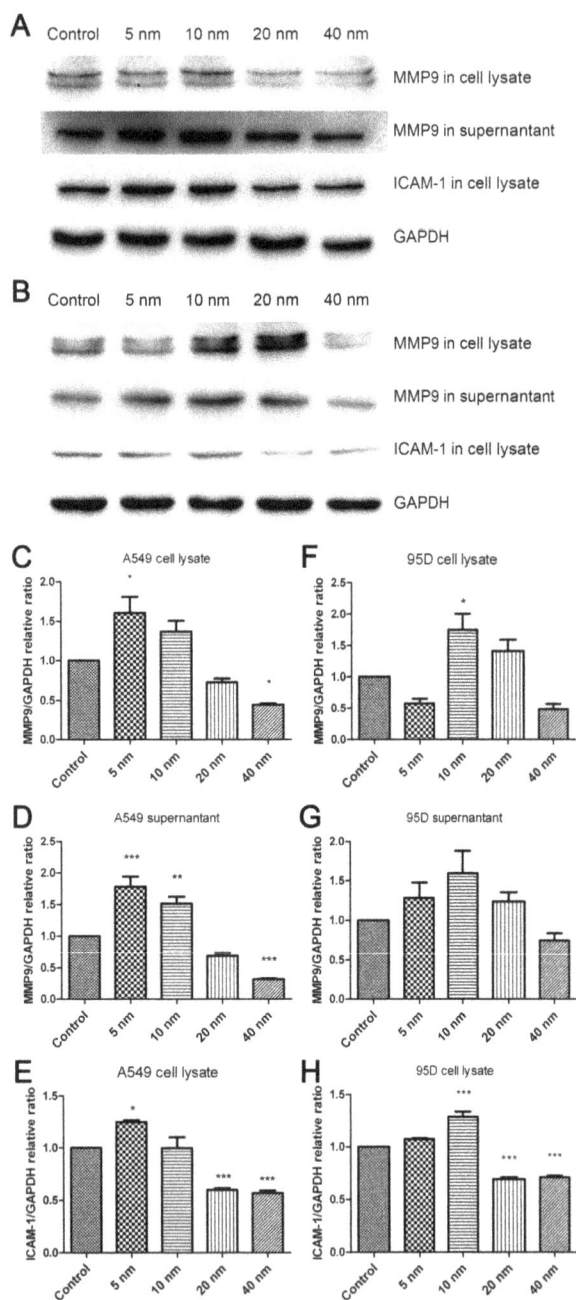

Figure 7. Western blot analysis of the expression of MMP9 and ICAM-1 in A549 and 95D cells. Representative results showing the effect of Au-NPs treatment on the expression of MMP9 and ICAM-1 in A549 (A) and 95D (B) cells. Relative band density of MMP9 and ICAM-1 to the mean value of the control group (C–H). Values are expressed as relative intensity normalized to GAPDH intensity and are given as mean ± SD from three independent experiments. *$P < 0.05$, **$P < 0.01$, ***$P < 0.001$ vs. control.

cores benign. A size larger than 6–8 nm is also optimal for precluding renal excretion and consequently diminishing the circulatory half-life [27]. Unexpectedly, our results showed that the proliferation of A549 cells was promoted after 24-h treatment with Au-NPs of 20 nm or 40 nm in diameter. This phenomenon was not observed in 95D cells. Moreover, 5-nm Au-NPs and 10-nm significantly promoted the invasion of A549 and 95D cells, respectively, while the invasive ability was not affected in the presence of particles with a size greater than 20 nm. These results support the view that Au-NPs do not universally target all cell types [28]. Coulter et al. found that the surviving fraction for Au-NPs–treated cells showed a strong dependence on the cell type compared with that of untreated cells in respect to radiosensitization potential [24].

In our study, the improved invasion ability was accompanied by a notable upregulation of ICAM-1 and MMP-9 expression. Invasion through the ECM is an important step in tumor metastasis. Cancer cells initiate invasion by adhering to and spreading along the blood vessel wall. Matrix metalloproteinases are endopeptidases that are able to degrade ECM components, which allows cancer cells to access the vasculature and lymphatic systems[29–31]. MMP-9 has attracted much attention for its ability to degrade type IV collagen, the basic component of the basement membrane [32]. Increased expression of MMP-9 in patients with non-small cell lung cancer has been reported [33,34]; therefore, agents suppressing the expression of the MMPs could inhibit cancer cell migration and invasion [35]. ICAM-1 is a representative adhesion molecule involved in the interaction among tumor cells, the endothelium, and ECM. High expression of ICAM-1 in human lung cancer specimens was correlated with a greater risk of advanced cancers (stages III and IV). A549/ICAM-1 cells were shown to induce in vitro cell invasion and in vivo tumor metastasis [36]. Denissenko et al. observed that ICAM-1 downregulation at the mRNA and protein levels led to strong suppression of human breast cell invasion through a Matrigel matrix [37]. We found that 5-nm and 10-nm Au-NPs effectively promoted the expression of ICAM-1and MMP9 in A549 and 95D cells, respectively, which partially explained the enhanced action of the particles on cancer cell invasion.

Since the upregulation effects of Au-NPs on MMP-9 and ICAM-1 expression in A549 and 95D cell suggested that small particles might possess the ability to facilitate the invasion of lung cancer cells, further in vivo studies are required to confirm the mechanisms. In summary, treatment with 5-nm Au-NPs effectively inhibited cells proliferation and promoted apoptosis, but it also upregulated the expression of ICAM-1 and MMP9, as well as increased the invasion of A549 cells. In contrast, Mukherjee et al. reported that Au-NPs (5 nm in diameter) exhibited anti-angiogenic properties (i.e., inhibited the tumorigenic growth of new blood vessels) both in vitro and in vivo [38]. Because of the complexity of this phenomenon and contradictory conclusions on the size of nanoparticles, further investigations are necessary to verify the effects and determine the appropriate size of Au-NPs for alleviating lung cancer progression.

This study provides new insights on the influence of Au-NPs of different sizes on cancer cell invasion, which could be of a great value for the application of Au-NPs to novel therapies in lung cancer. Our research indicated that the biological function of unmodified Au-NPs depended strongly on the particle size and cell type and that the particle size should be selected carefully for biological applications. There are both opportunities and challenges ahead in developing Au-NPs with suitable diameters that will be able to deliver novel anti-tumor agents or with intrinsic therapeutic potential for clinical use. In any case, the path toward

observed in 10 nm, 20 nm, and 40 nm Au-NP–treated cells, which is in agreement with the results of other research groups. Previous studies showed that surface size, and not surface charge, plays a large role in the therapeutic effect of Au-NPs [26]. Generally, Au-NPs for drug delivery and photothermal therapy applications have dimensions larger than 10 nm, which is sufficient to decrease the surface reactivity and thus make gold the development of a feasible nanomedicine is still long and material safety and long-term bioeffects should be carefully considered.

Conclusions

In this study, we have demonstrated the effects of citrate-capped Au-NPs of different sizes (5 nm, 10 nm, 20 nm, 40 nm) on the cytotoxicity and invasion in lung cancer cells. Based on the results, the following conclusions were drawn: First, nanoparticle size is an essential variable affecting cell proliferation, apoptosis, cell cycle, and cell invasion. Small particles endocytosed into the cells could cause great cytotoxicity, whereas large particles have no significant cytotoxicity. Second, in addition to particle size, cell type is also an important factor affecting the interaction between the Au-NPs and cells. Third, small Au-NPs upregulate the expression of MMP9

and ICAM-1, which may be associated with the increased invasion activity of A549 and 95D cells. This study provides useful information on cell cytotoxicity and invasion, although the molecular mechanisms need further investigation.

Acknowledgments

The authors gratefully acknowledge Chunsun Dai for the useful discussions and Zhiyang Li and Brian Jia for the technical assistance.

Author Contributions

Conceived and designed the experiments: XL ZL ZG. Performed the experiments: YW ZG YL PZ YS. Analyzed the data: ZL YW PZ. Contributed reagents/materials/analysis tools: XL ZL ZG. Wrote the paper: ZL XL.

References

1. Connor EE, Mwamuka J, Gole A, Murphy CJ, Wyatt MD (2005) Gold nanoparticles are taken up by human cells but do not cause acute cytotoxicity. Small 1: 325–327.
2. Shukla R, Bansal V, Chaudhary M, Basu A, Bhonde RR, et al. (2005) Biocompatibility of gold nanoparticles and their endocytotic fate inside the cellular compartment: a microscopic overview. Langmuir 21: 10644–10654.
3. Dreaden EC, Alkilany AM, Huang X, Murphy CJ, El-Sayed MA (2012) The golden age: gold nanoparticles for biomedicine. Chem Soc Rev 41: 2740–2779.
4. El-Sayed IH, Huang X, El-Sayed MA (2005) Surface plasmon resonance scattering and absorption of anti-EGFR antibody conjugated gold nanoparticles in cancer diagnostics: applications in oral cancer. Nano Lett 5: 829–834.
5. Weissleder R (2006) Molecular imaging in cancer. Science 312: 1168–1171.
6. Veiseh O, Gunn JW, Kievit FM, Sun C, Fang C, et al. (2009) Inhibition of tumor-cell invasion with chlorotoxin-bound superparamagnetic nanoparticles. Small 5: 256–264.
7. Huang X, Jain PK, El-Sayed IH, El-Sayed MA (2007) Gold nanoparticles: interesting optical properties and recent applications in cancer diagnostics and therapy. Nanomedicine (Lond) 2: 681–693.
8. Dreaden EC, Mackey MA, Huang X, Kang B, El-Sayed MA (2011) Beating cancer in multiple ways using nanogold. Chem Soc Rev 40: 3391–3404.
9. Giljohann DA, Seferos DS, Daniel WL, Massich MD, Patel PC, et al. (2010) Gold nanoparticles for biology and medicine. Angew Chem Int Ed Engl 49: 3280–3294.
10. Arvizo RR, Rana S, Miranda OR, Bhattacharya R, Rotello VM, et al. (2011) Mechanism of anti-angiogenic property of gold nanoparticles: role of nanoparticle size and surface charge. Nanomedicine 7: 580–587.
11. Alkilany AM, Murphy CJ (2010) Toxicity and cellular uptake of gold nanoparticles: what we have learned so far? J Nanopart Res 12: 2313–2333.
12. Pan Y, Leifert A, Ruau D, Neuss S, Bornemann J, et al. (2009) Gold nanoparticles of diameter 1.4 nm trigger necrosis by oxidative stress and mitochondrial damage. Small 5: 2067–2076.
13. Chithrani BD, Ghazani AA, Chan WC (2006) Determining the size and shape dependence of gold nanoparticle uptake into mammalian cells. Nano Lett 6: 662–668.
14. Eccles SA, Box C, Court W (2005) Cell migration/invasion assays and their application in cancer drug discovery. Biotechnol Annu Rev 11: 391–421.
15. Molina JR, Yang P, Cassivi SD, Schild SE, Adjei AA (2008) Non-small cell lung cancer: epidemiology, risk factors, treatment, and survivorship. Mayo Clin Proc 83: 584–594.
16. Manegold C (2001) Chemotherapy for advanced non-small cell lung cancer: standards. Lung Cancer 34 Suppl 2: S165–170.
17. Ju D, Sun D, Xiu L, Meng X, Zhang C, et al. (2012) Interleukin-8 is associated with adhesion, migration and invasion in human gastric cancer SCG-7901 cells. Med Oncol 29: 91–99.
18. Han S, Ritzenthaler JD, Sitaraman SV, Roman J (2006) Fibronectin increases matrix metalloproteinase 9 expression through activation of c-Fos via extracellular-regulated kinase and phosphatidylinositol 3-kinase pathways in human lung carcinoma cells. J Biol Chem 281: 29614–29624.
19. Dowlati A, Gray R, Sandler AB, Schiller JH, Johnson DH (2008) Cell adhesion molecules, vascular endothelial growth factor, and basic fibroblast growth factor in patients with non-small cell lung cancer treated with chemotherapy with or without bevacizumab–an Eastern Cooperative Oncology Group Study. Clin Cancer Res 14: 1407–1412.
20. Li Y, Liang J, Liu X, Liu H, Yin B, et al. (2012) Correlation of polymorphisms of the vascular endothelial growth factor gene and the risk of lung cancer in an ethnic Han group of North China. Exp Ther Med 3: 673–676.
21. Fajersztajn L, Veras M, Barrozo LV, Saldiva P (2013) Air pollution: a potentially modifiable risk factor for lung cancer. Nat Rev Cancer 13: 674–678.
22. Slot JW, Geuze HJ (1985) A new method of preparing gold probes for multiple-labeling cytochemistry. Eur J Cell Biol 38: 87–93.
23. Frens G (1973) Controlled nucleation for the regulation of the particle size in monodisperse gold suspensions. Nature 241: 20–22.
24. Coulter JA, Jain S, Butterworth KT, Taggart LE, Dickson GR, et al. (2012) Cell type-dependent uptake, localization, and cytotoxicity of 1.9 nm gold nanoparticles. Int J Nanomedicine 7: 2673–2685.
25. Tsai SW, Liaw JW, Kao YC, Huang MY, Lee CY, et al. (2013) Internalized gold nanoparticles do not affect the osteogenesis and apoptosis of MG63 osteoblast-like cells: a quantitative, in vitro study. PLoS One 8: e76545.
26. Arvizo R, Bhattacharya R, Mukherjee P (2010) Gold nanoparticles: opportunities and challenges in nanomedicine. Expert Opin Drug Deliv 7: 753–763.
27. Longmire M, Choyke PL, Kobayashi H (2008) Clearance properties of nano-sized particles and molecules as imaging agents: considerations and caveats. Nanomedicine (Lond) 3: 703–717.
28. Patra HK, Banerjee S, Chaudhuri U, Lahiri P, Dasgupta AK (2007) Cell selective response to gold nanoparticles. Nanomedicine 3: 111–119.
29. Hamano Y, Zeisberg M, Sugimoto H, Lively JC, Maeshima Y, et al. (2003) Physiological levels of tumstatin, a fragment of collagen IV alpha3 chain, are generated by MMP-9 proteolysis and suppress angiogenesis via alphaV beta3 integrin. Cancer Cell 3: 589–601.
30. Hanahan D, Weinberg RA (2000) The hallmarks of cancer. Cell 100: 57–70.
31. Egeblad M, Werb Z (2002) New functions for the matrix metalloproteinases in cancer progression. Nat Rev Cancer 2: 161–174.
32. Pritchard SC, Nicolson MC, Lloret C, McKay JA, Ross VG, et al. (2001) Expression of matrix metalloproteinases 1, 2, 9 and their tissue inhibitors in stage II non-small cell lung cancer: implications for MMP inhibition therapy. Oncol Rep 8: 421–424.
33. Guo CB, Wang S, Deng C, Zhang DL, Wang FL, et al. (2007) Relationship between matrix metalloproteinase 2 and lung cancer progression. Mol Diagn Ther 11: 183–192.
34. Hung WC, Tseng WL, Shiea J, Chang HC (2010) Skp2 overexpression increases the expression of MMP-2 and MMP-9 and invasion of lung cancer cells. Cancer Lett 288: 156–161.
35. Lin SS, Lai KC, Hsu SC, Yang JS, Kuo CL, et al. (2009) Curcumin inhibits the migration and invasion of human A549 lung cancer cells through the inhibition of matrix metalloproteinase-2 and -9 and Vascular Endothelial Growth Factor (VEGF). Cancer Lett 285: 127–133.
36. Lin YC, Shun CT, Wu MS, Chen CC (2006) A novel anticancer effect of thalidomide: inhibition of intercellular adhesion molecule-1-mediated cell invasion and metastasis through suppression of nuclear factor-kappaB. Clin Cancer Res 12: 7165–7173.
37. Rosette C, Roth RB, Oeth P, Braun A, Kammerer S, et al. (2005) Role of ICAM1 in invasion of human breast cancer cells. Carcinogenesis 26: 943–950.
38. Mukherjee P, Bhattacharya R, Wang P, Wang L, Basu S, et al. (2005) Antiangiogenic properties of gold nanoparticles. Clin Cancer Res 11: 3530–3534.

2

Nanopore Fabrication by Controlled Dielectric Breakdown

Harold Kwok[Ↄ], **Kyle Briggs**[Ↄ], **Vincent Tabard-Cossa***

Department of Physics, University of Ottawa, Ottawa, Ontario, Canada

Abstract

Nanofabrication techniques for achieving dimensional control at the nanometer scale are generally equipment-intensive and time-consuming. The use of energetic beams of electrons or ions has placed the fabrication of nanopores in thin solid-state membranes within reach of some academic laboratories, yet these tools are not accessible to many researchers and are poorly suited for mass-production. Here we describe a fast and simple approach for fabricating a single nanopore down to 2-nm in size with sub-nm precision, directly in solution, by controlling dielectric breakdown at the nanoscale. The method relies on applying a voltage across an insulating membrane to generate a high electric field, while monitoring the induced leakage current. We show that nanopores fabricated by this method produce clear electrical signals from translocating DNA molecules. Considering the tremendous reduction in complexity and cost, we envision this fabrication strategy would not only benefit researchers from the physical and life sciences interested in gaining reliable access to solid-state nanopores, but may provide a path towards manufacturing of nanopore-based biotechnologies.

Editor: Adam Hall, Wake Forest University School of Medicine, United States of America

Funding: This work was supported by the Natural Sciences and Engineering Research Council of Canada, the Canada Foundation for Innovation, and Ontario Network of Excellence. The funders had no role in study design, data collection and analysis, decision to publish, or preparation of the manuscript.

Competing Interests: A patent application was filed on the content of the work presented. "Fabrication of nanopores using high electric fields", with publication number: WO2013167955 A1, and application number: PCT/IB2013/000891). Since this study, the authors have been awarded a research grant from NSERC in collaboration with Abbott Laboratories, but that funding was not used to support the work presented here. There are currently no further patents, products in development or marketed products to declare.

* E-mail: tcossa@uOttawa.ca

[Ↄ] These authors contributed equally to this work.

Introduction

Nanopore sensing relies on the electrophoretically driven translocation of biomolecules through nanometer-scale holes embedded in thin insulating membranes to confine, detect and characterize the properties or the activity of individual biomolecules electrically, by monitoring transient changes in ionic current [1–4]. The field was initially shaped by the ability of researchers to exploit biological channels to translocate single molecules [5,6]. It rapidly expanded when new techniques to fabricate individual molecular-sized holes in thin solid-state materials were developed over the last decade [7–12]. These techniques, based on beams of high-energy particles, either produced by a dedicated ion beam machine (i.e ion-beam sculpting) or a transmission electron microscope (i.e TEM drilling), allowed researchers to control the nanopore size at the sub-10-nm length scale with single nanometer precision, thus greatly diversifying the breadth of applications. Since then, a host of applications for DNA, RNA and proteins analysis using both biological and solid-state nanopores have been demonstrated [4,13,14]. Compared to their organic counterparts, solid-state nanopores were expected to emerge as an essential component of any practical nanopore-based instrumentation due to the size control, increased robustness of the membrane, and their natural propensity for integration with wafer-scale technologies, including CMOS and microfluidics [15,16]. Yet, this prospect is significantly hindered due to the constraints and limitations imposed by ion beam sculpting and transmission

electron microscopy-based drilling, which, to this date, remain the only viable tools for achieving nanopores fabrication with dimensional control at the 1-nm scale. The complexity, low-throughput, and high-cost associated with these techniques restrict accessibility of the field to many researchers, greatly limit the productivity of the community, and prevent mass production of nanopore-based devices. Alternative nanofabrication strategies are therefore needed for the field to continue to thrive, and for the promised health-related applications to be successfully commercialized (including single-molecule DNA sequencing). Here, we introduce a fabrication technique based on the use of high electric fields to control dielectric breakdown in solution. The method is automated, simple, and low-cost, allowing nanopores to be created directly in aqueous solution with sub-nm precision, greatly facilitating use and improving yield of functional solid-state nanopore devices. We envision this fabrication strategy will not only provide a path towards nanomanufacturing of nanopore-based devices for a wide range of biotechnology applications, but will democratize the use of solid-state nanopores, while offering researchers new strategies for designing nanofluidics devices, as well as integrating nanopores with CMOS and microfluidics technologies.

Results and Discussion

We fabricate individual nanopores on thin insulating solid-state membranes directly in solution. A thin silicon nitride (SiN_x)

membrane, supported by a silicon frame, is mounted in a liquid cell and separates two reservoirs containing an aqueous solution of 1M KCl. Ag/AgCl electrodes immersed on both sides of the membrane are connected to a custom-built resistive feedback current amplifier, which allow trans-membrane potentials of up to ± 20 V to be applied. The setup shown in Figure 1 is otherwise identical to what is commonly used for biomolecular detection [17], which greatly facilitates the transition to sensing experiments, eliminating further handling of membranes. See Material and Methods section and Section S1 for more detail.

A single nanopore is fabricated by applying a constant potential difference, ΔV, across a $t = 10$-nm or 30-nm thick SiN_x membrane, to produce an electric field, $E = \Delta V/t$ in the dielectric membrane in the range of 0.4-1 V/nm (Figure 2a). At these high field strengths, a sustainable leakage current, $I_{leakage}$, is observed through the membrane, which remains otherwise insulating at low fields. $I_{leakage}$ rapidly increases with electric field strength, but is typically tens of nanoamperes for our operating conditions. We attribute the dominant conduction mechanism to a form of trap-assisted tunneling of electrons, supplied by ions in solution [18–21] (Figure 2b and 2c), since the membrane is too thick for significant direct tunnelling [18], and migration of impurities cannot produce lasting currents [22]. Direct migration of electrolyte ions is also unlikely, or negligible, since for a given electric field strength, a

higher $I_{leakage}$ is observed in thicker membranes (Figure 2e). A larger current is observed on thicker membranes since the number of charge traps (defects) per unit area is greater, as their number in the material increases with volume. We provide additional discussion on the characteristics of the leakage current in Section S2.

We observe the creation of a single nanopore (i.e. fluidic channel spanning the membrane) by a sudden irreversible increase in $I_{leakage}$, which is attributed to the onset of ionic current (Figure 2d and 2f) due to a discrete dielectric breakdown event. As the current continues to increase, the nanopore further enlarges (Figure 2g). We use a feedback control mechanism to rapidly terminate the trans-membrane potential when the current exceeds a pre-determined threshold, I_{cutoff}. A threshold, set as $I_{cutoff}/I_{leakage} < 1.2$, which is generally sufficient to terminate ΔV within ~0.1 s of the breakdown event, can produce nanopores on the order of 2-nm in diameter as shown by the I-V curves in Figure 2 h (see Section S3 for additional results). In addition, following the nanopore fabrication event, we can continue to enlarge its size with sub-nm precision by applying moderate AC electric field square pulses in the range of ± 0.2-0.3 V/nm, similar to Beamish et al. [23,24]. This allows the nanopore size to be precisely tuned, for a particular sensing application, directly in neutral KCl solution.

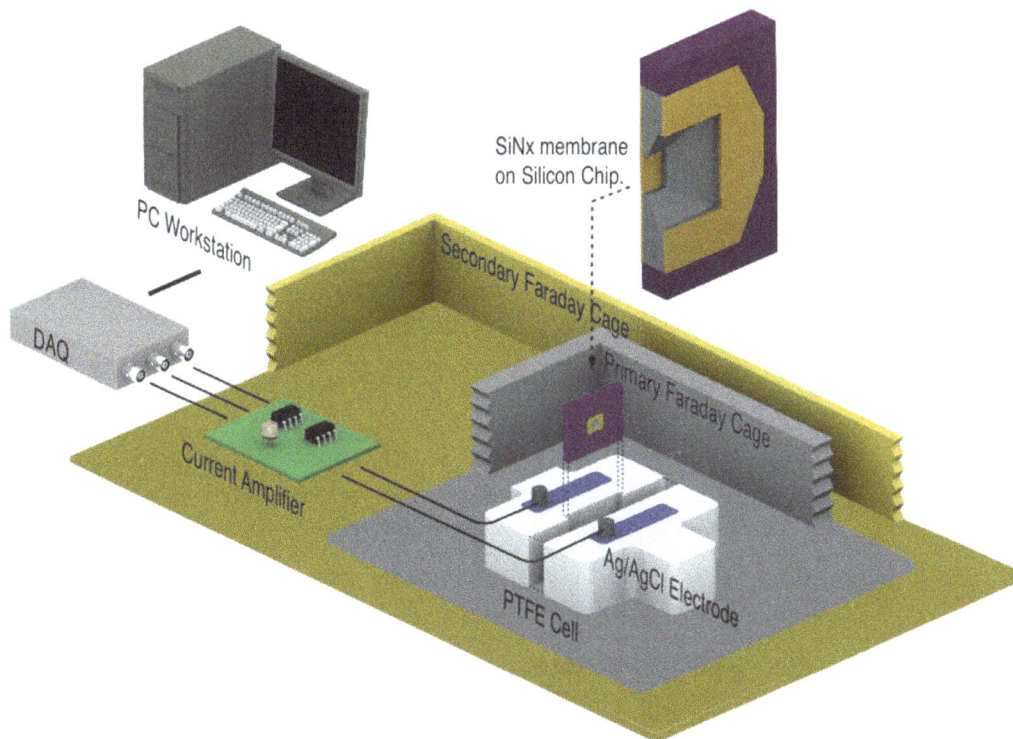

Figure 1. Schematic of the fabrication setup. A computer-controlled custom current amplifier is used to apply voltages up to ± 20 V and measure the current with sub-nA sensitivity from one of the two Ag/AgCl electrodes positioned on either sides of the membrane. It is noteworthy to realize that this experimental setup is identical (with the exception of the custom current amplifier replacing the commonly used Axopatch 200B) to the instrumentation used to study DNA or proteins translocation through nanopores.

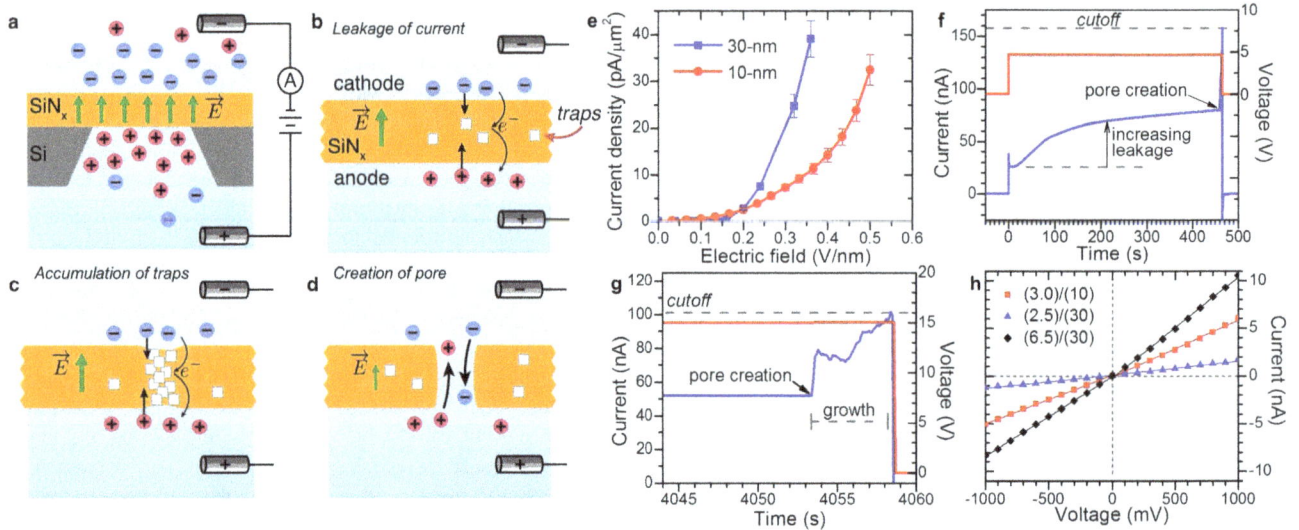

Figure 2. Nanopore formation by dielectric breakdown. a) Application of a trans-membrane potential generates an electric field inside the SiN_x, and charges the interfaces with opposite ions. b) Leakage current through the membrane follows a trap-assisted tunneling mechanism. Free charges (electrons or holes) can be produced by redox reactions at the surface or by field ionization of incorporated ions. The number of available charged traps (structural defects) sets the magnitude of the observed leakage current. c) Accumulation of charge traps produced by electric field-induced bond breakage or energetic charges carries leads to a highly localized conductive path, and a discrete dielectric breakdown event. d) A nanopore is formed following removal of the defects. e) Leakage current density for SiN_x membranes (50-μm $\times$ 50-μm). The leakage current is fully reversible and stable, unless high fields are sustained, see Section S2 f) Leakage current at 5 V, on a 10-nm-thick SiN_x membrane, in 1 M KCl at pH13.5. Pore created is ~5-nm (18 nS). The slowly increase leakage current, following the capacitive spike, is a result of the accumulation of traps in the membrane. g) Experiment performed at 15 V, on a 30-nm-thick SiN_x membrane, in 1 M KCl pH10. The nanopore is allowed to grow until a pre-determined threshold current is reached, at which point the voltage is turned off. The observed current fluctuations at the onset of pore formation are attributed to significant low-frequency noise at this voltage. Pore created is ~3-nm (2.9 nS). h) Current-to-voltage curves for 3 nanopores fabricated on different membranes. The legend indicates the (pore diameter)/(membrane thickness) in nm. Measurements performed in 1 M KCl pH8, with an Axopatch 200B.

I-V Characteristics and Noise

To infer the nanopore size upon fabrication, we measure its ionic conductance, G, and relate it to an effective diameter, d, assuming a cylindrical geometry and accounting for access resistance [25,26], using $G = \sigma \left[\dfrac{4t}{\pi d^2} + \dfrac{1}{d} \right]^{-} 1$, where σ is the bulk conductivity of the solution. This method, practical for nanopores fabricated in liquids, provides a reasonable first order estimate of the pore size [26,27] as confirmed by DNA translocations, and compares well with actual dimensions obtained from TEM images (see Sections S4 and S8). I-V curves are performed in a ± 1 V window, where the leakage current can safely be ignored. Figure 2 h reveals an ohmic electric response in 1 M KCl. The majority of our nanopores exhibit linear I-V curves upon fabrication. The remaining nanopores that show signs of self-gating or rectification can be conditioned, by applying moderate electric field pulses [23], to slightly enlarge them until an ohmic behaviour is attained in high salt. Such I-V characteristics imply a relatively symmetric internal electric potential pore profile [28] which supports the symmetrical geometry with a uniform surface charge distribution assumed by our pore conductance model. Otherwise, one would expect significant rectification from multiple $\leq$1-nm fluidic paths or from a single narrow nano-crack of similar conductance, due to strong electrostatic double layer overlap. To further characterize the nanopores, we examined the noise in the ionic current by performing power spectral density measurements. Our fabrication method consistently produces nanopores with low-$1/f$ noise levels, comparable to fully wetted TEM-drilled nanopores (see Section S5)[29,30]. This may be attributed to the fact that nanopores are created directly in liquid rather than in

vacuum. Thus far, we have successfully fabricated hundreds of individual nanopores ranging from 1 to 25-nm in size with comparable electrical characteristics that are stable for days, 66 of which are included in Figure 2. The success rate for fabricating a nanopore under the experimental parameters presented here is estimated at >99%.

Dielectric Breakdown Mechanism

In order that a single, well-defined nanopore be created each time, we postulate that the leakage current must be highly localized on the insulating membrane, since for conductive substrates (semiconductors or metals) anodic oxidation leads to an array of nanopores[31–33]. The leakage spot(s) must also modify the membrane at the nanoscale since an aqueous KCl solution at neutral pH is not an active etchant of SiN_x. To elucidate the mechanism leading to the formation of a nanopore, we investigate the fabrication process as a function of applied voltage, membrane thickness, electrolyte composition, concentration, and pH. Figure 3a shows the time-to-pore creation, τ, as a function of the trans-membrane potential for 30-nm-thick membranes, in 1 M KCl buffered at various pHs. Interestingly, τ scales exponentially with the applied voltage irrespective of other conditions, and can be as short as a few seconds. For a given voltage, pH has a strong effect. As seen in Figure 3b, τ can be reduced by 1,000-fold when lowering the pH from 7 to 2. We have also observed that lower salt concentrations increase the fabrication time (see Section S6). Overall, for a given fabrication condition τ is relatively consistent, though variations by a factor of 4 are common, and is uncorrelated with the size of the fabricated pore. Figure 3c shows τ for 10-nm-thick SiN_x membranes, buffered at pH10 in various 1 M Cl-based aqueous solutions. The fabrication time in these thinner membranes also decreases exponentially with potential, but the value required for forming a nanopore is now reduced by $\sim 1/3$ compared to 30-nm-thick membranes, irrespective of the different cations (K^+, Na^+, Li^+) tested. This observation indicates that the applied electric field in the membrane is the main driving force for initiating the fabrication of a single nanopore. Fields in the range of 0.4-1 V/nm are close to the dielectric breakdown strength of low-stress SiN_x films[19], and are key for intensifying the leakage current, which is thought to ultimately cause breakdown in thin insulating layers[34]. The exponential dependence of τ on potential implies the same electric field dependency, which is reminiscent of the time-to-dielectric breakdown in gate dielectrics[34]. According to the current understanding, dielectric breakdown mechanisms proceed as follows[34–36]: (i) probabilistic accumulation of charge traps (i.e. structural defects) by electric field-induced bond breakage or generated by charge injection from the anode or cathode, (ii) increasing up to a critical density forming a highly localized conductive path, and (iii) causing physical damage due to substantial power dissipation and the resultant heating. We propose that the process by which we fabricate a nanopore in solution is similar, though we control the damage to the nanoscale by limiting the localized leakage current, at the onset of the first, discrete breakdown event. As indicated by Figure 3, the likelihood of defect formation within the silicon nitride membrane increases with the applied voltage and the strength of the electric field. At low values, the accumulation of charge traps is accomplished with relatively low efficiency compared to the amount of charge carriers traversing the membrane, since a leakage current of tens of nanoamperes can be sustained for hours or days. Given the stochastic nature of the pore creation process, multiple simultaneous nanoscale breakdown events are unlikely. Termination of the applied voltage following the occurrence of the first breakdown event, observed by the sudden irreversible increase in $I_{leakage}$, ensures that ultimately a single nanopore is created. Moreover, the fabrication of a single nanopore may result from the fact that the formation path of a nanopore experiences increased electric field strength during growth, which locally reinforces the rate of defect generation. The process by which material is removed from the membrane remains unclear, but broken bonds could be chemically dissolved by the electrolyte or following a conversion to oxides/hydrides [37,38]. Another possibility is shearing due to localized plasticity of the membrane as a result of heating at the breakdown spot, but the efficiency of heat dissipation at the nanoscale, resulting from high surface-area-to-volume ratios, makes this less likely [22]. We explain the pH dependency on the fabrication time, for 30-nm thick SiN_x membranes, by the fact that breakdown at low pH is amplified by impact ionization producing an avalanche, due to the increased likelihood of hole injection or H^+ incorporation from the anode (see Section S6 for more detail). To support the general character of nanofabrication by dielectric breakdown, we created nanopores in a different material (silicon dioxide) and present the data in Section S7.

DNA Translocations

We performed DNA translocation experiments to demonstrate that these nanopores can be leveraged for the benefit of single-molecule detection. Electrophoretically driven passage of a DNA molecule across a membrane is expected to transiently block the flow of ions in a manner that reflects the molecule length, size, charge and shape. The results using a $\sim$6.4-nm-diameter pore, as estimated from conductance measurements, in a 10-nm thick SiN_x membrane are shown in Figure 4. The scatter plot shows event

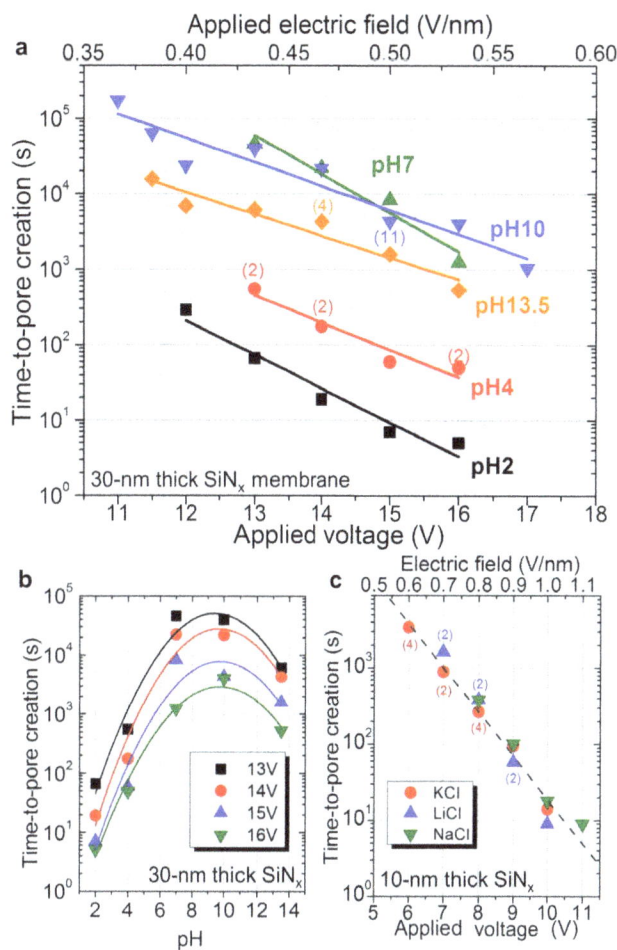

Figure 3. Time-to-pore creation as a function of experimental conditions. a) Semi-log plot of fabrication time of individual nanopores created in 30-nm-thick SiN_x membranes in 1 M KCl buffered as indicated, versus the applied voltage and the calculated applied electric field. The number of separate nanopores each data point is averaged over is indicated in parentheses. The vast majority of nanopores plotted are sub-5-nm in size (i.e. <7 nS). b) Semi-log plot of fabrication time versus pH for the data plotted in a). c) Semi-log plot of fabrication time of individual nanopores created in 10-nm-thick SiN_x membranes in 1 M Cl-based electrolyte buffered at pH 10 for different cationic species versus the applied voltage and the calculated applied electric field. All 66 nanopores plotted are sub-5-nm in size (i.e. <20 nS).

duration and average current blockage of over 2,400 single-molecule translocations events of 5-kb dsDNA. The characteristic shape of the events is indistinguishable to data obtained on TEM-drilled nanopores [26,39–41]. The observed quantized current blockades strongly support the presence of a single nanopore spanning the membrane. Using dsDNA ($\sim$2.2 nm in diameter) as a molecular-sized ruler, the value of the single-level blockage events, $\Delta G = 7.4 \pm 0.9$ nS, provides an effective pore diameter of 6.0 ± 0.5-nm consistent with the size extracted from the pore conductance model [26]. This result also suggests that the membrane thickness at the vicinity of the nanopores has not been significantly altered. We observed similar DNA translocation signatures from most nanopores tested (revealing >80% success rate in detecting DNA for N = 19 nanopores tested), and provide further discussion and additional translocation data in Section S8.

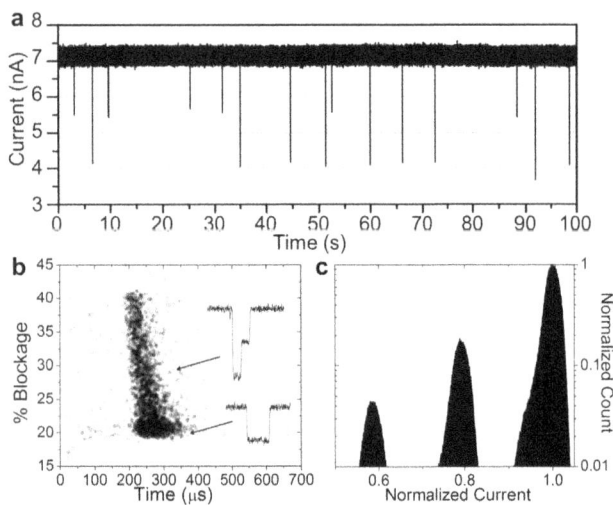

Figure 4. DNA Translocations. a) Ionic current trace showing multiple DNA translocation events through a ~6.4-nm pore in a 10-nm-thick SiN$_x$ membrane. Experiments performed with 10μg/mL of 5-kb DNA fragments in 3.6 M LiCl pH8, at 200 mV using an Axopatch 200B. Data sampled at 250 kHz and low-pass filtered at 100 kHz. b) Scatter plot of the normalized average current blockade (0% presenting a fully opened pore, and 100% a fully blocked pore) versus the total translocation time of a single-molecule event. Each data point represents a single DNA translocation event. The majority of the events are unfolded. There are very few anomalously long events, indicating weak DNA-pore interactions. The inset shows ionic current signatures of two single-molecule translocation events, passing in a linear and partially folded conformation. c) Histogram of the current level revealing the expected quantization of the amplitude of current blockades. Quantized levels corresponding to zero, one, two dsDNA strands in the nanopore are clearly observed.

Outlook

Nanopore fabrication by controlled dielectric breakdown in solution represents a major reduction in complexity and cost over current fabrication methods, which will greatly facilitate accessibility to the field to many researchers, and provides a path to commercialize nanopore-based technologies. While we attribute the nanopore creation process to an intrinsic property of the dielectric membrane, such that the nanopore can form anywhere on the surface, our current understanding strongly suggests that the position of the pore can be determined by locally controlling the electric field strength or the material dielectric strength. This could be achieved, for instance, by nanopatterning or locally thinning the membrane, by positioning of a nanoelectrode, or by confining the field to specific areas on the membrane via micro- or nanofluidic channel encapsulation (see Section S9). The latter would also allow for the simple integration of independently addressable nanopores in an array format on a single chip.

Materials and Methods

Dielectric Membranes

Silicon Nitride (SiN$_x$) membranes used in our experiments are commercially available as transmission electron microscope (TEM) windows (Norcada product # NT005X and NT005Z). Each membrane is made of 10-nm or 30-nm thick low-stress (<250 MPa) SiN$_x$, deposited on 200-μm thick lightly doped silicon (Si) substrate by low-pressure chemical vapour deposition

(LPCVD). A 50-μm × 50-μm window on the backside of the Si substrate is opened by a KOH anisotropic chemical etch. Prior to mounting into liquids, SiN$_x$ membranes can be cleaned in oxygen plasma for 30 s at 30 W to facilitate wetting of the membrane surface, though this is not a requirement. All solutions used were filtered and degassed prior to use. The absence of pre-existing structural damages (e.g. pinholes, nano-cracks) is inferred by the fact that no current (<pA) is measured across a membrane at low voltages (<±1 V) prior to nanopore fabrication. The membrane side opposing the Si etch pit is the reference point for all applied voltages in this article. Silicon dioxide membranes were also purchased from TEMWindows (product# SO100-A20Q33). Note that we have also successfully fabricated nanopores on SiN$_x$ membranes purchased from TEMWindows, and on custom fabricated SiN$_x$ membranes.

Instrumentation and Data Acquisition

A schematic of the experimental setup is shown in Figure 1. A silicon chip with an intact silicon nitride membrane is sandwiched between two silicone gaskets (shown in purple on the figure). It is then positioned between the two electrolyte reservoirs in a PTFE (polytetrafluoroethylene) or a PEEK (polyether ether ketone) fluidic cell. The two reservoirs filled with liquid electrolyte are electrically connected to a current amplifier by two Ag/AgCl electrodes. The entire system is encapsulated in a grounded faraday cage to isolate from electromagnetic interference. Data acquisition and measurement automation were performed using custom-designed LabVIEW software controlling a National Instruments USB-6351 or PXIe-6366 DAQ card. The value of the trans-membrane potential is set by the DAQ card. Leakage current is digitized at 250 kHz and the signal is filtered at 10 Hz. When a current exceed a pre-set threshold, the voltage bias is immediately ceased by the software (response time is ~100 ms). I-V measurements and ionic current signal during DNA translocations are recorded using an Axopatch 200B with a 4-pole Bessel filter set at 100 kHz, with at 250 kHz sampling rate. Data analysis was carried out using custom-designed LabVIEW software to measure the duration and depth of each current blockade events.

DNA Studies

We performed DNA translocation studies, using dsDNA fragments of 100 bp, 5 kbp, 10 kbp purchased from Fermantas (NoLimits products) in 1 M KCl pH8 or in 3.6 M LiCl pH8 at a final concentration of 10μg/mL. Lambda DNA (48.5 kbp) purchased from NewEngland BioLabs was also used.

Supporting Information

Section S1   Section S1 contains additional information regarding the experimental setup.

Section S2   Section S2 provides further discussion on the leakage current.

Section S3   Section S3 shows I-V curves of 8 nanopores <2.5-nm in diameter.

Section S4   Section S4 shows TEM images of nanopores fabricated by controlled dielectric breakdown.

Section S5   Section S5 presents the ionic current noise characteristics of 4 nanopores.

Section S6   Section S6 presents additional data of the time-to-pore creation versus trans-membrane potential and electric field strength.

Section S7   Section S7 demonstrates fabrication of a nanopore on a silicon dioxide membrane.

Section S8   Section S8 presents additional DNA translocation data and their analysis.

Section S9   Section S9 discusses strategies for localizing nanopores on a membrane.

Acknowledgments

The authors would like to thank Y. Liu for aid in TEM imaging and L. Andrzejewski for valuable technical support.

Author Contributions

Conceived and designed the experiments: VTC HK. Performed the experiments: HK KB. Analyzed the data: KB HK VTC. Contributed reagents/materials/analysis tools: HK. Wrote the paper: VTC. Discovered the fabrication process: HK.

References

1. Venkatesan BM, Bashir R (2011) Nanopore sensors for nucleic acid analysis. Nature nanotechnology 6: 615–624. doi:10.1038/NNANO. 2011.129.
2. Dekker C (2007) Solid-state nanopores. Nature nanotechnology 2: 209–215. doi:10.1038/nnano.2007.27.
3. Branton D, Deamer DW, Marziali A, Bayley H, Benner S a, et al. (2008) The potential and challenges of nanopore sequencing. Nature biotechnology 26: 1146–1153. doi:10.1038/nbt.1495.
4. Kasianowicz JJ, Robertson JWF, Chan ER, Reiner JE, Stanford VM (2008) Nanoscopic porous sensors. Annual review of analytical chemistry (Palo Alto, Calif) 1: 737–766. doi:10.1146/annurev.anchem.1.031207.112818.
5. Bezrukov SM, Vodyanoy I, Parsegian VA (1994) Counting polymers moving through a single ion channel. Nature 370: 279–281.
6. Kasianowicz JJ, Brandin E, Branton D, Deamer DW (1996) Characterization of individual polynucleotide molecules using a membrane channel. Proceedings of the National Academy of Sciences of the United States of America 93: 13770–13773.
7. Li J, Stein D, McMullan C, Branton D, Aziz MJ, et al. (2001) Ion-beam sculpting at nanometre length scales. Nature 412: 166–169. doi:10.1038/35084037.
8. Storm AJ, Chen JH, Ling XS, Zandbergen HW, Dekker C (2003) Fabrication of solid-state nanopores with single-nanometre precision. Nature materials 2: 537–540. doi:10.1038/nmat941.
9. Storm a J, Chen JH, Ling XS, Zandbergen HW, Dekker C (2005) Electron-beam-induced deformations of SiO[sub 2] nanostructures. Journal of Applied Physics 98: 014307. doi:10.1063/1.1947391.
10. Kuan AT, Golovchenko JA (2012) Nanometer-thin solid-state nanopores by cold ion beam sculpting. Applied physics letters 100: 213104–2131044. doi:10.1063/1.4719679.
11. Russo CJ, Golovchenko JA (2012) Atom-by-atom nucleation and growth of graphene nanopores. Proceedings of the National Academy of Sciences of the United States of America 109: 5953–5957. doi:10.1073/pnas.1119827109.
12. Yang J, Ferranti DC, Stern L a, Sanford C a, Huang J, et al. (2011) Rapid and precise scanning helium ion microscope milling of solid-state nanopores for biomolecule detection. Nanotechnology 22: 285310. doi:10.1088/0957-4484/22/28/285310.
13. Miles BN, Ivanov AP, Wilson KA, Doğan F, Japrung D, et al. (2013) Single molecule sensing with solid-state nanopores: novel materials, methods, and applications. Chemical Society reviews 42: 15–28. doi:10.1039/c2cs35286a.
14. Oukhaled A, Bacri L, Pastoriza-Gallego M, Betton J-M, Pelta J (2012) Sensing Proteins through Nanopores: Fundamental to Applications. ACS chemical biology 7: 1935–1949. doi:10.1021/cb300449t.
15. Rosenstein JK, Wanunu M, Merchant CA, Drndic M, Shepard KL (2012) Integrated nanopore sensing platform with sub-microsecond temporal resolution. Nature methods 9: 487–492. doi:10.1038/nmeth.1932.
16. Jain T, Guerrero RJS, Aguilar CA, Karnik R (2013) Integration of solid-state nanopores in microfluidic networks via transfer printing of suspended membranes. Analytical chemistry 85: 3871–3878. doi:10.1021/ac302972c.
17. Tabard-Cossa V (2013) Instrumentation for Low-Noise High-Bandwidth Nanopore Recording. In: Edel J, Albrecht T, editors. Engineered Nanopores for Bioanalytical Applications. Elsevier. pp. 59–88.
18. Frenkel J (1938) On Pre-Breakdown Phenomena in Insulators and Electronic Semi-Conductors. Physical Review 54: 647–648. doi:10.1103/PhysRev.54.647.
19. Habermehl S, Apodaca RT, Kaplar RJ (2009) On dielectric breakdown in silicon-rich silicon nitride thin films. Applied Physics Letters 94: 012905. doi:10.1063/1.3065477.
20. Jeong DS, Hwang CS (2005) Tunneling-assisted Poole-Frenkel conduction mechanism in HfO[sub 2] thin films. Journal of Applied Physics 98: 113701. doi:10.1063/1.2135895.
21. Kimura M, Ohmi T (1996) Conduction mechanism and origin of stress-induced leakage current in thin silicon dioxide films. Journal of Applied Physics 80: 6360. doi:10.1063/1.363655.
22. Lee S, An R, Hunt AJ (2010) Liquid glass electrodes for nanofluidics. Nature nanotechnology 5: 412–416. doi:10.1038/nnano.2010.81.
23. Beamish E, Kwok H, Tabard-Cossa V, Godin M (2012) Precise control of the size and noise of solid-state nanopores using high electric fields. Nanotechnology 23: 405301. doi:10.1088/0957-4484/23/40/405301.
24. Beamish E, Kwok H, Tabard-Cossa V, Godin M (2013) Fine-tuning the Size and Minimizing the Noise of Solid-state Nanopores. Journal of visualized experiments: JoVE: e51081. doi:10.3791/51081.
25. Vodyanoy I, Bezrukov SM (1992) Sizing of an ion pore by access resistance measurements. Biophysical journal 62: 10–11. doi:10.1016/S0006-3495(92)81762-9.
26. Kowalczyk SW, Grosberg AY, Rabin Y, Dekker C (2011) Modeling the conductance and DNA blockade of solid-state nanopores. Nanotechnology 22: 315101. doi:10.1088/0957-4484/22/31/315101.
27. Frament CM, Dwyer JR (2012) Conductance-Based Determination of Solid-State Nanopore Size and Shape: An Exploration of Performance Limits. The Journal of Physical Chemistry C 116: 23315–23321. doi:10.1021/jp305381j.
28. Kosińska ID (2006) How the asymmetry of internal potential influences the shape of I-V characteristic of nanochannels. The Journal of chemical physics 124: 244707. doi:10.1063/1.2212394.
29. Tabard-Cossa V, Trivedi D, Wiggin M, Jetha NN, Marziali A (2007) Noise analysis and reduction in solid-state nanopores. Nanotechnology 18: 305505. doi:10.1088/0957-4484/18/30/305505.
30. Smeets RMM, Keyser UF, Dekker NH, Dekker C (2008) Noise in solid-state nanopores. Proceedings of the National Academy of Sciences of the United States of America 105: 417–421. doi:10.1073/pnas.0705349105.
31. Thompson GE, Wood GC (1981) Porous anodic film formation on aluminium. Nature 290: 230–232. doi:10.1038/290230a0.
32. Létant SE, Hart BR, Van Buuren AW, Terminello LJ (2003) Functionalized silicon membranes for selective bio-organism capture. Nature materials 2: 391–395. doi:10.1038/nmat888.
33. Tseng AA, Notargiacomo A, Chen TP (2005) Nanofabrication by scanning probe microscope lithography: A review. Journal of Vacuum Science & Technology B: Microelectronics and Nanometer Structures 23: 877. doi:10.1116/1.1926293.
34. Lombardo S, Stathis JH, Linder BP, Pey KL, Palumbo F, et al. (2005) Dielectric breakdown mechanisms in gate oxides. Journal of Applied Physics 98: 121301. doi:10.1063/1.2147714.
35. McPherson JW, Mogul HC (1998) Underlying physics of the thermochemical E model in describing low-field time-dependent dielectric breakdown in SiO[sub 2] thin films. Journal of Applied Physics 84: 1513. doi:10.1063/1.368217.
36. DiMaria DJ, Cartier E, Arnold D (1993) Impact ionization, trap creation, degradation, and breakdown in silicon dioxide films on silicon. Journal of Applied Physics 73: 3367. doi:10.1063/1.352936.
37. Liu H, Steigerwald ML, Nuckolls C (2009) Electrical double layer catalyzed wet-etching of silicon dioxide. Journal of the American Chemical Society 131: 17034–17035. doi:10.1021/ja903333s.
38. Jamasb S, Collins S, Smith RL (1998) A physical model for drift in pH ISFETs. Sensors and Actuators B: Chemical 49: 146–155.
39. Chen P, Gu J, Brandin E, Kim Y-R, Wang Q, et al. (2004) Probing Single DNA Molecule Transport Using Fabricated Nanopores. Nano Letters 4: 2293–2298. doi:10.1021/nl048654j.
40. Fologea D, Brandin E, Uplinger J, Branton D, Li J (2007) DNA conformation and base number simultaneously determined in a nanopore. Electrophoresis 28: 3186–3192.
41. Li J, Gershow M, Stein D, Brandin E, Golovchenko J a (2003) DNA molecules and configurations in a solid-state nanopore microscope. Nature materials 2: 611–615. doi:10.1038/nmat965.

3

Inhibiting the Growth of Pancreatic Adenocarcinoma *In Vitro* and *In Vivo* through Targeted Treatment with Designer Gold Nanotherapeutics

Rachel A. Kudgus[1], Annamaria Szabolcs[1], Jameel Ahmad Khan[1], Chad A. Walden[2], Joel M. Reid[2], J. David Robertson[3], Resham Bhattacharya[1], Priyabrata Mukherjee[1,2,4]*

1 Department of Biochemistry and Molecular Biology, College of Medicine, Mayo Clinic, Rochester, Minnesota, United States of America, **2** Department of Physiology and Biomedical Engineering, College of Medicine, Mayo Clinic, Rochester, Minnesota, United States of America, **3** Department of Chemistry and University of Missouri Research Reactor, University of Missouri, Columbia, Missouri, United States of America, **4** Mayo Clinic Cancer Center, College of Medicine, Mayo Clinic, Rochester, Minnesota, United States of America

Abstract

Background: Pancreatic cancer is one of the deadliest of all human malignancies with limited options for therapy. Here, we report the development of an optimized targeted drug delivery system to inhibit advanced stage pancreatic tumor growth in an orthotopic mouse model.

Method/Principal Findings: Targeting specificity *in vitro* was confirmed by preincubation of the pancreatic cancer cells with C225 as well as Nitrobenzylthioinosine (NBMPR - nucleoside transporter (NT) inhibitor). Upon nanoconjugation functional activity of gemcitabine was retained as tested using a thymidine incorporation assay. Significant stability of the nanoconjugates was maintained, with only 12% release of gemcitabine over a 24-hour period in mouse plasma. Finally, an *in vivo* study demonstrated the inhibition of tumor growth through targeted delivery of a low dose of gemcitabine in an orthotopic model of pancreatic cancer, mimicking an advanced stage of the disease.

Conclusion: We demonstrated in this study that the gold nanoparticle-based therapeutic containing gemcitabine inhibited tumor growth in an advanced stage of the disease in an orthotopic model of pancreatic cancer. Future work would focus on understanding the pharmacokinetics and combining active targeting with passive targeting to further improve the therapeutic efficacy and increase survival.

Editor: Soumitro Pal, Children's Hospital Boston & Harvard Medical School, United States of America

Funding: Supported by National Institutes of Health CA135011, CA136494 (PM). The funders had no role in study design, data collection and analysis, decision to publish, or preparation of the manuscript.

* E-mail: Mukherjee.priyabrata@mayo.edu

Introduction

Pancreatic cancer is the fourth leading cause of cancer deaths in America [1]. It continues to have a less than 5% survival rate over 5 years, with a median survival of only six months [2,3]. Pancreatic cancer is an aggressive and illusive cancer that is typically diagnosed at the late stages of the disease where surgical intervention is no longer an option and traditional chemotherapeutics have minimal therapeutic effects.

Gemcitabine is the standard of care for pancreatic cancer treatment [4–8]. The therapeutic efficacy of gemcitabine is governed by the triple phosphorylation within the cell by deoxycytidine kinase (dCK) to an active form; followed by subsequent intercalation into the DNA of the cell leading to the inhibition of DNA synthesis and hence, inhibition of cellular proliferation [9,10]. Despite this being the current protocol, gemcitabine continues to have a modest beneficial outcome on its own in a clinical setting [11,12]. There have been a number of combination therapies that utilize gemcitabine and other drugs or

antibodies in an attempt to enhance the therapeutic effects of the gemcitabine, but all have shown dismal outcomes [4,13–17].

Nanotechnology has the potential to overcome the limitations in current cancer therapeutic options [18–23]. The adverse effects of chemotherapies are an enormous problem in the current treatment of cancer in general, causing systemic toxicity leading to severe side effects. The utilization of monoclonal antibodies conjugated to gold nanoparticles has proven to be effective in targeting cancer cells with an over expression of EGFR (epidermal growth factor receptor) [24,25]. Gold nanoparticles have been shown to be biologically viable and highly adaptable for conjugation with nearly any compound having an amine or thiol functionality utilizing Au-SH, Au-NH$_2$ interactions [18,26–30].

In this study we utilized cetuximab, an anti-EGFR monoclonal antibody, as a targeting agent. Cetuximab was approved by the FDA for the treatment of colorectal cancer, as well as head and neck cancer in 2004 [31–38]. Our group previously reported the effective targeting of EGFR-overexpressing pancreatic cancer cells both *in vitro* and *in vivo* with gold nanoparticles conjugated with C225 as a targeting agent [24]. Utilizing these findings we now incorporated gemcitabine in the nanoformulation to create an optimized targeted drug delivery vehicle to inhibit the growth of pancreatic cancer cells simulating an advanced stage of the disease in an orthotopic model.

The aim of this current study was to develop a gold nanoparticle-based therapeutic with enhanced efficacy to inhibit pancreatic cancer growth in an advanced stage of the disease. The designer therapeutic introduced in this paper is a novel approach to increasing the efficacy of gemcitabine, or any chemotherapy, with the utilization of a targeted delivery system that employs gold nanoparticles as the delivery vehicle. This study was aimed to evaluate the *in vitro* and *in vivo* anti-tumor effect of a gold nanoparticle based targeted drug delivery system that inhibits pancreatic tumor growth in an advanced stage of the disease.

Materials and Methods

Materials

Tetrachloroauric acid trihydrate and sodium borohydride were from Sigma-Aldrich, St. Louis, MO. ^{3}H-thymidine was from Perkin-Elmer, (Waltham, MA). Media and PBS was purchased from Mediatech (Manassas, VA). Scintillation cocktail was purchased from Fisher Scientific.

Synthesis and Characterization of Au-antibody and Au-antibody-gemcitabine Nanoconjugates

The core gold nanoparticles (GNPs) were synthesized by reduction of 1200 ml of 0.1 mM tetrachloroauric acid trihydrate ($HAuCl_4$) solution with 600 ml of a freshly prepared aqueous solution containing 51.6 mg of sodium borohydride ($NaBH_4$) under vigorous and constant stirring, overnight at ambient temperature. Upon addition of the sodium borohydride, the pale yellow solution becomes orange and then turns to a wine red color within minutes. The GNPs were characterized by UV-visible spectrometry, scanning from 400–800 nm and transmission electron microscopy (TEM) after drop-coating 10 μl of the sample on a 400 mesh carbon-coated copper grid followed by side blotting. The size of the nanoparticles was determined from analysis of the TEM images and Dynamic Light Scattering (DLS) (Malvern Zetasizer Nano ZS). Zeta potential measurements were done using a clear zeta disposable capillary (Malvern DTS1061).

The GNP-antibody conjugates (AC4 and AI4) were synthesized by mixing 4 μg/ml of antibody (C225 or IgG, respectively) with the core GNP solution as previously reported. Cetuximab (C225) was purchased as a solution of 2 mg/ml (Erbitux[TM] Injection, ImClone Inc and Bristol-Myers Squibb Co.) and whole molecule human IgG was purchased as a solution of 10.0–11.2 mg/ml (Jackson Immuno Research Laboratories, Inc.). After dilution in 1 ml of water each antibody was added dropwise to the GNP solutions. These solutions were stirred vigorously at ambient temperature for 1 hr. One half of these solutions were centrifuged at 20,000 rpm in a Beckman Ultracentrifuge in a 50.2 Ti rotor to separate AC4 and AI4 nanoconjugates from unconjugated antibody.

The other half of the solution was subjected to further conjugation with various concentrations of gemcitabine (Eli Lilly, Indianapolis, IN) to generate ACG4X and AIG4X (X = 1, 2, 4, 6 and 8 μg/ml). These solutions were also stirred vigorously at ambient temperature for 1 hr and then centrifuged at 20,000 rpm in a Beckman Ultracentrifuge in a 50.2 Ti rotor to separate ACG4X and AIG4X from unbound antibody and gemcitabine. All conjugates formed a loose pellet at the bottom of the centrifuge

tube and were collected after careful aspiration of the supernatant. The gold concentration of the nanoconjugates was determined from absorbances obtained by UV-visible spectrometry (Spectra-Max M5e) at 500 nm (A_{500}) and 800 nm (A_{800}), taken before and after centrifugation and by instrumental neutron activation analysis (INAA). The antibody loading was previously determined [24] and the gemcitabine concentration in the nanoconjugates was determined through high-performance liquid chromatography (HPLC) analysis of the supernatant and subtracted from the total μg added to determine the bound concentration as previously reported [25]. The size and hydrodynamic diameter of the nanoparticle conjugates was determined from analysis of the TEM images and DLS, respectively. Zeta potential measurements were done using a clear zeta disposable capillary (Malvern DTS1061).

The stability of the ACG4X and AIG4X conjugates was tested against 150 mM sodium chloride (NaCl) solution. The absorbance spectrum was taken for all nanoconjugates before and after incubation with NaCl solution for 15 minutes.

Release Study

The release profile of gemcitabine in different biological fluids was characterized following incubation of 100 μl of nanoconjugates having a gemcitabine concentration of 24.4 μg/ml with 100 μl of PBS or mouse plasma and incubated for different time periods. For a control, nanoconjugates were incubated in water, this sample was used to determine the baseline (free) gemcitabine concentration at the zero time point. The samples were centrifuged at 100,000 g for 1 hr and the supernatants were collected. The concentration of gemcitabine in the supernatants was determined by HPLC analysis. The total concentration of gemcitabine at different time points was plotted after subtracting the baseline concentration.

Cell Culture

Pancreatic cancer cell lines: AsPC-1, PANC-1 and MiaPaca-2 were grown in RPMI and Dulbecco's Modified Eagles medium (Gibco). All media was supplemented with 10% fetal bovine serum (Gibco) and 1% antibiotics (Penicillin/Streptomycin) and the cell lines were maintained at 37°C, in a humidified atmosphere under 20% O_2 and 5% CO_2.

[3]H-thymidine Incorporation Assay

For each cell line, (3×10^4) cells were seeded in 24-well culture plates in their respective media and allowed to incubate overnight (16 hrs). After incubation for 2 hrs with ACG44, AIG44 and no gold controls at 0.1, 1 and 10 μM concentrations of gemcitabine cells were washed with PBS to remove unbound nanoconjugates and replaced with fresh media followed by additional incubation for another 48 hrs. At the end of 48-hr incubation the experiment, the media in the wells was replaced with ^{3}H-thymidine containing media (1 μCi/mL) and incubated for an additional 4 hrs at 37°C and processed for the assay as described previously [39]. Experiments were repeated at least three times, in triplicate each time, averages and standard deviations are reported.

In vitro Targeting Studies

To determine the effect of serum components on the targeting efficacy of the nanoconjugates, we preincubated AC4, AI4, ACG44 and AIG44 in RPMI containing fetal bovine serum for 15 mins. After incubation we studied cellular uptake in AsPC-1 cells in 100 mm tissue culture dishes. Both preincubated and as synthesized nanoconjugates were added to the cells in separate culture dishes and incubated at 37°C for 2 hrs. After the treatment the media was removed and the cells were washed once with PBS to remove excess nanoconjugates and trypsinized to obtain a cell pellet for gold content determination by INAA.

Inhibitor Studies

Cells were pre-treated with either C225 or NBMPR (Nitro-benzylthioinosine) to determine the specificity of uptake of ACG44 and AIG44 nanoconjugates. AsPC-1 cells were grown in 100 mm tissue culture dishes and pre-treated with either 50 µg/mL of C225 or 100 nM NBMPR for 1 hr followed by the addition of 2 µg/mL ACG44 and AIG44, respectively. After 2 hrs, the culture medium was removed, the plates were washed once with PBS to remove any unbound nanoconjugates and cells were trypsinized to obtain a cell pellet to determine gold content by INAA.

Measurement of Gold Content by Instrumental Neutron Activation Analysis (INAA)

Samples were analyzed by instrumental neutron activation analysis at the University of Missouri Research Reactor Center as previously described [24,25,40–43]. Briefly, cell pellets/tissues were prepared by weighing the samples into high-density poly-ethylene irradiation vials and lyophilized to a dry weight. Solution samples were prepared by gravimetrically transferring 100 µl to an irradiation vial followed by lyophilization. All samples were loaded in polyethylene transfer "rabbits" in sets of nine and irradiated for 90 sec in a thermal flux density of approximately 5×10^{13} n cm^{-2} s^{-1}. The samples were then allowed to decay for 24 to 48 hrs and counted on a high-purity germanium detector for 3600 sec at a sample-to-detector distance of approximately 5 cm. The mass of gold in each sample was quantified by measuring 411.8 keV gamma ray from the β^- decay of ^{198}Au ($t_{1/2} = 2.7$ days). The area of this peak was determined by the Genie ESP spectroscopy package from Canberra. A minimum of six geo-metrically equivalent comparator standards were also run. The standards were prepared by aliquoting approximately 0.1 (n = 3) and 0.01 (n = 3) µg of gold from a (10.0±0.5) µg/mL certified standard solution (High-Purity Standards) in the polyethylene irradiation vials, and were used with each sample set.

Transmission Electron Microscopy (TEM)

TEM samples (cell pellets and tissues) were fixed in trumps solution and processed as previously described [24]. Micrographs were taken on a TECNAI 12 operating at 120 KV.

Animal Handling and in vivo Tumor Uptake

Male athymic nude mice (NCr-nu; 4–6 weeks old) were purchased from the National Cancer Institute-Frederick Cancer Research and Development Center (Frederick, MD). All mice were housed and maintained under specific pathogen-free conditions in facilities approved by the American Association for Accreditation of Laboratory Animal Care and in accordance with current regulations and standards of the U.S. Department of Agriculture, U.S. Department of Health and Human Services, and NIH. All studies were approved and supervised by the Mayo Clinic Institutional Animal Care and Use Committee.

For the generation of orthotopic pancreatic tumor models, before injection, tumor cells were washed twice with PBS, lifted with 0.25% trypsin, centrifuged for 5 minutes, and reconstituted in PBS. AsPC-1 cells (1.5×10^6) were implanted into the pancreas of nude mice. The mice were imaged non-invasively every week under isoflurane anesthesia using a Xenogen-IVIS-cooled CCD optical system (Xenogen-IVIS) as previously described [24]. Seven

days after tumor cell implantation, the mice were randomized into 6 groups (n = 5). All nanoconjugates were normalized to a gold concentration of 450 µg/mouse (1.8 mg/kg of gemcitabine) and injected i. p. thrice for week one and twice for weeks two and three. The mice were sacrificed on day 28 and the tumors, kidneys, liver, lungs, spleen, pancreas and blood were collected and analyzed for gold content through INAA.

Immunohistochemistry

Tumor samples were fixed in 4% paraformaldehyde and stained for hemotoxylin and eosin (H&E) and Ki-67. The percentage of Ki-67 positive cells in 5 high-powered fields at 20X was determined through visual inspection and counting.

Statistical Analysis

Statistical analysis was done by a two-tailed student t-test and a value of $P < 0.05$ was considered to be significant.

Results and Discussion

Synthesis and Characterization of Gold Nanoparticles and Nanoconjugates

Physicochemical characterization of the unmodified GNPs and the nanoconjugates were performed by UV-visible spectroscopy (UV-vis), transmission electron microscopy (TEM), dynamic light scattering (DLS) and zeta potential (ζ-Potential) measurements.

The UV-visible spectrum of unmodified GNPs exhibits a characteristic surface plasmon resonance (SPR) band of spherical gold nanoparticle at 512 nm as previously reported [24,25]. Addition of an anti-EGFR antibody cetuximab (C225), or its non-targeted counterpart immunoglobulin G (IgG) at a concentration of 4 µg/ml to the GNP solution increases the absorbance of the solution with a simultaneous red shift in the λ_{max} value from 512 nm of unmodified GNP to 518 nm for Au-C225 (AC4) and Au-IgG (AI4) conjugates [24,25], respectively. Such a red shift in the SPR band suggests binding of C225 and IgG to the GNP surface as previously reported [24,44]. The rationale for selecting a concentration of 4 µg/ml of C225/IgG was based on our previous report where we demonstrated that the AC4 conjugates have the highest ability to target EGFR-overexpressing pancreatic cancer cells both in vitro and in vivo in an orthotopic pancreatic cancer model [24]. Based on this information, we next designed a targeted drug delivery system with the incorporation of gemcitabine (Gem) on the AC4 and AI4 nanoconjugates. The addition of different concentrations of gemcitabine (1, 2, 4, 6, and 8 µg/ml) to the AC4 and AI4 solutions caused a further red-shift in the λ_{max} from 518 nm to 520 nm for the ACGs and AIGs, suggesting binding of gemcitabine to the available reactive surface on the gold nanoparticle (Figures S1A,B).

Stability of the nanoconjugates in 150 mM sodium chloride (NaCl) further supports the binding of both C225/IgG and gemcitabine on the GNPs. It is evident from Figure 1A and B that the addition of 150 mM NaCl to unmodified GNPs shifted the λ_{max} from 512 nm to 562 nm with a strong decrease in absorbance, suggesting significant aggregation of uncovered GNPs by NaCl in the absence of surface protection by C225/IgG and Gem. However, the addition of NaCl to the ACGs and AIGs did not alter the absorbance and λ_{max} value, thereby confirming the stabilization of GNPs by conjugation with C225/IgG and gemcitabine. TEM analysis further confirms the absence of aggregation as well as the formation of ~5 nm spherical nanoparticles (Figure 1C, D). Dynamic light scattering and zeta potential results were used to further characterize the hydrody-

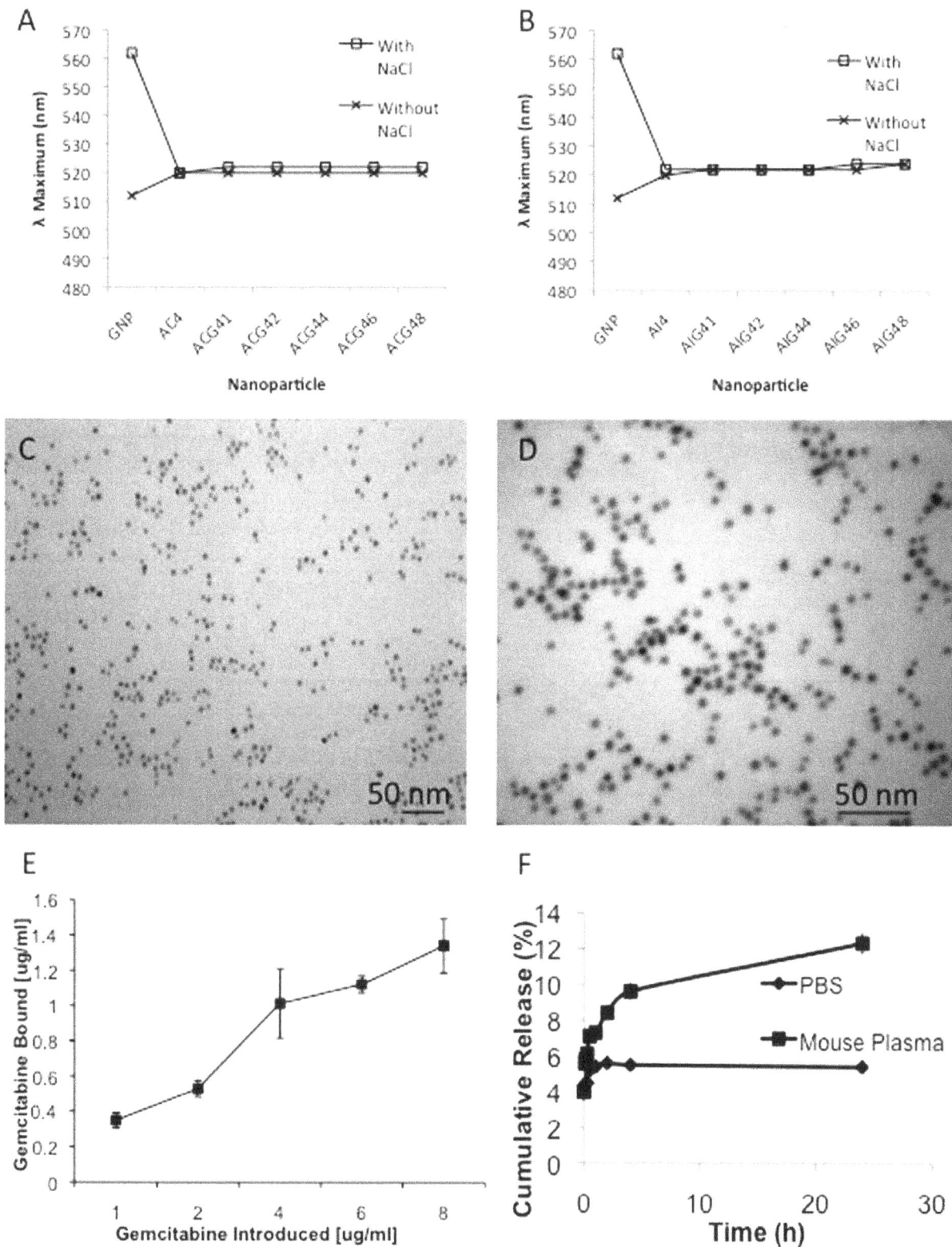

Figure 1. Physicochemical characterization of gold nanoconjugates. Figure 1A and 1B describes the changes in the λ_{max} value of GNP and different ACG44 and AIG44 nanoconjugates, respectively, with/without incubation with NaCl for 15 minutes. Figure 1C and 1D exhibits the transmission electron micrographs (TEM) of ACG44 and AIG44 nanoconjugates, drop coated after synthesis without alteration. Figure 1E describes the amount of gemcitabine bound to AC4 nanoconjugates analyzed by HPLC; the x-axis shows the µg/ml used to synthesize the conjugates and the y-axis represents the µg/ml bound to the particle. Figure 1F describes the release of gemcitabine from ACG44 nanoconjugates when incubated in PBS and mouse plasma over time.

Table 1. Dynamic Light Scattering and Zeta Potential Measurements of GNP (the core particle), ACG44 and AIG44.

Sample	DLS (d.nm)	Zeta Potential (mV)
GNP*	5.3	−29.6
ACG44	32.09	−20.2
AIG44	88.95	−20.6

namic diameter (HD) as well as the charge of the particles at each step of conjugation (Table 1 and Table S1).

A MELVERN Zetasizer Nano ZS instrument was used to measure both the DLS and ζ-Potentials of all the nanoconjugates. The average of 5 independent runs is presented in Table 1. As expected, the HD increases from 5 nm for the unmodified GNP, to approximately 38 nm for AC4 and 56 nm for AI4, further demonstrating the binding of C225 and IgG to GNP. The addition of gemcitabine, at all concentrations does not alter the HD of the nanoconjugates; this observation was expected due to the small molecule size of gemcitabine. The ζ-potential measurements also followed a similar trend; as expected, the GNP was most negative, with an average of approximately −30 mV. The ζ-Potential starts to become less negative with the addition of C225 and IgG due to the antibody binding. In turn, there is a minor increase in ζ-potential with the addition of gemcitabine. These results clearly suggest the binding of C225/IgG and gemcitabine to the nanoparticle to form ACGs and AIGs.

Quantifying the Gemcitabine Loading on AC4 and AI4

Using varying concentrations of ^{125}I-labeled C225 and IgG we previously demonstrated that nearly 90% of 4 µg/ml antibody was bound to the GNP [24]. To determine the loading of gemcitabine on AC4 and AI4, gemcitabine was added to the antibody covered particles in various concentrations (X = 1, 2, 4, 6 and 8 µg/ml) under vigorous stirring at room temperature. After one hour, the nanoconjugates containing gemcitabine were purified by ultracentrifugation. The ACG4X and AIG4X pellets were collected at the bottom of the centrifuge tubes and the supernatant, containing the unbound gemcitabine was analyzed using HPLC. The bound fraction was calculated by subtracting the concentration of gemcitabine present in the supernatant from the original concentration introduced during the nanofabrication process. The concentration of bound gemcitabine increases from approximately 0.4 to 1 µg/ml for the ACG4X particles with the introduction of 1–4 µg/ml (Figure 1E). However, the addition of 6 and 8 µg/ml of gemcitabine does not significantly increase the amount gemcitabine in the nanoconjugate. Therefore, the optimum conjugation reaction with the highest percent yield for gemcitabine bound to the particle was determined to be 4 µg/ml. As expected, the AIG4X particles showed a similar trend for gemcitabine binding (data not shown). Subsequently, the ACG44 and AIG44 nanoconjugates were employed for all further experiments.

The ACG44 and AIG44 nanoconjugates were synthesized using the AC4 and AI4 solutions, respectively, as described in the materials and methods section. This synthesis was possible due to the spontaneous binding of the antibodies through their cysteine/lysine residues utilizing the Au-S/Au-NH$_2$ bond as previously described [24]. It has been well described in the literature that interactions of proteins, antibodies and small molecules with gold could be due to electrostatics, covalent bonding or hydrophobic

interactions. The addition of gemcitabine exploits the Au-NH$_2$ binding through the amine moiety on gemcitabine to the gold particle surface. Our previous data suggests the initial binding is due to electrostatics and then covalent, investigated with XPS and TGA analysis and previously reported by our group [25,45].

Mechanism and Targeting Efficacy of ACG44 and AIG44 in vitro

Studies were performed to ascertain the targeting efficacy of C225 and the functional activity of gemcitabine in the nanoconjugated form against various pancreatic cancer cell lines *in vitro*. The targeting experiments were performed in two ways to address the role of serum components on targeting efficacy; (i) the nanoconjugates were preincubated with cell growth media for 15 minutes before adding to the cells in tissue culture dishes; (ii) as synthesized nanoconjugates were directly added to the cells in tissue culture dishes. It is evident from the Figure 2A and Figure S2 that the addition of C225/IgG increases the absorbance of the GNPs followed by a red shift in the λ$_{max}$. Subsequent addition of gemcitabine further shifts the λ$_{max}$ while simultaneously increasing the absorbance, suggesting binding of both the components to the GNPs. Interestingly, preincubation with cell growth media decreases the absorbance of all the nanoconjugates probably due to the formation of a protein corona around the GNPs [46]. Additionally, TEM analysis showed efficient intracellular uptake of the ACG44 nanoconjugates in AsPC-1 cells (Figure 2B) and no significant aggregation of the particles in the presence of serum was observed (Figure S3) further supporting the formation of protein-corona after pre-incubation with the serum.

The effect of preincubation on the intracellular uptake of different nanoconjugates (AC4, ACG44, AI4 and AIG44) with AsPC-1 cells *in vitro* was determined. Uptake efficiency was determined by measuring the gold content in the cell pellet through INAA analysis (Figure 2C). It is evident from the Figure that AC4 and ACG44 are far more effective in targeting AsPC-1 cells than AI4 and AIG44. It is also interesting to note that cellular uptake of ACG44 is higher than AC4. This enhanced uptake of ACG44 could be due to the presence of gemcitabine, mediating uptake through nucleoside transporters (NTs), or due to an alternative path because of the available reactive surface area on gold nanoparticles. The non-specific uptake appears to be minimized after preincubation, which could be due to the coating of serum components on the available gold surface. Non-specific uptake of AI4 and AIG44 was completely inhibited by preincubation with the cell growth media. These data clearly suggest that C225 retains its specific targeting ability to EGFR-expressing cancer cells *in vitro*. The ability of C225 in the nanoconjugated form to effectively target EGFR over-expressing AsPC-1 cells was further confirmed by preincubation with C225. Preincubating AsPC-1 cells with C225 greatly diminished the uptake of ACG44, whereas there was no intracellular uptake of AIG44. These results further confirm that the endocytosis of ACG44 is via the EGFR pathway. Similarly, no difference in intracellular uptake was observed when cells were pre-treated with a nucleoside transporter blocker (NBMPR) [47–49], suggesting the absence of nucleoside transporter mediated uptake of the nanoconjugates (Figure 2E).

Testing Functional Activity of Gemcitabine in the Nanoconjugated Form with Various Pancreatic Cancer Cell Lines in vitro

We utilized three different pancreatic cancer cell lines, AsPC-1, PANC-1 and MiaPaca-2, all having variable EGFR expression, to test whether the activity of gemcitabine had been retained in the

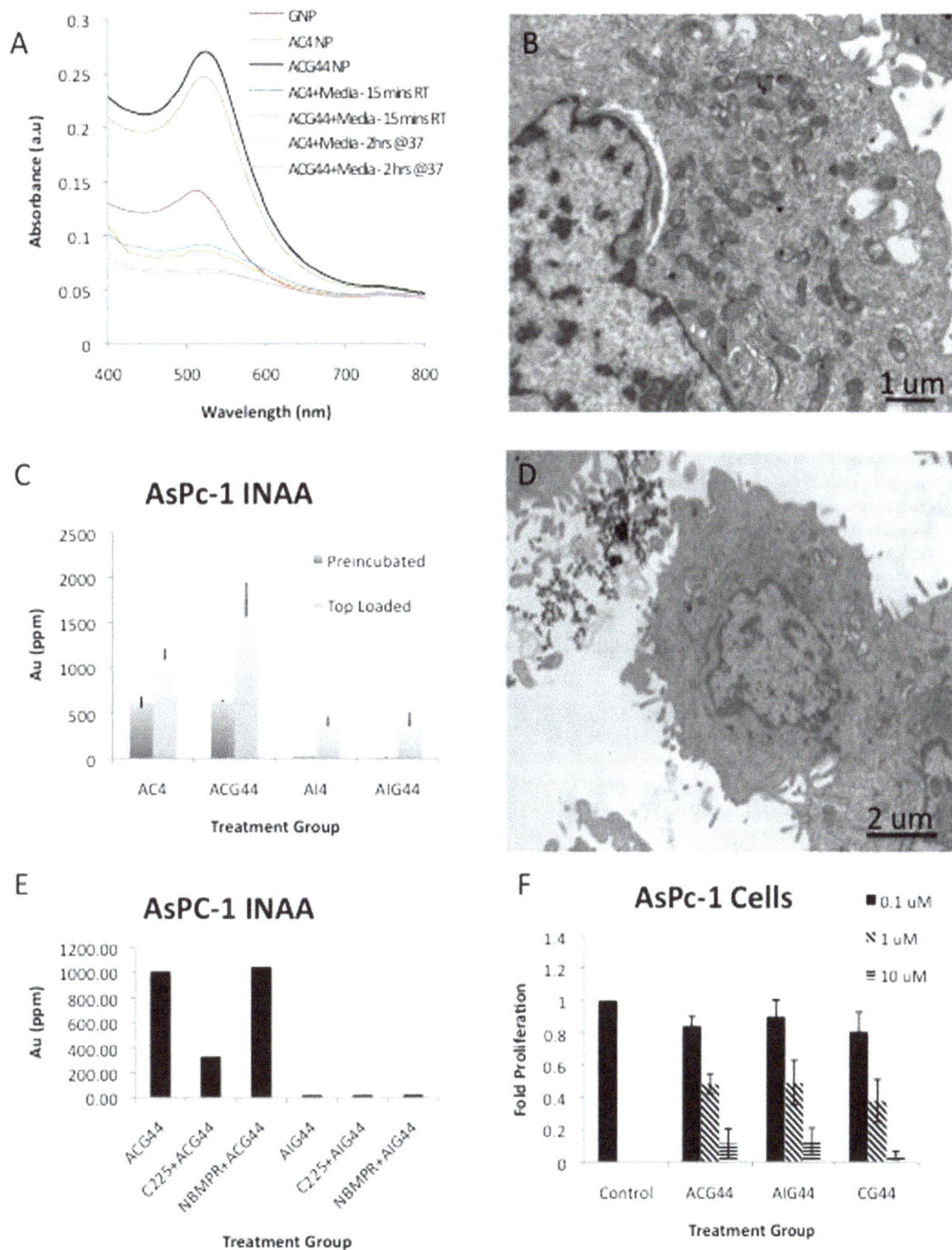

Figure 2. Role of pre-incubation with serum, C225 and NBMPR on targeting efficacy of gold nanoconjugates and their *in vitro* biological function. Figure 2A depicts the absorbance spectrums of GNP, AC4 and ACG44 before and after pre-incubation with serum either 15 minutes at room temperature or 2 hrs at 37°C. Figure 2B and 2D are transmission electron microscopy images of the *in vitro* uptake of ACG44 and AIG44 in AsPC-1 cells, respectively. Figure 2C describes the effect of pre-incubation with serum on the cellular uptake of the nanoconjugates into AsPC-1 cells analyzed for gold content utilizing INAA. Figure 2E also depicts INAA analysis of cellular uptake of ACG44 and AIG44 in AsPC-1 cells, both with/without pre-incubation with C225 or NBMPR to demonstrate possible uptake mechanisms. Figure 2F shows the anti-proliferative effect of ACG44, AIG44 and CG44 on AsPC-1 cells determined through ^{3}H-thymidine incorporation.

nanoconjugated form. It is evident from the Figures that ACG44, AIG44 and CG44 (a no gold control containing gemcitabine and C225) substantially inhibited proliferation of pancreatic cancer cells in a concentration dependent manner as determined by ^{3}H-thymidine assays. It is also apparent from the Figure that maximum inhibition of AsPC-1 cells was observed at the highest dose of both ACG44 and AIG44, which is comparable to the same dose of C225 and gemcitabine in the no gold control, CG44. A similar trend was observed with PANC-1 and MiaPaca-2 cells (Figure 2F and Figure S4). These results clearly demonstrate that gemcitabine retains its functional activity in the nanoconjugated form.

Stability of the Nanoconjugates in Biological Fluids

It is important to test the stability of the nanoconjugates in biological fluids before *in vivo* applications. The stability of the nanoconjugates in terms of gemcitabine release was performed in phosphate buffered saline (PBS) and in mouse plasma. It is evident from the Figure 1F that the ACG44 nanoconjugates are very stable both in PBS (only 5% of bound gemcitabine is released over a period of 24 hrs) as well as in mouse plasma (~12% of gemcitabine released over a period of 24 hrs). Furthermore, for future clinical application it is important to determine the stability of the nanoconjugates after long-term storage. To address this concern, we flash froze the nanoconjugates in liquid nitrogen and lyophilized them to a powder form. The powder was easily reconstituted in water, with no noticeable aggregation observed upon visual inspection. The solution was then tested for functional efficacy to inhibit the proliferation of AsPC-1 cells. Similar inhibition in proliferation of pancreatic cancer cells was observed by the reconstituted nanoconjugates, confirming the stability of the nanoconjugates under long-term storage conditions (Figure S5).

Therapeutic Efficacy of the Nanoconjugates to Inhibit Tumor Growth in an Orthotopic Model of Pancreatic Cancer

To test the therapeutic efficacy of the nanoconjugates to inhibit tumor growth *in vivo*, we generated an orthotopic model of pancreatic cancer by implanting 2×10^6 AsPC-1 cells directly into the pancreas of 4–6 week old nude male mice. To simulate an advanced stage of the disease, tumors were allowed to grow for 7 days before initiation of the treatment (Figure 3A). One week after the tumor cell implantation, the mice were imaged non-invasively through bioluminescence for tumor growth and randomized into 6 groups (n = 5) before initiation of the treatment. The animals were injected in the intraperitoneal (i.p.) cavity thrice a week for the first week and twice for the next two weeks with 450 μg of gold nanoconjugates, 1.8 mg/kg of gemcitabine. The tumor progression was monitored non-invasively once a week with bioluminescence imaging as discussed above (Figure 3B). Bioluminescence measurements clearly demonstrate a significant reduction of tumor growth in ACG44 group as compared to the other groups. As expected, AC4 and AI4 had the least effect on decreasing the tumor growth while the AIG44 and CG44 groups had a moderate response due to the presence of gemcitabine. These observations were further confirmed by directly measuring tumor weight and volume after sacrificing the mice at the end of the experiment (Figure 3C and D). The ACG44 group showed the tightest cluster of data points as well as the highest therapeutic effect of all groups with an average tumor mass of 0.3 grams and tumor volume of 275 cm^3 (0.55 g and 940 cm^3 for the PBS group). Likewise, this therapeutic effect was also shown in the tumor gold content for the ACG44 group (Figure 3E).

Further TEM analysis confirmed the uptake of ACG44 into the tumor tissue (Figure 3F), also show AIG44 internalization in Figure S7. Additionally, the biodistribution of gold in the liver, lungs, kidneys and spleen was also analyzed with INAA (Figure S6). As expected, the organs with the greatest uptake of gold were the spleen and the liver in all of the treatment groups. These results were further confirmed at the molecular level by quantifying the number of proliferating cells in different treatment groups using hematoxilin-eosin (H&E) and Ki-67 staining (Figure 4). Figure 4 exhibits H&E and Ki-67 for tumor tissue from the PBS (4A and B) and ACG44 groups (Figure 4C and D), respectively. The number of cancer cells as demonstrated by H&E staining are far less in ACG44 group than in the control PBS group. Furthermore, the number of proliferating cell nuclei as demonstrated by Ki-67 staining, is significantly decreased (~60%) in the ACG44 group as opposed to the control PBS group (Figure 4E). Interestingly, treatment with ACG44 also revealed the presence of gold nanoparticles in the form of black specs inside the tumor tissue, suggesting better tumor penetration with the ACG44 nanoconjugates compared to the other groups. Additionally, it is evident from Figure 4 that the number of Ki-67 positive cells is greatly diminished surrounding the nanoparticles, suggesting a better therapeutic effect correlating with proximity to the nanoparticles. Taken together, these data indicate that the targeted delivery of a low dose of gemcitabine via gold nanoparticles can significantly inhibit the tumor growth in an advanced model of orthotopic pancreatic cancer *in vivo*.

Discussion

Pancreatic cancer is one of the deadliest among human malignancies with no effective therapies currently available [11,15]. Pancreatic cancer is typically detected in the late stage of the disease and the overall survival from this late diagnosis is commonly just a few months [1]. Surgery is the only option when detected early and gemcitabine is the front line chemotherapy for the advanced stage of the disease. However, chemotherapy with gemcitabine has only a limited therapeutic effect due to severe dose limiting toxicity. Thus, the overall survival is only 5–6 months. For this reason, effective therapeutic strategies are urgently required to combat this deadly disease. Therefore, any approach that will reduce the systemic toxicity and increase the efficacy of gemcitabine will have a significant impact on the therapeutic management of pancreatic cancer.

It has been recently recognized that nanotechnology has great potential to improve the quality of lives of cancer patients [22,50,51]. Major areas where nanotechnology can impact significantly in cancer are; i) Detection/diagnosis; (ii) Imaging and (iii) Therapeutics [52,53]. Specifically, targeted drug delivery could significantly impact the therapeutic management of cancer by increasing the efficacy and reducing systemic toxicity of many chemotherapeutics. Development of a nanoparticle formulation to effectively target tumors is an area of active investigation. Among the inorganic nanomaterials, gold nanoparticles have generated considerable interest in various biomedical applications including targeting [27]. Several advantages that GNPs have over other nanomaterials are; (i) ease of synthesis; (ii) ease of characterization due to the presence of the SPR band; (iii) ease in binding of biomolecules such as peptides, proteins and antibodies exploiting the gold-thiol, gold-amine interactions and most importantly, (iv) biocompatibility compared to other inorganic nanomaterials currently being investigated [30,54–57]. Previously we demonstrated that the targeted delivery of a low dose of gemcitabine using cetuximab bound to a gold nanoparticle in a "2 in 1" fashion

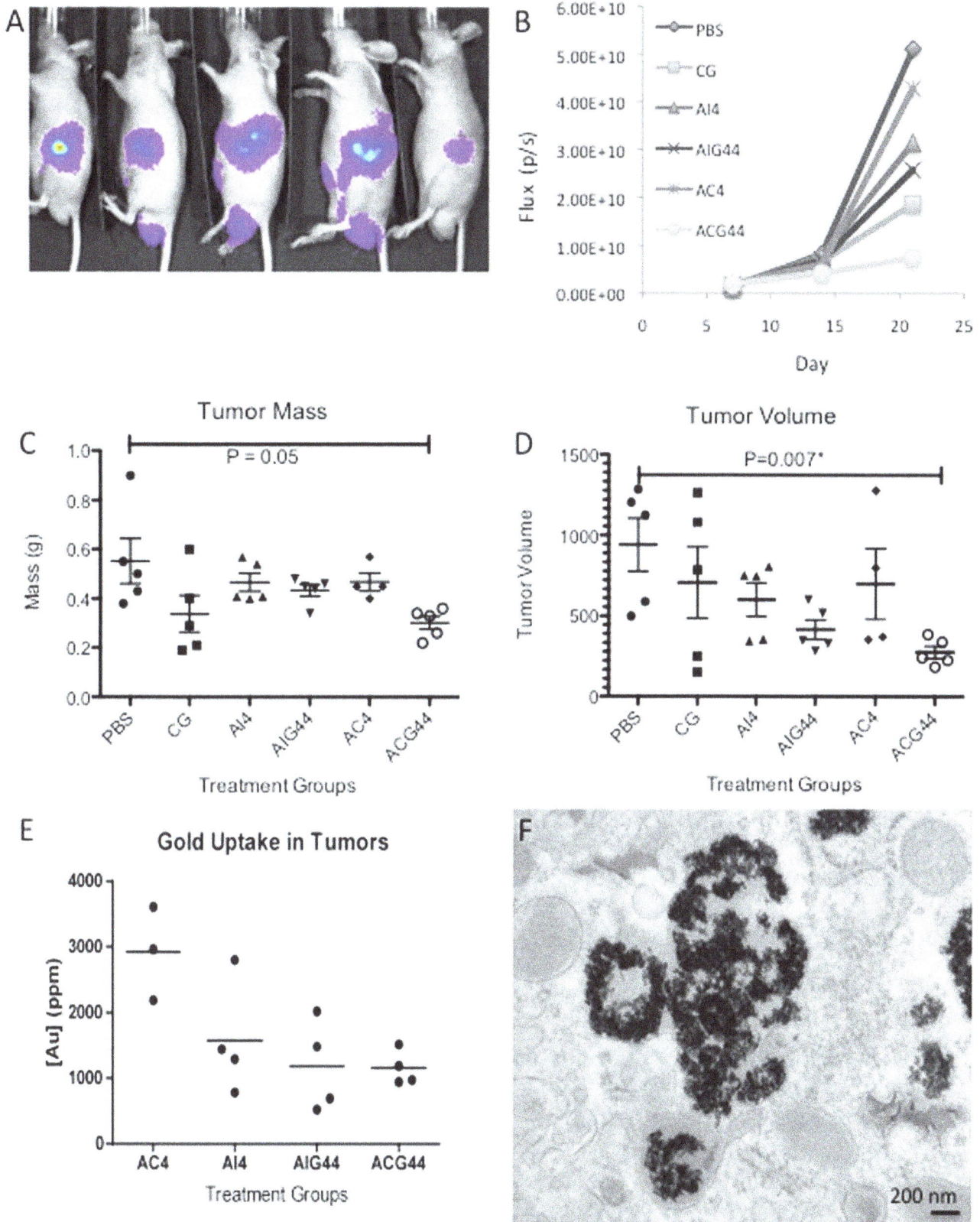

Figure 3. *In vivo* effects of gold nanoconjugates in an orthotopic model of pancreatic cancer. Figure 3A shows a representative bioluminescence image of 5 mice, 7 days after orthotopic implantation of AsPC-1 cells into the pancreas. Figure 3B is flux quantification from the bioluminescence imaging taken every 7 days. Figure 3C, 3D and 3E all show scatter plots of tumor analysis of each animal per group, post study termination. Figure 3C shows tumor mass, 3D shows tumor volume determined through caliper measurements and 3E shows total gold uptake in each tumor determined by INAA. Figure 3F is a TEM micrograph showing ACG44 conjugates in a cross section of tumor tissue.

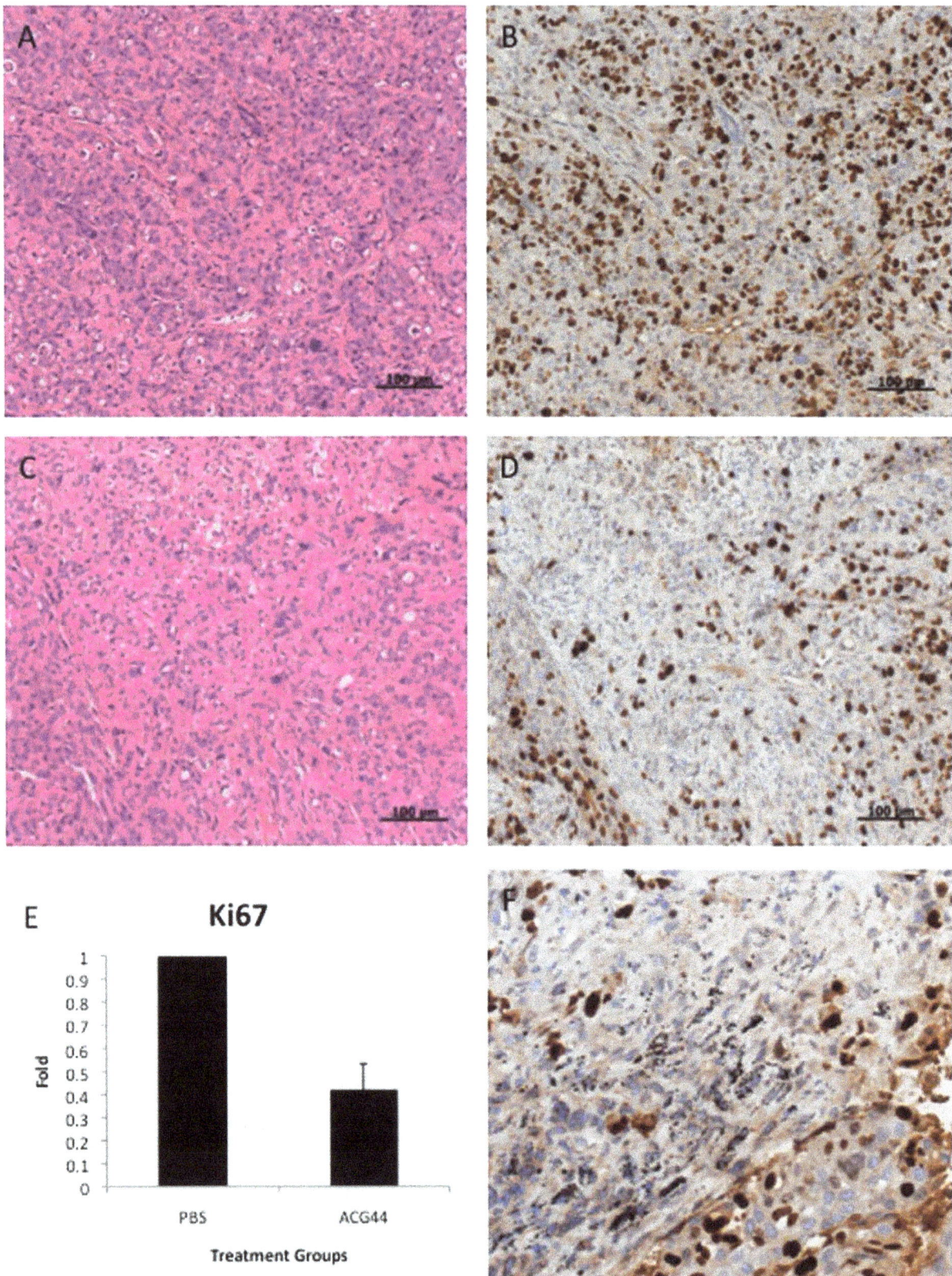

Figure 4. Immunohistochemistry Analysis of Tumors from the PBS and ACG44 groups. Figure 4A and 4B show representative images of H&E and Ki-67 stained tumor tissues, respectively, from the PBS treated group whereas Figure 4C and 4D show images of H&E and Ki-67 staining of tumor tissue from the ACG44 treated group. All images were taken with 20× magnification. Figure 4E is quantification of the Ki-67 positive proliferative nuclei shown in Figures 4B and 4D. Figure 4F is a tumor image of Ki-67 staining from the ACG44 treated group, taken at 100 X to show gold accumulation (black spots) at a high magnification.

resulted in significant inhibition of tumor growth at an early stage of an orthotopic model of pancreatic cancer [25]. EGFR is overexpressed in a number of human malignancies. The FDA approved cetuximab, an anti-EGFR antibody, for the use alone or in combination therapies to treat various malignancies. Therefore, the successful development of a targeted drug delivery system containing C225, gemcitabine and gold nanoparticles could be widely applicable to a variety of malignances including pancreatic cancer. Recently, we defined the design criteria to effectively target pancreatic cancer cells in an orthotopic model of pancreatic cancer *in vivo* [24]. Based on this prior work, we report here the design of an optimized targeted drug delivery system that inhibits pancreatic tumor growth in an advanced stage of the disease in an orthotopic model.

The binding of C225 and gemcitabine to the gold nanoconjugates was demonstrated by an increase in absorbance and a red shift of the absorption maxima in the SPR band of gold nanoparticles. These findings were further confirmed by stability testing in 150 mM NaCl. Uncovered nanoparticles undergo rapid aggregation in the presence of high salt concentration. However, such aggregation is prevented through surface coverage of the gold nanoparticles by a combination of cetuximab and gemcitabine.

Gemcitabine, being a purine nucleoside requires nucleoside transporters, such as human equilibrative nucleoside transporters (hENT) or concentrative nucleoside transporters (hCNT) for intracellular uptake [58–61]. Therefore, ACG44 may be taken up by the cells either via EGFR endocytosis through active targeting with C225 or via nucleoside transporters (NTs). In general, nanoconjugates having C225 exhibit much higher cellular uptake compared to nanoconjugates having the non-targeting antibody, IgG. Furthermore, preincubating pancreatic cancer cells that overexpress EGFR, such as AsPC-1, with C225 significantly reduces the uptake of ACG44 nanoconjugates, while the uptake of AIG44 remains unaffected. Additionally, treatment with a NT inhibitor, NBMPR, does not reduce the uptake of either of the nanoconjugates. Together, these results support the conclusion that ACG44 enters into the cells through EGFR mediated endocytosis and that NTs do not contribute significantly to the uptake of the nanoconjugates. These findings could prove beneficial in combating gemcitabine resistance in cancer cells with a low expression of NTs. It is also important to note that preincubating the nanoconjugates in serum, decreases the uptake of both AC4 and ACG44, suggesting the adsorption of serum proteins blocks the available reactive surface on the gold particle and thereby increases the specificity of targeting. Likewise, preincubation of AI4 and AIG44 in serum reduced the intracellular uptake, presumably by blocking the available surface area on the gold particle that is involved in non-specific uptake. The fact that AI4 and AIG44 uptake is reduced to a non-detectable level, further confirms the targeting specificity of AC4 and ACG44.

Stability of the nanoconjugates under the physiological salt concentrations (150 mM NaCl) and in biological fluids (PBS and mouse plasma) demonstrates the suitability of these nanoconjugates for *in vivo* use. Only 12% of gemcitabine was released over a 24 hr period in mouse plasma. Previously, we demonstrated a similar amount of C225 released under various physiological environments [24].

Finally, the efficacy of the nanoconjugates *in vivo* in an aggressive orthotopic model of pancreatic cancer was demonstrated. It is important to note here, that treatments in many studies are typically initiated in this model on day 3 or 4 after tumor cell implantation in the pancreas. However, to simulate an advanced stage of the disease, we allowed the tumor cells to grow in the pancreas for 7 days before initiating the treatment. Typically, all the animals would die in a model like this within 3–4 weeks of implantation. However, our *in vivo* data clearly demonstrate that a low dose of gemcitabine delivered in a targeted fashion significantly reduced the tumor growth in this advanced stage model. This study highlights the potential of a gold nanoparticle based targeted drug delivery system to inhibit tumor growth in a orthotopic model of pancreatic cancer in advanced stage of the disease.

Conclusion

In conclusion, we demonstrated that a low dose of gemcitabine delivered in the form of a targeted drug delivery system inhibits tumor growth in an advanced stage of an orthotopic model of pancreatic cancer. Using different cellular uptake path inhibitors we demonstrated that the uptake of the nanoconjugates is specific to the EGFR pathway. We also demonstrated significant stability of the nanoconjugates under different biological environments, as well as long-term stability of the nanoconjugates after lyophilization and storage. The potential impact of this study to inhibit pancreatic tumor growth at the advanced stage is significant, as no therapy is currently available for this deadly disease. Future work will focus on further improving the therapeutic efficacy, understanding the pharmacokinetics and combining active targeting with passive targeting.

Supporting Information

Figure S1 Representative absorption spectrums of GNP, AC4, AI4 and the addition of various loadings of gemcitabine. Figure S1A represents the absorption spectra of ACG4X nanoconjugates, after incubation of AC4 with gemcitabine (X = 1, 2, 4, 6 and 8 µg/ml) for 1 h. Figure S1B represents the absorption spectra of AIG4X conjugates.

Figure S2 Absorption spectrum showing the role of pre-incubation with serum of AIG44. Figure S2 depicts the absorbance spectrums of GNP, AI4 and AIG44 before and after pre-incubation with serum either 15 minutes at room temperature or 2 hrs at 37°C.

Figure S3 Transmission electron microscopy images showing the effect of pre-incubating with serum on conjugate shape and size. Figure S3A is AC4 pre-incubated with serum. Figure S3B is AI4 pre-incubated with serum. Figure S3C is ACG44 pre-incubated with serum and Figure S3D is AIG44 pre-incubated with serum.

Figure S4 *In Vitro* effect of nanoconjugates on pancreatic cancer cell lines. Figure S4A and S4B shows the anti-proliferative effect as determined by ^{3}H-thymidine incorporation assay, of ACG44, AIG44 and CG44 on MiaPaCa-2 and Panc-1 cells, respectively.

Figure S5 Effect of lyophilizing the nanoconjugates on *in vitro* proliferation with AsPc-1 cells as determined by ^{3}H-thymidine incorporation assay.

Figure S6 Tissue distribution of the nanoconjugates in vital organs. Cumulative gold uptake concentrations in the lung, kidney, liver and spleen. Shown in ppm and measured by INAA.

Figure S7 *In vivo* **internalization of AIG44 in tumor tissue depicted with a TEM image.**

Table S1 **Dynamic Light Scattering and Zeta Potential Analysis of all Nanoconjugates.**

Author Contributions

Conceived and designed the experiments: RB JMR PM. Performed the experiments: RK AS JAK JDR CW. Analyzed the data: RK AS JAK JDR CW RB JMR PM. Contributed reagents/materials/analysis tools: JMR JDR. Wrote the paper: RK AS JAK JDR CW RB JMR PM.

References

1. Jemal A, Siegel R, Xu J, Ward E (2010) Cancer statistics, 2010. CA Cancer J Clin 60: 277–300.
2. Shore S, Vimalachandran D, Raraty MG, Ghaneh P (2004) Cancer in the elderly: pancreatic cancer. Surg Oncol 13: 201–210.
3. Maitra A, Hruban RH (2008) Pancreatic cancer. Annu Rev Pathol 3: 157–188.
4. Burris HA 3rd, Moore MJ, Andersen J, Green MR, Rothenberg ML, et al. (1997) Improvements in survival and clinical benefit with gemcitabine as first-line therapy for patients with advanced pancreas cancer: a randomized trial. J Clin Oncol 15: 2403–2413.
5. Burris H, Storniolo AM (1997) Assessing clinical benefit in the treatment of pancreas cancer: gemcitabine compared to 5-fluorouracil. Eur J Cancer 33 Suppl 1: S18–22.
6. Rathos MJ, Joshi K, Khanwalkar H, Manohar SM, Joshi KS (2012) Molecular evidence for increased antitumor activity of gemcitabine in combination with a cyclin-dependent kinase inhibitor, P276-00 in pancreatic cancers. J Transl Med 10: 161.
7. Wong A, Soo RA, Yong WP, Innocenti F (2009) Clinical pharmacology and pharmacogenetics of gemcitabine. Drug Metab Rev 41: 77–88.
8. Bayraktar S, Bayraktar UD, Rocha-Lima CM (2010) Recent developments in palliative chemotherapy for locally advanced and metastatic pancreas cancer. World J Gastroenterol 16: 673–682.
9. Plunkett W, Huang P, Xu YZ, Heinemann V, Grunewald R, et al. (1995) Gemcitabine: metabolism, mechanisms of action, and self-potentiation. Semin Oncol 22: 3–10.
10. Hidalgo M (2010) Pancreatic cancer. N Engl J Med 362: 1605–1617.
11. Kleeff J, Michalski C, Friess H, Buchler MW (2006) Pancreatic cancer: from bench to 5-year survival. Pancreas 33: 111–118.
12. Philip PA (2010) Novel targets for pancreatic cancer therapy. Surg Oncol Clin N Am 19: 419–429.
13. Moore MJ, Goldstein D, Hamm J, Figer A, Hecht JR, et al. (2007) Erlotinib plus gemcitabine compared with gemcitabine alone in patients with advanced pancreatic cancer: a phase III trial of the National Cancer Institute of Canada Clinical Trials Group. J Clin Oncol 25: 1960–1966.
14. Conroy T, Desseigne F, Ychou M, Bouche O, Guimbaud R, et al. (2011) FOLFIRINOX versus gemcitabine for metastatic pancreatic cancer. N Engl J Med 364: 1817–1825.
15. Philip PA, Benedetti J, Corless CL, Wong R, O'Reilly EM, et al. (2010) Phase III study comparing gemcitabine plus cetuximab versus gemcitabine in patients with advanced pancreatic adenocarcinoma: Southwest Oncology Group-directed intergroup trial S0205. J Clin Oncol 28: 3605–3610.
16. Kindler HL, Niedzwiecki D, Hollis D, Sutherland S, Schrag D, et al. (2010) Gemcitabine plus bevacizumab compared with gemcitabine plus placebo in patients with advanced pancreatic cancer: phase III trial of the Cancer and Leukemia Group B (CALGB 80303). J Clin Oncol 28: 3617–3622.
17. Bramhall SR, Schulz J, Nemunaitis J, Brown PD, Baillet M, et al. (2002) A double-blind placebo-controlled, randomised study comparing gemcitabine and marimastat with gemcitabine and placebo as first line therapy in patients with advanced pancreatic cancer. Br J Cancer 87: 161–167.
18. Paciotti GF, Myer L, Weinreich D, Goia D, Pavel N, et al. (2004) Colloidal gold: a novel nanoparticle vector for tumor directed drug delivery. Drug Deliv 11: 169–183.
19. Dreaden EC, Alkilany AM, Huang X, Murphy CJ, El-Sayed MA (2012) The golden age: gold nanoparticles for biomedicine. Chem Soc Rev 41: 2740–2779.
20. Brigger I, Dubernet C, Couvreur P (2012) Nanoparticles in cancer therapy and diagnosis. Adv Drug Deliv Rev.
21. Saha K, Bajaj A, Duncan B, Rotello VM (2011) Beauty is skin deep: a surface monolayer perspective on nanoparticle interactions with cells and bio-macromolecules. Small 7: 1903–1918.
22. Kudgus RA, Bhattacharya R, Mukherjee P (2011) Cancer nanotechnology: emerging role of gold nanoconjugates. Anticancer Agents Med Chem 11: 965–973.
23. Arvizo RR, Bhattacharyya S, Kudgus RA, Giri K, Bhattacharya R, et al. (2012) Intrinsic therapeutic applications of noble metal nanoparticles: past, present and future. Chem Soc Rev 41: 2943–2970.
24. Khan JA, Kudgus RA, Szabolcs A, Dutta S, Wang E, et al. (2011) Designing nanoconjugates to effectively target pancreatic cancer cells in vitro and in vivo. Plos One 6: e20347.
25. Patra CR, Bhattacharya R, Wang E, Katarya A, Lau JS, et al. (2008) Targeted delivery of gemcitabine to pancreatic adenocarcinoma using cetuximab as a targeting agent. Cancer Res 68: 1970–1978.
26. Qian X, Peng XH, Ansari DO, Yin-Goen Q, Chen GZ, et al. (2008) In vivo tumor targeting and spectroscopic detection with surface-enhanced Raman nanoparticle tags. Nat Biotechnol 26: 83–90.
27. Perrault SD, Walkey C, Jennings T, Fischer HC, Chan WC (2009) Mediating tumor targeting efficiency of nanoparticles through design. Nano Lett 9: 1909–1915.
28. Han G, Ghosh P, Rotello VM (2007) Functionalized gold nanoparticles for drug delivery. Nanomedicine (Lond) 2: 113–123.
29. Khan JA, Pillai B, Das TK, Singh Y, Maiti S (2007) Molecular effects of uptake of gold nanoparticles in HeLa cells. Chembiochem 8: 1237–1240.
30. Daniel MC, Astruc D (2004) Gold nanoparticles: Assembly, supramolecular chemistry, quantum-size-related properties, and applications toward biology, catalysis, and nanotechnology. Chemical Reviews 104: 293–346.
31. Rocha-Lima CM, Soares HP, Raez LE, Singal R (2007) EGFR targeting of solid tumors. Cancer Control 14: 295–304.
32. Herbst RS, Kim ES, Harari PM (2001) IMC-C225, an anti-epidermal growth factor receptor monoclonal antibody, for treatment of head and neck cancer. Expert Opin Biol Ther 1: 719–732.
33. Kim ES, Lu C, Khuri FR, Tonda M, Glisson BS, et al. (2001) A phase II study of STEALTH cisplatin (SPI-77) in patients with advanced non-small cell lung cancer. Lung Cancer 34: 427–432.
34. Kim ES, Khuri FR, Herbst RS (2001) Epidermal growth factor receptor biology (IMC-C225). Curr Opin Oncol 13: 506–513.
35. Liao HJ, Carpenter G (2009) Cetuximab/C225-induced intracellular trafficking of epidermal growth factor receptor. Cancer Res 69: 6179–6183.
36. Li S, Schmitz KR, Jeffrey PD, Wiltzius JJ, Kussie P, et al. (2005) Structural basis for inhibition of the epidermal growth factor receptor by cetuximab. Cancer Cell 7: 301–311.
37. Orth JD, Krueger EW, Weller SG, McNiven MA (2006) A novel endocytic mechanism of epidermal growth factor receptor sequestration and internalization. Cancer Res 66: 3603–3610.
38. Schreiber AB, Libermann TA, Lax I, Yarden Y, Schlessinger J (1983) Biological role of epidermal growth factor-receptor clustering. Investigation with monoclonal anti-receptor antibodies. J Biol Chem 258: 846–853.
39. Arvizo RR, Giri K, Moyano D, Miranda OR, Madden B, et al. (2012) Identifying new therapeutic targets via modulation of protein corona formation by engineered nanoparticles. Plos One 7: e33650.
40. Arvizo RR, Rana S, Miranda OR, Bhattacharya R, Rotello VM, et al. (2011) Mechanism of anti-angiogenic property of gold nanoparticles: role of nanoparticle size and surface charge. Nanomedicine 7: 580–587.
41. Patra CR, Bhattacharya R, Mukherjee P (2010) Fabrication and functional characterization of goldnanoconjugates for potential application in ovarian cancer. J Mater Chem 20: 547–554.
42. Patra CR, Bhattacharya R, Mukhopadhyay D, Mukherjee P (2010) Fabrication of gold nanoparticles for targeted therapy in pancreatic cancer. Adv Drug Deliv Rev 62: 346–361.
43. Bhattacharya R, Mukherjee P (2008) Biological properties of "naked" metal nanoparticles. Adv Drug Deliv Rev 60: 1289–1306.
44. Mangeney C, Ferrage F, Aujard I, Marchi-Artzner V, Jullien L, et al. (2002) Synthesis and properties of water-soluble gold colloids covalently derivatized with neutral polymer monolayers. J Am Chem Soc 124: 5811–5821.
45. Mukherjee P, Bhattacharya R, Lee YK, Patra CR, et al. (2007) Potential therapeutic application of gold nanoparticles in B-chronic lymphocytic leukemia (BCLL): enhancing apoptosis. J Nanobiotechnology 5: 4.
46. Walczyk D, Bombelli FB, Monopoli MP, Lynch I, Dawson KA (2010) What the cell "sees" in bionanoscience. J Am Chem Soc 132: 5761–5768.
47. Yao SY, Ng AM, Muzyka WR, Griffiths M, Cass CE, et al. (1997) Molecular cloning and functional characterization of nitrobenzylthioinosine (NBMPR)-sensitive (es) and NBMPR-insensitive (ei) equilibrative nucleoside transporter proteins (rENT1 and rENT2) from rat tissues. J Biol Chem 272: 28423–28430.
48. Griffiths M, Beaumont N, Yao SY, Sundaram M, Boumah CE, et al. (1997) Cloning of a human nucleoside transporter implicated in the cellular uptake of adenosine and chemotherapeutic drugs. Nat Med 3: 89–93.
49. Paproski RJ, Ng AM, Yao SY, Graham K, Young JD, et al. (2008) The role of human nucleoside transporters in uptake of 3'-deoxy-3'-fluorothymidine. Mol Pharmacol 74: 1372–1380.
50. Cobley CM, Chen J, Cho EC, Wang LV, Xia Y (2011) Gold nanostructures: a class of multifunctional materials for biomedical applications. Chem Soc Rev 40: 44–56.
51. Bhattacharyya S, Kudgus RA, Bhattacharya R, Mukherjee P (2011) Inorganic nanoparticles in cancer therapy. Pharm Res 28: 237–259.

52. Giljohann DA, Mirkin CA (2009) Drivers of biodiagnostic development. Nature 462: 461–464.

53. Sardar R, Funston AM, Mulvaney P, Murray RW (2009) Gold Nanoparticles: Past, Present, and Future. Langmuir 25: 13840–13851.

54. Ferrari M (2005) Cancer nanotechnology: opportunities and challenges. Nat Rev Cancer 5: 161–171.

55. Burda C, Chen X, Narayanan R, El-Sayed MA (2005) Chemistry and properties of nanocrystals of different shapes. Chem Rev 105: 1025–1102.

56. Bhattacharya R, Mukherjee P (2008) Biological properties of "naked" metal nanoparticles. Advanced Drug Delivery Reviews 60: 1289–1306.

57. Whitesides GM (2003) The 'right' size in nanobiotechnology. Nat Biotechnol 21: 1161–1165.

58. Mackey JR, Baldwin SA, Young JD, Cass CE (1998) Nucleoside transport and its significance for anticancer drug resistance. Drug Resist Updat 1: 310–324.

59. Mackey JR, Mani RS, Selner M, Mowles D, Young JD, et al. (1998) Functional nucleoside transporters are required for gemcitabine influx and manifestation of toxicity in cancer cell lines. Cancer Res 58: 4349–4357.

60. Young JD, Yao SY, Sun L, Cass CE, Baldwin SA (2008) Human equilibrative nucleoside transporter (ENT) family of nucleoside and nucleobase transporter proteins. Xenobiotica 38: 995–1021.

61. Giovannetti E, Del Tacca M, Mey V, Funel N, Nannizzi S, et al. (2006) Transcription analysis of human equilibrative nucleoside transporter-1 predicts survival in pancreas cancer patients treated with gemcitabine. Cancer Res 66: 3928–3935.

4

Resveratrol Protects Chondrocytes from Apoptosis via Altering the Ultrastructural and Biomechanical Properties: An AFM Study

Hua Jin[1]♦, Qian Liang[1]♦, Tongsheng Chen[2], Xiaoping Wang[1]*

1 Department of Pain Management, The First Affiliated Hospital of Jinan University, Guangzhou, China, **2** MOE Key Laboratory of Laser Life Science & Institute of Laser Life Science, South China Normal University, Guangzhou, China

Abstract

Osteoarthritis (OA), a degenerative joint disease with high prevalence among older people, occurs from molecular or nanometer level and extends gradually to higher degrees of the ultrastructure of cartilage, finally resulting in irreversible structural and functional damages. This report aims to use atomic force microscopy (AFM) to investigate the protective effects of resveratrol (RV), a drug with good anti-inflammatory properties, on cellular morphology, membrane architecture, cytoskeleton, cell surface adhesion and stiffness at nanometer level in sodium nitroprusside (SNP)-induced apoptotic chondrocytes, a typical cellular OA model. CCK-8 assay showed that 100 µM RV significantly prevented SNP-induced cytotoxicity. AFM imaging and quantitative analysis showed that SNP potently induced chondrocytes changes including shrunk, round, lamellipodia contraction and decrease in adherent junctions among cells, as well as the destruction of biomechanics: 90% decrease in elasticity and 30% decrease in adhesion. In addition, confocal imaging analysis showed that SNP induced aggregation of the cytoskeleton and decrease in the expression of cytoskeletal proteins. More importantly, these SNP-induced damages to chondrocytes could be potently prevented by RV pretreatment. Interestingly, the biomechanical changes occurred before morphological changes could be clearly observed during SNP-induced apoptosis, indicating that the biomechanics of cellular membrane may be a more robust indicator of cell function. Collectively, our data demonstrate that RV prevents SNP-induced apoptosis of chondrocytes by regulating actin organization, and that AFM-based technology can be developed into a powerful and sensitive method to study the interaction mechanisms between chondrocytes and drugs.

Editor: Etienne Dague, LAAS-CNRS, France

Funding: This work was supported by the National Natural Science Foundation of China (Grant No. 81071491 and 61178078), and Key Project of the Department of Education and Finance of Guangdong Province (cxzd115). The funders had no role in study design, data collection and analysis, decision to publish, or preparation of the manuscript.

Competing Interests: The authors have declared that no competing interests exist.

* E-mail: txp2938@jnu.edu.cn

♦ These authors contributed equally to this work.

Introduction

Osteoarthritis (OA) is known as a degenerative arthritis or degenerative joint disease, which affects 20 million people in U.S. [1]. At present, the treatment for OA mainly focuses on relieving pains and symptoms, and improving function of cartilage. However, there are no treatments to cure OA or reduce the degradation of cartilage. Current treatments for OA are restricted to anti-inflammatory drugs which bring numerous side effects and are only temporarily effective to the patients. To find safe and highly effective drugs for OA treatment are therefore very urgent.

Resveratrol (3,5,4′ -trihydroxystilbene, RV), a polyphenol derived from grapes, berries, peanuts and other plants, has been shown to possess anti-proliferative, anti-oxidative and anti-inflammatory properties [2], and these effects are associated with the suppression of inflammation, arthritis and cardiovascular diseases [3]. It is reported that RV protects chondrocytes from apoptosis via preventing mitochondrial depolarization and ATP consumption [4] or suppressing ROS and p53-production [5]. RV also can be as a potent safe drug for OA treatment, but the mechanisms are still unclear.

Apoptosis of chondrocytes is regarded as a feature of progressive cartilage degeneration in OA [6]. Sodium nitroprusside (SNP) was widely used as the donor of nitric oxide (NO) to study the molecular mechanism of NO-induced chondrocytes apoptosis [7,8]. Although NOC-12 may be a more effective NO donor in OA metastasis [9], it could not effectively induce apoptosis of chondrocytes [10]. Eo and co-workers [11] reported that RV could rescue SNP-induced degradation of I-kappa B alpha mainly through SN50 peptide-mediated inhibition of NF-kappa B activity, thus blocking SNP-induced caspase-3 activation and apoptosis. However, the effects of RV on morphological and biomechanical properties of chondrocytes at subcelluar or nanometer-level have not been studied.

Nanobiomechanics of cells have been identified as a vital characteristic to distinguish normal cells from diseased cells which differ physically from healthy cells [12]. Diseases can not only cause biological and functional alterations but also induce abnormalities in physical and structural characteristics of cells.

Therefore, research into biomechanics at the cellular and molecular levels of some human diseases can provide a better elucidation on the mechanisms behind disease progression [13], thereby providing important information for treatment of these diseases as well. Due to the nanometer resolution, AFM has been extensively used in detection of some diseases at cellular or subcellular level [14]. The ultrastructural and biomechanical properties have been altered a lot in disease or cancerous cells, and these alterations can be used as target to diagnose or distinguish diseased cells from healthy cells [15,16]. Furthermore, biomechanics of cell membrane is always changed in the context of drugs. Therefore, detecting these changes at nanometer level is very important for evaluating curative effect and elucidating mechanisms of drugs.

In this work, we used the rabbit chondrocytes as the cell model to detect the protective effects of RV on SNP-induced chondrocytes apoptosis. Rabbit chondrocytes have been extensively used in the basic research of mechanisms of chondrocytes or OA [17–24], and we have gained ripe experimental experiences [25]. Alterations in ultrastructure and biomechanics of cellular membrane of chondrocytes with or without RV pretreatment were investigated using AFM at nanometer scale. Our results showed that RV could effectively protect chondrocytes from apoptosis through altering the cytoskeleton arrangements and biomechanical properties including cellular stiffness and adhesion force.

Materials and Methods

Materials

Trypsin and type II collagenase, DMEM, fetal bovine serum, Cell Counting Kit-8 were purchased from Invitrogen (California, USA), Hyclone (Logan, Utah, USA), Sijiqing (Hangzhou, China) and Dojindo (Kumamoto, Japan), respectively. Actin-Tracker Green (phalloidin-FITC) and Tubulin-Tracker Red (α-Tubulin-Alexa Fluor 555) were both obtained from Beyotime Institute of Biotechnology (Naijing, China). Dulbecco's modified Eagle medium (DMEM) was from Gibco (Carlsbad, California, USA), fetal bovine serum (FBS) was from Sijiqing (Hangzhou, China).

Isolation and culture of chondrocytes

New Zealand rabbits were purchased from Experimental Animal Center of Guangzhou (China). As Tonomura, et al [26] described, articular cartilage was derived from knee, hip and shoulder joints of 6-week-old New Zealand white rabbits. The utilization of rabbit articular cartilage has been approved by the Animal Ethics Committee of Guangdong province, China. The extracted cartilages were firstly minced into small pieces and Chondrocytes were isolated by enzymatic digestion of 0.25% Trypsin in phosphate buffered solution (PBS) for 1 h and 0.2% type II collagenase in DMEM for 4–6 h. After collection by centrifugation, chondrocytes were resuspended in DMEM supplemented with 10% FBS and antibiotics (100 U/ml penicillin and 100 U/ml streptomycin) and 4.5% glucose. The cells were transferred when confluent monolayer cells reached to 85–90%, the transferred density was 5×10^4 cells/cm^2. The growth medium was changed every other day. The second and third generations of chondrocytes were used in our study.

CCK-8 assay to analyze cell viability

Chondrocytes were cultured in 96-well plates for 24 h, and then exposed to different concentrations of SNP for different periods. Cell viability was assessed using Cell Counting Kit assay according to the manufacturer's instructions. All experiments were performed three times.

AFM measurements of cell morphology

For all topographic images, the cells were fixed with 2.5% paraformaldehyde, and imaged by a tapping mode AFM (Park Scientific Instruments) in air. The silicon nitride tips (UL20B) used in all AFM measurements were irradiated with ultraviolet in air for 15 min to remove any organic contaminates prior to use. The curvature radius of the tips is less than 10 nm, and the length, width and thickness of the cantilevers are 115, 30, and 3.5 μm, respectively, with the oscillation frequency of 255 kHz and a force constant of 0.03 N/m.

Surface roughness of cell membrane

The average surface roughness (Ra) is defined as the arithmetic mean of the deviations in height from the line mean value, and Rq is the root mean square. As the roughness has a dependence on the sampling size, Ra and Rq were analyzed in two different areas: 10 randomly selected 4 μm^2 (2 μm×2 μm) and 10 randomly selected 25 μm^2 (5 μm×5 μm). $P<0.05$ were considered as statistical significance.

Determination of nanomechanical properties of chondrocytes

The force spectroscopy of cells was detected using an AFM (Agilent 5500) in the near physiological environment. The methods here were according to Kim's procedure [27]. In brief, the cells were firstly fixed with 2.5% glutaraldehyde and kept in PBS (PH = 7.4) during AFM tip indentation. All force measurements were performed at the same loading rate (1.2×10^5 pN/s). The deflection-vs-displacement curves were obtained by the instrument, and to convert the deflection-vs-displacement curves into the force-vs-distance curves, we adopted Cappella's method [28]. In each group, over 20 cells were measured. The data of stiffness and adhesion forces were processed using SPSS13.0 to gain the Gaussian distribution (or normal distribution) histograms.

The Young's modulus was calculated using Hertz model which shows the relationship between the applied force F and the indentation δ:

$$F = \frac{4}{3} \frac{ER^{1/2}\delta^{3/2}}{(1-\upsilon^2)}$$

In the equation, υ is the poisson ratio, F the loading force, δ the indentation, E the Young's modulus, and R the curvature radius of the AFM tip, respectively. A Poisson ratio of 0.5 is appropriate for cells [29–31]. Young's modulus during the calculations to obtain the best fit to the model considering the least-squares method as proposed by Dimitriadis et al [32].

Immunofluorescence staining

The characterizations of cytoskeleton were evaluated by staining with phalloidin-FITC and Tubulin-Tracker, separately. The chondrocytes were fixed with 4% paraformaldehyde for 30 min and incubated with 1 μM phalloidin-FITC or 1 μM Tubulin-Tracker for 60 min in dark at room temperature, separately, and then washed twice with PBS. After that, the cytoskeleton organization was imaged by a laser scanning confocal microscope (LCM 510 Meta Duo Scan, Carl Zeiss, Germeny). The resulting fluorescence was also measured by flow cytometer at excitation wavelength 488 nm, emission wavelength 530 nm to quantitatively elucidate the alterations of cytoskeleton proteins.

Results and Discussions

The changes in cell viability induced by SNP and RV

To detect the protecting effects of RV on chondrocytes, we firstly established the OA model by exposure of chondrocytes to SNP, an inorganic compound with the formula $Na_2[Fe(CN)_5-NO] \cdot 2H_2O$. SNP has been used as anti-hypertensive treatments for decades and it has not obvious side effects. In vitro, SNP could be as an external NO donor to induce apoptosis.

As shown in Figure 1A, SNP induced a dose-dependent cytotoxicity in chondrocytes, and treatment with 1.5 mM SNP for 24 h induced an over 85% of decrease of cell viability. Treatment with 1.5 mM of SNP for different time induced a time-dependent cytotoxicity (Fig.1B). The results indicated that SNP induced dose- and time-dependent cytotoxicity in chondrocytes. Based on these data, chondrocytes treated with 1.5 mM SNP for 24 h were used as the in vitro OA model.

To investigate the effects of RV on chondrocytes, different concentrations of RV were used to pretreat chondrocytes for 24 h before SNP treatment, and then the cell viability was measured. As shown in Figure 2, RV not only did not significantly induce cytotoxicity but also potently prevented SNP-induced cytotoxicity (Fig. 2), indicating that RV could protect chondrocytes from SNP-induced apoptosis and could be used as a potent drug to treat OA.

AFM detects morphological changes of chondrocytes

The specific shape of cells plays vital roles in maintaining specific functions of cells. Cellular shapes and morphology determine the interaction extent between cells and their environment. If cellular shapes were changed, the physiological and functional situations of cells would be damaged or disturbed [33,34]. Therefore, detecting the structural details of chondrocytes is very helpful for understanding their functions. Morphological data of cells obtained using AFM could be an important index to evaluate the effects of drugs [35].

As shown in Figure 3, control chondrocytes were spindle and elongated shapes (A1, A2), and lots of lamellipodia and filopodia were observed around the cells (A3, A4, A5, A6). In addition, the adherent junctions between/among cells were connected by lamellipodia (A7, A8), which provided essential condition for the exchange of energy and matter among chondrocytes. After treatment with 1.5 mM of SNP for 12 h, chondrocytes become shrunk and round, the lamellipodia contracted, and the adherent junctions among cells significantly decreased or even diminished (shown by B1–B4), the characterizations of apoptosis. These SNP-

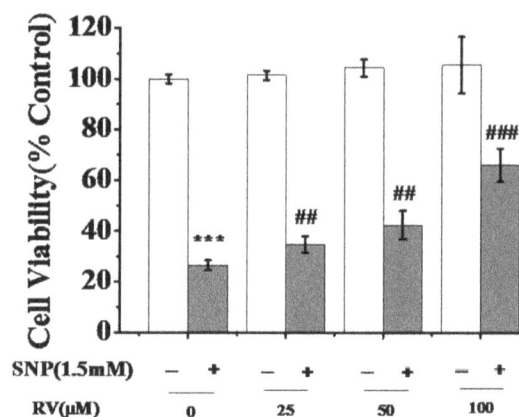

Figure 2. Protection effects of RV on SNP-induced apoptosis of chondrocytes. Cells were pretreated with different concentrations (0, 25, 50 and 100 mM) of RV for 24 h, and then treated with 1.5 mM of SNP for 12 h. After that, the cell viability was assayed using CCK-8 (comparing with control group, *P<0.05, **P<0.01; comparing with SNP treated group, #P<0.05, ##P<0.01, ###P<0.001).

induced changes were potently prevented by pretreatment with 100 μM of RV for 24 h (shown by images C1–C4), demonstrating that RV pretreatment could markedly prevent SNP-induced morphological changes and apoptosis of chondrocytes.

Besides, quantitative morphological data were also compared for accurate assessment of the effects of RV. As shown in Figure 3D1, the average length and width of control chondrocytes were 62.8±5.2, 33.5±2.6 μm, and they decreased to 27.8±2.1 and 23.6±3.3 μm after treatment with SNP. In addition, SNP treatment induced a significant increase in the average height of chondrocytes from 427.4±44.6 nm (control) to 774.8±85.6 nm (Fig. 3D2), and also induced a significant decrease in the ratio of major and minor cell axis from 1.9±0.2 (control) to 1.2±0.1 (Fig.3D2). Both the decreased width and increased height indicated that the cells were tending to detach from the cellular matrix and became shrank round, even apoptosis. More importantly, all these morphological changes induced by SNP were potently prevented by RV pretreatment (shown by Fig. 3D1, D2).

Taken together, RV pretreatment could significantly prevented SNP-induced apoptosis of chondrocytes by protecting cellular structure, shape and biomechanics.

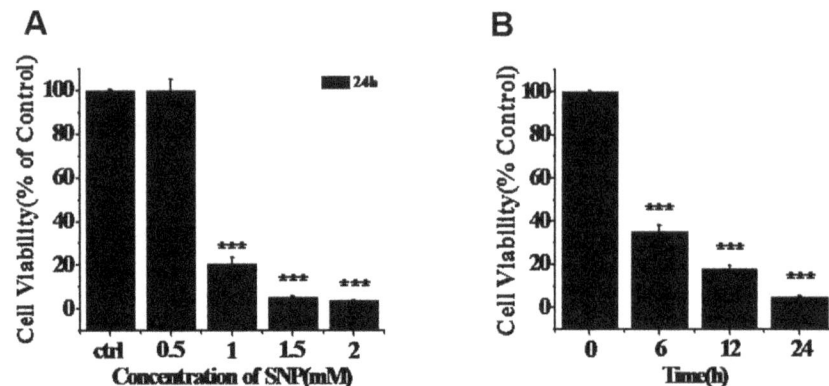

Figure 1. Cytotoxicity of SNP in chondrocytes. (A) Cell viability of chondrocytes treated by different concentrations of SNP for 24 h. (B) Cell viability of chondrocytes treated by 1.5 mM of SNP for different time periods (comparing with control group, *P<0.05, **P<0.01, ***P<0.001). The results indicated the killing effects of SNP on chondrocytes were in a dose- and time-dependent manner.

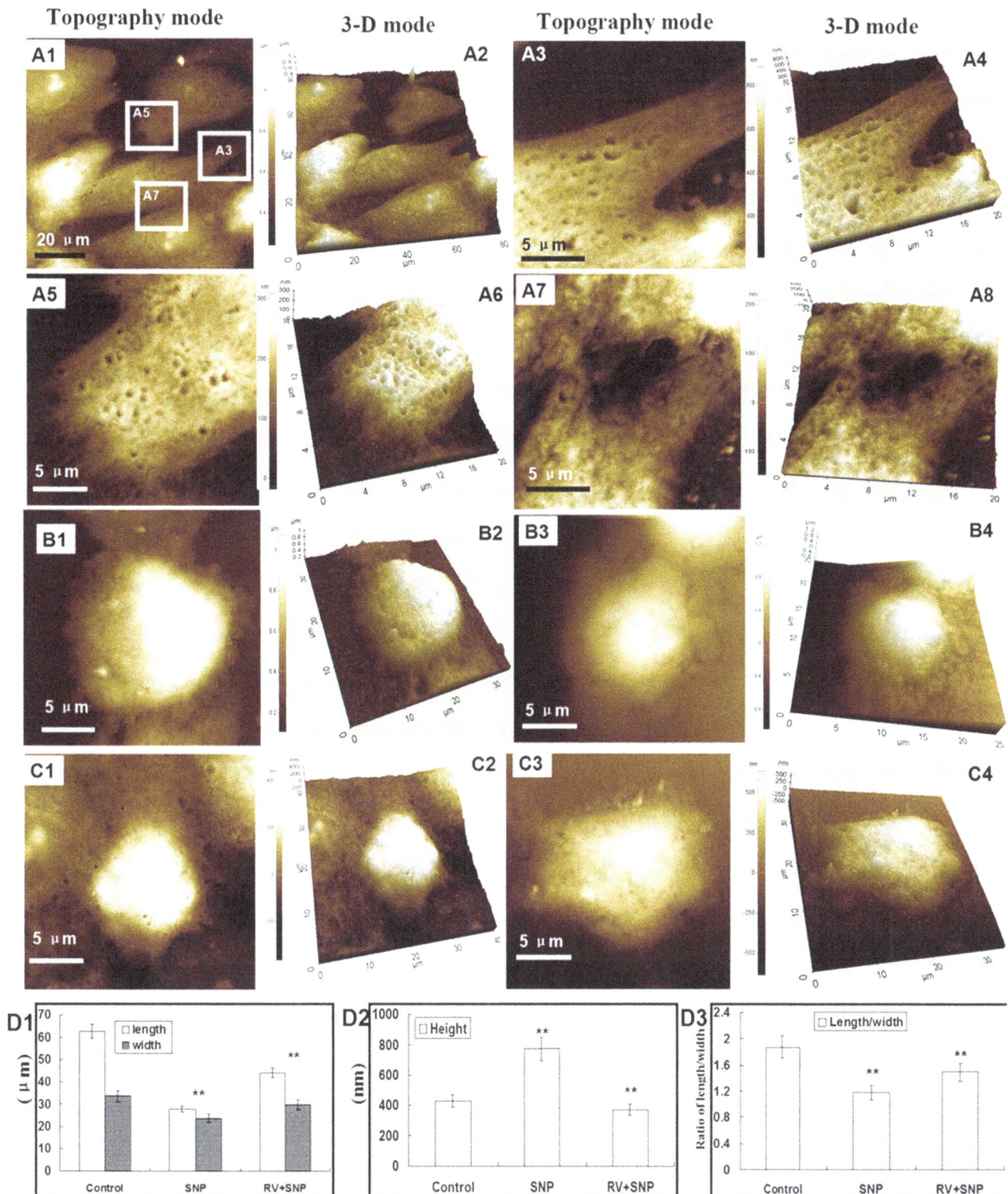

Figure 3. Morphological data of chondrocytes. (A1–A2) Control chondrocytes. (A3–A8) The enlargement images of white panes in A1. (B1–B4) Chondrocytes treated with 1.5 mM SNP for 12 h. (C1–C5) The chondrocytes were pretreated with 100 mM of RV for 24 h, and then treated with 1.5 mM of SNP for 12 h. (D1–D3) Histograms of average length, width (D1), height (D2) and ratio of length/width of cells in three groups. In D1–D3, more than ten cells in each group were selected to measure the values. *$P < 0.05$ was regarded as statistically significant.

Topography Mode   3-D Mode   Contour Map

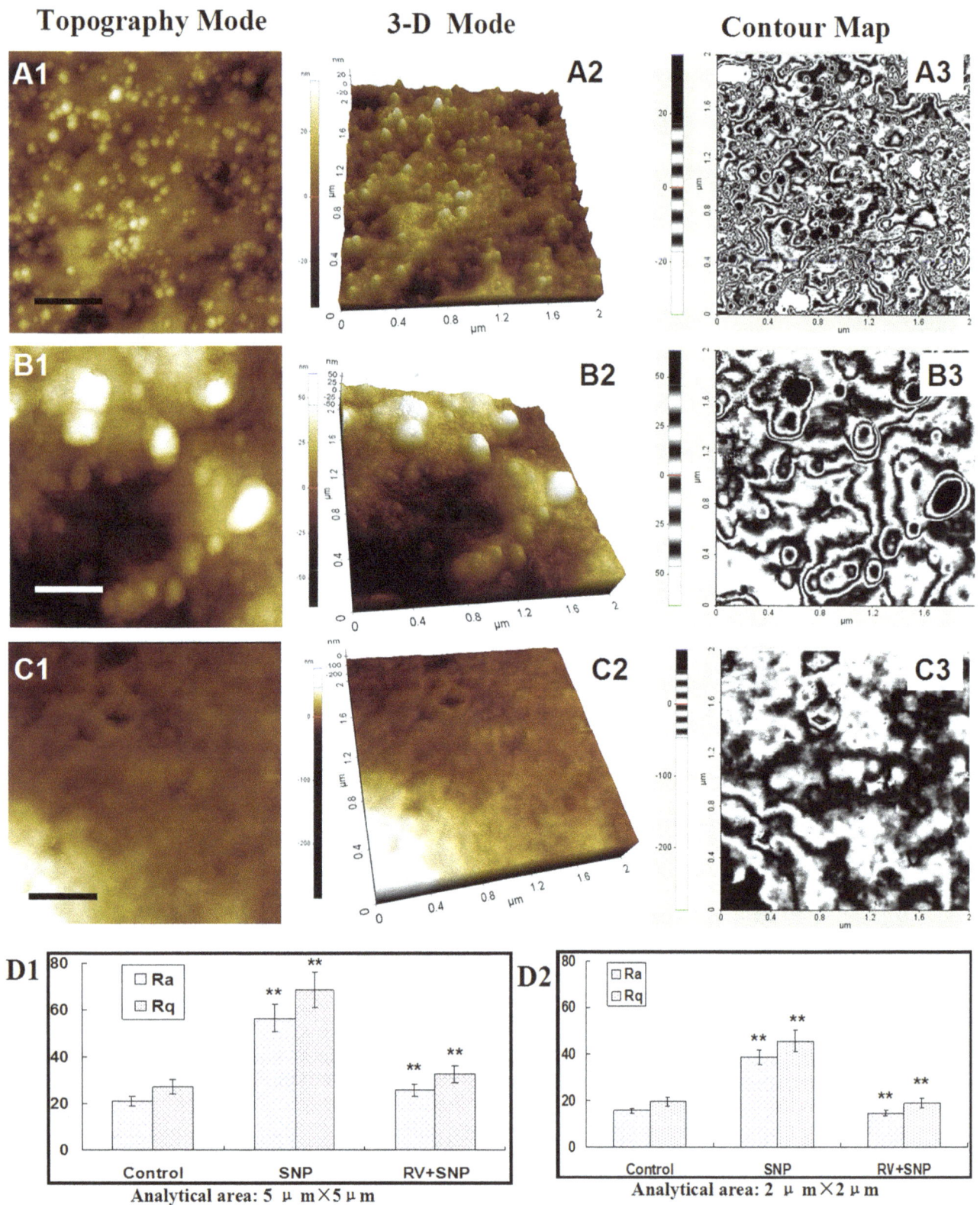

Figure 4. AFM ultrastructural data of control chondrocytes. (A1–A3) Control chondrocytes. (B1–B3) Chondrocytes treated with 1.5 mM SNP for 12 h. (C1–C3) The chondrocytes were pretreated with 100 μM of RV for 24 h, and then treated with 1.5 mM of SNP for 12 h. Scanning area: 2×2 μm^2. (A1), (B1), (C1) was topography mode. (A2), (B2), (C2) 3-D mode of (A1), (B1) and (C1), respectively. (A3), (B3), (C3) was contour map of (A1), (B1) and (C1), respectively. (D1) and (D2) were histograms of average roughness (Ra) of chondrocytes which were analyzed in 5×5 μm^2 and 2×2 μm^2, respectively. In (D1) and (D2), ten cells in each group were selected to measure the values of Ra, statistical analysis was performed using Student's t-test. $P<0.05$ was regarded as statistically significant.

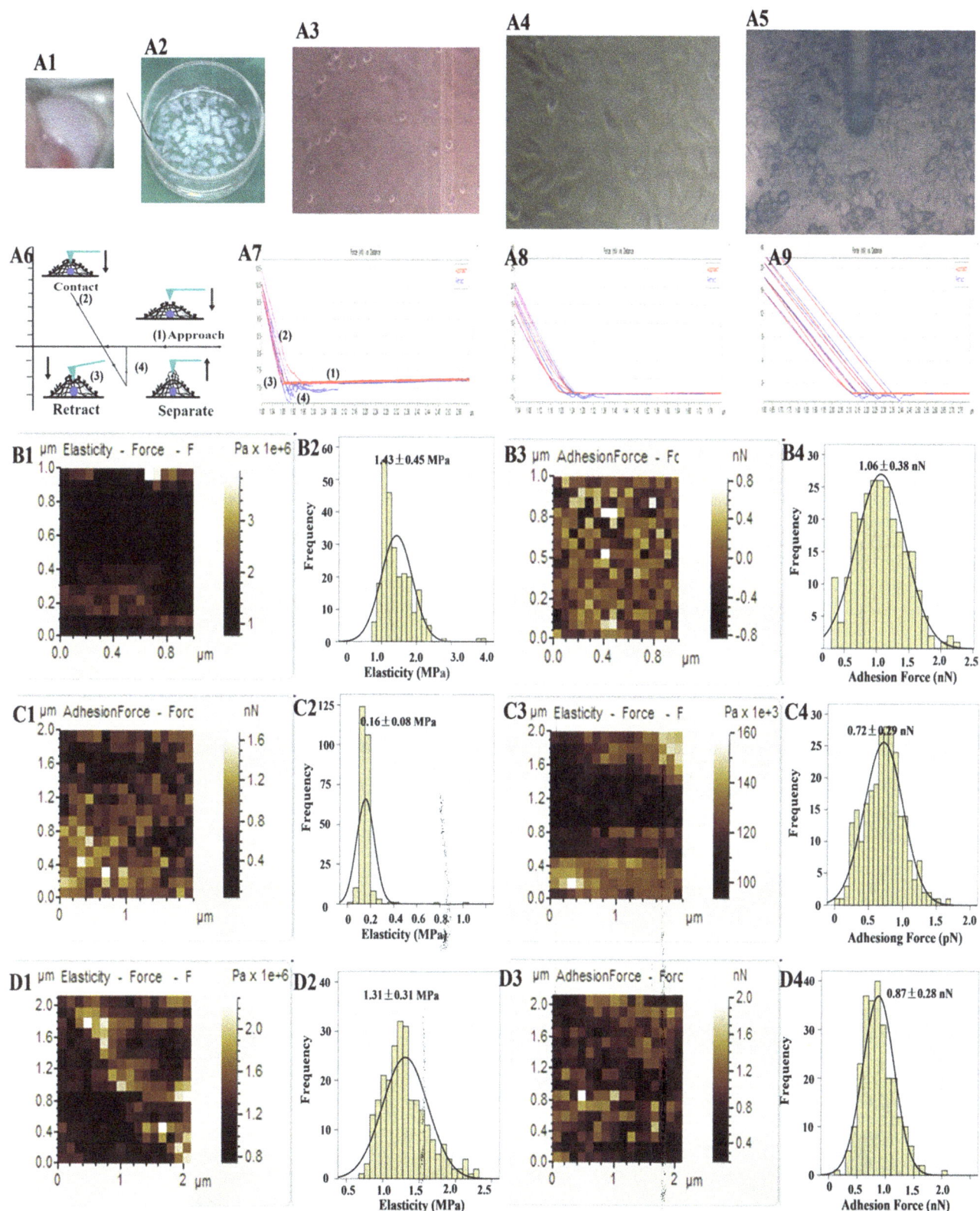

Figure 5. Alterations in nanobiotechnology of chondrocytes detected by AFM. (A1–A5) isolation of chondrocytes: (A1) Cartilage collected from the bilateral joints of the knees, hips, and shoulders. (A2) The joints were minced into small pieces, treated with 0.015% trypsin for 30 min, and subsequently digested. (A3) Morphology of primary joint chondrocytes. (A4) The morphology of primary joint chondrocytes cultured for 7 days. (A5) The AFM tip was employed to detect the morphology and biomechanics of chondrocytes. (A6) Typical force-distance curve detected using AFM: (1) The tip is approaching the surface of sample, (2) the tip is just in contact with the surface of cells, (3) the tip is further put into repulsive contact with the cellular surface, (4) lastly, the tip-sample contact is retracted. (A7–A9) are the representative force-distance curves obtained on control chondrocytes (A7), chondrocytes treated with 1.5 mM SNP for 12 h (A8), and chondrocytes pretreated with RV and the induce with SNP (A9),

respectively. The elasticity maps, histogram of elasticity, adhesion force map and histogram of adhesion force of control chondrocytes (B1–B4), chondrocytes treated with 1.5 mM SNP for 12 h (C1–C4), and chondrocytes pretreated with RV and then cotreated with SNP (A4), respectively.

Alterations in cellular membrane architecture detected at nanometer level

Figure 4 showed the ultrastructural data of chondrocytes. The membrane architecture of control chondrocytes (Figs.4A1–A3) showed uniform structures and granular morphology with the surface particles of 50~100 nm. Figures 4B1–B3 showed the surface architecture of SNP-induced chondrocytes which became heterogeneous, and the sizes of the membrane particles increased to 150~200 nm. The ultrastructure of SNP-treated chondrocytes pretreated by RV became smooth and homogeneous but the granular morphology on cellular membrane diminished (Figs.4C1–C3), implying that RV could significantly protect chondrocytes from SNP-induced apoptosis and changes in morphological properties, but could not completely prevented SNP-induced changes in nanostructure of cellular membrane. Therefore, AFM with nanometer-scale resolution provides us new insights about cellular structure-function.

Additionally, the average roughness of cell membrane is directly or indirectly sensitive to the membrane-skeleton integrity [36]. The average roughness (Ra) and root mean square roughness (Rq) were measured for comparison. As shown in Figures 4D1 and D2, the values of both Ra and Rq of SNP-induced chondrocytes increased significantly compared with that of control chondrocytes. While Ra and Rq of the -chondrocytes pretreated with RV decreased 50% than that of SNP group, and their values were similar to that of control chondrocytes. These data suggested that

RV protected chondrocytes from SNP-induced damages via altering the membrane architecture.

Taken together, all these morphological data revealed that SNP could successfully induced apoptosis in chondrocytes, and RV could protect the chondrocytes from damaging or apoptosis via changing their morphological properties and architectures.

Alterations in nanomechanical properties of chondrocytes

Since the biomechanical properties of cells can potentially indicate their function and health, it is therefore very important to study the cellular biomechanics. Lots of literatures have shown that study on cellular mechanics is very helpful for clinical diagnostics and even the formulation of suitable strategies towards effective therapeutic treatments of human diseases. Although AFM measures the nanobiomechanics of single cell, it has been used to diagnose some diseases [16,37]. Although the potent protection of RV on chondrocytes has been reported [38,39], the effects on the nanobiomechanics, particularly on cell function and growth, is poor.

Here, the nanobiomechanical properties including elasticity and adhesion force were detected at levels of nanometer and pN, respectively. Figures 5A1–A4 showed the isolation of chondrocytes. Figures A5 and A6 indicated that the AFM tip was employed to detect the morphology and biomechanics of chondrocytes. After positioning the AFM tip over the cell center (Fig.5A5), the tip was brought to contact and pressed against the

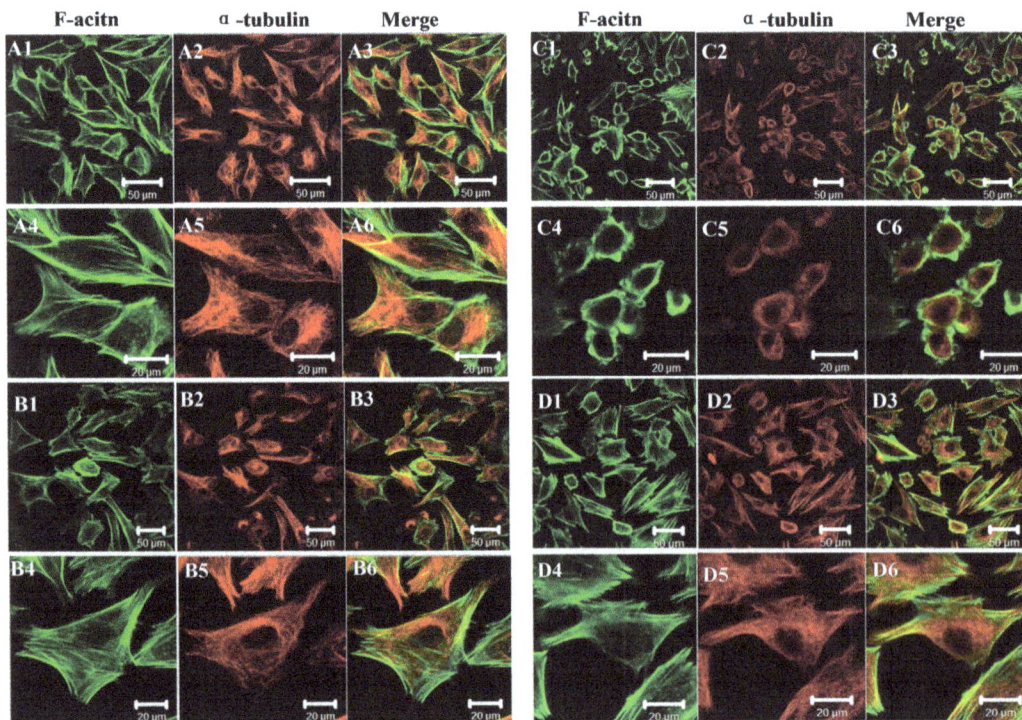

Figure 6. Organization of cytoskeleton chondrocytes. Control chondrocytes (A1–A6), chondrocytes treated with 100 μM RV for 12 h (B1–B6), chondrocytes treated with 1.5 mM SNP for 24 h (C1–C6) and chondrocytes pretreated with RV and then treated with SNP (D1–D6), respectively. A4–A6, B4–B6, C4–C6, D4–D6 were enlarged images of cellular cytoskeleton in A1–A3, B1–B3, C1–C3, D1–D3, respectively. Bars in A1–A3, B1–B3, C1–C3, D1–D3 and A4–A6, B4–B6, C4–C6, D4–D6 were 50 and 20 μm, respectively.

cell surface (Fig.5A6). During the tip retraction from the cell surface rupture events, the retraction events in the force-distance curves revealed the general tip-cell-surface adhesive interactions. Figs 5A7–A9 showed the typical force-distance curves obtained from control chondrocytes, SNP and RV+SNP groups, respectively.

As shown in Figs 5B1 and B2, the elasticity/stiffness of control chondrocytes was 1.43 ± 0.45 MPa. After treatment with 1.5 mM SNP for 12 h, the elasticity of the chondrocytes decreased to 0.16 ± 0.08 MPa (Figs.5C1 andC2), indicating that SNP obviously destructed the rigidity and chemical compositions of chondrocytes membrane. Nevertheless, if the chondrocytes pretreated with 100 μM of RV for 24 h before exposure to 1.5 mM SNP for 12 h, the elasticity was 1.31 ± 0.31 MPa (Fig.5D1,D2), indicating that RV could effectively protect the elasticity/stiffness of chondrocytes. The damage to the cell envelope and the changes in the composition of chondrocytes induced by SNP were suspected to be the major causes of the decreases in the cell stiffness/elasticity. As reported by Cai and co-workers [40], the damage and destruction of the cytoskeleton directly led to the decrease of cellular rigidity.

Moreover, adhesion of cellular membrane plays a very important role in cell physiological and pathological processes [41]. Here, we employed force spectroscopy of AFM to measure the non-specific adhesion force between AFM tip and cellular membrane as a function of the nanomechanical properties of the existing surface adhesive molecules. The adhesion force of control chondrocytes was 1.06 ± 0.38 pN (Figs.5B3 and B4), but it decreased to 0.72 ± 0.29 pN after SNP treatment (Figs.5C3 and C4), indicating that membrane proteins were damaged by SNP treatment. However, the adhesion force of SNP-treated chondrocytes pretreated with RV only increased to 0.87 ± 0.28 pN, demonstrating that RV could partially protect the membrane proteins. Furthermore, RV significantly increased the number of actins (A1–A4), and the particles with nano-meter scale were mainly distributed on/around actins (B1–B3) (Fig.S1). Notably, the nanomechanical properties of cellular membrane, including adhesion force and stiffness, were both enhanced with RV treatment (Figs.S1 C1 and C2).

Taken together, these results showed that SNP induced the destruction of biomechanics in chondrocytes, including 90% decrease in elasticity and 30% decrease in adhesion. Notably, RV pretreatment could recover the elasticity/stiffness closely to that of control chondrocytes, but RV possessed only a little protecting effect on membrane proteins, implying that RV could not entirely protect the physiological and functional properties of chondrocytes.

Alterations in cytoskeletal proteins F-actin and α-tubulin

From the biomechanical data, we can see that RV protects chondrocytes from SNP-induced apoptosis mainly through maintaining their elasticity/stiffness. As cytoskeleton is very important to maintain the biomechanics of cells, we further qualitatively investigated the organization of cytoskeleton, including F-actin and α-tubulin using confocal microscopy. The cytoskeleton is the mesh-like structure beneath the cell membrane, which is an important reflection of cellular structure organization [42]. Particularly, F-actin cytoskeleton is extensively regarded as the key factor to regulate the shapes and generate the mechanical forces of cells, and it plays vital roles during cellular physiological and pathological behaviors.

As shown in Figure 6, control chondrocytes presented well-spreading shapes, and F-actin and α-Tubulin were well-organized uniformly assembly (Figs.6A1–A6). The F-actins were of paralleled-like organization and α-Tubulin microtubules were of mesh-like alignment, which showed a well-grown station of control chondrocytes. After exposure of cells to 100 μM RV for 24 h, the fluorescence intensity increased significantly (Figs.6B1–B6) indicative of the increase of both F-actins and α-Tubulin, suggesting that RV treatment may protect the cellular cytoskeleton and promote the expression of cytoskeletal proteins. While after treatment with SNP, the chondrocytes became shrunk and round, and both the F-actin and α-Tubulin cytoskeleton were reorganized and polymerized (Figs.6C1–C6 showed). Interestingly, in the RV-pretreated chondrocytes, we found that the cytoskeleton could recover to some extent and represented well-spreading organizations (shown by Figs.6D1–D6), indicating that RV protected chondrocytes from SNP-induced apoptosis mainly through altering the organization of cytoskeleton. For further detecting the expression of cytoskeletal proteins induced by SNP and RV, we also measured the MFI of F-actin and α-tubulin using fluorescence-based flow cytometry, and found that SNP induced apoptosis and stiffness decrease mainly through the reorganization and decrease the cytoskeletal proteins—F-actin instead of α-tubulin (Fig.S2).

All these data demonstrated that the biomechanical properties were heavily influenced by the organization and expression levels of actin microfilaments instead of α-tubulin. SNP induced aggregation of cytoskeleton, decrease in the expression of cytoskeletal proteins, 90% decrease of elasticity and 30% decrease of adhesion force in chondrocytes. While RV pretreatment could protect the elasticity/stiffness close to that of control chondrocytes through protecting the integrity and organization of cell cytoskeleton. These data demonstrate that AFM can be as a promising nanodevice to study cells cytoskeleton integrity and arrangement in vitro models of apoptosis and migration.

It is very imperative to evaluate the effects of RV on chondrocytes when it is used in tissue engineering or OA treatment. The evidence provided in this study showed that RV could protect chondrocytes from SNP-induced apoptosis by regulating actin organization, but it had only a little effect on adhesive molecules or proteins in/on cell membrane, and that AFM-based technique provides us an effective and feasible tool to detect the changes and underlying mechanism of cells induced by drugs at nanometer level.

Supporting Information

Figure S1 The morphological and nanomechanical properties of chondrocytes treated with 100 nM of RV for 24 h, which detected using AFM. A1–A2 Morphology of chondrocytes. A3–A4 the enlargement of white frame in A1. B1–B3 The ultrastructure of membrane on chondrocytes. C1,C2 The adhesion force and stiffness modulus of cellular membrane was 2.13 ± 0.66 nN and 5.05 ± 2.43 MPa.

Figure S2 The alterations in expression of F-actin andα-tubulin in chondrocytes treated with SNP, RV and SNP+RV, respectively, which detected by fluorescence-based flow cytometry. The F-actins and microtubule were stained with FITC- phalloidine and Alexa Fluor 555-α-Tubulin antibody, respectively.

Author Contributions

Conceived and designed the experiments: XPW. Performed the experiments: HJ QL. Analyzed the data: HJ AL TSC. Contributed reagents/materials/analysis tools: HJ. Wrote the paper: HJ TSC.

References

1. Anderson AS, Loeser RF (2010) Why is osteoarthritis an age-related disease? Best Pract Res Cl Rh 24(1):15–26.
2. Wadsworth TL, Koop DR (1999) Effects of the wine polyphenolics quercetin and resveratrol on pro-inflammatory cytokine expression in RAW 264.7 macrophages. Biochem Pharmacol 57(8): 941–949.
3. Shakibaei M, Harikumar KB, Aggarwal BB (2009) Reseratrol addiction: to die or not to die. Mol Nutr Food Res 53(1): 115–128.
4. Dave M, Attur M, Palmer G, Al-Mussawir HE, Kennish L, et al. (2008) The antioxidant resveratrol protects against chondrocyte apoptosis via effects on mitochondrial polarization and ATP production. Arthritis Rheum 58(9): 2786–2797.
5. Csaki C, Keshishzadeh N, Fischer K, Shakibaei M (2008) Regulation of inflammation signalling by resveratrol in human chondrocytes in vitro. Biochem Pharmacol 75(3): 677–687.
6. Del Carlo M Jr, Loeser RF (2008) Cell death in osteoarthritis. Curr Rheumatol Rep 10(1): 37–42.
7. Wu GJ, Chen TG, Chang HC, Chiu WT, Chang CC, et al. (2007) Nitric oxide from both exogenous and endogenous sources activates mitochondria-dependent events and induces insults to human chondrocytes. J Cell Biochem 101(6):1520–1531.
8. Nakagawa S, Arai Y, Mazda O, Kishida T, Takahashi KA, et al. (2010) N-acetylcysteine prevents nitric oxide-induced chondrocyte apoptosis and cartilage degeneration in an experimental model of osteoarthritis. J Orthopaed Res 28(2):156–163.
9. Andrés MC de, Maneiro E, Martín MA, Arenas J, Blanco FJ (2013) Nitric oxide compounds have different effects profiles on human articular chondrocyte metabolism. Arthritis Res Ther 15(5):R115.
10. Carlo MD, Loeser RF (2003) Increased oxidative stress with aging reduces chondrocyte survival: correlation with intracellular glutathione levels. Arthritis Rheum 48(12):3419–3430.
11. Eo SH, Cho H, Kim SJ (2013) Resveratrol Inhibits Nitric Oxide-Induced Apoptosis via the NF-Kappa B Pathway in Rabbit Articular Chondrocytes. Biomol Ther 21(5):364–370.
12. Lee GY, Lim CT (2007) Biomechanics approaches to studying human diseases. Trends Biotechnol 25(3): 111–118.
13. Stolz M, Gottardi R, Raiteri R, Miot S, Martin I, et al. (2009) Early detection of aging cartilage and osteoarthritis in mice and patient samples using atomic force microscopy. Nat nanotechnol 4(3): 186–192.
14. Stolz M, Raiteri R, Daniels AU, VanLandingham MR, Baschong W, et al. (2004) Dynamic elastic modulus of porcine articular cartilage determined at two different levels of tissue organization by indentation-type atomic force microscopy. Biophys J 86(5): 3269–3283.
15. Iyer S, Gaikwad RM, Subba-Rao V, Woodworth CD, Sokolov I (2009) Atomic force microscopy detects differences in the surface brush of normal and cancerous cells. Nat Nanotechnol 4(6): 389–393.
16. Cross SE, Jin YS, Tondre J, Wong R, Rao J, et al. (2008) AFM-based analysis of human metastatic cancer cells. Nanotechnology 19: 384003–384011.
17. Kim SJ, Ju JW, Oh CD, Yoon YM, Song WK, et al. (2002) ERK-1/2 and p38 kinase oppositely regulate nitric oxide-induced apoptosis of chondrocytes in association with p53, caspase-3, and differentiation status. J Biol Chem 277(2):1332–1339.
18. Kim SJ, Kim HG, Oh CD, Hwang SG, Song WK, et al. (2002) p38 kinase-dependent and -independent inhibition of protein kinase C ζ and -α regulates nitric oxide-induced apoptosis and dedifferentiation of articular chondrocytes. J Biol Chem 277(33):30375–30381.
19. Kim SJ, Hwang SG, Shin DY, Kang SS, Chun JS (2002) p38 Kinase Regulates Nitric Oxide-induced Apoptosis of Articular Chondrocytes by Accumulating p53 via NFkappaB-dependent Transcription and Stabilization by Serine 15 Phosphorylation. J Biol Chem 277(36):33501–33508.
20. Oh CD, Chun JS (2003) Signaling mechanisms leading to the regulation of differentiation and apoptosis of articular chondrocytes by insulin-like growth factor-1. J Biol Chem 278(38):36563–36571
21. Kim SJ, Hwang SG, Kim IC, Chun JS (2003) Actin cytoskeletal architecture regulates nitric oxide-induced apoptosis, dedifferentiation, and cyclooxygenase-2

expression in articular chondrocytes via mitogen-activated protein kinase and protein kinase C pathways. J Biol Chem 278(43):42448–42456.
22. Yoon JB, Kim SJ, Hwang SG, Chang S, Kang SS, et al. (2003) Non-steroidal anti-inflammatory drugs inhibit nitric oxide-induced apoptosis and dedifferentiation of articular chondrocytes independent of cyclooxygenase activity. J Biol Chem 278(17):15319–15325.
23. Hwang SG, Ryu JH, Kim IC, Jho EH, Jung HC, et al. (2004) Wnt-7a causes loss of differentiated phenotype and inhibits apoptosis of articular chondrocytes via different mechanisms. J Biol Chem 279(25):26597–26604.
24. Kim JS, Park ZY, Yoo YJ, Yu SS, Chun JS (2005) p38 kinase mediates nitric oxide-induced apoptosis of chondrocytes through the inhibition of protein kinase C by blocking autophosphorylation. Cell Death Differ 12(3):201–212.
25. Zhuang CP, Wang XP, Chen TS (2013) H_2O_2 induces apoptosis of rabbit chondrocytes via both the extrinsic and the caspase-independent intrinsic pathways. J Innov Opt Health Sci 6(3):1350022.
26. Tonomura H, Takahashi KA, Mazda O, Arai Y, Inoue A, et al. (2006) Glutamine protects articular chondrocytes from heat stress and NO-induced apoptosis with HSP70 expression. Osteoarthritis Cartilage 14(6): 545–553.
27. Kim KS, Cho CH, Park EK, Jung MH, Yoon KS, et al. (2012) AFM-detected apoptotic changes in morphology and biophysical property caused by paclitaxel in Ishikawa and HeLa cells. PloS one 7(1):e30066.
28. Cappella B, Baschieri P, Frediani C, Miccoli P, Ascoli C (1997) Force-distance curve by AFM. IEEE Eng Med Biol 16(2): 58–65.
29. Laney DE, Garcia RA, Parson SM, Hansma HG (1997) Changes in the Elastic Properties of Cholinergic Synaptic Vesicles as Measured by Atomic Force Microscopy. Biophys J 72: 806–813.
30. Radmacher M (2002). Measuring the elastic properties of living cells by the atomic force microscope. Methods Cell Biol 68: 67–90.
31. Liang X, Mao G, Simon Ng KY (2004) Probing small unilamellar EggPC vesicles on mica surface by atomic force microscopy. Colloids Surf B: Biointerfaces 34(1): 41–5.
32. Dimitriadis EK, Horkay F, Maresca J, Kachar B, Chadwick RS (2002) Determination of Elastic Moduli of Thin Layers of Soft Material Using the Atomic Force Microscope. Biophys J 82(5): 2798–2810.
33. Yin Z, Sadok A, Sailem H, McCarthy A, Xia X, et al. (2013) A screen for morphological complexity identifies regulators of switch-like transitions between discrete cell shapes. Nat Cell Biol 15(7): 860–871.
34. McBeath R, Pirone DM, Nelson CM, Bhadriraju K, Chen CS (2004) Cell shape, cytoskeletal tension, and RhoA regulate stem cell lineage commitment. Dev Cell 6(4): 483–495.
35. Cai X, Yang X, Cai J, Wu S, Chen Q (2010) Atomic force microscope-related study membrane-associated cytotoxicity in human pterygium fibroblasts induced by mitomycin C. J Phys Chem B 114(11): 3833–3839.
36. Girasole M, Pompeo G, Cricenti A, Congiu-Castellano A, Andreola F, et al. (2007) Roughness of the plasma membrane as an independent morphological parameter to study RBCs: A quantitative atomic force microscopy investigation. Biochim Biophys Acta 1768(5): 1268–1276.
37. Teck Chwee Lim (2005) Single Cell Mechanics and its connections to Human Diseases. Asia-Pacific Biotech News 09: 674–675.
38. Csaki C, Mobasheri A, Shakibaei M (2009) Synergistic chondroprotective effects of curcumin and resveratrol in human articular chondrocytes: inhibition of IL-1beta-induced NF-kappaB-mediated inflammation and apoptosis. Arthritis Res Ther 11(6): R165.
39. Liu FC, Hung LF, Wu WL, Chang DM, Huang CY, et al. (2010) Chondroprotective effects and mechanisms of resveratrol in advanced glycation end products-stimulated chondrocytes. Arthritis Res Ther 12(5): R167.
40. Cai X, Yang X, Cai J, Wu S, Chen Q (2010) Atomic force microscope-related study membrane-associated cytotoxicity in human pterygium fibroblasts induced by mitomycin C. J Phys Chem B 114(11):3833–3839.
41. Geiger B (2001) Cell biology: encounters in space. Science 294(5547):1661–1663.
42. Hsieh CH, Lin YH, Lin S, Tsai-Wu JJ, Herbert Wu CH, et al. (2008) Surface ultrastructure and mechanical property of human chondrocyte revealed by atomic force microscopy. Osteoarthritis Cartilage 16(4):480–488.

5

Foreign Body Reaction Associated with PET and PET/Chitosan Electrospun Nanofibrous Abdominal Meshes

Beatriz Veleirinho[1,2]*, **Daniela S. Coelho**[3], **Paulo F. Dias**[3], **Marcelo Maraschin**[4], **Rúbia Pinto**[5], **Eduardo Cargnin-Ferreira**[6], **Ana Peixoto**[7], **José A. Souza**[7], **Rosa M. Ribeiro-do-Valle**[2], **José A. Lopes-da-Silva**[1]

1 QOPNA Research Unit, Department of Chemistry, University of Aveiro, Aveiro, Portugal, 2 Biotechnology and Biosciences Post-Graduation Program, Federal University of Santa Catarina, Florianópolis, Brazil, 3 Department of Cell Biology, Embryology, and Genetics, Federal University of Santa Catarina, Florianópolis, Brazil, 4 Plant Morphogenesis and Biochemistry Laboratory, Federal University of Santa Catarina, Florianópolis, Brazil, 5 Epagri, Florianopolis, Brazil, 6 Federal Institute of Education, Science, and Technology of Santa Catarina, Garopaba, Brazil, 7 Department of Pediatrics, Federal University of Santa Catarina, Florianópolis, Brazil

Abstract

Electrospun materials have been widely explored for biomedical applications because of their advantageous characteristics, i.e., tridimensional nanofibrous structure with high surface-to-volume ratio, high porosity, and pore interconnectivity. Furthermore, considering the similarities between the nanofiber networks and the extracellular matrix (ECM), as well as the accepted role of changes in ECM for hernia repair, electrospun polymer fiber assemblies have emerged as potential materials for incisional hernia repair. In this work, we describe the application of electrospun non-absorbable mats based on poly(ethylene terephthalate) (PET) in the repair of abdominal defects, comparing the performance of these meshes with that of a commercial polypropylene mesh and a multifilament PET mesh. PET and PET/chitosan electrospun meshes revealed good performance during incisional hernia surgery, post-operative period, and no evidence of intestinal adhesion was found. The electrospun meshes were flexible with high suture retention, showing tensile strengths of 3 MPa and breaking strains of 8–33%. Nevertheless, a significant foreign body reaction (FBR) was observed in animals treated with the nanofibrous materials. Animals implanted with PET and PET/chitosan electrospun meshes (fiber diameter of 0.71 ± 0.28 µm and 3.01 ± 0.72 µm, respectively) showed, respectively, foreign body granuloma formation, averaging 4.2-fold and 7.4-fold greater than the control commercial mesh group (Marlex). Many foreign body giant cells (FBGC) involving nanofiber pieces were also found in the PET and PET/chitosan groups (11.9 and 19.3 times more FBGC than control, respectively). In contrast, no important FBR was observed for PET microfibers (fiber diameter $= 18.9\pm0.21$ µm). Therefore, we suggest that the reduced dimension and the high surface-to-volume ratio of the electrospun fibers caused the FBR reaction, pointing out the need for further studies to elucidate the mechanisms underlying interactions between cells/tissues and nanofibrous materials in order to gain a better understanding of the implantation risks associated with nanostructured biomaterials.

Editor: Mário A. Barbosa, Instituto de Engenharia Biomédica, University of Porto, Portugal

Funding: Authors acknowledge Fundação para a Ciência e a Tecnologia (FCT, Portugal), the European Union, QREN, FEDER, COMPETE, for funding the Organic Chemistry Research Unit (QOPNA) (project PEst-C/QUI/UI0062/2013; FCOMP-01-0124-FEDER-037296) and for the PhD grant (SFRH/BD/38881/2007). Authors also acknowledge CNPq (Brazil) for a young talent researcher grant (375205/2012-8) and the financial support from FAPESC (Brazil) (PRONEX 17420/2011/3). The funders had no role in study design, data collection and analysis, decision to publish, or preparation of the manuscript.

Competing Interests: The authors have declared that no competing interests exist.

* E-mail: bveleirinho@ua.pt

Introduction

Electrospinning has attracted the interest of researchers from many fields as a versatile technique to produce nanofibers from synthetic and naturally derived polymers. With typical diameters ranging from 10 nm to a few micrometers, these fibers are usually collected continuously as nonwoven fibrous mats. These mats usually show a tridimensional nanostructure with high surface-to-volume ratio, high porosity, and interconnectivity, and they have demonstrated high potential for biomedical applications, such as tissue engineering scaffolds, vascular grafts, and drug delivery systems [1–3].

In the past decade, our research group has been exploring the potential of electrospinning for different applications [4]. Cell culture studies revealed that hybrid nanofibers of poly(ethylene terephthalate) (PET) and chitosan provide a good substratum for fibroblast adhesion, proliferation, extracellular matrix secretion, and three-dimensional colonization [5], in addition to their interesting surface and mechanical properties [6,7]. The promising results obtained *in vitro* prompted us to test these nondegradable electrospun mats as abdominal meshes for incisional hernia repair.

Incisional hernia is a frequent complication of laparotomy resulting from the decrease of abdominal strength in the injured tissue. The incidence of incisional hernia after abdominal surgery depends on the pattern of the incision performed. For a midline incision, the preferred incision for the upper abdominal surgery, the incidence lies around 10–14% [8–10]. Different patterns of incision, however, such as the transverse incision, yield much lower rates of hernia formation (2%) [10]. In comparison to the traditional hernia repair strategy by primary closure, the implementation of a tension-free repair by using a prosthetic biomaterial, i.e., nonresorbable abdominal mesh, to substitute or reinforce abdominal strength at the damaged area has decreased the recurrence rates significantly [8–12]. Nevertheless, serious complications have also been associated with this procedure, including infection, visceral adhesions to the mesh, seroma, mechanical failure of the mesh, and foreign body reaction [13,14]. Bowel adhesion to the implanted mesh is a major concern as it causes serious complications, including bowel obstruction, enterocutaneous fistula, and chronic pain [15].

Absorbable meshes are also used as scaffolds for abdominal defect correction. In this case, the biodegradable mesh provides support for cell growth and for extracellular matrix secretion, promoting the tissue repair process. A few recent reports have shown successful results in exploring partially degradable or absorbable electrospun mats as abdominal meshes. An electrospun blended fiber mesh prepared from biodegradable poly(ester urethane) urea and poly(lactide-co-glycolide), latter loaded with an antibiotic, was shown to provide good mechanical properties, while imparting antibacterial activity and, hence, reducing the risk of infection during application of the composite material to abdominal wall closure [16]. A similar approach used electrospun poly(ester urethane) urea fibers deposited with electrosprayed serum-based culture medium [17] or porcine dermal extracellular matrix digest [18]. When these materials were tested as abdominal wall repair meshes, they were demonstrated to provide adequate mechanical properties and, at the same time, enhanced bioactivity, biocompatibility and cell infiltration, with no herniation, infection, or tissue adhesion. Other examples include the application of electrospun absorbable polycaprolactone scaffolds [19], which were also evaluated for their suitability in hernia repair.

Despite the advantages of absorbable materials, the newly formed tissue has a decreased tensile strength and thereby, re-herniation is a frequent problem after the absorption of the prosthetic material. In this context, electrospun mats of nonde-gradable polymers have emerged as potential alternative meshes for abdominal defect repair. Because of its unique properties, i.e., unaligned nanofibrous arrangement, microporosity, and high hydrophobicity, the PET electrospun mat has emerged as a potential candidate for abdominal wall repair [6]. In addition to these advantages, PET is a highly biocompatible, biostable, and nondegradable polymer which possesses the mechanical features required for this application. Moreover, the low density (~ 0.091 g/cm^{-3}) and the high malleability of this material may promote enhanced adaptation and, consequently, patient comfort [7]. On the other hand, as a hydrophobic material with small pores, electrospun PET meshes may restrict the integration of parietal conjunctive tissue. McGinty et al. have demonstrated that a better incorporation of the mesh in the parietal side reduces the number and the severity of adhesion formations on the visceral side [20]. Hence, with the aim of enhancing the interaction of the mesh with the parietal conjunctive tissue, a hybrid mat of PET/chitosan (PET/C) was developed. Chitosan has shown ideal properties for biomedical applications, including the anti-inflam-matory and wound healing effects, which may attenuate the typical symptoms of the post-surgery period and prevent adhesiogenesis [21,22]. Additionally, in the current study, a double-layered mesh (DL), containing one layer of PET (turned to the visceral side) and one layer of PET/C mat (turned to the parietal side), was also developed.

In this paper, we describe the application of three electrospun nonabsorbable mats, including PET, PET/C and DL, in the repair of abdominal defects, comparing the performance of these nanofibrous meshes with that of a commercial polypropylene mesh (Marlex) (control) and a multifilament microfibrous PET mesh. An in vivo study with an abdominal hernia Wistar rat model was performed to evaluate the clinical and histological aspects of using these meshes for abdominal hernia repair.

Materials and Methods

Animal experiments were approved by the Animal Ethics Committee of the Federal University of Santa Catarina (PP0406/2009).

Materials

PET pellets and PET woven fabric were kindly supplied by Flexitex (Portugal). Marlex (Intracorp) was purchased from Cirurgica Passos, Brazil. Chitosan medium molecular weight (15% acetylation degree) was purchased from Sigma-Aldrich Chemical Company. The molecular weight of the initial chitosan sample (1500 kDa) was reduced to 15 kDa by oxidative depolymerization [23]. All chemicals were of analytical grade and obtained from Sigma-Aldrich Chemical Company.

Mesh Fabrication by Electrospinning

Thirty percent (w/v) PET solution was prepared in a blend of trifluoroacetic acid and dichloromethane [8:2 v/v] by moderate stirring for 2 hours at room temperature. PET/C blend was prepared by adding chitosan (6 wt. %) to the PET 30 wt. % solution and stirring for 3 hours at room conditions.

Electrospinning was performed using a typical experimental setup previously described [6]. The process was conducted at 26 kV of applied voltage, with a flow rate of 0.08 mL/min (V = 20 mL) and a needle tip-to-collector distance of 12 cm.

The double-layer (DL) mesh was fabricated by electrospinning 10 mL of PET solution, followed by 10 mL of PET/C solution. Fibers were collected as a nonwoven fibrous mat on the rotating drum (900 rpm), in air and at room conditions ($20\pm2°$C, 45–50% RH), and the mat was dried at 35°C for 24 hours.

Morphological Analysis

The morphology of the fibrous scaffolds was investigated by scanning electron microscopy (SEM). Small sections of the scaffolds were sputter-coated with gold and analyzed using a scanning electron microscope (Hitachi S4100) at an accelerating voltage of 25 kV.

Image processing was performed using ImageJ - 1.37c software (Wayne Rasband, National Institutes of Health, USA). Five random images (1000 X magnification) were obtained for each sample. Fiber diameters were calculated from at least 100 measurements of the sample fibers. SEM images were thresholded, and pore areas were automatically calculated using the "analyze particle" tool of ImageJ (n >200).

Mechanical Properties

Mechanical properties in tension were evaluated using texture analyzer equipment (Model TA HDi, Stable Micro Systems, England) equipped with fixed grips lined with thin rubber on the ends. Test specimens 90 mm long×10 mm wide were obtained perpendicular to the axis of the collector rotation, and the ends were mounted on the grips using sticky tape. The thickness of the test samples was measured at different locations on each sample using a digital micrometer (Model MDC-25S, Mitutoya Corp., Tokyo, Japan). The initial grip separation was set at 50 mm, and the crosshead speed was 0.5 mm/s. At least eight samples of each mat were tested.

Water Contact Angle (WCA) Measurements

The wettability of scaffolds was assessed by the sessile drop method using an OCA-20 contact angle system (DataPhysics Instruments). A drop of distilled water (1 μL) was automatically dispensed on the scaffold surface, and the WCA and drop life times were calculated using the SCA 20 software. At least 10 measurements were taken for each sample.

Animal Model

Male Wistar rats 3 months of age and weighing 250–300 g were obtained from the Central Biotery of the Federal University of Santa Catarina. Animals were randomly distributed among the 7 treatments (Marlex30, PET30, PET/C, DL, Woven-PET, Marlex90, and PET90), according to Table 1 (n≥8).

Before biological assay, all meshes were sterilized under UV light for 1 hour of exposure (both faces), immersed in ethanol 70% (v/v) for 10 minutes, and washed with sterile physiologic solution.

A graphical illustration of the surgical procedure is provided (Figure S1). After an intramuscular injection of a mixture of ketamine (90 mg/kg) and xylazine (15 mg/kg) and abdominal shaving, a 5 cm paramedian skin incision was made at the left side, using a sterile scalpel blade. Skin was dissected to expose the underlying abdominal fascia, and a 1.5×1.5 cm defect of anterior abdominal wall was created by the complete resection of abdominal layers. The edges of the meshes (2.0×2.0 cm) were sutured to the remaining muscle of the abdominal wall with interrupted suture and also with simple running suture all over the borders, using 5–0 polypropylene. The skin was closed with intradermal suture with nylon 4–0 monofilament. Animals were allowed to recover from anesthesia, housed in individual cages, and observed daily for evidence of wound complications, such as redness, infection, seroma, abcess, hematoma, or skin dehiscence.

On day 30 and 90 post-surgery, animals were sacrificed in a carbon dioxide chamber, and the presence of intestinal adhesions was analyzed. The abdominal wall was carefully excised well away from the mesh (see Figure S2) to preserve any adherence to the bowel or omentum. After adhesion analysis, tissue was completely excised and collected for histopathological analysis.

Histopathological Analysis

Tissues were excised from the animals and fixed in phosphate buffered formaldehyde solution (4%, pH 7.2, 0.1 M), embedded in paraffin, and sectioned at 4 μm thickness. Giemsa, hematoxylin and eosin (HE), and Garvey's staining were performed [24]. For immunohistochemical analysis, sections were deparaffinized in xylene and rehydrated in a graded ethanol series. Antigen retrieval was performed with 0.05% trypsin and 0.1% calcium chloride (20 minutes, 37°C). Endogenous peroxidase activity was blocked by incubation in a hydrogen peroxide solution. Following, sections were incubated with a monoclonal antibody directed against CD68, clone KP1, dilution 1:100 (Zeta Corporation, CA). Antibody detection was performed using Histofine Simple Stain Max-Po Multi (Nichirei Biosciences, Tokyo, Japan) and 3,3′-diaminobenzidine tetrahydrochloride (Spring Bioscience, CA). Samples were analyzed under a microscope (Nikon Eclipse 50i equipped with a Nikon Digital Sight DS-Fi2). The thickness of the foreign body granuloma was measured and the absolute number of foreign body giant cells (FBGC) per granuloma section was determined by manual counting (n≥8). A detailed description of morphometric procedures is given in Figure S2.

Statistics

Statistical analysis was carried out using Instat 3.0 software. Results were expressed as the mean ± standard deviation and compared through one-way ANOVA and Tukey-Kramer.

Results

Characterization of Meshes

Figure 1 displays SEM images of PET, PET/C, and DL mats. The electrospun mats showed a typical nonwoven fibrous structure with random fiber orientation and high porosity. Table 2 summarizes some morphological, mechanical, and surface properties of the meshes used in this study. The average diameter of PET fibers was 0.71±0.28 μm, and the average pore area was 9.4 μm^2. Addition of chitosan promoted a substantial increase in fiber diameter and pore area to 3.01±0.72 μm and 89.3 μm^2, respectively. Compared to PET/C, PET mesh showed superior mechanical properties with a higher tensile strength (3.17±0.23 MPa compared to 2.89±0.27 MPa), Young's modulus (120±10 compared to 70±10 MPa), and elongation (32.8±5.7% compared to 8.2±1.3%). Also, a decrease in the hydrophobicity of the mesh was observed by the presence of chitosan (WCA decreased from 133.2±2.9° to 125.2±4.6° with the addition of chitosan).

Surgical Procedure and Post-operative Period

The electrospun mats were evaluated in an incisional hernia experiment with Wistar rats and compared to Marlex, the control. Following the creation of the abdominal defect by resection of 1.5×1.5 cm of the Wistar rats' abdominal muscle, the prosthetic meshes were implanted and fixed through the borders to the remaining muscle (see Figure S1). Suture of electrospun meshes was easily performed without breakage. In fact, electrospun meshes were more suitable and more resistant to suture than

Table 1. Experimental groups of incisional hernia repair.

Group	Mesh chemical composition	Duration of experiment (days)	Mesh thickness (mm)
Marlex30	Polypropylene	30	0.22±0.07
Marlex90	Polypropylene	90	0.22±0.07
PET30	PET	30	0.31±0.02
PET90	PET	90	0.31±0.02
Woven-PET	PET	30	0.49±0.09
PET/C	PET/C 5:1 (w/w)	30	0.52±0.05
DL	PET + PET/C 5:1 (w/w)	30	0.46±0.11

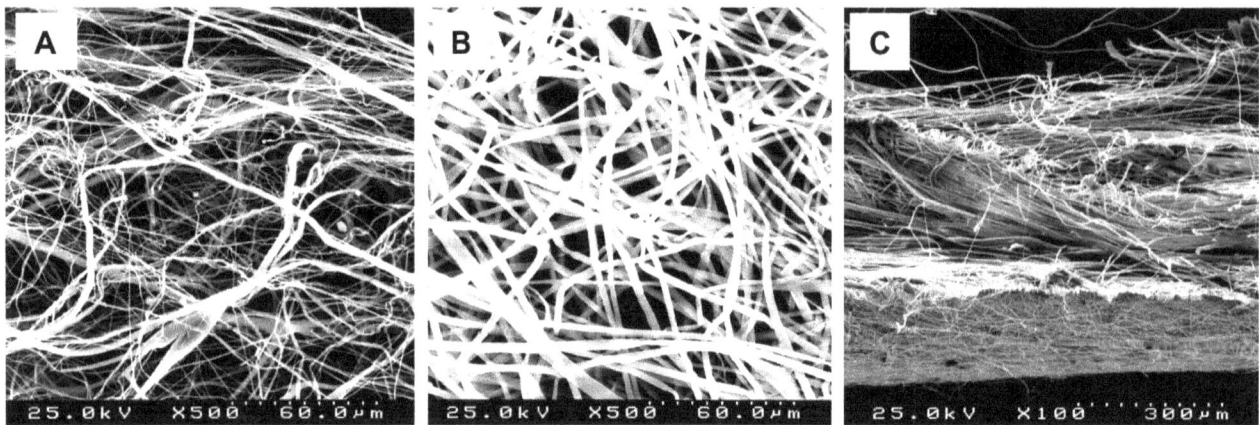

Figure 1. Morphological analysis of the meshes. SEM images of (A) PET, (B) PET/C, and (C) DL meshes. The transversal section of the DL mesh shows PET mat (bottom) and PET/C mat (top). *Bars*: 60 μm (A, B); 300 μm (C).

control, where an extra margin of around 5 mm was used to avoid breaking of filaments.

During the post-operative period, some complications, such as local redness, abscess, or skin dehiscence, were registered. Table 3 displays macroscopic complications found in the post-operative period. No complications were found during the post-operative period for the woven-PET group. In the other experimental groups, local redness, characterized by redness or swelling, was the most common complication, with higher incidence of these complications observed for PET/C and DL groups, occurring in 50% and 75% of the animals, respectively. Redness tended to decrease with time, while abscess and skin dehiscence persisted until the end of the experimental period. Omentum adhesions to the mesh were observed in all animals (see Figure S3), but no visceral adhesions were found in the experimental groups.

Histological Analysis

Histological analysis was performed to evaluate the cellular response to the prosthetic biomaterials. Figure 2 and Figure 3 display representative images of HE-stained sections of animals treated with electrospun meshes and Marlex (low and high magnification, respectively). All animals showed typical nonimmunogenic granulomas (foreign body granulomas) surrounding the mesh structure placed below the abdominal subcutaneous tissue. Representative photomicrographs from Garvey's staining are provided in Figure 4. The foreign body granuloma were mostly composed of macrophages, foreign body giant cells (FBGC), and fibroblasts. Multinucleated cells frequently involved

one or more fiber segments. Immunohistochemical analysis confirmed the high density of both macrophages and FBGC (CD68+) in animals treated with the electrospun materials (Figure 5). The mean thickness of the granuloma and the average number of FBGC in the granuloma are plotted in Figure 6A and Figure 6B, respectively. Animals treated with electrospun meshes showed significantly thicker granulomas and a higher number of FBGC compared to Marlex and the woven-PET group. The mean granuloma thickness induced by PET nanofibers was 4-fold higher than control (Marlex) and 10-fold higher than in the woven-PET group. Also a 10-fold increase in the number of FBGC was observed in the PET group compared to control. Hybrid meshes showed even thicker granulomas (1522±277 μm and 1211±547 μm for PET/C and DL, respectively), as well as a large number of FBGC comprising one or more fibrous structures. The woven-PET group produced the weakest inflammatory response with an average granuloma thickness of 87±35 μm and FBGC rarely observed.

Long-term inflammatory response was evaluated for two selected groups (Marlex and electrospun PET) 90 days after mesh implantation. Both groups showed a decrease in inflammation with time. The average granuloma thickness of the electrospun PET group decreased from 959±473 μm to 513±217 μm, and the number of FBGC decreased from 106±30 to 89±12.

Table 2. Fiber diameter, pore area, mechanical properties, and WCA of abdominal meshes.

Group	Fiber diameter (μm)	Pore area (μm²)	Tensile strength (MPa)	Percentage elongation (%)	Young's modulus (MPa)	WCA (°)
Marlex	177±24	31400	N.A.	N.A.	N.A.	N.A.
PET	0.71±0.28	9.4	3.17±0.23	32.8±5.7	1.2±0.1	133.2±2.9
PET/C	3.01±0.72	89.3	2.89±0.27	8.2±1.3	0.7±0.1	125.2±4.6
Woven-PET	18.9±2.1[a]	N.A.	N.A.	N.A.	N.A.	N.A.
	342±68[b]					

N.A. not available
[a]Average filament diameter.
[b]Average diameter of the multifilament yarn.

Table 3. Occurrence rate (%) of complications during post-operative period (PO) and day euthanized (E).

Mesh	Period	Local redness	Dehiscence	Seroma or abscess	Adhesion
Marlex	PO	37.5	12.5	12.5	N.A.
	E	12.5	12.5	12.5	omentum
PET	PO	37.5	0	12.5	N.A.
	E	25	0	12.5	omentum
PET/C	PO	50	0	0	N.A.
	E	0	0	0	omentum
DL	PO	75	12.5	12.5	N.A.
	E	37.5	12.5	12.5	omentum
Woven- PET	PO	0	0	0	N.A.
	E	0	0	0	omentum

N.A. not available.

Discussion

Several biomaterials have been used as prosthetic meshes for abdominal wall repair, in particular, incisional hernia repair. Among them, nonabsorbable polymers with recognized biocompatibility, such as polypropylene, PET or polytetrafluoroethylene, are the most common. Although a considerable improvement in the recurrence rate has been achieved with these materials in comparison to the traditional suture technique, several problems are still associated with this procedure. Among them, the formation of adhesions between the mesh and the bowel results in several complications, such as chronic pain, bowel obstruction or enterocutaneous fistula. Also, nonresorbable meshes have been associated with a high occurrence of chronic foreign body response and increased risk of infection [25,26]. Improvements have been achieved by using polymer meshes that gradually degrade *in vivo*, promoting improved tissue integration and rapid resorption [19]. However, as scaffolds to support native tissue ingrowth, resorbable synthetic meshes are still limited by their loss of strength. Considering the attractive properties of electrospun polymer meshes, we have developed, for the first time, electrospun mats of nondegradable polymers for the repair of abdominal defects.

Chemical composition, weight, pore size, and filament structure, represent critical parameters employed in surgical mesh design [27]. Apart from preventing adhesion, it is generally accepted that the ideal abdominal mesh should be chemically inert and stable for long periods, promote tissue regeneration to form an adequate barrier against protrusion, cause no immune or inflammatory response, and fulfill the required mechanical needs for the application. The studied PET electrospun mats are nonwoven meshes of nanofibers (average diameter $= 0.71 \pm 0.28$ μm) thought to be ideal for the prevention of visceral adhesion by their highly hydrophobic microporous structure (WCA $= 133.2 \pm 2.9°$; average pore area $= 9.4$ μm^2). Indeed, according to Mathews *et al.*, biomaterials with pores smaller than 75 μm reduce the occurrence of bowel adhesion [28].

With the purpose of manipulating the architecture of PET mesh and also taking advantage of the anti-inflammatory and wound healing effects of chitosan, a hybrid fibrous mat (PET/C) was also developed, with higher fiber diameter and pore area. These morphological differences can have significant effects on cell-biomaterial interactions, as previously demonstrated [5]. The decreased stiffness of the hybrid mesh in comparison to the PET mesh may be attributed to a heterogeneous polymer distribution within fibers as a result of phase separation during the

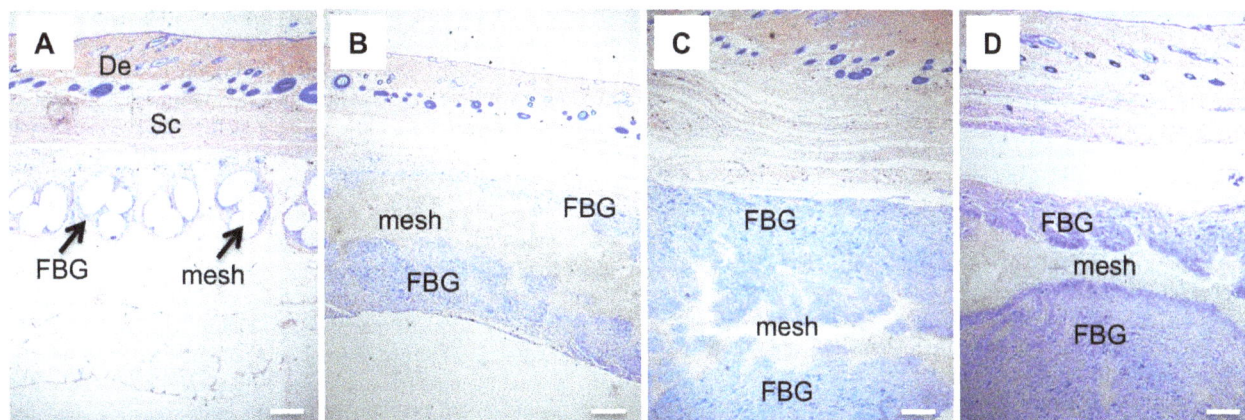

Figure 2. Foreign body granuloma induced by the abdominal meshes. Histological sections evidencing foreign body granuloma (Giemsa staining) of (A) Marlex30, (B) PET30, (C) PET/C, and (D) DL groups. (De) dermis, (Sc) subcutaneous tissue, and (FBG) foreign body granuloma. Bar = 220 μm.

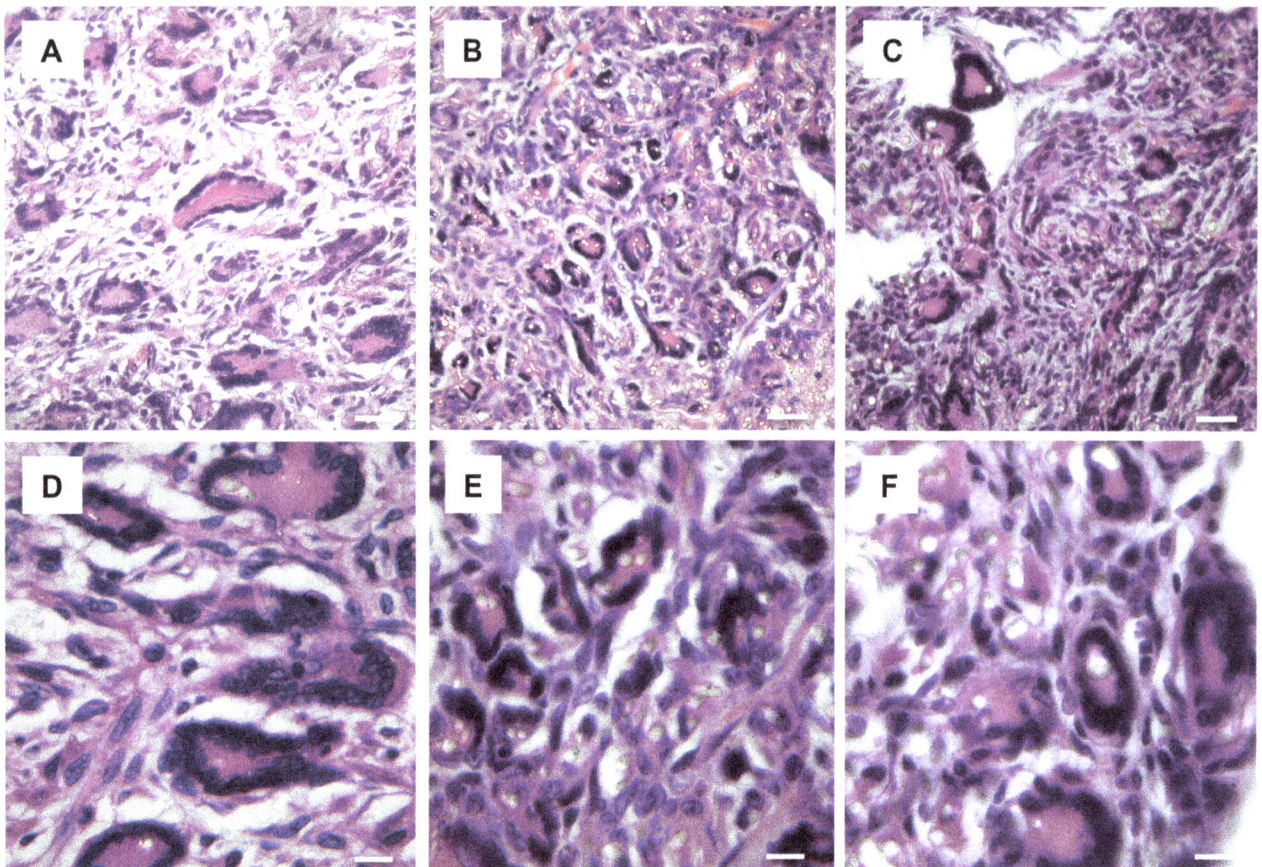

Figure 3. Foreign body giant cells covering nano- and microfibers of electrospun materials. Histological sections (HE staining) evidencing FBGC in (A and D) PET30, (B and E) PET/C, and (C and F) DL groups. *Bars*: 25 μm (A–C); 10 μm (D–F).

electrospinning process [7]. The increased hydrophilic character observed for the PET/C hybrid may be advantageous for mesh integration on the parietal side. A double-layer mesh was also developed, comprising one layer of PET to prevent formation of bowel adhesion and one layer of PET/C nanofibers to stimulate integration of the mesh in the subcutaneous tissue, thus reinforcing the mechanical strength of the prosthetic wall.

In the rat abdominal hernia model, electrospun meshes were demonstrated to be adequate for the surgical procedure, i.e., easy to suture, and as a soft and flexible material, the electrospun meshes adapted well to the abdominal tissues. In contrast, the stiffness of the Marlex material may affect surrounding tissues and cause discomfort. This material was even perceived through the animal's skin by touching. Electrospun meshes, on the other hand, were much softer, as well as more malleable and adaptable, while effectively performing their role of containing visceral components without mesh failure during the experimental period.

In contrast to bowel adhesion that can lead to serious complications, the observed omentum adhesion to the meshes is considered clinically irrelevant [29]. In fact, some authors have even suggested that the interposition of omentum between the prosthetic mesh and viscera is effective in restricting omentum adhesion, both in preclinical and clinical studies [29,30].

Importantly, we observed a significant foreign body reaction associated with the electrospun nanofibrous meshes. This was an unexpected result since the chemical composition of the electrospun mesh is 100% PET, a recognized biocompatible polymer, and many studies have reported the absence of inflammatory

response to polymer electrospun fibers by different cell types, both *in vitro* and *in vivo* [31–34]. Numerous studies have reported on the anti-inflammatory and wound healing effects of chitosan; however, in this specific application, chitosan had a negative impact, as an increased inflammatory response was observed in animals treated with chitosan- containing meshes. Similar results were obtained by Barbosa *et al.* [35] when testing the inflammatory response to chitosan scaffolds with different acetylation degree. In fact, the chitosan scaffold with an acetylation degree of 15% induced the formation of a thick granuloma with high infiltration of inflammatory cells, after subcutaneous implantation in mice [35]. Another study also showed that chitosan a scaffold with an acetylation degree of 15% caused a macrophage M1 pro-inflammatory response [36].

The foreign body granulomas were mostly composed of macrophages, FBGC, and fibroblasts. An abundance of FBGC was found in tissues surrounding the electrospun meshes, evidencing a typical foreign body reaction. FBGC were most often found to be enclosing one or more nanofibers as an attempt to isolate the foreign material (*cf.* Figure 3). Extending the experimental time seemed to result in an attenuation of the foreign body reaction; nonetheless, a large FBR persisted 90 days after mesh implantation. Although foreign body reaction has few clinical implications and is usually limited to the close periphery of the implanted material, certain clinical disadvantages are always present as an associated risk condition. Indeed, chronic inflammation and the related proangiogenic process have been assumed

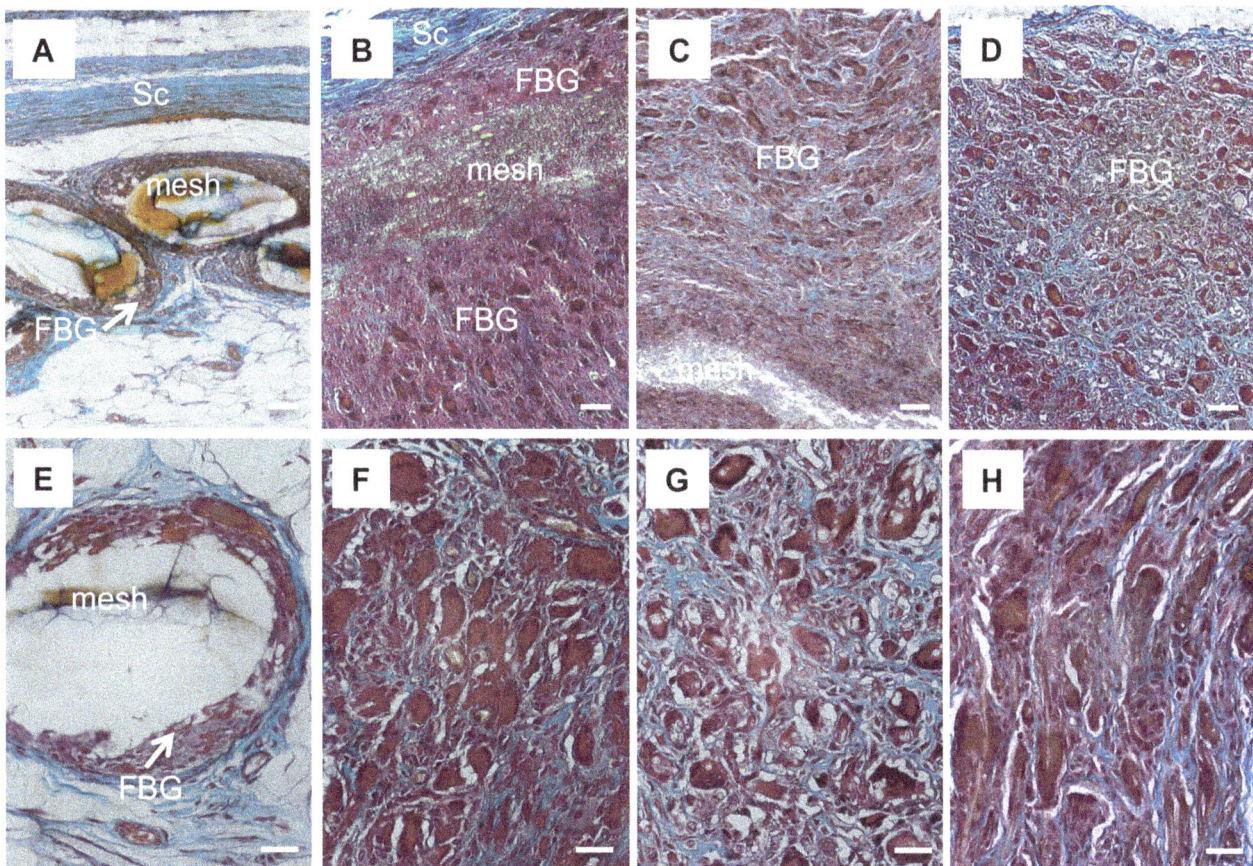

Figure 4. Garvey staining of foreign body granulomas. Histological sections stained with Garvey's staining (nuclei stained black/dark red, cytoplasmic elements stained red and collagen fibers stained light blue). (A,E) Marlex (B,F) PET (C,G) PET/C (D,H) DL. (De) dermis, (Sc) subcutaneous tissue, and (FBG) foreign body granuloma. Bars = 50 μm (A–D); 25 μm (E–H).

to underlie most chronic diseases, including cancer, cardiovascular diseases, and diabetes [37,38].

It is well established that polymer type, mesh construction, fiber size, mesh porosity and contact surface, as well as the specific characteristics of the tissue where the biomaterial is implanted, play important roles in biocompatibility and induced tissue reactions [39,40]. Still, no consensus has been reached with respect to the effects of implanted mesh and the development of inflammation [41–43]. Indeed, cell/tissue-mesh interactions still require further elucidation.

Considering the high biocompatibility of bulk PET and the minimum foreign body reaction found for woven-PET mesh (PET-woven microfibers), we hypothesize that the nanostructure of the electrospun materials underlies the huge foreign body reaction found in animals implanted with electrospun meshes. The reduced diameter of the electrospun fibers and pore size of the meshes, combined with the high surface-to-volume ratio of the electrospun materials, may therefore have important effects on the inflammatory reaction. Among surface properties, the material's ability to adsorb proteins plays a key role, as those proteins, not the material's composition itself, are major contributors of FBR [40]. It is well documented that the high surface-to-volume ratios of electrospun nanofibrous materials contribute to their high protein adsorption capability [44,45] and likely explain the high foreign body reaction observed in electrospun PET meshes. Indeed, higher foreign body reactions are often found for biomaterials, in both in micro- and nanoscale dimensions, with large surface areas

[46,47]. Specifically, in abdominal defect repair, Conze et al. [48] have verified a pronounced foreign body reaction for a multifilament small-diameter polypropylene mesh. In fact, the diameter of foreign body granuloma 90 days after mesh placement decreased from 106.5 μm to 70.9 μm by increasing the filament diameter from 0.6 mm to 2.5 mm. On the other hand, many reports show a decrease of inflammatory response with the decrease of fiber diameter. Saino et al. [49] showed that the decrease of fiber diameter of electrospun polylactic acid fibrous mats reduced in vitro macrophage activation and the secretion of proinflammatory molecules. Similarly, Cao et al. [50] demonstrated the importance of the nanofibrous scaffold architecture and topography on the in vivo and in vitro foreign body reaction and showed a decrease of granuloma thickness of subcutaneous implants from 38 μm for a PCL film to ~8 μm and 4 μm for PCL-aligned electrospun nanofibers or nonwoven electrospun nanofibers, respectively.

Other aspects that are thought to have influenced the observed intensive foreign body reaction are related to the extension of the trauma and the specific characteristics of the tissues where the materials had been implanted. Abdominal mesh implantation involves the creation of a 1.5×1.5 cm defect by complete resection of the abdominal wall. This is a severe trauma, considering the relative size of the animal, and, consequently, a large inflammatory response may be induced. Furthermore, specific characteristics of the implantation tissues, such as cell composition and function, vascularization, extracellular matrix composition, and

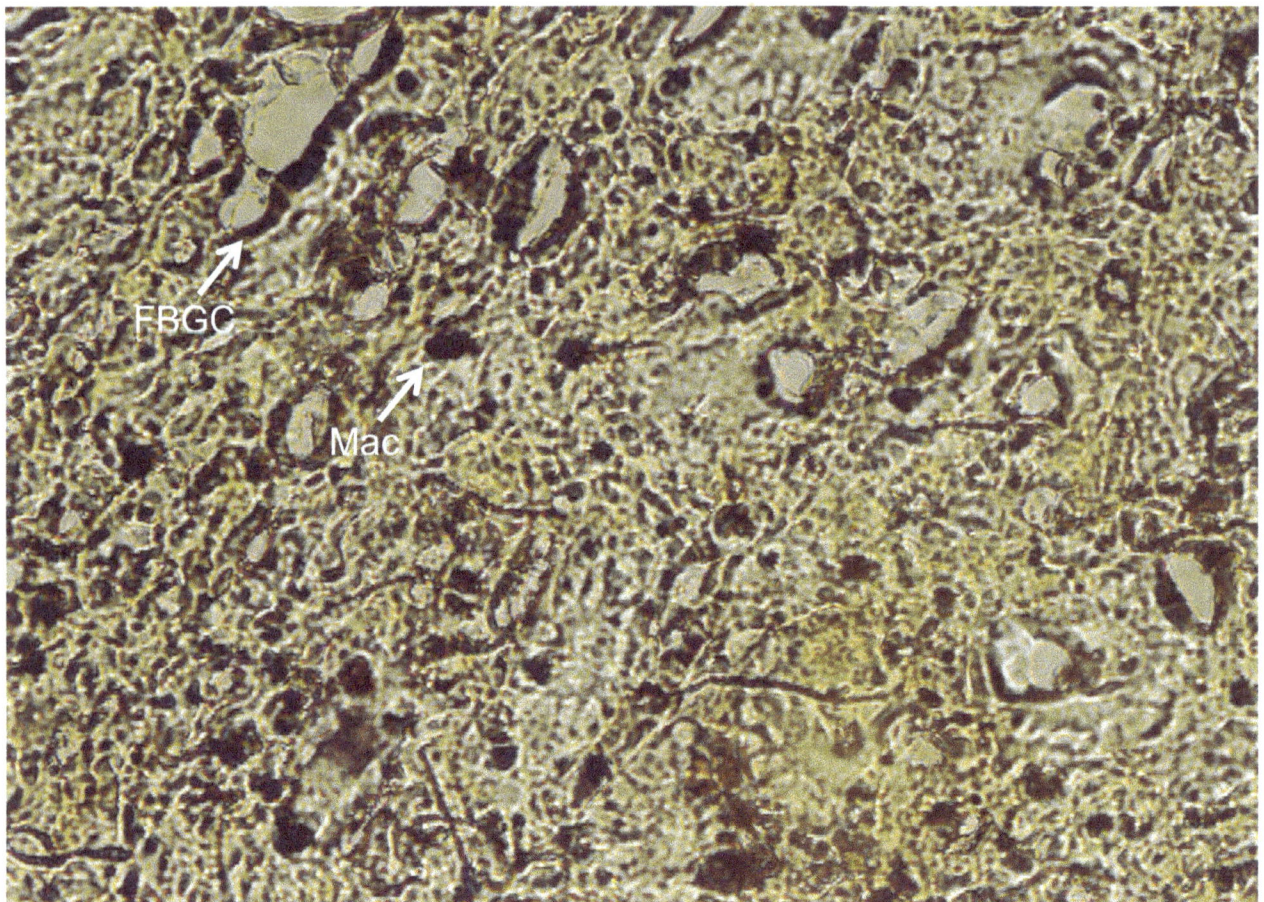

Figure 5. Immunohistochemistry for CD68. Immunohistochemical staining (CD68+) evidencing macrophages (Mac) and foreign body giant cells (FBGC) around a PET electrospun mat implanted in rat as abdominal mesh. Sections were incubated with a monoclonal antibody directed against CD68, clone KP1 (Zeta Corporation, CA), Histofine Simple Stain Max-Po Multi and 3,3′-diaminobenzidine tetrahydrochloride.

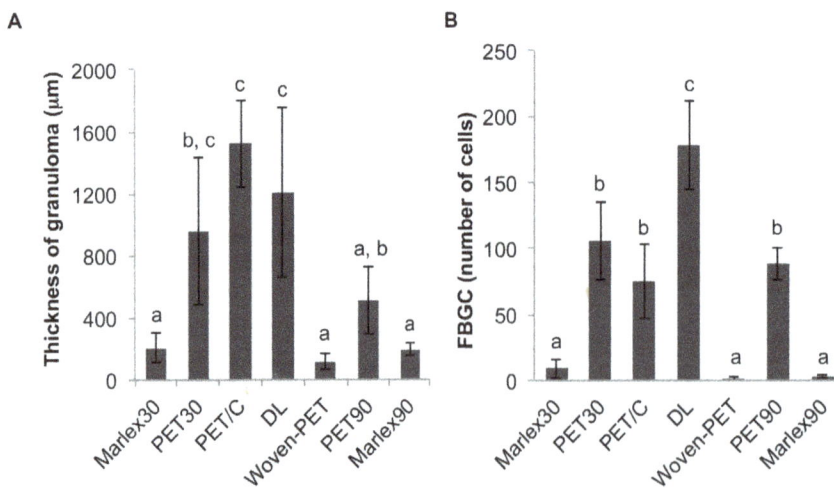

Figure 6. Analysis of the foreign body granuloma induced by abdominal meshes. (A) Granuloma thickness (mean ± standard deviation) and (B) absolute number of FBGC per section of granuloma (mean ± standard deviation) in animals implanted with different abdominal meshes. All experimental groups were compared to each other using Tukey-Kramer multiple comparison tests. For each chart, bars with different letters are significantly different at $P < 0.05$.

contact with ascitic fluid, for example, have important effects on foreign body reaction.

Materials are becoming smaller than the basic body unity, i.e. the cell. Despite the enormous progress of the electrospinning technique over the past decade, cellular response and the associated risks involved in the use of nanostructure fibrous biomaterials are still poorly understood [51,52], and further studies are needed to gain more insight.

Conclusions

PET and PET/chitosan electrospun meshes demonstrated good performance during the implantation surgery, adequate mechanical attributes, and no evidence of intestinal adhesion. Nevertheless, a large foreign body reaction was found in animals treated with the electrospun mats. Indeed, the reduced dimension of nanofibers and the high surface-to-volume ratio of electrospun nonwoven materials may induce a high foreign body reaction, depending on the extent and location of the lesion. Refinement may be achieved by the inclusion of biological components on the fiber's surface to enhance bioactivity and biocompatibility, thus increasing the potential of these nondegradable electrospun fiber scaffolds for abdominal wall replacement. Nevertheless, these results demonstrate the need for more studies to elucidate the mechanisms underlying cell/tissue-nanomaterial interactions in order to gain a better understanding of the risks involved in implantation of nanostructured biomaterials.

Acknowledgments

Authors acknowledge Central Laboratory of Electron Microscopy (LCME) and Fluorbeg (UFSC) for microscopic analysis.

Author Contributions

Conceived and designed the experiments: BV JALdS JAS. Performed the experiments: BV DSC RP AP. Analyzed the data: BV MM ECF PFD RMRdV JALdS JAS. Contributed reagents/materials/analysis tools: MM ECF PFD RMRdV JALdS. Wrote the paper: BV JALdS.

References

1. Schofer MD, Roessler PP, Schaefer J, Theisen C, Schlimme S, et al. (2011) Electrospun PLLA nanofiber scaffolds and their use in combination with BMP-2 for reconstruction of bone defects. PLoS ONE 6(9): e25462.
2. Vasita R, Katti DS (2006) Nanofibers and their applications in tissue engineering. Int J Nanomed 1(1): 15–30.
3. Cui W, Zhou Y, Chang J (2010) Electrospun nanofibrous materials for tissue engineering and drug delivery. Sci Technol Adv Mater 11: 014108 (11pp).
4. Veleirinho B, Coelho DS, Dias PF, Maraschin M, Ribeiro-do-Valle RM, et al. (2012) Nanofibrous poly(3-hydroxybutyrate-co-3-hydroxyvalerate)/chitosan scaffolds for skin regeneration. Int J Biol Macromol 51: 343–350.
5. Veleirinho B, Berti FV, Dias PF, Maraschin M, Ribeiro-do-Valle RM, et al. (2013) Manipulation of chemical composition and architecture of non-biodegradable poly(ethylene terephthalate)/chitosan fibrous scaffolds and their effects on L929 cell behavior. Mat Sci Eng C-Biomim 33: 37–46.
6. Veleirinho B, Rei MF, Lopes-da-Silva JA (2008) Solvent and concentration effects on the properties of electrospun poly(ethylene terephthalate) nanofiber mats. J Polym Sci Pol Phys 46(5): 460–471.
7. Lopes-da-Silva JA, Veleirinho B, Delgadillo I (2009) Preparation and characterization of electrospun mats made of pet/chitosan hybrid nanofibers. J Nanosci Nanotechnol 9(6): 3798–3804.
8. Anthony T, Bergen PC, Kim LT, Henderson M, Fahey T, et al. (2000) Factors affecting recurrence following incisional herniorrhaphy. World J Surg 24(1): 95–101.
9. Gecim IE, Kocak S, Ersoz S, Bumin C, Aribal D (1996) Recurrence after incisional hernia repair. Results and risk factors. Surg Today 26(8): 607–609.
10. Halm JA, Lip H, Schmitz PI, Jeekel J (2009) Incisional hernia after upper abdominal surgery: a randomised controlled trial of midline versus transverse incision. Hernia 13: 275–280.
11. Luijendijk RW, Hop WCJ, van den Tol P, de Lange DCD, Braaksma MMJ, et al. (2000) A comparison of suture repair with mesh repair for incisional hernia. New Engl J Med 343(6): 392–398.
12. Kingsnorth A, LeBlanc K (2003) Hernias: inguinal and incisional. Lancet 362: 1561–1571.
13. Carbajo MA, del Olmo JCM, Blanco JI, de la Cuesta C, Toledano M, et al. (1999) Laparoscopic treatment vs open surgery in the solution of major incisional and abdominal wall hernias with mesh. Surg Endosc-Ultras 13(3): 250–252.
14. Robinson TN, Clarke JH, Schoen J, Walsh MD (2005) Major mesh-related complications following hernia repair - Events reported to the Food and Drug Administration. Surg Endosc-Ultras 19(12): 1556–1560.
15. Menzies D, Ellis H (1990) Intestinal-obstruction from adhesions - how big is the problem? Ann Roy Coll Surg 72(1): 60–63.
16. Hong Y, Fujimoto K, Hashizume R, Guan J, Stankus JJ, et al. (2008) Generating elastic, biodegradable polyurethane/poly(lactide-co-glycolide) fibrous sheets with controlled antibiotic release via two-stream electrospinning. Biomacromolecules 9: 1200–1207.
17. Hashizume R, Kazuro L, Fujimoto KL, Hong Y, Amoroso NJ, et al. (2010) Morphological and mechanical characteristics of the reconstructed rat abdominal wall following use of a wet electrospun biodegradable polyurethane elastomer scaffold. Biomaterials 31: 3253–3265.
18. Hong Y, Takanari K, Amoroso NJ, Hashizume R, Brennan-Pierce EP, et al. (2012) An elastomeric patch electrospun from a blended solution of dermal extracellular matrix and biodegradable polyurethane for rat abdominal wall repair. Tissue Eng Pt C-Meth 18: 122–132.
19. Ebersole GC, Buettmann EG, MacEwan MR, Tang ME, Frisella MM, et al. (2012) Development of novel electrospun absorbable polycaprolactone (PCL) scaffolds for hernia repair applications. Surg Endosc 26: 2717–2728.
20. McGinty JJ, Hogle NJ, McCarthy H, Fowler DL (2005) A comparative study of adhesion formation and abdominal wall ingrowth after laparoscopic ventral hernia repair in a porcine model using multiple types of mesh. Surg Endosc 19(6): 786–790.
21. Ueno H, Mori T, Fujinaga T (2001) Topical formulations and wound healing applications of chitosan. Adv Drug Deliver Rev 52(2): 105–115.
22. VandeVord PJ, Matthew HWT, DeSilva SP, Mayton L, Wu B, et al. (2002) Evaluation of the biocompatibility of a chitosan scaffold in mice. J Biomed Mater Res 59(3): 585–590.
23. Tommeraas K, Varum KM, Christensen BE, Smidsrod O (2001) Preparation and characterisation of oligosaccharides produced by nitrous acid depolymerisation of chitosans. Carbohydr Res 333: 137–144.
24. Garvey W (1984) Modified Elastic Tissue-Masson Trichrome Stain. Stain Technol. 59(4): 213–216.
25. Robinson TN, Clarke JH, Schoen J, Walsh MD (2005) Major mesh-related complications following hernia repair: events reported to the Food and Drug Administration. Surg Endosc 19: 1556–1560.
26. Welty G, Klinge U, Klosterhalfen B, Kasperk R, Schumpelick V (2001) Functional impairment and complaints following incisional hernia repair with different polypropylene meshes. Hernia 5: 142–147.
27. Voskerician G, Jin J, White MF, Williams CP, Rosen MJ (2010) Effect of biomaterial design criteria on the performance of surgical meshes for abdominal hernia repair: a pre-clinical evaluation in a chronic rat model. J Mater Sci-Mater M 21(6): 1989–1995.
28. Matthews BD, Pratt BL, Pollinger HS, Backus CL, KercherKW, etal. (2003) Assessment of adhesion formation to intra-abdominal polypropylene mesh and polytetrafluoroethylene mesh. J Surg Res 114(2): 126–132.
29. Karabulut B, Sonmez K, Turkyilmaz Z, Demiroğullari B, Karabulut R, et al. (2006) Omentum prevents intestinal adhesions to mesh graft in abdominal infections and serosal defects. Surg Endosc 20(6): 978–982.
30. Bingener J, Kazantsev GB, Chopra S, Schwesinger WH (2004) Adhesion formation after laparoscopic ventral incisional hernia repair with polypropylene mesh: a study using abdominal ultrasound. JSLS-J Soc Laparoend 8(2): 127–131.
31. Nisbet DR, Rodda AE, Horne MK, Forsythe JS, Finkelstein DI (2009) Neurite infiltration and cellular response to electrospun polycaprolactone scaffolds implanted into the brain. Biomaterials 30: 4573–4580.
32. Pan H, Jiang HL, Kantharia S, Chen WL (2011) A fibroblast/macrophage co-culture model to evaluate the biocompatibility of an electrospun dextran/PLGA

scaffold and its potential to induce inflammatory responses. Biomedical Mat 6(6): 065002.

33. Ni PY, Fu SZ, Fan M, Guo G, Shi S, et al. (2011) Preparation of poly(ethylene glycol)/polylactide hybrid fibrous scaffolds for bone tissue engineering. Int J Nanomed 6: 3065–3075.

34. Bergmeister H, Grasl C, Walter I, Plasenzotti R, Stoiber M, et al. (2011) Electrospun small-diameter polyurethane vascular grafts: Ingrowth and differentiation of vascular-specific host cells. Artificial Organs 36(1): 54–61.

35. Barbosa JN, Amaral IF, Aguas AP, Barbosa MA (2010) Evaluation of the effect of the degree of acetylation on the inflammatory response to 3D porous chitosan scaffolds. J Biomed Mater Res-A 93(1): 20–28.

36. Vasconcelos DP, Fonseca AC, Costa M, Amaral IF, Barbosa MA, et al. (2013) Macrophage polarization following chitosan implantation. Biomaterials 34: 9952–9959.

37. Bartsch H, Nair J (2006) Chronic inflammation and oxidative stress in the genesis and perpetuation of cancer: role of lipid peroxidation, DNA damage, and repair. Langenbeck Arch Surg 391(5): 499–510.

38. Manabe I (2011)Chronic inflammation links cardiovascular, metabolic and renal diseases. Circ J 75(12): 2739–2748.

39. Kamath S, Bhattacharyya D, Padukudru C, Timmons RB, Tang L (2008) Surface chemistry influences implant-mediated host tissue responses. J Biomed Mater Res-A 86(3): 617–626.

40. Hu WJ, Eaton JW, Tang LP (2001) Molecular basis of biomaterial-mediated foreign body reactions. Blood 98(4): 1231–1238.

41. Klosterhalfen B, Klinge U, Schumpelick V, Tietze L (2000) Polymers in hernia repair–common polyester vs. polypropylene surgical meshes. J Mater Sci 35: 4769–4776.

42. Brown CN, Finch JG (2010) Which mesh for hernia repair? Ann R Coll Surg Engl 92(4): 272–278.

43. Binnebösel M, von Trotha KT, Jansen PL, Conze J, Neumann UP, et al. (2011) Biocompatibility of prosthetic meshes in abdominal surgery. Semin Immuno-pathol 33: 235–243.

44. Shalumon KT, Binulal NS, Deepthy M, Jayakumar R, Manzoor K, et al. (2011) Preparation, characterization and cell attachment studies of electrospun multi-scale poly(caprolactone) fibrous scaffolds for tissue egineering. J Macromol Sci Pure 48(1): 21–30.

45. Leong MF, Chian KS, Mhaisalkar PS, Ong WF, Ratner BD (2009) Effect of electrospun poly(D, L-lactide) fibrous scaffold with nanoporous surface on attachment of porcine esophageal epithelial cells and protein adsorption. J Biomed Mater Res-A 89(4): 1040–1048.

46. Voskerician G, Gingras PH, Anderson JM (2006) Macroporous condensed poly(tetrafluoroethylene). I. In vivo inflammatory response and healing characteristics. J Biomed Mater Res-A 76(2): 234–242.

47. Sanchez VC, Weston P, Yan A, Hurt RH, Kane AB (2011) A 3-dimensional in vitro model of epithelioid granulomas induced by high aspect ratio nanomaterials. Part Fiber Toxicol 8(17): 1–18.

48. Conze J, Rosch R, Klinge U, Weiss C, Anurov M, et al. (2004) Polypropylene in the intra-abdominal position: influence of pore size and surface area. Hernia 8(4): 365–372.

49. Saino E, Focarete ML, Gualandi C, Emanuele E, Cornaglia AI, et al. (2011) Effect of electrospun fiber diameter and alignment on macrophage activation and secretion of proinflammatory cytokines and chemokines. Biomacromole-cules 12(5): 1900–1911.

50. Cao H, McHugh K, Chew SY, Anderson JM (2010) The topographical effect of electrospun nanofibrous scaffolds on the in vivo and in vitro foreign body reaction. J Biomed Mater Res-A 93(3): 1151–1159.

51. Soto K, Garza KM, Murr LE (2007) Cytotoxic effects of aggregated nanomaterials. Acta Biomater 3(3): 351–358.

52. Stern ST, McNeil SE (2008) Nanotechnology safety concerns revisited. Toxicol Sci 101(1): 4–21.

6

Nano-Tubular Cellulose for Bioprocess Technology Development

Athanasios A. Koutinas[1]*, Vasilios Sypsas[1], Panagiotis Kandylis[1], Andreas Michelis[1], Argyro Bekatorou[1], Yiannis Kourkoutas[2], Christos Kordulis[3,4], Alexis Lycourghiotis[3], Ibrahim M. Banat[5], Poonam Nigam[5], Roger Marchant[5], Myrsini Giannouli[6], Panagiotis Yianoulis[6]

1 Food Biotechnology Group, Department of Chemistry, University of Patras, Patras, Greece, 2 Department of Molecular Biology, Democritus University of Thrace, Alexandroupolis, Greece, 3 Group of Catalysis and Interfacial Chemistry for Environmental Applications, University of Patras, Patras, Greece, 4 Institute of Chemical Engineering and High Temperature Chemical Processes (FORTH/ICE-HT), Patras, Greece, 5 School of Biomedical Sciences, University of Ulster, Coleraine, United Kingdom, 6 Department of Physics, University of Patras, Patras, Greece

Abstract

Delignified cellulosic material has shown a significant promotional effect on the alcoholic fermentation as yeast immobilization support. However, its potential for further biotechnological development is unexploited. This study reports the characterization of this tubular/porous cellulosic material, which was done by SEM, porosimetry and X-ray powder diffractometry. The results showed that the structure of nano-tubular cellulose (NC) justifies its suitability for use in "cold pasteurization" processes and its promoting activity in bioprocessing (fermentation). The last was explained by a glucose pump theory. Also, it was demonstrated that crystallization of viscous invert sugar solutions during freeze drying could not be otherwise achieved unless NC was present. This effect as well as the feasibility of extremely low temperature fermentation are due to reduction of the activation energy, and have facilitated the development of technologies such as wine fermentations at home scale (in a domestic refrigerator). Moreover, NC may lead to new perspectives in research such as the development of new composites, templates for cylindrical nano-particles, etc.

Editor: Vipul Bansal, RMIT University, Australia

Funding: The authors have no support or funding to report.

Competing Interests: The authors have declared that no competing interests exist.

* E-mail: a.a.koutinas@upatras.gr

Introduction

Research on bioprocess technology development has been extensive during the last decades; however most of these technologies have not been used at industrial level. The main problems associated with these technologies are related to productivity, ease of industrial application and production cost. However there are still opportunities to use abundant, low cost materials with specific chemical or nano-mechanical properties to create multiple new, effective bioprocessing systems.

One of the major problems of food and drug production is the use of the relatively costly pasteurization processes. In addition to high operational cost, pasteurization reduces the nutritional quality of food products. However due to its importance, pasteurization is extensively used despite these disadvantages. Only in some food bioprocesses like wine production, membranes [1] are used to remove e.g. *Leuconostoc oenos* cells but they have a high cost and are not easy to handle at industrial level. Alternatively, if a nano-tubular solid of food grade purity, could be used to remove microbial cells through cell entrapment and immobilization at ambient and low temperatures, it could have wide applications in the food and pharmaceutical industries, such as the cold pasteurization of liquid foods and the cold sterilization of liquids that are used in drug production plants, respectively.

Porous materials like γ-alumina pellets [2] and mineral kissiris [3] have been used for cell entrapment and immobilization; however

their composition (that liberates aluminium in the processed media) limits their applications. Especially their use in food and drug production is not recommended. On the other hand delignified cellulosic material is of food grade purity, abundant in nature and therefore of low cost. This material has been successfully used as support for cell immobilization [4]. The produced biocatalyst was used for extremely low temperature fermentations in order to improve the quality of the products [5], and it was found able to promote the alcoholic fermentation of molasses [6].

Nanotechnology is one of the most growing areas of research the last decades. Especially the subarea of nano-cellulosic materials has attracted considerably high-level attention from researchers. The applications of this technology are numerous and in many fields. For example cellulosic nano-fibrils coated with SnO_2 have been used for the production of membranes [7], while nano-porous cellulose combined with carbon nano-tubes have been used in batteries and supercapacitors [8].

Following that trend the main goal of the present work was to use nano-tubular cellulose (NC) for the development of bioprocesses related to food industries. More specifically the work addresses the characterization of the tubular structure of NC and its suitability, due to that structure, for (i) "cold pasteurization" processes, (ii) promotion of alcoholic fermentation, (iii) extremely low temperature fermentations, and (iv) home-scale (domestic refrigerator) bioprocesses. Finally the perspectives for NC

a

b	NC sample 1	NC sample 2
BJH Adsorption Cumulative Surface area of pores between 17000 and 3000000 Å diameter	0.8016 m² g⁻¹	0.8946 m² g⁻¹
BJH Adsorption Average Pore diameter	80.875 Å	85.366 Å

d	
Crystallinity (%)	**Crystallite size (nm)**
65	3.0

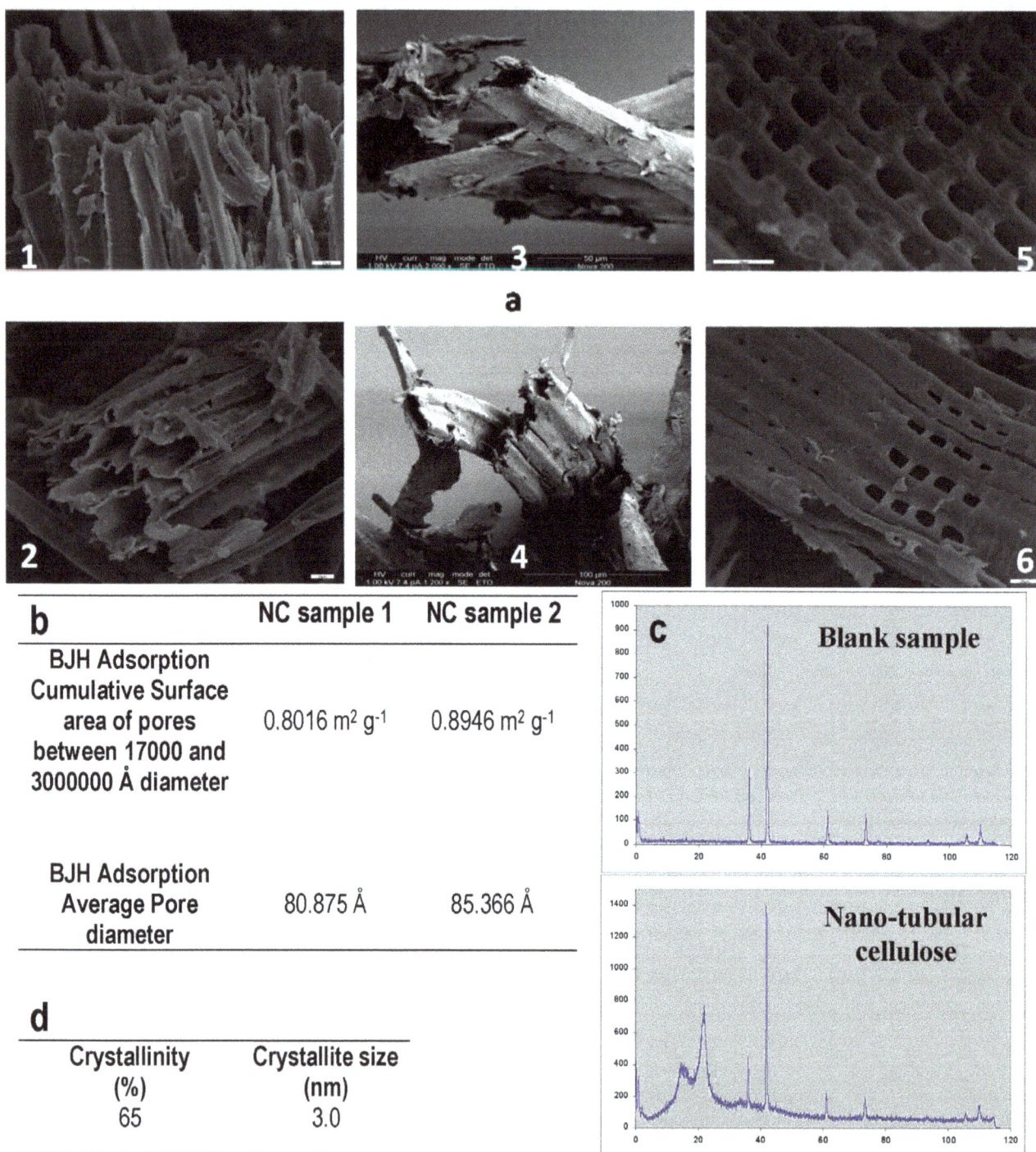

Figure 1. Characterization of nano-tubular cellulose (NC). (a) SEM micrographs of NC produced by delignification of sawdust with NaOH (scale bar in 1, 2, 5, 6 corresponds to 20 µm, while in 3 to 50 µm and in 4 to 100 µm). (b) NC average pore diameter and cumulative surface area of the pores. (c) Spectra from X-ray diffractometry of blank sample and NC sample. (d) Crystallinity and crystallite size of NC calculated using spectra from Figure 2C.

applications in nanobiotechnology are discussed, taking into account the results of the present study.

Results and Discussion

Nanostructure and microstructure of tubular cellulose

In the context of limited availability of complex materials for advanced bioprocessing, this investigation initially examines (i) the characterization of the structure of the NC material, (ii) the crystallinity of NC, and (iii) the degree of its crystallinity. Figure 1A shows the tubular structure in the remaining mass of cellulose after removal of lignin, while Figure 1B illustrates that the tubes have nano-scale dimensions. The cumulative surface area of the tubes is in the range of 0.8–0.89 m² g⁻¹ as indicated by porosimetry analysis. This surface is relatively small compared with other porous materials such as γ-alumina [9]. However, using a natural organic tubular biopolymer is attractive from the point of view that it is safer for bioprocess applications. The relatively small surface can be attributed to the fact that the lignin content in sawdust is about 16% and therefore its removal leads to limited tubing in the remaining mass of cellulose. However, this tubing is enough to encourage application for bioprocess development. Furthermore, single tubes (Fig. 1A, panel 3) and small groups of 3–4 tubes (Fig. 1A, panel 4) are observed. The tubes are horizontal (Fig. 1A, panel 2) and vertical (Fig. 1A, panel 1) and contain holes of varying diameters covering a relatively small area of the tubes (Fig. 1A, panel 6). Tubes with big differences of diameter are shown in the SEM photo of Figure 1A, panel 1. Likewise, tubes forming normal patterns such as a normal square are shown in Figure 1A, panel 5. However, Figure 1A, panel 1, 1A, panel 2 and 1A, panel 6 show tubes of small and bigger diameters. In the SEM micrographs it is shown that the bigger tubes are in micro-scale dimension and the small ones in nano-scale. Therefore, there is a mixture of micro and nano-tubing. The cumulative surface area and average pore diameter of the nano-tubes was measured by porosimetry. Figure 1D illustrates that NC has 65% degree of crystallinity indicating that the treatment with sodium hydroxide applied for delignification did not destroy its crystal structure. The size of the crystallites is 3 nm, which is big enough to contain a part of the tubes. Therefore, NC provides the possibility of cell entrapment and immobilization. These three different potential properties of NC make this nano-material suitable for different bioprocesses development as described above. Furthermore, this natural structure facilitates various new perspectives in research, which are discussed in this paper and which point to new nano-structured materials development. NC was produced by delignification of softwood sawdust and the observed microstructures reflect the architecture of the tracheids in the wood. The use of other starting materials would produce slightly different end products. In any case it is necessary to increase the nano-tubing and microtubing of cellulose. Pure microbial cellulose containing higher percentages of lignin could be examined, that would lead to more extended tubing after lignin removal.

Cold pasteurization of liquid foods

NC can be used to remove microorganisms from liquid foods or liquids used in the pharmaceutical industries at ambient and low temperatures. The aim is to develop a novel technology for cold pasteurization of drinking water, fruit juices, milk, wine, beer and other beverages, which could also be applied in pharmaceutical processing liquids. To test this possibility, NC was selected as a suitable new material due to its nano-tubular structure, ideal for the entrapment and immobilization of cells [4]. Water contaminated with bacteria and yeasts was passed continuously through the nano-tubular NC filter and the filtrate was analyzed for viable cells. The experiments were carried out at ambient temperature. There was an obvious difference in turbidity among the contaminated influent and the effluent water. Figure 2A shows that the microbiological load removal was 100%. The operational stability at this level of microbiological load removal remained constant for two days and then dropped. It was restored to 100% after regeneration of the NC filter achieved by washing it with hot water. The regeneration was successfully repeated every two days for more than one month, keeping the microbial removal at 100%. After the success of this experiment, cold pasteurization of the contaminated flow was carried out at 3–5°C since food processing is preferred at low temperatures to maintain the nutritional value and quality of the food. However, the removal of cells was greatly reduced and the effluent water was turbid. This unsuccessful result

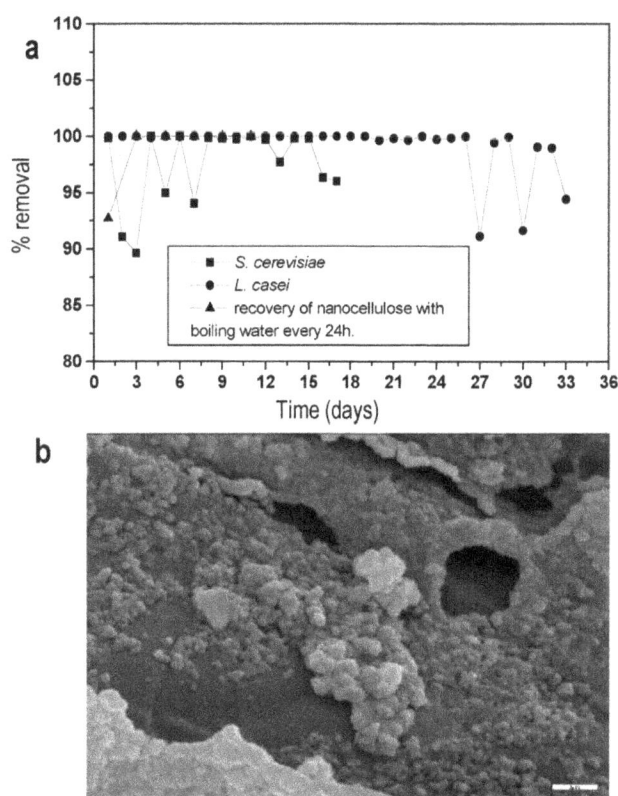

Figure 2. Nano-tubular cellulose (NC) in food, environmental and health care applications. (a) Removal of bacteria and yeasts from water (22–26°C). (b) SEM micrograph of L. casei flocculation in NC that was sampled from the bioreactor, which was pumped at low temperature (5°C) with its liquid culture for cold pasteurization (scale bar corresponds to 10 μm).

was attributed to the crystallization of sodium chloride added to the water to maintain the osmotic balance of cells. The crystals seemed to block the NC tubes preventing cell entrapment. Additionally, cell flocculation caused by low temperature stress [10], could also block the tubes. Figure 2B illustrates the formation of microbial flocks and proves that the reduction in microbial removal can be attributed to cell flocculation.

Although low temperature cold pasteurization of the model system (contaminated water) failed, the experimental work involving a real food, i.e. orange juice, was attempted. Cold pasteurization of orange juice should be performed at low temperatures to preserve the quality characteristics, and it could be presumed successful since components in the juice (like ascorbic acid) are essential for the synthesis of substances like carnitine that reduce the stress of cells [11]. Indeed, the low temperature cold pasteurization of orange juice resulted in 100% microbial load removal, as demonstrated by plate counting of influent and effluent samples (data not shown).

Nano-tubular cellulose as promoted of the alcoholic fermentation

Enhancement of the alcoholic fermentation of molasses by delignified cellulosic material has been exhibited [6]. This observation motivated further research to examine the physico-chemical structure of delignified cellulosic material and correlate it with its promotional effect on the rate of alcoholic fermentation. The characterization showed that cellulose has nano-scale tubular pores. This nano-tubular material (NC) was used as yeast immobilisation support to examine the activation energy E_a of the alcoholic fermentation and the reaction speed constant k versus temperature, in comparison with free yeast cells, in order to prove if NC affects the catalytic activity and therefore acts as promoter of the alcoholic fermentation. Fermentations were performed at different temperatures in the presence of NC and with free cells, separately. Figure 3A shows that the NC biocatalyst increased the fermentation rate and was more effective as the temperature was reduced compared with free cells. Furthermore, Figure 3B and 3C shows that the activation energy E_a of the NC biocatalyst was 28% lower than that of free cells. Likewise, the reaction speed constant k was higher for the NC biocatalyst. These results indicate that NC is an excellent carrier to promote the catalytic action of cells for the fermentation of molasses, as it was observed for delignified cellulosic material by Iconomou et al. [6].

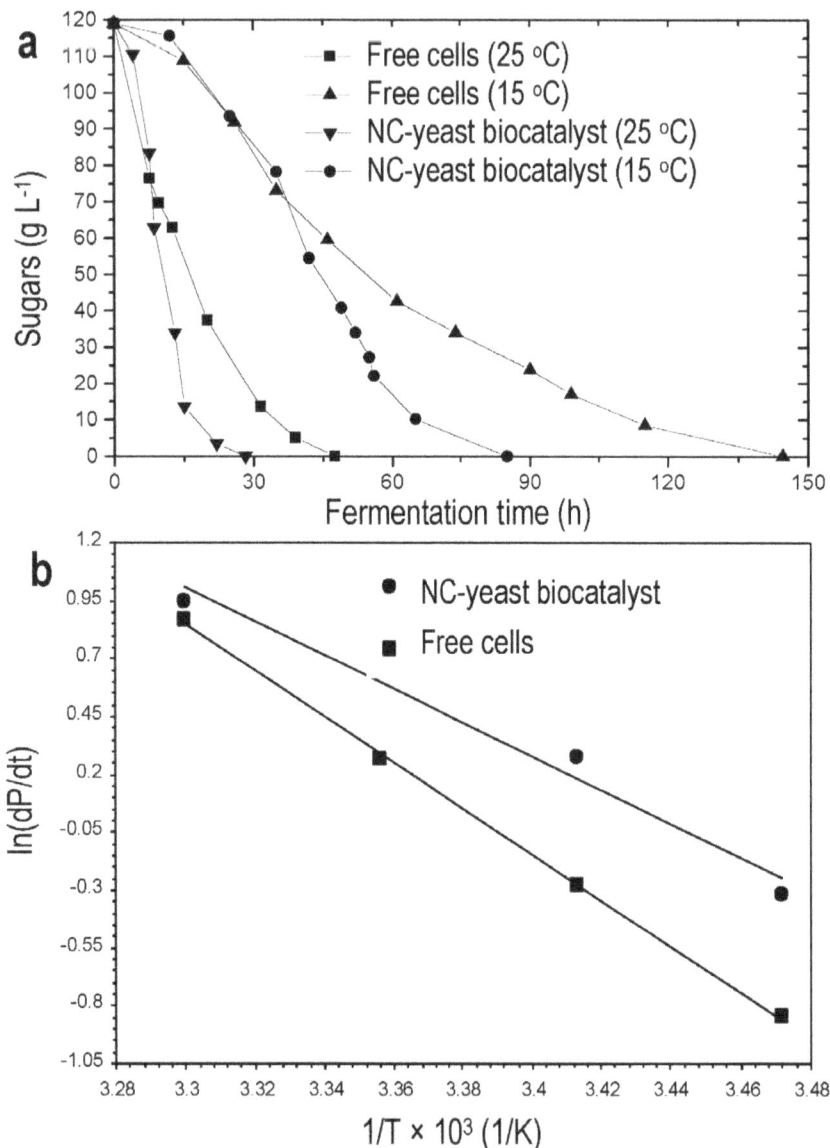

Support	Activation energy (Ea, KJ mol⁻¹)	Reaction rate constant (h⁻¹)			
		k_{30}	k_{25}	k_{20}	k_{15}
FC	82.6	0.239	0.137	0.078	0.043
IC	60.6	0.281	0.187	0.123	0.080

Figure 3. Nano-tubular cellulose (NC) and promotion of the alcoholic fermentation. (a) Fermentation kinetics observed at 25°C and 15°C by free cells and cells immobilised on NC (NC-yeast biocatalyst). (b) Arrhenius plot for evaluation of the activation energy and the pre-exponential factor of alcoholic fermentation performed with free and NC-yeast biocatalyst. (c) Activation energies and reaction rate constants of the fermentations made using free cells (FC) and cells immobilised on NC (IC).

It seems that this catalytic behaviour of NC is strongly related to its tubular structure, and can find application in fuel-grade ethanol production by increasing productivity and allowing a reduction in the size of the production plant [12].

NC is the first natural nano-tubular material characterized as a promoter of a bioconversion and the first biomolecule that promotes the alcoholic fermentation. The effect on the catalytic activity by NC that contains functional active hydroxyl groups is in agreement with a recent study, which proposed that effective catalysts can be created by adding functional hydroxyl groups in materials with nano-tubes, for example carbon nano-tubes [13]. This creates a new opportunity in research to examine this material as a possible catalytic promoter of other biochemical reactions like the reaction of lactic acid fermentation, malolactic fermentation, or the oxidation of alcohol by acetic acid bacteria to acetic acid. Also, its chemical and physical structure that has been studied in the context of this investigation can provide the base for the synthesis of new promoters of bioprocesses, having similar polymeric structures.

Glucose pump theory for the promotion of fermentation by nano-tubular cellulose

The catalytic effect of NC on the fermentation rate can be attributed to a glucose pump operating among the system of cellulose tubes that increase the surface, cells and glucose. The operation and rationale of a glucose pump is presented in Figure 4. According to this theory cells are encapsulated and joined by hydrogen bonding with hydroxyl groups between the surface of NC and the cell wall. Likewise, glucose molecules are joined with cellulose at the surface of the tubes through their hydroxyl group hydrogen bonds. Therefore, high cell and glucose concentrations are created mainly on the surface of the cellulose nano-tubes, increasing the rate of the biochemical reaction. Another factor that may affect the rate of the reaction is that glucose molecules and cells that are attached on the surface of the nano-tubes come very close to each other, increasing the glucose uptake rate. The bioconversion of glucose by the cells liberates alcohol, leading to continuous glucose pumping from the solution, mainly to the surface of the nano-tubes.

Nano-tubular cellulose for domestic refrigerator bioprocess development

The catalytic effect and reduction of the activation energy Ea of the alcoholic fermentation by NC, explains the high increase of the fermentation rate at low temperatures that was observed when delignified cellulosic materials where used as yeast immobilisation supports [4]. The feasibility of low temperature fermentation led to the idea of developing technologies for producing foods at home scale. Domestic wine making was selected to show the possibility for the development of bioprocesses that the consumers could apply in their own refrigerators. To develop this technology using NC, the strategy adopted was the preparation of a powder that could be preserved for a long period and consists of freeze dried nano-tubular cellulose supported biocatalyst and solidified grape must. The process is described in detail in the Methods section. The problem was that after freeze drying of grape must, the product was a very viscous liquid that could not be solidified by further freeze drying. However, carrying out freeze drying of grape must in the presence of NC a crystalline solid mass was formed containing crystals of glucose and fructose. Figure 5D shows an electron micrograph of this solid product with crystals of sugars derived from the viscous liquid grape must. The crystals are formed inside a nano-compartment. Crystal formation in the presence of NC may be explained by considering that some of the

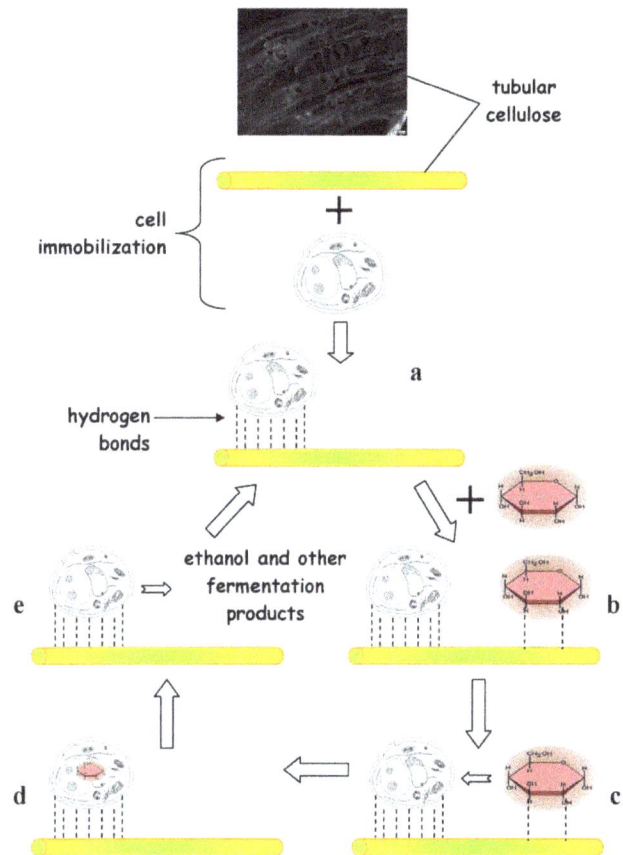

Figure 4. "Glucose pump". (a) Cell is immobilized on cellulose fiber by hydrogen bonding. (b) Glucose also attached on the surface of the cellulose fiber by hydrogen bonding. (c) Glucose is transferred inside the cell. (d) Glucose is biochemically converted inside the cell. (e) After glycolysis, alcoholic fermentation and other processes, ethanol and other fermentation products are produced and transferred to the solution, and another cycle begins.

adsorption sites for the sugar molecules inside the tubes may act as crystallization centres promoting interfacial heterogeneous crystallization [14]. Another observation is that cells encapsulated in the NC tubes were not destroyed and remained alive, although they were in a very viscous solution and therefore under increased osmotic stress. This protection of freeze dried cells attached in the tubes of NC is attributed to the water content of the viscous sugar solution, which, even though low, forms hydrogen bonds in the tubes with the hydroscopic cellulose macromolecules and is retained by it. In addition, the fact that crystallization of the sugars did not damage the cell walls of the yeast cells may be due to the protective effect of dehydration by freeze-drying.

Figure 5A shows the kinetics of the model fermentation system using commercial invert sugar powder. The NC-yeast biocatalyst resulted in a dramatic reduction of the fermentation time as compared with free cells. The results of the model fermentation provided a basis for the second experiment using powdered grape must, the results of which are presented in Figure 5B. These results show that both fermentations, which were carried out at 1°C, were completed and the NC resulted in an increase of the fermentation rate and reduction of the fermentation time in comparison with free cells. As an alternative strategy in the development of the technology, dried grapes were used to substitute solid freeze dried grape must. Figure 5C shows that once again there was a

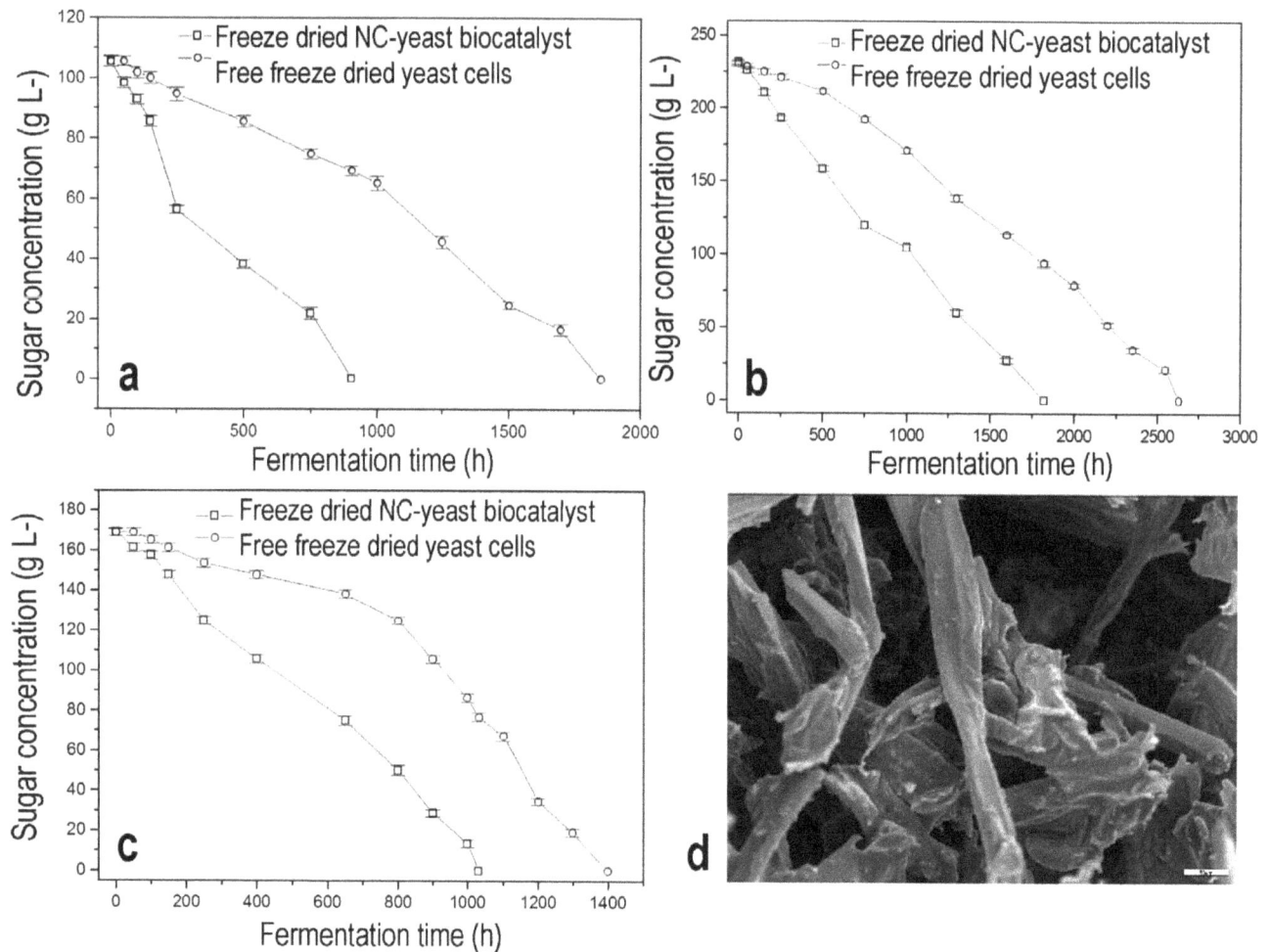

Figure 5. Domestic refrigerator wine making. (a) Fermentation kinetics observed at 1°C of freeze dried mixture of invert sugar and NC-yeast biocatalyst. (b) Fermentation kinetics at 1°C of freeze dried mixture of grape must and NC-yeast biocatalyst. (c) Fermentation kinetics observed at 1°C of freeze dried mixture of raisins and NC-yeast biocatalyst. Fermentation kinetics using free freeze dried cells are also shown in all Figures (a), (b) and (c). (d) SEM micrograph of the freeze dried mixture of grape must and NC-yeast biocatalyst (scale bar corresponds to 20 μm).

reduction in the fermentation time in the case of the NC biocatalyst.

Using NC, the improvement of wine quality is shown by the increase of esters and reduction of higher-alcohols as compared with free cells (Figure 6A). Moreover, Figure 6B and 6C reveals the reduction of amyl alcohols and increase of ethyl acetate on total volatiles as the temperature was reduced. These results lead to improvement of wine quality that has also demonstrated by sensory testing (Fig. 6D).

Nano-tubular cellulose for creation of new trends in research

NC may be tested (a) as template for the production of cylindrical nano-particles [15]. The precipitation or gel formation inside the tubular pores may be followed by thermal treatment to stabilize the fibres whereas the removal of the template could be obtained by combustion or hydrolysis of cellulose using cellulases [16]. The production of semiconductor particles with particular schemes is urgent for water splitting to produce hydrogen taking advantage of the visible range of the solar spectrum [17]. A second opportunity (b) lies in further investigation of the deadly disease of septicaemia, through nano-encapsulation of microbes, when

antibiotics are not effective. It could be achieved with research using nano-tubular cellulose as a nano-filter or alternatively nano-tubular natural or synthetic polymeric material. Further research is required on these approaches.

NC is the first nano-tubular material characterized as a promoter of a bioconversion (alcoholic fermentation). This creates a new research opportunity (c) to examine this material as a possible promoter of other productive biochemical reactions such as lactic acid fermentation and malolactic fermentation, or the oxidation of alcohol by acetic acid bacteria to acetic acid. Also, its physicochemical structure has been studied in this investigation and could form the basis (d) for synthesis of new catalytic promoters of bioprocesses, having similar polymeric structures. NC can also be employed (e) to produce new composites containing different materials with different properties [18] after introducing inorganic or organic nano-particles in its nano-tubes, as described. In addition, it could be used to strengthen other nano-composites [19]. Nano-tubing in cellulose (f) could also be examined to increase the rate of cellulose hydrolysis by nano-tubes and micro-tubes facilitating the diffusion of cellulases. NC can be employed to accelerate the chemical hydrolysis of cellulose by reversing the technology, from passing cellulose through pores of a

a

RI	Compounds	20 °C FDNCB	20 °C FFDC	5 °C FDNCB	5 °C FFDC	1 °C FDNCB	1 °C FFDC
	ESTERS						
1040	Isoamyl acetate	7.803	3.604	8.607	1.611	8.609	0.802
1143	Ethyl hexanoate	1.740	0.403	1.550	0.142	1.571	0.120
1270	Hexyl acetare	0.002	0.001	0.002	0.001	0.002	Tr
1390	Ethyl octanoate	3.503	0.028	2.709	0.092	1.831	0.082
1412	3-OH-ethyl-butanoate	0.024	0.002	0.027	0.002	0.038	0.002
1524	2-OH-ethyl propanoate	0.017	0.001	0.001	nd	Tr	0.004
1592	Ethyl decanoate	0.204	0.032	0.531	0.072	1.171	0.082
1637	9-Ethyl decenoate	1.550	0.054	1.652	0.052	2.624	0.060
1740	2-phenylethyl acetate	0.905	0.303	1.920	0.943	1.982	0.909
1789	Ethyl dodecanoate	0.401	0.204	0.090	0.023	0.311	0.071
	Sum of esters	**16.149**	**4.632**	**17.089**	**2.938**	**18.139**	**2.132**
	ALCOHOLS						
1060	Butanol	0.221	0.056	0.121	0.113	0.031	0.005
1107	2-hexanol	0.809	0.731	Tr	Tr	0.308	0.201
1255	2-heptanol	0.271	0.821	0.121	0.413	Tr	0.441
1320	1-hexanol	0.042	0.072	0.048	0.061	0.048	0.069
1535	2-nonanol	0.142	0.372	0.008	0.005	0.005	0.050
1542	2,3-butanediol	0.024	3.809	0.034	3.373	0.005	3.230
1560	2-decanol	0.463	0.132	0.181	0.014	0.001	0.013
1570	1,3-butanediol	0.012	1.209	0.018	0.733	0.001	0.122
1613	Nonanol	0.085	2.109	0.004	2.420	Tr	2.809
1810	2-phenylethanol	13.108	12.420	10.301	11.521	8.306	10.412
	Sum of alcohols	**15.177**	**21.731**	**10.836**	**18.653**	**8.705**	**17.352**

d	Ferm. Temp. (°C)	Wine from freeze dried NC-yeast biocatalyst	Wine from free freeze dried yeast cells	Commercial wine
Taste score	20	7.1±0.97	7.0±0.94	4.9±1.74
	5	7.3±0.7	7.2±0.92	
	1	7.4±0.9	7.0±0.77	
Aroma score	20	6.9±0.84	6.2±1.31	6.2±1.6
	5	7.5±0.64	6.7±0.96	
	1	7.8±0.47	6.8±0.87	

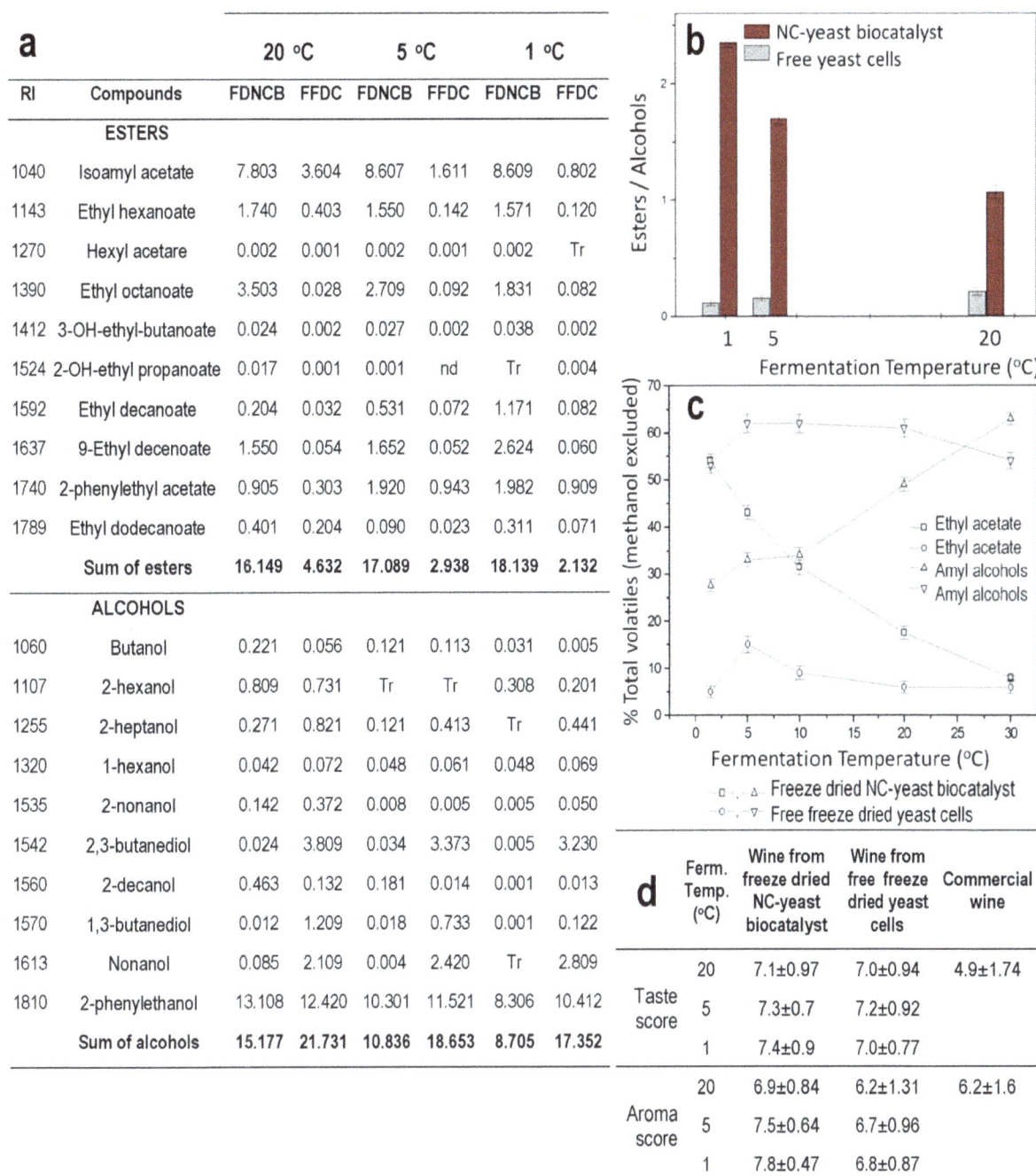

Figure 6. Quality of the wine produced in the refrigerator using mixture of freeze dried grape must and freeze dried biocatalysts. (a) Effect of temperature on esters and alcohols formation during alcoholic fermentation of mixtures of freeze dried grape must and freeze dried biocatalysts. Tr: compounds <1 µg/L (traces), nd: not detected. FDNCB: Freeze dried NC-yeast biocatalyst, FFDC: Free freeze dried cells. (b) Effect of temperature on the ratio of esters-to-alcohols formed during alcoholic fermentation of mixtures of freeze dried grape must and freeze dried biocatalysts. (c) Effect of temperature on the (%) percentage of ethyl acetate and amyl alcohols on total volatiles formed during alcoholic fermentation of mixtures of freeze dried grape must and freeze dried biocatalysts. Plots show means of three replicates with standard errors. (d) Sensory evaluation of commercial wine and wines produced by freeze dried free cells and freeze dried NC-yeast biocatalyst.

resin that contains sulphonic groups [20] to introducing in the tubes of cellulose compounds with sulphonic groups. Furthermore, (g) NC can also be used in the development of novel fermentation methods such as three-phase fermentation, which can be developed by introducing in its tubes non-polar organic solvents that could dissolve the product of fermentation. Therefore,

product recovery with lower energy demand could be obtained during distillation, e.g. in fuel-grade alcohol production.

Conclusions

The produced nano-tubular cellulose (NC) is a suitable new material for multiple applications in bioprocessing. Specifically, it was proved to be effective for removal of microbes in cold pasteurization processes of liquid foods. Also, it has been demonstrated as the first nano-tubular promoter that reduces the activation energy Ea in a bioprocess. Taking into account the chemical and physical structure of NC, the theory of a glucose pump is proposed to explain its promotional activity. The reduction of the activation energy Ea explains the low temperature performance and in combination with the property of NC to aid crystallization of viscous sugar solutions, allows home scale, domestic refrigerator bioprocessing for food production. Such a bioprocess as wine making in a domestic refrigerator was found to improve the quality of the wine produced.

Materials and Methods

Ethics Statement

Due to the nature of the research and because no human or animal experimentation was conducted, no ethics statement is required.

Characterization of the Nano-tubular cellulose structure

The studied nano-tubular cellulose material was produced by delignification of softwood sawdust with NaOH [4]. The average pore diameter and cumulative surface are of pores were measured using a Micromeritics TriStar 3000 porosimeter. For the X-ray studies a ENRAF NONIUS FR 590 X-ray diffractometer was used. The crystallinity [21] and crystallite size [22,23] of the material was calculated using the X-ray spectra. The scanning electron micrographs were taken using JOEL JSM-6300 and FEI Nova 200 SEM microscopes.

Cold pasteurization

Cold pasteurization was carried out on aqueous suspensions of *Lactobacillus casei* and the yeast *Saccharomyces cerevisiae* and on liquid orange and apple juices contaminated with *L. casei*. In the case of aqueous microbial suspensions, the experiments were run in a 1.5 L continuous bioreactor system. The cylindrical glass bioreactor (1.5 L) was filled and packed with NC and pumped daily with 1 L microbial suspension of 8 log cfu mL^{-1} at 26°C using a high accuracy peristaltic pump. When the removal fell from 100 to 90%, 5 L of boiling water were pumped through the bioreactor to regenerate the surface of NC. In the case of fruit juices, orange juice contaminated with *L. casei* at 7 log cfu mL^{-1} was used in the same experimental unit. Orange juice was produced by squashing

oranges, pasteurization at 62°C for 30 minutes and cold filtration in the laboratory. The pH was adjusted to 3.2 with NaHCO$_3$. The bioreactor was fed for 6 days at 3–5°C in a domestic refrigerator. Daily estimations of cell counts at the outlet were performed using the Standard Plate Counting method.

Solid crystalline grape must production

Thermally dried NC (27.5 g) was mixed with 190 mL grape must of 20° Be density. This mixture was agitated overnight and freeze dried in a Labconco Freezone 4.5 Freeze Dry system. This produced a crystalline mass of grape must.

Bioconversion in the refrigerator

Cell immobilization of the strain *S. cerevisiae* AXAZ-1 on NC was performed according to a previous study [4]. The solid biocatalyst was mixed separately with solid crystalline grape must, invert sugar and raisins and was freeze dried. Storage of these mixtures for one week at 4°C followed and then tap water was added to obtain the initial sugar concentrations. The mixtures were placed in a refrigerator and fermentation was done at 1°C. Kinetics of bioconversion were measured by analyzing sugar at various time intervals using a Shimadzu HPLC system [24]. Samples were also analyzed for volatile by-products by GC and GC-MS analysis [25]. Sensory evaluation was performed by 14 laboratory members (7 previously trained and 7 untrained) using locally approved protocols. The panel was asked to score the wines based on a 0 to 10 scale (0 unacceptable, 10 exceptional).

Calculation of activation energy

Fermentations of 400 mL of 12% (w/v) glucose medium were carried out with immobilized and free cells at various temperatures. The activation energies of the fermentation systems were calculated based on the Arrhenius equation according to a previous study [24] by a curve obtained by plotting ln(dP/dt) versus T^{-1}.

Author Contributions

Conceived and designed the experiments: AAK. Performed the experiments: VS PK AM. Analyzed the data: AAK PK YK CK. Contributed reagents/materials/analysis tools: CK AL. Wrote the paper: AAK PK AB YK CK AL IMB PN RM MG PY.

References

1. Girard B, Fukumoto LR, Sefa Koseoglu S (2000) Membrane processing of fruit juices and beverages: A review. Crit Rev Biotechnol 20: 109–175.
2. Kana K, Kanellaki M, Papadimitriou A, Psarianos C, Koutinas AA (1989) Immobilization of *Saccharomyces cerevisiae* on γ-alumina pellets and its ethanol production in glucose and raisin extract fermentation. J Ferment Bioeng 68: 213–215.
3. Kana K, Kanellaki M, Psarianos C, Koutinas AA (1989) Ethanol production by *Saccharomyces cerevisiae* immobilized on mineral Kissiris. J Ferment Bioeng 68: 144–147.
4. Bardi EP, Koutinas AA (1994) Immobilization of yeast on delignified cellulosic material for room temperature and low-temperature wine making. J Agric Food Chem 42: 221–226.
5. Bardi EP, Koutinas AA, Soupioni M, Kanellaki M (1996) Immobilization of yeast on delignified cellulosic material for low temperature brewing. J Agric Food Chem 44: 463–467.
6. Iconomou L, Psarianos C, Koutinas AA (1995) Ethanol fermentation promoted by delignified cellulosic material. J Ferment Bioeng 79: 294–296.
7. Huang J, Matsunaga N, Shimanoe K, Yamazoe N, Kunitake T (1998) Process for the stabilization of the properties of cellulosic membranes. U.S. Patent 5853647.
8. Scrosati B (2007) Nanomaterials: Paper powers battery breakthrough. Nature Nanotechnol 2: 598–599.
9. Palkar VR (1999) Sol-gel derived nanostructured γ-alumina porous spheres as an adsorbent in liquid chromatography. Nanostr Mater 11: 369–374.
10. Gonzalez MG, Fernandez S, Sierra JA (1996) Effect of temperature in the evaluation of yeast flocculation ability by the Helm's method. J Am Soc Brew Chem 54: 29–31.
11. Franken J, Kroppenstedt S, Swiegers JH, Bauer FF (2008) Carnitine and carnitine acetyltransferases in the yeast *Saccharomyces cerevisiae*: a role for carnitine in stress protection. Curr Genet 53: 347–360.
12. Balat M, Balat H (2009) Recent trends in global production and utilization of bioethanol fuel. Appl Energ 86: 2273–2282.
13. Resasco DE (2008) Nanotubes: Giving catalysis the edge. Nat Nanotech 3: 708–709.
14. Bourikas K, Kordulis C, Lycourghiotis A (2006) The role of the liquid-solid interface in the preparation of supported catalysts. Catal Rev 48: 363–444.
15. Mehdaoui S, Benslim N, Aissaoui O, Benabdeslem M, Bechiri L, et al. (2009) Study of the properties of CuInSe$_2$ materials prepared from nanoparticle powder. Mater Charact 60: 451–455.
16. Zhang YHP, Lynd LR (2004) Toward an aggregated understanding of enzymatic hydrolysis of cellulose: Noncomplexed cellulase systems. Biotechnol Bioeng 88: 797–824.
17. Navarro RM, Alvarez-Galvan MC, del Valle F, Villoria de la Mano JA, Fierro JLG (2009) Water splitting under visible light irradiation on semiconductor catalysts. Chem Sus Chem 2: 471–485.
18. McClory C, Chin SJ, McNally T (2009) Polymer/Carbon Nanotube Composites. Aust J Chem 62: 762–785.

19. Beecer JF (2007) Organic materials: Wood, trees and nanotechnology. Nat Nanotech 2: 466–467.

20. Buick R (2008) Biofuel acid test. Nature 455: 569.

21. Mihranyan A, Llagostera AP, Karmhag R, Stromme M, Ek R (2004) Moisture sorption by cellulose powders of varying crystallinity. Int J Pharm 269: 433–442.

22. Farrauto RJ, Bartholomew CH (2003) Fundamentals of Industrial Catalytic Processes. London: Blackie Academic and Professional.

23. Arinstein A, Burman M, Gendelman O, Zussman E (2007) Effect of supramolecular structure on polymer nanofibre elasticity. Nat Nanotech 2: 59–62.

24. Kandylis P, Goula A, Koutinas AA (2008) Corn starch gel for yeast cell entrapment. A view for catalysis of wine fermentation. J Agr Food Chem 56: 12037–12045.

25. Kandylis P, Koutinas AA (2008) Extremely low temperature fermentations of grape must by potato-supported yeast, strain AXAZ-1. A contribution is performed for catalysis of alcoholic fermentation. J Agr Food Chem 56: 3317–3327.

7

Catalytic Nanoceria Are Preferentially Retained in the Rat Retina and Are Not Cytotoxic after Intravitreal Injection

Lily L. Wong[1]*, Suzanne M. Hirst[2], Quentin N. Pye[1], Christopher M. Reilly[2], Sudipta Seal[3], James F. McGinnis[1,4]*

1 Department of Ophthalmology, University of Oklahoma Health Sciences Center, College of Medicine, and Dean McGee Eye Institute, Oklahoma City, Oklahoma, United States of America, 2 Biomedical Sciences and Pathobiology, Virginia Polytechnic Institute and State University, and Via College of Osteopathic Medicine, Blacksburg, Virginia, United States of America, 3 Advanced Materials Processing Analysis Center, Mechanical Materials Aerospace Engineering, Nanoscience and Technology Center, University of Central Florida, Orlando, Florida, United States of America, 4 Department of Cell Biology and Oklahoma Center for Neuroscience, University of Oklahoma Health Sciences Center, Graduate College, Oklahoma City, Oklahoma, United States of America

Abstract

Cerium oxide nanoparticles (nanoceria) possess catalytic and regenerative radical scavenging activities. The ability of nanoceria to maintain cellular redox balance makes them ideal candidates for treatment of retinal diseases whose development is tightly associated with oxidative damage. We have demonstrated that our stable water-dispersed nanoceria delay photoreceptor cell degeneration in rodent models and prevent pathological retinal neovascularization in *vldlr* mutant mice. The objectives of the current study were to determine the temporal and spatial distributions of nanoceria after a single intravitreal injection, and to determine if nanoceria had any toxic effects in healthy rat retinas. Using inductively-coupled plasma mass spectrometry (ICP-MS), we discovered that nanoceria were rapidly taken up by the retina and were preferentially retained in this tissue even after 120 days. We also did not observe any acute or long-term negative effects of nanoceria on retinal function or cytoarchitecture even after this long-term exposure. Because nanoceria are effective at low dosages, nontoxic and are retained in the retina for extended periods, we conclude that nanoceria are promising ophthalmic therapeutics for treating retinal diseases known to involve oxidative stress in their pathogeneses.

Editor: Tailoi Chan-Ling, University of Sydney, Australia

Funding: This work was supported in part by grants from the National Institutes of Health R21EY018306 (JFM), R01EY018724 (JFM) (http://www.nei.nih.gov/) and R15AI072756 (CMR) (http://www.fda.gov/Drugs/ScienceResearch/default.htm); the Foundation Fighting Blindness (FFB) C-NP-0707-0404-UOK08 (JFM) (http://www.blindness.org/); the National Science Foundation: Chemical, Bioengineering, Environmental, and Transport Systems (NSF:CBET) 0708172 (SS and JFM) (http://www.nsf.gov/div/index.jsp?org = CBET); Oklahoma Center for the Advancement of Science and Technology (OCAST) HR06-012 (JFM) (http://www.ok.gov/ocast/), unrestricted funds from the Presbyterian Health Foundation (PHF) (http://www.phfokc.com/) and RPB to JFM (http://www.rpbusa.org/rpb/), and an Research to Prevent Blindness (RPB) SSI award to JFM. This work was also supported in part by the DMEI/NEI Imaging Core Facility at OUHSC (NIH: P30-EY021725, COBRE-P20 RR017703) (http://www.nigms.nih.gov/). The funders had no role in study design, data collection and analysis, decision to publish, or preparation of the manuscript.

Competing Interests: LLW, SS, and JFM are co-inventors of US Patents: 7727559 and 7347987, European Patent: 1879570, and Australian Patent: 2006242541. SS is a co-inventer of US Patent: 7504356. JFM is Chief Scientist of a startup company, Nantiox, which is licensed to use nanoceria to destroy reactive oxygen species but no corporate funding was used to support this research.

* E-mail: lily-wong@ouhsc.edu (LLW); james-mcginnis@ouhsc.edu (JFM)

Introduction

Nanomaterials which include nano-sized and nano-structured objects, have gained importance in biomedical research and medicine in recent years. Because of the dramatic increase in surface area when synthesized in the nanometer range, nanomaterials exhibit enhanced or unique reactivity that is not found in their macroscopic counterparts. Many promising nanomaterials are currently under investigation for drug or nucleic acid delivery to target specific organ/tissue for therapy. Others are tested for diagnostic, imaging, tissue healing, and surgical aids [1]. Another unique class of nanomaterials, namely the redox-active radical scavenging nanoparticles including fullerenes and cerium oxide nanoparticles (nanoceria or CeNPs), is being developed as bona fide antioxidants for treatment of neurodegenerative diseases [2–4].

The oxides of cerium, a rare earth element, have unique physical and chemical properties. The cerium ions have both the 3+ and 4+ valence states and therefore can act as electron donors or acceptors. Oxygen defects or vacancies on the surface or subsurface of the lattice crystals act as sites for radical scavenging [5,6]. When synthesized in the 3–5 nm range, nanoceria possess enhanced catalytic activities that mimic superoxide dismutase and catalase [7–9], two major anti-oxidative enzymes, to neutralize superoxide anions and hydrogen peroxides, respectively. The enhanced redox capacity of nanoceria is most likely due to the dramatically increased surface to volume ratio of these nanoparticles.

Accumulating evidence has shown that the disease progression of many neurodegenerative conditions such as Alzheimer's, Parkinson's and retinal degenerative diseases including age-related macular degeneration, diabetic retinopathy, and various forms of retinitis pigmentosa, are tightly associated with oxidative damage due to either chronically or acutely increased reactive oxygen species [10–16]. During the past few years, we have focused on developing our stable water-dispersed nanoceria as ophthalmic therapeutics for treatment of retinal diseases. We showed that these nanoceria increased the lifespan of retinal neurons in culture and protected them from oxidative damage when challenged with hydrogen peroxide [17]. Nanoceria synthesized using the same methodology also protected photoreceptor cells in a light-induced retinal degeneration model [17]. They inhibited the development and caused the regression of pathologic retinal neovascularization in the *very low density lipoprotein receptor* (*vldlr*) knockout mouse [18]. These nanoceria also delayed the degeneration of photoreceptor cells in a retinal degeneration mouse carrying the *tubby* mutation [19].

Despite the well-documented ability of nanoceria to reduce oxidative damage, retinal degeneration, and inflammation [17–20], the mechanisms of radical scavenging by nanoceria in biological systems are still unclear [2,21]. Furthermore, the bio-distribution and pharmacokinetics of nanoceria in ocular tissues after a single intravitreal injection (the optimal route for nanoceria delivery) are unknown. Nanoceria appear to have differential effects in different cell types. From cell culture studies, certain cell types exhibited enhanced longevity and protection from oxidative insults while a few showed reduced viability when exposed to nanoceria at specific dosages [2,22]. Currently, a systematic study of nanoceria cytotoxicity *in vivo* in ocular tissues is lacking. We therefore carried out a detailed study to specifically address these fundamental questions to characterize the interactions of nanoceria in the unique biological environment of the eye. We used inductively-coupled plasma mass spectrometry (ICP-MS), a highly sensitive method for trace element detection in biological samples, to study the bio-distribution of nanoceria after a single intravitreal injection in the rat eye. We discovered that nanoceria were rapidly and preferentially retained in the retina for at least 120 days. We also showed that nanoceria were not toxic to retinal cells over a range of dosages applied. Our study is the first to demonstrate that nanoceria are retained in the retina for an extended period after a single intravitreal injection and that nanoceria do not have toxic side effects in retinal cells *in vivo* at the dosage levels applied.

Materials and Methods

1. Animal
We kept a breeding colony of Sprague-Dawley (SD) albino rats in the Dean McGee Eye Institute (DMEI) vivarium under cyclic light conditions (12 h on/12 h off, 5–20 lux).

2. Ethics Statement
Animals were cared for and handled according to the Association for Research in Vision and Ophthalmology statement for the use of animals in vision and ophthalmic research. The study was approved by the University of Oklahoma Health Sciences Center Institutional Animal Care and Use Committee (OUHSC IACUC) and the DMEI IACUC. The approved protocol numbers were 10-087 and 10-088 from the OUHSC IACUC, and D-10-087 and D-10-088 from the DMEI IACUC.

3. Synthesis of Nanoceria
Cerium oxide nanoparticles were synthesized using simple wet chemistry methods as described previously [23]. Briefly, stoichiometric amounts of cerium nitrate hexahydrate (99.999% from Sigma Aldrich) were dissolved in deionized water. The solution was oxidized using an excess amount of hydrogen peroxide. After the synthesis of nanoparticles, the pH of the suspension was maintained below 3.0 using nitric acid (1M) to keep the synthesized nanoceria in suspension. These 3–5 nm particles were thoroughly characterized using transmission electron microscopy (size and shape determination), dynamic light scattering (zeta potential measurement), and X-ray photoelectron spectroscopy (estimating the oxidation state of cerium) as described in [18]. Each batch was validated for the abundance of catalytically active Ce3+ oxidation state and stable aqueous dispersion. More importantly, we did not use hexamethylenetetramine (HMT) in the synthesis of nanoceria due to cytotoxic effects exhibited by nanoparticles prepared in this manner (unpublished observation).

4. Intravitreal Injection of Nanoceria
Adult SD rats (8 weeks or older) were selected for intravitreal injection. Animals were anesthetized by intramuscular injection of a mixture of ketamine (80 mg/kg) and xylazine (4 mg/kg). Pupils were dilated by application of a drop of phenylephrine (10% solution) to the cornea before the delivery of 2 µl of either CeNPs of varying dosages (1 µM or 0.344 ng to 1 mM or 344 ng in saline solution) or saline alone into the vitreous with the aid of an ophthalmic operating microscope. Both eyes of each animal received the same treatment.

5. Sample Collection for Inductively Coupled Plasma Mass Spectrometry (ICP-MS)
Eyes were harvested at designated times post injection. Enucleated eyes were fixed in 4% (para-formaldehyde) PFA (in 0.1M phosphate buffer, pH 7.4) at 4°C until ICP-MS analysis. If further dissection was performed, eyes were fixed at room temperature for 30 minutes before dissection. Eyes were dissected into component parts in cold PBS, pH 7.4. Ocular components: retina (R), lens (L), rest of eyecup including cornea, iris, retinal pigment epithelium, choroid, and sclera (EC) or whole eyes were kept in individual Eppendorf tubes containing 1 ml 4% PFA and stored at 4°C until processing for ICP-MS. In most cases (80%), the vitreous body was found associated with the lens tissue and was included in the lens component. When the vitreous body was not associated with the lens tissue, it was discarded in the dissecting buffer. We confirmed that inclusion of the vitreous body in the lens component did not alter the amount of nanoceria in the lens component.

6. ICP-MS
Tissues in 1 ml fixative were mixed in 10 ml 70% nitric acid overnight to start the digestion process. Samples were then microwave-digested in an Xpress Microwave Digester. The temperature was ramped to 200°C over a span of 20 minutes and held there for another 20 minutes. Samples were then boiled down to less than 1 ml each and reconstituted in water to 10 ml exactly. The Ce levels were assessed using a 7700 Series ICP-MS from Agilent Technologies (Santa Clara, CA). The level of Ce was converted to CeO_2 for data presentation. Each time point represented the average from at least four eyes from two individual rats.

7. Electroretinogram Recordings (ERG)
Animals were dark adapted overnight and manipulated under dim red light. Anesthesia was induced with a mixture of ketamine (80 mg/kg i.m.) and xylazine (4 mg/kg i.m.) and the animal was placed on a thermal water heated pad with the temperature controlled (38°C) by a circulating water pump (GAYMAR T/PUMP, Orchard Park, NY). Pupils were dilated at least five minutes before testing using a topical phenylephrine (10%) solution (AK-DILATE, Akorn, Inc., Lake Forest, IL). A drop of 2.5% hypromellose solution (GPS, Wilson Ophthalmic, Mustang, OK) was applied to the cornea to prevent dehydration and allow for electrical contact with the gold recording electrode (Goldring

Electrode, 4 mm, Roland Consult, Stasche & Finger GmbH, Brandenburg, Germany). A platinum subdermal needle (Grass Technologies, West Warwick, RI) hooked into the mouth, right side of the cheek, served as the reference electrode. To complete the circuit, a platinum subdermal needle electrode was placed under the skin at the base of the tail. The Espion system from Diagnosys LLC (Lowell, MA) provided amplification (at 0.30 to 300 Hz bandpass, with notch filtering), stimulus presentation, and data acquisition.

Scotopic. The scotopic stimuli consisted of white flashes provided by a xenon bulb projected on a ganzfeld. The intensity of stimuli evaluated and specifics of the protocol are shown in Table 1. The amplitude of each A-wave response corresponded to the maximum negative deflection found between 6 and 35 ms after the stimulus. The amplitude of each B-wave response corresponded to the difference between the maximum negative deflection determined and the maximum positive deflection found between 30 and 130 ms after the stimulus.

Photopic. The source of photopic stimuli was the same as the one presented for scotopic stimuli and flashes were presented consecutively at 0.06, 0.6, 3.0, 30.0, 300.0 and 600.0 cd.s/m2. Prior to the first flash, a five minute light adaptation sequence was used at a luminance of 30 cd/m^2. A total of 10 responses were averaged for each intensity tested with an inter trial delay of 1 s. We determined the amplitudes of each A-, and B-wave response as described under the Scotopic condition. Only B-wave amplitudes are presented.

Flicker. The flicker stimuli at 30 $cd.s/m^2$ flash intensity with a background luminance of 30 cd/m^2 were presented at 3 Hz, 5 Hz and then at 10–40 Hz in 10 Hz steps. For each flicker frequency tested, the stimulus was presented followed by a 5 s delay prior to data collection. This guaranteed that the first few responses (not preceded by repeated stimuli and of potentially greater amplitude) were excluded. The wave trains were examined to ensure consistency of responses over the duration of stimuli (500 ms). Amplitude from the first wave form from each frequency was determined. Finally, presentation of the 3 Hz stimulus was repeated to ascertain the stability of the responsiveness, i.e. confirming that amplitudes obtained with this last trial were comparable to those obtained with the first trial. The amplitude of each flicker response corresponded to the difference between the first maximum negative and the first positive deflections. These protocols were modified from the ones described in [24,25]. Measurements from these ERG recordings enable us to evaluate the functions of several classes of retinal neurons. Scotopic a-wave amplitude reflects primarily the function of rod cells. Scotopic b-wave amplitude reflects the function of neurons in the inner retina, post-synaptic to the rod photoreceptor cells. Photopic b-wave amplitude and flicker ERG reflect the function of cone cells [26]. Each data point represented the average from at least three individual rats or six eyes.

8. Histology and Morphometric Analysis

After ERG recordings, the rats were euthanized with carbon dioxide, the eyes enucleated and fixed in Prefer fixative (Anatech Ltd., Battle Creek, MI) for 45 min. Five μm thick paraffin sections were obtained along the superior/inferior axis of the globe through the central retina. Hematoxylin and Eosin (H&E) stained sections were used for morphometric analyses. Detailed description of each measurement is provided in Figure 1. Each data point was from 2–5 eyes from different rats; all but 2 data points had samples of at least 3 eyes.

9. Statistical Analysis

For ICP-MS data, each data point was averaged from 4–8 eyes whereas ERG data were averaged from 6–10 eyes. Values were expressed as means ± SEM. Statistical analyses were performed using one-way ANOVA followed by the Tukey multiple comparison tests comparing every group with every other group using GraphPad Prism version 5.00 for Windows (GraphPad Software, San Diego CA USA, www.graphpad.com). A P value less than 0.05 was considered significant and is indicated by an asterisk.

Results

1. Characterization of Nanoceria

Each batch of synthesized nanoceria was thoroughly characterized using (1.) transmission electron microscopy for size and shape determination, (2.) dynamic light scattering for zeta potential measurement, and (3.) X-ray photoelectron spectroscopy for determination of the relative abundance of 3+ and 4+ oxidation state of cerium. The detailed characterization results can be found in the supplemental materials in [18]. Our stable water-dispersed nanoceria are 3–5 nm in size. The size of these particles remains the same in a wide range of pH buffers and upon aging [27]. The surface charge on nanoceria is an important consideration for their interactions with the biological environment. We established that the zeta potential of our synthesized nanoceria at 1 mM in saline solution to be +10±3 mV [18].

Table 1. Parameters of the nine steps used for Scotopic ERG recording.

Step	Light intensity (cd.s/m²)	Light intensity Log scale (cd.s/m²)	# trials per result	Inter trial delay (s)
1	0.006	−2.22184875	4	10
2	0.009	−2.045757491	4	10
3	0.03	−1.522878745	3	15
4	0.06	−1.22184875	3	15
5	0.3	−0.522878745	3	20
6	3	0.477121255	3	25
7	30	1.477121255	3	30
8	300	2.477121255	2	50
9	600	2.77815125	2	60

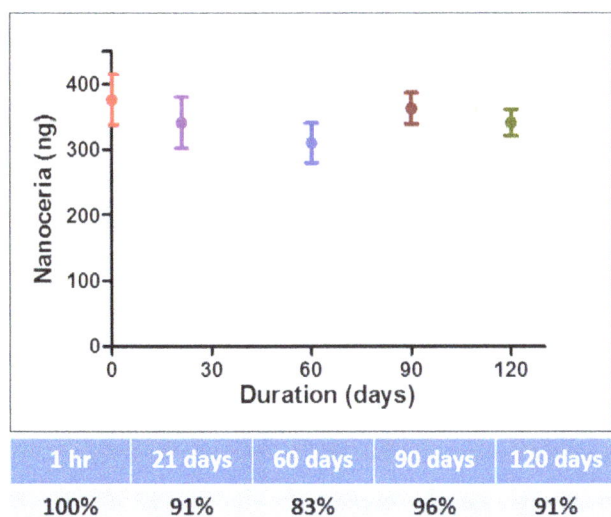

1 hr	21 days	60 days	90 days	120 days
100%	91%	83%	96%	91%

Figure 1. Parameters for evaluation of effects of nanoceria on retinal cytoarchitecture. A. Schematic diagram of a cross section of a rodent eye cut through the optic nerve head. The thick black line in the posterior part of the globe represents the retina. Measurements were taken from i) the central portion of the retina: 960 µm from the optic nerve head (ONH), and ii) the peripheral portion of the retina: 960 µm from the ora serrata, along the superior and inferior axis of the globe. Each marked interval represents ~960 µm. **B.** Photomicrograph of an H&E stained retinal section to illustrate the measurements taken from different retinal layers for quantitative analyses. Inner retina is up. 1 = retina thickness (RT) = inner limiting membrane (ILM) to outer limiting membrane (OLM), 2 = inner nuclear layer (INL), 3 = outer nuclear layer (ONL), 4 = inner and outer segments (IOS) of rod cells. To minimize the effects of uneven tissue shrinkage among eye samples, we normalized the thickness of INL, ONL, and IOS by comparing these measurements with the overall thickness of the retina (RT). Measurements from a single retinal section were taken using a calibrated reticle on one of the binoculars of a Nikon E400 microscope under a 20X objective. Averages were from eyes of different animals.

2. Ninety Percent Of Injected Nanoceria Stayed in the Eye for 120 Days

Since we did not know how quickly nanoceria were cleared in the eye, we injected 1000X the effective dose previously used in our light-damage model in rats [17]. Each eye received 2 µl of 1 mM nanoceria (344 ng). We harvested the eyes at different times after injection: from 1 hour to 120 days. The Ce levels were assessed using a 7700 Series ICP-MS from Agilent technologies that was capable of detecting Ce at a minimum range of 50–100 parts per trillion. **Figure 2** shows the retention of nanoceria in the eye after a single intravitreal injection. This data set shows the combined amount of nanoceria from each eye that had been dissected into component parts: retina, lens, and rest of eye cup. Another experiment, in which whole eyes were analyzed at 1 hour and 30 days post injection, showed similar results (data not shown). Surprisingly, we found that 90% of the injected nanoceria remained in the eye after 120 days. Unlike many ophthalmic agents such as peptides, nucleic acids or organic compounds that are cleared from the eye within hours or days, we observed that nanoceria were not actively eliminated from the eye. Additional retention data from 8- and 12-month time points became available after our submission. We determined that the elimination half-life of nanoceria in the eye and in the retina to be 525 days and 414 days, respectively. These data are presented in **Figures S1 and S2**.

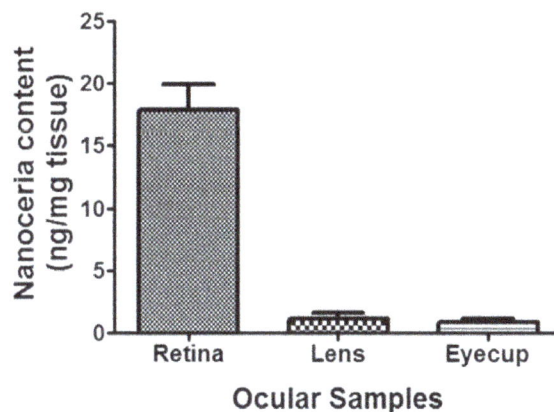

Figure 2. Nanoceria were retained in the eye for months. The amount of nanoceria retained in the eye was determined by ICP-MS. The elimination was extremely slow with approximately 90% of the injected nanoceria still retained in the eye at 120 days or four months post injection.

3. Nanoceria were Rapidly and Preferentially Taken up by Retinal Cells

To determine where nanoceria were distributed after injection, we dissected the eye into component parts. We separated the retina and the lens from the rest of the eyecup before ICP-MS analysis. **Figure 3** shows the bio-distribution of nanoceria one hour post injection. During this initial hour, retinal tissue accumulated the highest concentration of nanoceria (17.89 ng/mg tissue, followed by the lens tissue (1.13 ng/mg tissue), and the lowest in the eyecup tissue (0.83 ng/mg tissue). Uninjected eyes contained negligible amount of nanoceria.

To determine if nanoceria were also preferentially retained in retinal tissue, we analyzed retinal samples from 1 hour to 120 days post injection. **Figure 4** shows the amount of nanoceria

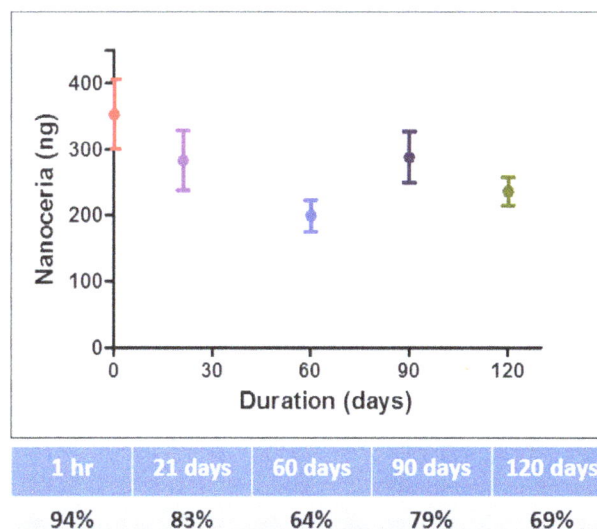

1 hr	21 days	60 days	90 days	120 days
94%	83%	64%	79%	69%

Figure 3. Bio-distribution of nanoceria in ocular tissues one hour post injection. We detected the highest concentration and amount of nanoceria in the retinal portion of the eye whereas the lens and the rest of the eye cup retained only small amounts of nanoceria one hour post injection.

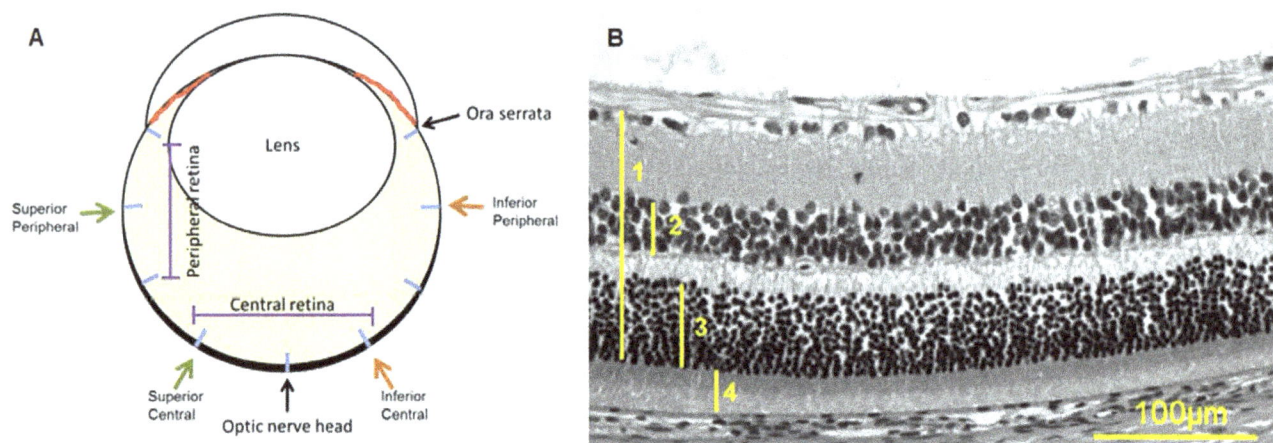

Figure 4. The retina retained nanoceria over prolonged periods of time following a single intravitreal injection. The injected nanoceria accumulated in the retina rapidly. We detected 94% of the injected nanoceria in the retina after one hour of injection and about 70% of the injected nanoceria was retained for the four months tested.

remaining in the retina over the 120-day period. We observed that about 70% of the injected nanoceria were retained in the retina even after 120 days. (New data from 8- to 12- month samples can be found in **Figure S2**.).

4. Nanoceria did not have Cytotoxic Effects in the Retina

To assess the safety of nanoceria for therapeutic treatment, we ascertained the effects of nanoceria in the retina after 9, 60 and 120 days post intravitreal injection. We administered a range of nanoceria dosages (1 μM, 100 μM, and 1 mM) to the rat eyes, and evaluated potential morphological and functional changes in the retina. For retinal cytoarchitecture assessment, we examined the thicknesses of the inner and outer retina (**Figure 1B**). The layer thickness reflects the number of neurons residing in these two layers, thus the overall health of the retina. We also included the thickness of the rod inner and outer segments (IOS) to further evaluate the health status of rod cells. We focus on the rod photoreceptor cells because they are one of the key light sensors and are the most abundant cell type in many mammalian retinas, including rodents and primates [28,29]. They are also exquisitely susceptible to oxidative damage due to the unusually high content of polyunsaturated fatty acids [30]. Evaluation of the thickness of IOS, therefore, can serve as an indicator of the health status of these cells. To evaluate functional changes, we conducted scotopic, photopic, and flicker full field ERG recordings. Scotopic a-wave amplitude reflects primarily the function of rod cells. Scotopic b-wave amplitude reflects the function of neurons in the inner retina, post-synaptic to the rod photoreceptor cells. Photopic b-wave amplitude and flicker ERG reflect the function of cone cells [26]. **Figure 5** shows the morphometric data from 9 days post injection. From the four surveyed areas: superior and inferior central retina, superior and inferior peripheral retina (see **Figure 1A** for orientation), we did not observe any reduction in thickness in the layers examined for nanoceria injected eyes. **Figure S3** shows representative H&E stained retinal sections from the inferior central portion of each eye for this set of the experiment. We also did not observe any changes in retinal functions among the nanoceria injected versus the saline injected animals (**Figure 6**). These results indicate that nanoceria did not cause acute negative side effects in the healthy retina. Similar results were obtained for the 60 days (data not shown) and the 120 days (**Figures 7, 8, and S4**) data sets. From these results, we

conclude that nanoceria generated according to our described formulation and procedure, are not toxic to the rat retina as measured by morphology or function, even when 344 ng are present for over 4 months.

Discussion

This is the first *in vivo* study to show that nanoceria are rapidly and preferentially taken up and retained in the retina after a single intravitreal injection. The retention of nanoceria in the retina does not have any short- or long-term cytotoxic effects. This study is a more targeted approach to understanding the bio-distribution and side-effects of nanoceria in rodent ocular tissues. The lack of toxic effects in the retina is consistent with our previous findings demonstrating that weekly systemic administration of nanoceria in mice did not have cytotoxic effects in the heart, kidney, brain, lungs, spleen, and liver for a 5-week period [31].

We do not yet know how nanoceria are taken up by retinal cells. However, in an *in vitro* uptake experiment, Singh and colleagues [32] showed that fluorescein-conjugated nanoceria were taken up by keratinocytes via clathrin-, and caveolae-dependent endocytic pathways. These fluorescently-labeled nanoceria were found in multiple cellular compartments including the mitochondria, lysosomes, endoplasmic reticulum, the nucleus, and the cytoplasm. At this moment, kinetic studies on elimination or exocytosis of nanoceria by cells are not available. However, in this *in vivo* study, we established that the elimination half-life in the rat retina to be 414 days (see Supplemental Materials). Our results suggest that the rate of elimination by cells is extremely slow. By 120 days, we observed 30% reduction of injected nanoceria in the retina. Could these nanoceria be accumulating in other organs? We detected trace amounts of nanoceria in liver and kidney tissues from 120 days animals from both nanoceria injected and uninjected animals (data not shown). Our findings suggest that nanoceria are removed from the eye eventually but the primary route of elimination is unknown. It can be by general circulation or locally in the eye, or both. The protracted retention of nanoceria in the eye could be beneficial if the self-regenerative property of nanoceria [33] is maintained. Presently, we have shown that nanoceria are not toxic to retinal neurons after 120 days of exposure and the prolonged retention in the retina should be an asset for the investigation and development of nanoceria as ophthalmic therapeutics.

A Superior Central

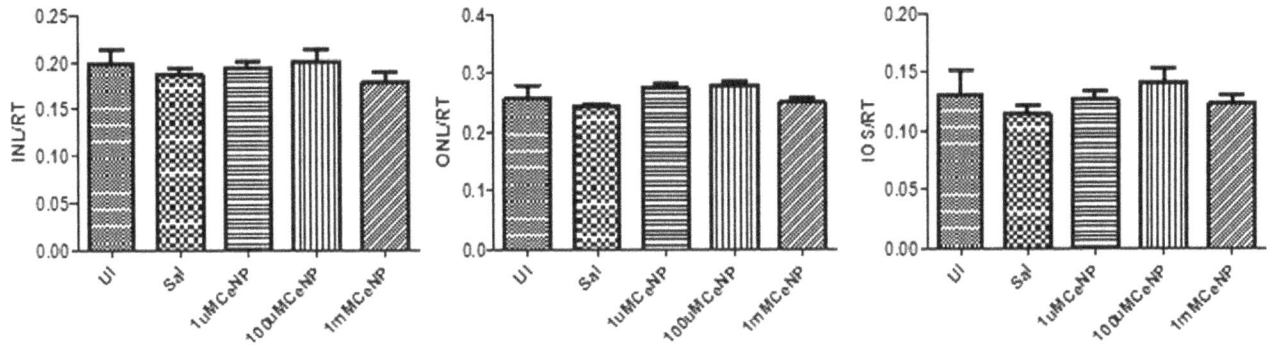

B Inferior Central

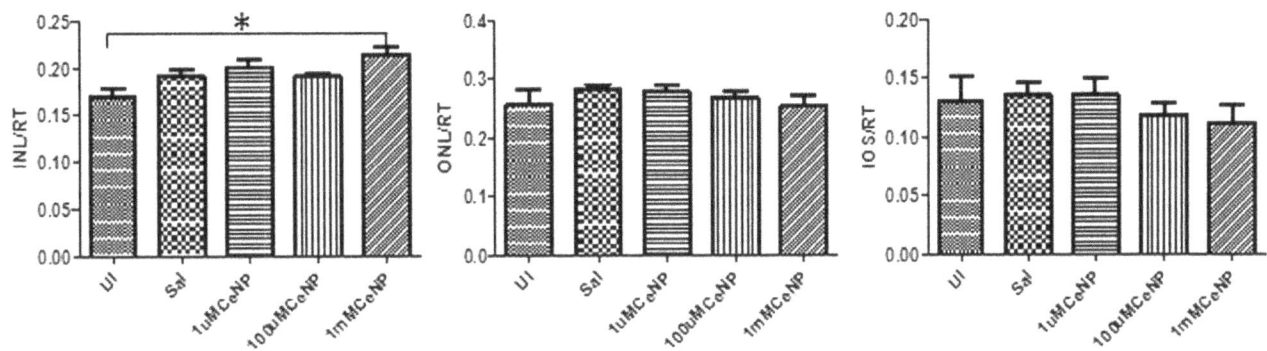

C Superior Peripheral

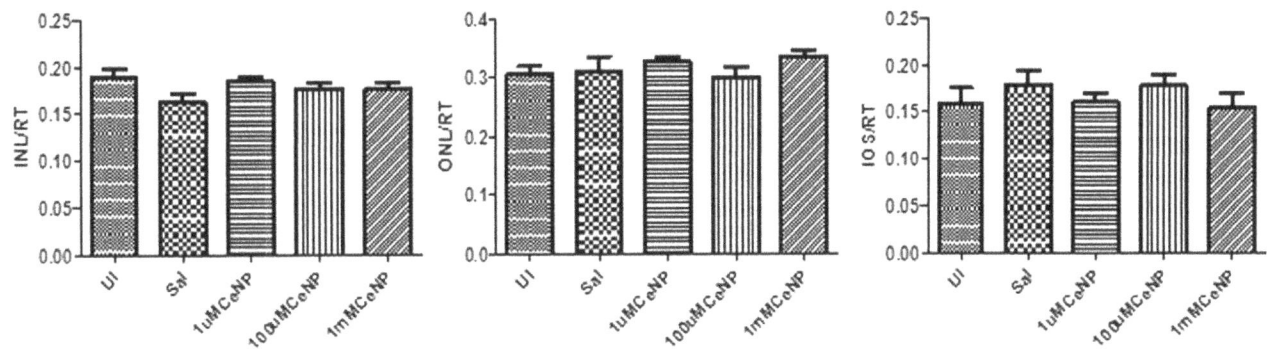

D Inferior Peripheral

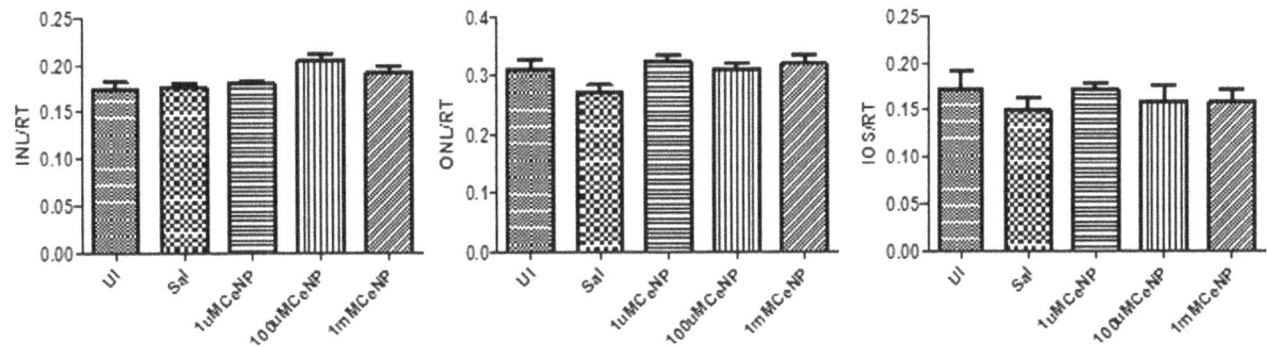

Figure 5. Nanoceria had no toxic effects on the morphological structures of the retina at nine days post injection. We performed quantitative morphometric analyses (see Figure 4 legend) on the central (A–B), and peripheral (C–D) portions of the retina of animals from different treatment groups (UI = uninjected, Sal = saline injected, or CeNP injected with the indicated dosage) and found no changes induced by nanoceria. Representative photomicrographs of retinal sections from this time point are shown in Figure S3.

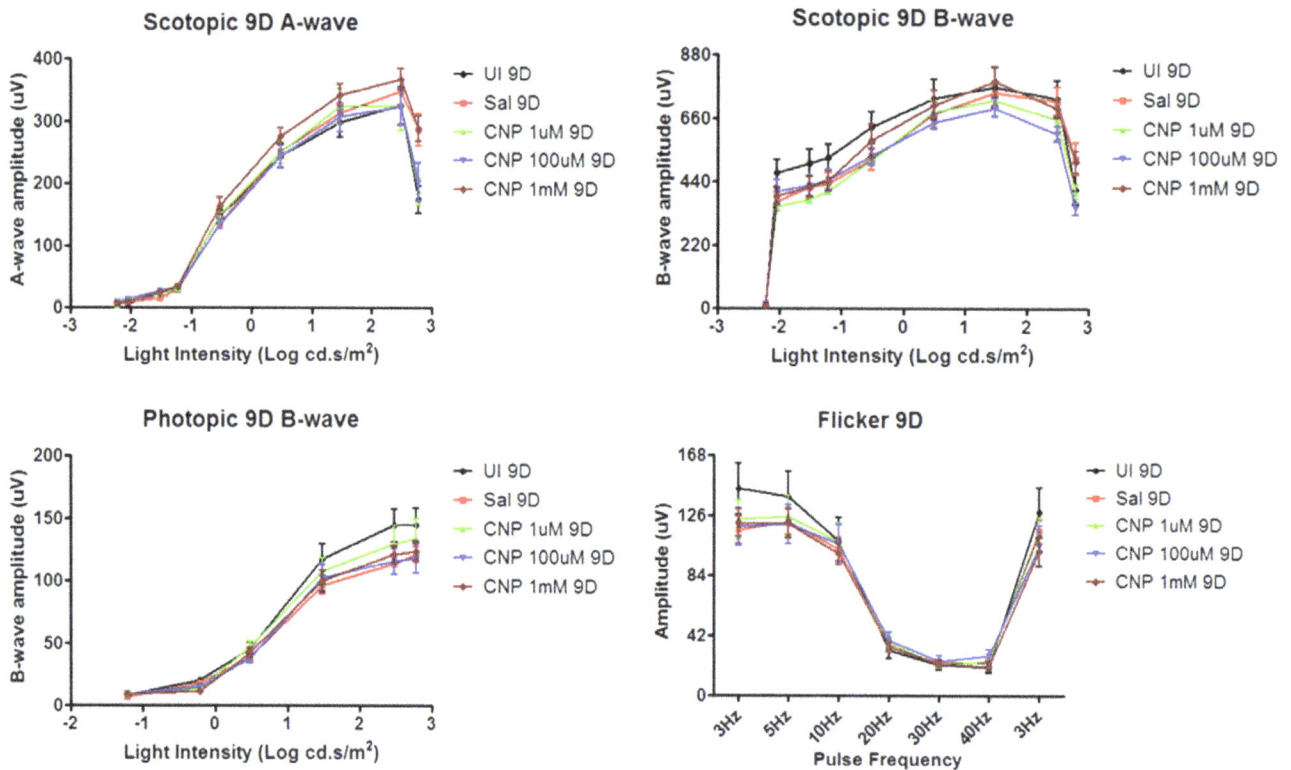

Figure 6. Nanoceria had no negative effects on retinal function as evaluated by ERG recordings nine days post injection. We did not observe changes in amplitudes of scotopic a- and b-waves, photopic b-wave, and flicker at the frequencies indicated.

We are just beginning to unravel the interactions of nanoceria in the complex biological environment. Reports of the differential effects of nanoceria on a variety of cell types in cell culture experiments add to the challenge of developing nanoceria for therapeutic treatments. One major defining feature in these cell culture studies that was seldom discussed was the synthesis and characterization of nanoceria administered [2]. For example, the usage of hexamethylenetetramine (HMT) in the synthesis of nanoceria might contribute to the cytotoxic effects of nanoceria (unpublished observations). The effects of particle aggregation might be another contributor to the negative effects of nanoceria observed in some cell culture studies [34,35]. The surface charge of nanoparticles plays a major role in their uptake by cells [34]. Asati and colleagues [36] showed that polymer-coated nanoceria with a positive charge were taken up by more cell types than the negatively-charged or neutrally-charged particles. They also demonstrated that the differently charged particles could be localized in the same or different cellular compartments depending on the cell type.

Delivery of nanoparticles to their site of action without dilution poses another challenge. Kim and colleagues [37] compared the movement of fluorescently-labeled human serum albumin nanoparticles with either positive or negative surface charge, in the vitreous and retina of the rat eye. Five hours after intravitreal administration, the authors found that the negatively-charged nanoparticles penetrated the vitreal barrier more easily and were found deep in the retina from the inner limiting membrane to all the nuclear layers and plexiform layers and even in the retinal pigment epithelium. The seemingly contradictory findings from cell culture versus *in vivo* studies with respect to nanomaterial surface charge and uptake illustrate the complexity of the

interactions between nanomaterials and biological environments. We caution that the observations made regarding interactions of nanomaterials on synthetic membranes or lipid bilayers and nanomaterials uptake in cell culture studies do not necessarily reflect the actual interactions in biological environments. For *in vivo* systemic delivery, the administered nanomaterials need to circulate in the blood to reach their target tissue. En route, the nanomaterials are likely to have serum proteins adsorbed on the surface through electrostatic or hydrophobic interactions. This adsorption may change the properties of how the nanomaterials will be taken up by target cells.

In this study, we demonstrated that nanoceria administered in the vitreous were able to penetrate through this complex structure and reach the underlying retinal cells. The mammalian vitreous body is a network of collagen fibers embedded in a matrix of highly water-bound glycosaminoglycans [38]. Albumin constitutes about 40% of the soluble proteins in the vitreous [38] and may adsorb to the injected positively-charged nanoceria by electrostatic interactions [39]. However, this potential interaction did not impede the uptake of nanoceria by retinal cells. One unusual feature of the vitreous is the high percentage of iron-binding proteins present compared to the plasma. Van Bockxmeer and coworkers [40] showed that about 35% of the soluble proteins in the monkey vitreous were composed of transferrin and/or lactoferrin. The high transferrin or transferrin-like proteins in the vitreous may contribute to the rapid uptake of nanoceria by retinal cells. We [41] previously showed that nanoceria with positive surface charges can bind to transferrin and were taken up by cells more efficiently than naked nanoceria at a number of concentrations. Transferrin receptors are present in all cell types in the body including retinal cells. We postulate that the rapid uptake

A Superior Central

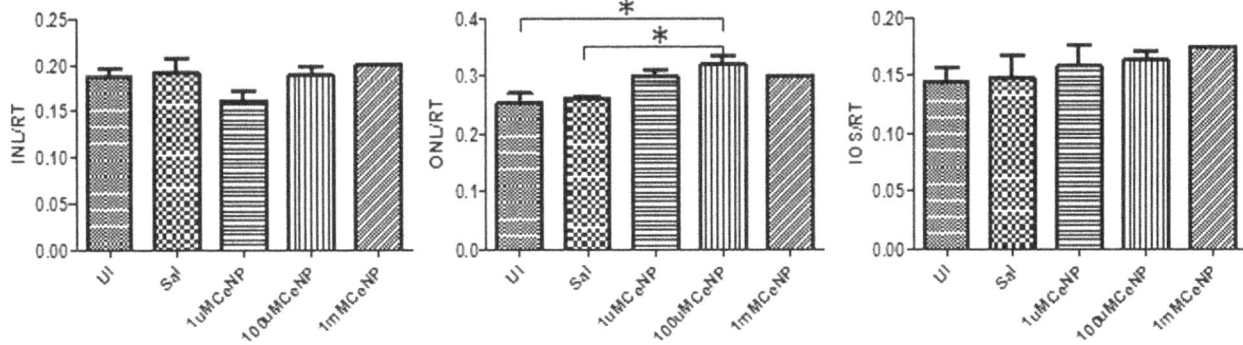

B Inferior Central

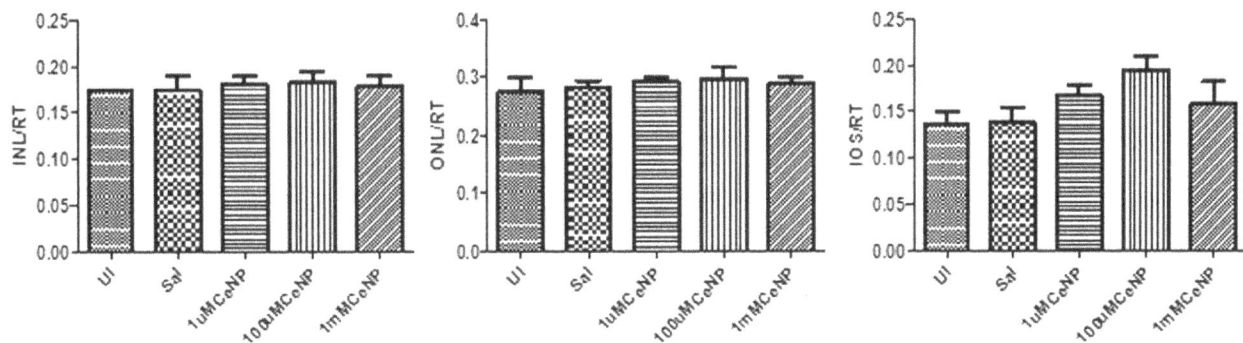

C Superior Peripheral

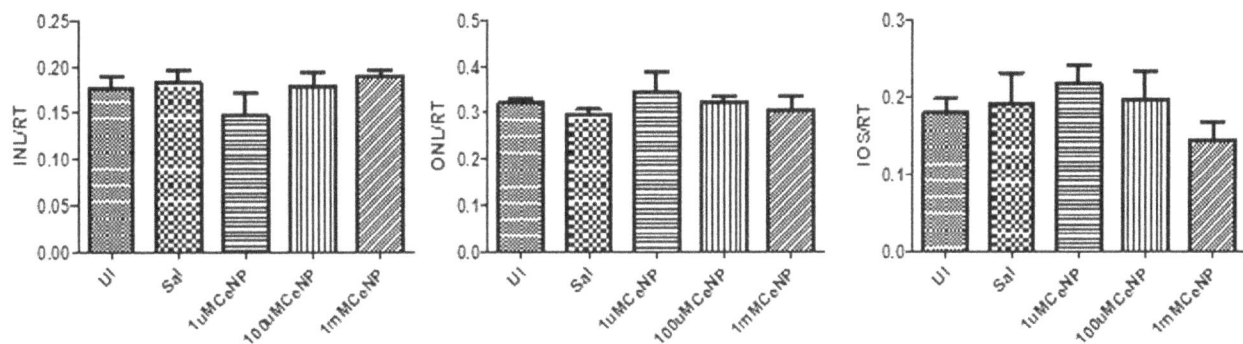

D Inferior Peripheral

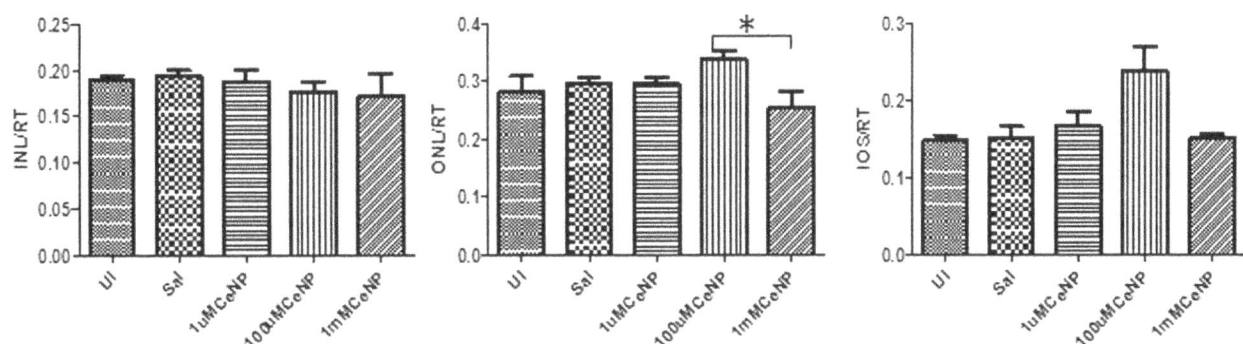

Figure 7. Nanoceria (1–1000 uM) had no negative effects on retinal morphology, even four months after injection, as measured by quantitative morphometric analyses. We examined the central (A–B), and peripheral (C–D) portions of the retina of animals from different treatment groups (UI = uninjected, Sal = saline injected, or CeNP injected with the indicated dosage), and found no changes induced by nanoceria. Measurements were performed as indicated in Figure 4.

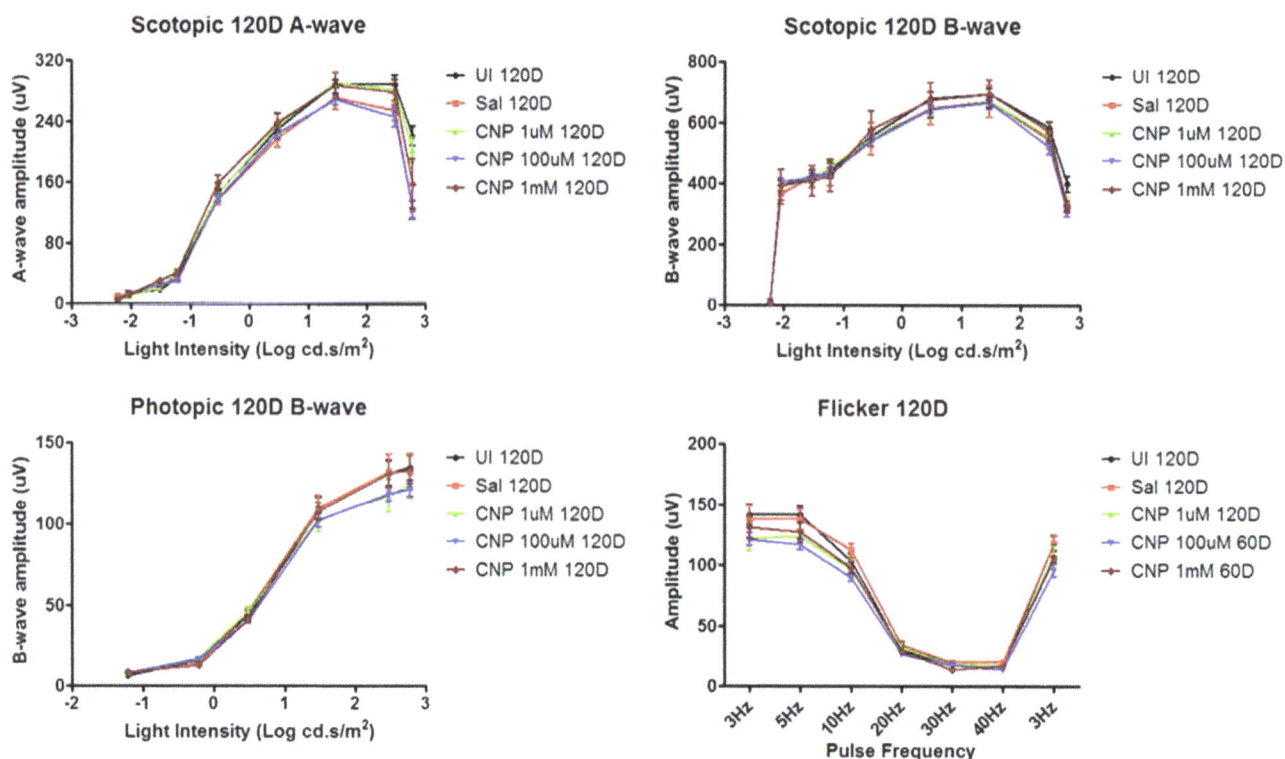

Figure 8. Nanoceria (1–1000 uM) had no negative effects on retinal function, even four months after injection, as measured by ERG recordings. No changes were detected in amplitudes of scotopic a- and b-waves, photopic b-wave, and flicker at the frequencies indicated.

of nanoceria may be facilitated by binding to transferrin and/or lactoferrin in the vitreous. This hypothesis is consistent with the observation that nanoceria-mediated photoreceptor cell protection is not focal but is found across the entire retina rather than being confined to the injection site [17].

In spite of the many unknown interactions of nanoceria in the vitreous and other ocular tissues, our results suggest that our stable water-dispersed nanoceria are likely to be taken up by one or more cell types in the retina by an active process. The enhanced redox capacity of nanoceria does not have negative effects in healthy retinal cells. Finally, nanoceria appear to have an extremely slow rate of removal once inside cells.

Currently, effective and comprehensive therapies for blinding diseases such as age-related macular degeneration, diabetic retinopathy, and retinitis pigmentosa are still unavailable. Targeting the reduction of reactive oxygen species by nanoceria in these diseased eyes may be one way to prolong the life and function of retinal cells before cures become a reality. Additionally, drug delivery to the posterior segment of the eye, such as the retina, is most effective when administered intravitreally. This procedure, though simple, carries the risk of complications such as infection and retinal detachment, especially when frequent, repeated applications are required. Taken together, nanoceria appear to be ideal candidates as ophthalmic antioxidants because the redox activity of nanoceria is regenerative [33] and frequent repetitive dosing may be avoided. We predict that nanoceria may become the "aspirin" of the 21st century for the therapeutic treatment of eye diseases whose pathologic progression associates tightly with oxidative damage.

Supporting Information

Figure S1   Detection of nanoceria in the eye after one year.

Figure S2   Detection of nanoceria in the retina after one year.

Figure S3   Photomicrographs of H&E stained retinal sections from adult SD rats nine days post nanoceria (CeNP) intravitreal injection.

Figure S4   Photomicrographs of H&E stained retinal sections from adult SD rats 120 days post nanoceria (CeNP) intravitreal injection.

Acknowledgments

We thank all the individuals at the Imaging and Animal Modules of the DMEI/OUHSC Vision Core Facility. We thank Dr Sukyung Woo at the Dept. of Pharmaceutical Sciences, College of Pharmacy, OUHSC for suggestions on refinement of the pharmacokinetic study, and Dr Yves Sauvé at the University of Alberta for sharing his ERG protocols.

Author Contributions

Conceived and designed the experiments: LLW JFM SS CMR. Performed the experiments: LLW SMH QNP CMR. Analyzed the data: LLW SMH QNP JFM CMR SS. Contributed reagents/materials/analysis tools: LLW CMR SS JFM. Wrote the paper: LLW SMH QNP JFM CMR SS.

References

1. Thomas DG, Klaessig F, Harper SL, Fritts M, Hoover MD, et al. (2011) Informatics and standards for nanomedicine technology. Wiley Interdiscip Rev Nanomed Nanobiotechnol.
2. Karakoti A, Singh S, Dowding JM, Seal S, Self WT (2010) Redox-active radical scavenging nanomaterials. Chem Soc Rev 39: 4422–4432.
3. McGinnis JF, Chen J, Wong L, Sezate S, Seal S, et al. (2010) Inhibition of reactive oxygen species and protection of mammalian cells. US Patent 7727559.
4. McGinnis JF, Chen J, Wong L, Sezate S, Seal S, et al. (2008) Inhibition of reactive oxygen species and protection of mammalian cells. US Patent 7347987.
5. Campbell CT, Peden CH (2005) Chemistry. Oxygen vacancies and catalysis on ceria surfaces. Science 309: 713–714.
6. Inerbaev TM, Seal S, Masunov AE (2010) Density functional study of oxygen vacancy formation and spin density distribution in octahedral ceria nanoparticles. J Mol Model 16: 1617–1623.
7. Korsvik C, Patil S, Seal S, Self WT (2007) Superoxide dismutase mimetic properties exhibited by vacancy engineered ceria nanoparticles. Chem Commun (Camb): 1056–1058.
8. Pirmohamed T, Dowding JM, Singh S, Wasserman B, Heckert E, et al. (2010) Nanoceria exhibit redox state-dependent catalase mimetic activity. Chem Commun (Camb) 46: 2736–2738.
9. Self WT, Seal S (2009) Nanoparticles of cerium oxide having superoxide dismutase activity. US Patent 7504356.
10. Onyango IG, Khan SM (2006) Oxidative stress, mitochondrial dysfunction, and stress signaling in Alzheimer's disease. Curr Alzheimer Res 3: 339–349.
11. Burn DJ (2006) Cortical Lewy body disease and Parkinson's disease dementia. Curr Opin Neurol 19: 572–579.
12. Hogg R, Chakravarthy U (2004) AMD and micronutrient antioxidants. Curr Eye Res 29: 387–401.
13. Hollyfield JG (2010) Age-related macular degeneration: the molecular link between oxidative damage, tissue-specific inflammation and outer retinal disease: the Proctor lecture. Invest Ophthalmol Vis Sci 51: 1275–1281.
14. Madsen-Bouterse SA, Mohammad G, Kanwar M, Kowluru RA (2010) Role of mitochondrial DNA damage in the development of diabetic retinopathy, and the metabolic memory phenomenon associated with its progression. Antioxid Redox Signal 13: 797–805.
15. Bhatti MT (2006) Retinitis pigmentosa, pigmentary retinopathies, and neurologic diseases. Curr Neurol Neurosci Rep 6: 403–413.
16. Hartong DT, Berson EL, Dryja TP (2006) Retinitis pigmentosa. Lancet 368: 1795–1809.
17. Chen J, Patil S, Seal S, McGinnis JF (2006) Rare earth nanoparticles prevent retinal degeneration induced by intracellular peroxides. Nat Nanotechnol 1: 142–150.
18. Zhou X, Wong LL, Karakoti AS, Seal S, McGinnis JF (2011) Nanoceria Inhibit the Development and Promote the Regression of Pathologic Retinal Neovascularization in the <italic>Vldlr</italic> Knockout Mouse. PLoS ONE 6: e16733.
19. Kong L, Cai X, Zhou X, Wong LL, Karakoti AS, et al. (2011) Nanoceria extend photoreceptor cell lifespan in tubby mice by modulation of apoptosis/survival signaling pathways. Neurobiol Dis 42: 514–523.
20. Hirst SM, Karakoti AS, Tyler RD, Sriranganathan N, Seal S, et al. (2009) Anti-inflammatory Properties of Cerium Oxide Nanoparticles. Small 5: 2848–2856.
21. Celardo I, De Nicola M, Mandoli C, Pedersen JZ, Traversa E, et al. (2011) Ce(3)+ ions determine redox-dependent anti-apoptotic effect of cerium oxide nanoparticles. ACS Nano 5: 4537–4549.
22. Celardo I, Pedersen JZ, Traversa E, Ghibelli L (2011) Pharmacological potential of cerium oxide nanoparticles. Nanoscale 3: 1411–1420.
23. Karakoti AS, Monteiro-Riviere NA, Aggarwal R, Davis JP, Narayan RJ, et al. (2008) Nanoceria as Antioxidant: Synthesis and Biomedical Applications. JOM (1989) 60: 33–37.
24. Pinilla I, Lund RD, Sauve Y (2004) Contribution of rod and cone pathways to the dark-adapted electroretinogram (ERG) b-wave following retinal degeneration in RCS rats. Vision Res 44: 2467–2474.
25. Sauve Y, Pinilla I, Lund RD (2006) Partial preservation of rod and cone ERG function following subretinal injection of ARPE-19 cells in RCS rats. Vision Res 46: 1459–1472.
26. Perlman I (2011) The Electroretinogram: ERG. In: Kolb H, editor. Webvision: The Organization of the Retina and Visual System. Salt Lake City, Utah, USA: University of Utah. Online Textbook of the Visual System.
27. Vincent A, Inerbaev TM, Babu S, Karakoti AS, Self WT, et al. (2010) Tuning hydrated nanoceria surfaces: experimental/theoretical investigations of ion exchange and implications in organic and inorganic interactions. Langmuir 26: 7188–7198.
28. Jeon CJ, Strettoi E, Masland RH (1998) The major cell populations of the mouse retina. J Neurosci 18: 8936–8946.
29. Masland RH (2011) Cell populations of the retina: the Proctor lecture. Invest Ophthalmol Vis Sci 52: 4581–4591.
30. Bazan NG (2003) Synaptic lipid signaling. Journal of Lipid Research 44: 2221–2233.
31. Hirst SM, Karakoti A, Singh S, Self W, Tyler R, et al. (2011) Bio-distribution and in vivo antioxidant effects of cerium oxide nanoparticles in mice. Environmental Toxicology: n/a-n/a.
32. Singh S, Kumar A, Karakoti A, Seal S, Self WT (2010) Unveiling the mechanism of uptake and sub-cellular distribution of cerium oxide nanoparticles. Mol Biosyst 6: 1813–1820.
33. Das M, Patil S, Bhargava N, Kang JF, Riedel LM, et al. (2007) Auto-catalytic ceria nanoparticles offer neuroprotection to adult rat spinal cord neurons. Biomaterials 28: 1918–1925.
34. Verma A, Stellacci F (2010) Effect of Surface Properties on Nanoparticle–Cell Interactions. Small 6: 12–21.
35. Murdock RC, Braydich-Stolle L, Schrand AM, Schlager JJ, Hussain SM (2008) Characterization of nanomaterial dispersion in solution prior to in vitro exposure using dynamic light scattering technique. Toxicol Sci 101: 239–253.
36. Asati A, Santra S, Kaittanis C, Perez JM (2010) Surface-charge-dependent cell localization and cytotoxicity of cerium oxide nanoparticles. ACS Nano 4: 5321–5331.
37. Kim H, Robinson SB, Csaky KG (2009) Investigating the movement of intravitreal human serum albumin nanoparticles in the vitreous and retina. Pharm Res 26: 329–337.
38. Kleinberg TT, Tzekov RT, Stein L, Ravi N, Kaushal S (2011) Vitreous Substitutes: A Comprehensive Review. Survey of Ophthalmology 56: 300–323.
39. Patil S, Sandberg A, Heckert E, Self W, Seal S (2007) Protein adsorption and cellular uptake of cerium oxide nanoparticles as a function of zeta potential. Biomaterials 28: 4600–4607.
40. Van Bockxmeer FM, Martin CE, Constable IJ (1983) Iron-binding proteins in vitreous humour. Biochim Biophys Acta 758: 17–23.
41. Vincent A, Babu S, Heckert E, Dowding J, Hirst SM, et al. (2009) Protonated Nanoparticle Surface Governing Ligand Tethering and Cellular Targeting. ACS Nano 3: 1203–1211.

8

Plasmonic Optical Trapping in Biologically Relevant Media

Brian J. Roxworthy[1], Michael T. Johnston[2], Felipe T. Lee-Montiel[3], Randy H. Ewoldt[2], Princess I. Imoukhuede[3], Kimani C. Toussaint Jr.[2]*

1 Department of Electrical and Computer Engineering, University of Illinois at Urbana-Champaign, Urbana, Illinois, United States of America, **2** Department of Mechanical Science and Engineering, University of Illinois at Urbana-Champaign, Urbana, Illinois, United States of America, **3** Department of Bioengineering, University of Illinois at Urbana-Champaign, Urbana, Illinois, United States of America

Abstract

We present plasmonic optical trapping of micron-sized particles in biologically relevant buffer media with varying ionic strength. The media consist of 3 cell-growth solutions and 2 buffers and are specifically chosen due to their widespread use and applicability to breast-cancer and angiogenesis studies. High-precision rheological measurements on the buffer media reveal that, in all cases excluding the 8.0 pH Stain medium, the fluids exhibit Newtonian behavior, thereby enabling straightforward measurements of optical trap stiffness from power-spectral particle displacement data. Using stiffness as a trapping performance metric, we find that for all media under consideration the plasmonic nanotweezers generate optical forces 3–4x a conventional optical trap. Further, plasmonic trap stiffness values are comparable to those of an identical water-only system, indicating that the performance of a plasmonic nanotweezer is not degraded by the biological media. These results pave the way for future biological applications utilizing plasmonic optical traps.

Editor: Giuseppe Chirico, University of Milano-Bicocca, Italy

Funding: This work was supported by the National Science Foundation (NSF ECCS 10-25868). The funders had no role in study design, data collection and analysis, decision to publish, or preparation of the manuscript.

Competing Interests: The authors have declared that no competing interests exist.

* E-mail: ktoussai@illinois.edu

Introduction

Optical tweezers, introduced by Ashkin in 1986 [1], have become an indispensable component in the biophysicists' toolkit, leading to breakthroughs in understanding DNA structure [2], RNA transcription [3], protein folding [4], cell motility [5,6], and single-molecule biophysics [7,8]. However, investigation of systems at increasingly smaller scales is hindered by optical diffraction, which limits the maximum optical forces that can be achieved in an optical tweezer for a given input power [9]. This is particularly salient for biological systems, wherein high input optical power can lead to specimen damage [10,11]. Recently, plasmonic optical tweezers have emerged as a promising avenue to circumvent this issue. Also known as plasmonic "nanotweezers", this architecture employs metallic nanoantennas to concentrate and enhance incident optical fields in deep-subwavelength gaps [12–17]. This yields large near-field intensity gradients that greatly amplify optical forces for a given input power [12,13], enabling strong optical trapping with input-power densities 2–3 orders of magnitude lower than the biological damage threshold [18].

Following this reasoning, there have been several studies employing plasmonic nanostructures to trap biological objects. For instance, Righini *et al.* showed that living *Escherichia coli* bacteria can be stably trapped in a plasmonic nanotweezer comprised of dipole nanoantennas for more than two hours without visible damage [10]. Similarly, studies by Huang *et al.* and Miao and Lin demonstrated plasmonic trapping of yeast cells using a microfluidic platform containing Au nanodisks [19] and a

spherical Au nanoparticle array [20], respectively. Despite these initial experimental demonstrations, no studies exist to date that systematically address the impact of biologically relevant buffers on the trapping capabilities of either standard or plasmonic-based tweezers. Biological buffers (media) are critical to *in vitro* studies in order to mimic the biological environment outside of a host organism. As a result, such buffers are often designed to operate, e.g., at specific atmospheric conditions (%CO_2), physiologically relevant temperature, and pH, and thus should not be ignored in calibration of optical trapping platforms used for biophysical assays.

In this paper, we investigate the effects of five widely used, biologically relevant media (3 cell growth media and 2 buffers) on the trapping performance of both plasmonic nanotweezers that are based on Au bowtie nanoantenna arrays (BNAs) and conventional high-numerical aperture (NA) optical tweezers. We perform high-precision, temperature-dependent rheological measurements on the media to determine their viscosity and assess trapping performance by measuring the optical trapping stiffness on 1.5-μm diameter polystyrene spheres in the various media. The effects of the medium pH and nanostructure geometry on trap stiffness are investigated. Our results show that the main contributor to the variation in performance of plasmonic nanotweezers in the biological media is the viscosity. Moreover, we show that in the biological media, plasmonic trapping strength is up to 4x that of conventional optical tweezers and is not mitigated compared to the water-only environment commonly used for trapping experiments. The cell growth solutions used in this study are utilized in cancer research, cardiovascular research, and common molecular biology assays. These findings have important implications for making plasmonic optical trapping more accessible to biological studies.

Experimental Methods

Biological Buffer Preparation

The breast cancer cell media (BC) is comprised of high-glucose Dulbecco's Modified Eagle Medium (DMEM) containing 10% fetal bovine serum (Invitrogen, Carlsbad, CA) and 1% Penicillin-Streptomycin (Invitrogen, Carlsbad, CA). The DMEM contains sodium pyruvate as an energy source and it contains sodium bicarbonate and sodium phosphate for buffering; such buffering is necessary for cellular growth in a 5% $CO2$ environment (incubator). The addition of 1% antibiotics prevents bacterial contamination and the serum contains biomolecules necessary for cell growth and cellular interactions, including: growth factors, enzymes, proteins, fatty acids and lipids, amino acids and carbohydrates [21]. This media is commonly utilized for the growth of human and mouse tumor cells, fibroblasts, macrophages, and other cell types. One of the co-authors (Imoukhuede) has recently employed this media in the growth of human breast cancer cell line MDA-MB-231 [22].

The endothelial growth medium (EGM2) is optimized for growth of human macrovascular endothelial cells in culture and is supplemented by the EGM-2 SingleQuot Kit, which contains FBS, growth factors and other ingredients for accelerated growth of healthy endothelial cells. This media is commonly used in cardiovascular research, including studies of angiogenesis. We have recently employed this media in the growth of human umbilical vein endothelial cells (HUVEC) [22–24]. The Lebovitz media (L15) contains glucose, free base amino acids and is buffered at pH 7.8 by salts. It is designed to be used with cells in a non-CO_2 atmospheric conditions (outside an incubator).

In addition to these media, we also use two buffers in this study: Phosphate buffered saline (PBS) and Flow Cytometry Stain Buffer (Stain). Phosphate buffered saline (1x PBS, Fisher Scientific 10x power concentrate) is an aqueous solution consisting of Sodium Chloride (81%), Sodium Phosphate Dibasic (14%), and trace amounts of Potassium Phosphate Monobasic and Potassium Chloride. The ion concentrations and osmolarity of PBS are based on those found in the human body and the phosphate helps to buffer cell pH at 7.4 outside of an incubator. The Stain buffer is utilized for immunofluorescent staining of suspended cells and is a PBS-based solution with 2% bovine serum albumin (BSA), to reduce non-specific antibody bonding, and 0.09% of the preservative sodium azide. These buffers have fewer ingredients than the growth media and are widely used in flow cytometry applications. Each solution is prepared with two different pH values, 7.4 and 8.0, and the pH of the individual solutions is measured with a FiveEasy FE20 pH meter (Mettier-Toledo AG). A digital photograph of the media used in this study is shown in the supporting information (Fig. S1 in File S1).

Viscosity Measurements

Viscosity measurements of the biological media are performed using a rotational rheometer (Discovery Series Hybrid Rheometer (DHR), model HR-3, TA Instruments). The geometry is a single-gap, concentric cylinder (DIN standard) with conical bottom on the inner rotor. A schematic diagram is shown in the Fig. 1 inset. This geometry has shown highly reproducible results for shear-rate dependent measurements of low viscosity liquids, specifically because it minimizes surface tension torque effects that can appear

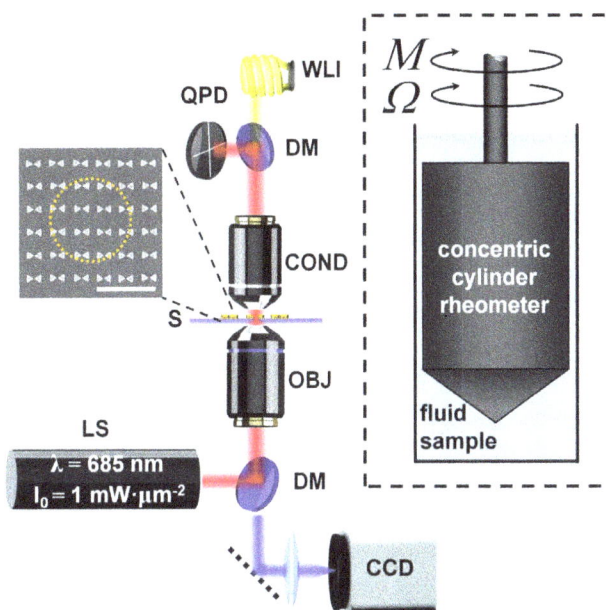

Figure 1. Schematic of the experimental setup. The experimental setup consists of a laser source (LS) coupled into the sample (S) by the microscope objective (OBJ) and dichroic mirror (DM). The sample inset shows a SEM image of the 425-nm array BNAs and the dotted-yellow line depicts the approximate focal spot diameter; scale bar is 1 μm. The condenser lens (COND) collects forward-scattered light from the trapped particle and the quadrant photodiode (QPD) detects Brownian fluctuations about the trap center. White-light illumination (WLI) provides visualization of particles on the CCD camera. The inset depicts the rotational rheometer geometry (not to scale) for the viscosity measurements. The measured torque M is due primarily to the simple shear flow in the thin gap between the inner rotor and outer stator. The shear viscosity η is calculated from the measured torque and angular velocity Ω.

inaccurately as shear-thinning [25]. The geometry has outer stator radius 30.35 mm, inner rotor radius 27.98 mm, and inner rotor working length 42.2 mm. A sample volume of 22.4 mL is used. Each sample is tested at temperatures of 20, 25, and 30 °C with Peltier temperature control at the outer surface. After loading, samples are held at the experimental temperature for 5 minutes prior to testing. Shear-rate sweeps are performed from 1 to 100 s^{-1} at T = 25 °C to determine the rate-dependent behavior of the biological media. Reported viscosity values are taken at 10 s^{-1} for the Newtonian samples and repeated in triplicate with separate sample loading to obtain precision error < 1%. For the measurably non-Newtonian Stain buffer at 8.0 pH, the reported viscosity is taken as the average from 2 to 50 s^{-1} with no repeated measurements.

Optical Trapping

The experimental optical trapping setup is built on an inverted microscope (Olympus IX-81) equipped with a 0.9-NA condenser (Olympus MPlanFL N 100x) that both provides white-light illumination for imaging trapped particles and collects the forward-scattered light from the trapping volume for trap stiffness measurements. The custom-built laser source is derived from a 685-nm wavelength laser diode that is spatially filtered and expanded to overfill the back-aperture of the microscope objective lens. For plasmonic optical trapping, a 0.6-NA objective (Olympus LUCPlanFLN 40x) is used to focus the incident beam onto the bowtie nanoantenna arrays (BNAs), which are fabricated onto a glass substrate with a 25-nm thick Indium-Tin-Oxide coating. The individual bowties comprising the BNAs are placed with two array spacings: 425 and 475 nm, which correspond to the center-to-center spacing between bowties along both x and y directions. Fabrication details can be found elsewhere [12]. The trapping chamber is formed using a 13-mm diameter gasket (Invitrogen) sandwiched between the BNA substrate and a rectangular #1 coverslip (Corning). The incident polarization is set parallel to the bowtie long axis in order to generate strong field concentration in the 20-nm gap. The chosen illumination wavelength is blue-detuned from the peak plasmon resonance of the BNAs, which produces strong optical forces without excessive plasmonic-absorption generated heating [15,26].

Conventional optical trapping is performed using a 1.4-NA, oil-immersion objective (Olympus UPlanSApo 100x). The trapping chamber for conventional tweezers is formed by replacing the BNA substrate with a standard #1-1/2 coverslip (Corning). In all cases, the input power is adjusted to achieve a focal power density $I_0 = P_0/A = 1$ mW·μm^{-2}, where the focal-spot area is given by $A = \pi w_0^2$, with focal-spot radius $w_0 = 0.61\lambda/\text{NA}$, λ is the free-space, input wavelength, and P_0 is the optical power measured at the focal plane. This process compensates for losses in the optical system. In the 0.6-NA case, P_0 is directly assessed by placing an optical power detector near the focal plane, whereas for the 1.4-NA case, P_0 is assessed by re-collimating the focused laser with an identical objective and placing the power detector in the back-focal-plane of the objective. A schematic diagram of the experimental setup is shown in Fig. 1.

Optical trapping experiments are performed on 1.5-μm diameter polystyrene particles (Thermo Scientific), and the trap stiffness is assessed via the power spectrum method [27]. Here, a quadrant photodiode (QPD, Thorlabs PDQ80A) placed in the back-focal-plane of the condenser measures the position fluctuations of the trapped particle. The power spectrum of these Brownian fluctuations about the trap center is given by the Lorentzian [27]

$$S_{xx}(f) = \frac{k_B T}{\pi^2 \gamma \left(f^2 + f_c^2\right)}, \qquad (1)$$

where k_B is Boltzmann's constant, T is the local temperature near the particle, f_c is the corner frequency, and $\gamma = 6\pi a \epsilon(a,h)\eta(T)$ is Stokes' drag coefficient with particle radius a and temperature-dependent viscosity $\eta(T)$ of the local fluid medium [27]. In order to account for the particle proximity to the substrate, we use the lubrication value of Faxen's correction

$$\epsilon(a,h) = \left| \frac{8}{15} \ln\left(\frac{h}{a}\right) - 0.9588 \right|, \qquad (2)$$

where h is the distance between the particle and the substrate [15,28,29]. In practice, it is difficult to determine h for a plasmonic trap, however, given the evanescent nature of plasmonic near-fields, particles must be within ~ 10–30 nm of the nanoantennas to experience enhanced optical forces [15,30]. Thus, we use $h = 15$ nm as an average value which gives $\epsilon(a,h) = 3.05$; this value is assumed for both plasmonic and conventional trapping experiments. In the latter case, the particle height is set to within ~ 15 nm using a precision closed-loop microscope stage. Here, the axial position of the stage is moved with 10-nm precision until trapped particles are observed to contact the surface of the coverslip. Then, the stage is moved a single step away from the

particle. The trap stiffness $\kappa = 2\pi\gamma f_c$ is then determined from the corner frequency obtained by fitting experimental power spectra to Eq. 1 via the Levenberg-Marquardt algorithm [27]. Position fluctuation signals are captured for 60 seconds using custom-written Labview software, and each corner frequency measurement represents the average of 15 independent measurements on the same particle, when possible.

Results

As a first step toward assessing the trap stiffness, we measure the steady-shear viscosities for the various media and the results are given in Fig. 2. Here, viscosity data are reported for T = 25 °C and a characteristic shear rate of $\dot{\gamma} = 10\ s^{-1}$; full temperature-dependent data are available in the supporting information (Fig. S2 in File S1). From the shear-rate dependent measurements, we find that all media (excluding Stain) exhibit Newtonian behavior for characteristic shear-rates $\dot{\gamma} \sim 2 - 50\ s^{-1}$. As a result, calculation of the trap stiffness utilizing the Stokes' drag coefficient γ is justified [27,29]. In contrast, the Stain media at pH = 8.0 showed measurable shear-thinning behavior (Fig. S3 in File S1). This buffer includes bovine serum albumin protein, which may be stretched and oriented by shear flow and cause non-constant shear viscosity. We observed an approximate plateau viscosity (within 6%) over the range of $\dot{\gamma} \sim 2 - 50\ s^{-1}$ for this particular case, and therefore calculated viscosity from the new average within this range. For all the fluids tested, the pH has little effect on the viscosity, with the only appreciable deviation occurring for L15 which shows a $\sim 1\%$ larger viscosity for pH = 7.4. We note that the BC media is unstable and phase separates at 8.0 pH, and therefore no data is reported for this particular case.

The measured trap stiffness using the plasmonic optical tweezers for all parameters considered in this study are shown in Fig. 3. Figures 3a and 3b show typical trap stiffness results comprising the time trace of the QPD voltage signal and the calculated power spectrum with a Lorentzian fit, respectively. The inset in Fig. 3b shows a particle displacement histogram overlaid with a Gaussian fit; the close-fit of the histogram with the Gaussian indicates that

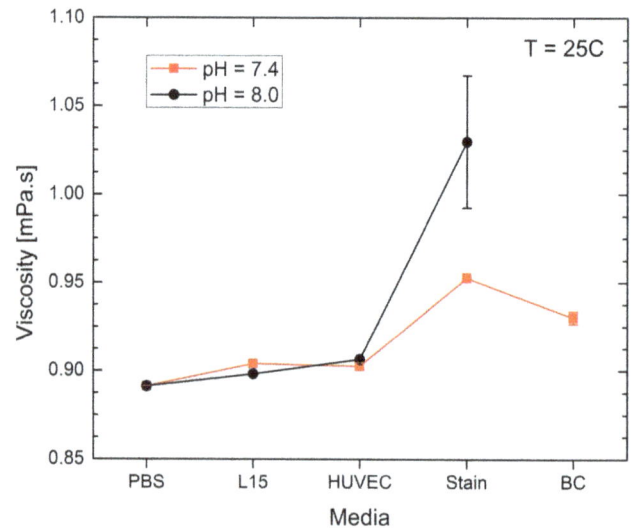

Figure 2. Viscosity measurement results. Experimentally measured viscosity data at 25°C for the various media at 7.4 and 8.0 pH (red and black curves, respectively). Note the BC media is unstable at 8.0 pH and is therefore not included.

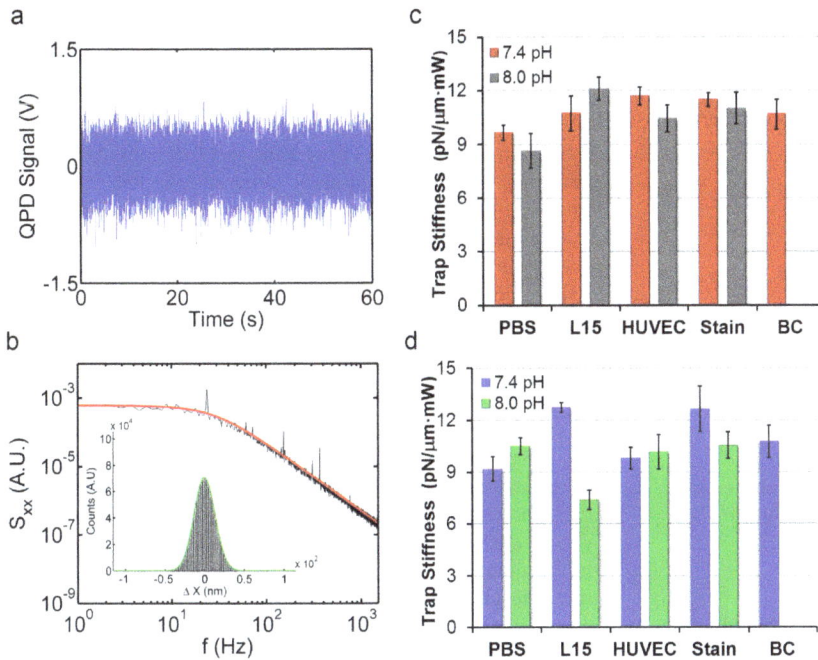

Figure 3. Plasmonic nanotweezer stiffness. Typical trap stiffness results showing (a) a time trace of the output from the quadrant photodiode and (b) the calculated power spectrum overlaid with a Lorentzian fit (red line). The inset shows a particle displacement histogram fit with a Gaussian curve. Measured trap stiffness for the biological media using plasmonic nanotweezers with (c) 425 and (d) 475-nm spaced BNAs. Error bars represent the standard error in stiffness measurements over 15 individual trials per data point.

the trapped particle experiences an approximate harmonic trapping potential, thereby validating the applicability of the trap stiffness model for plasmonic nanotweezers [15,18]. Representative power-spectral particle displacement data for all medias are available in the supporting information (Fig. S4 in File S1). Figures 3c and 3d show the stiffness of the plasmonic optical traps using 425- and 475-nm spaced BNAs, respectively. It can be seen that for all cases, the plasmonic trapping stiffness varies between $\sim 7 - 12$ pN$\cdot\mu$m$^{-1}\cdot$mW^{-1}, which is comparable to previously reported values in aqueous media [15,18]. In calculating the stiffness, we use viscosity data taken at 25 °C (Fig. 2) due to heating effects by the plasmonic nanoantennas, which for the given input intensity results in an approximately 2–5°C temperature rise of the illuminated bowties [15,26]. This indicates that the trapping performance of plasmonic nanotweezers is not significantly reduced in biologically relevant media. In most cases, there is no significant difference in the stiffness for the two pH values for a given media and the overall trend in trap stiffness follows that of the media viscosity reasonably well. This suggests that the most prominent cause for variation in trapping strength is the 5–10% variation in viscosity for the different media. Furthermore, the fact that κ does not change as a function of pH implies that free ions in solution do not significantly alter the optical forces generated by the nanoantennas.

The minimal difference in stiffness between the two array spacings (for most cases) can be understood by comparing the relative near-field intensity enhancement, $|E/E_0|^2$ where E (E_0) is the magnitude of the electric field generated by the nanoantennas (magnitude of the input electric field), and the absorption cross section data (σ_{abs}) computed via Finite-Difference Time-Domain calculations [18]. Here, the intensity enhancement and absorption cross section serve as proxies for the maximum optical force and local heating, respectively. Comparing these values, we see that

$|E/E_0|^2 \sim$ 310 (200) for the 425 (475) array, whereas $\sigma_{abs} \sim 0.0225$ μm^2 (0.015 μm^2) for the 425 (475) array. Thus, the 43% larger intensity enhancement, viz. optical force, for the 425 array is offset by a $\sim 40\%$ larger absorption cross section, which translates into higher local heating and thus enhanced Brownian perturbation to the trapped particle, i.e., lower trap stiffness. This effect has been previously observed in similar systems based on an aqueous solution [15,18], which further indicates that general performance of the plasmonic system is retained when using biological media.

It is useful to compare the trap stiffness of the plasmonic nanotweezers with a conventional optical trap. Figure 4 depicts κ for a conventional optical trap based on a 1.4-NA objective. The overall lower stiffness obtained using conventional tweezers is clear, with $\kappa \sim 3 - 5$ pN$\cdot\mu$m$^{-1}\cdot$mW^{-1}. Interestingly, conventional tweezers display a stronger variation in trap strength as the pH is varied in contrast to the plasmonic case. A potential reason for this may be that the overall lower error in conventional stiffness measurements, which itself is due to reduced heating in this case, exposes more clearly the differences in optical force for the different pH values. Notwithstanding these differences, the benefit of using plasmonic nanotweezers compared to conventional tweezers in biological media is clear: the former produces larger trapping forces with lower input powers, thereby reducing potential phototoxic effects. Furthermore, these results suggest that the apparent higher sensitivity of standard optical tweezers to specific buffers is an important design criterion when choosing a platform for optical trapping-based biological studies.

Discussion

Human physiological systems, along with almost all living things, are generally alkaline, water-based systems heavily reliant on acid-base equilibrium [31]. For this study, we choose the

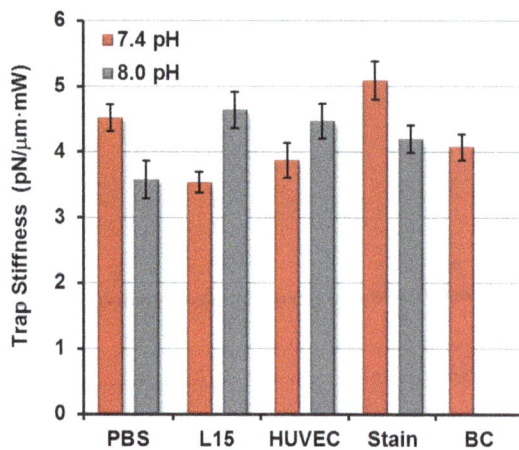

Figure 4. Conventional optical tweezer stiffness. Measured trap stiffness for the various media using a conventional optical tweezer. Error bars represent standard error in κ determination.

biologically relevant pH values 7.4 (typical) and 8.0 (maximal) due to the fact that optimal growth of mammalian cells is obtained at pH 7.2–7.4, human blood pH is regulated within the narrow range of 7.35 to 7.45, and mammalian cells are supported in the range of pH 6.6–7.8 [31,32]. The presence of bovine serum albumin in the Stain, HUVEC, and BC trapping media has the potential to alter the dynamics of the trapped particle, given that BSA readily adsorbs onto many different surfaces due to the ease with which its structure changes [33]. Evidence for such an adsorption event would manifest as a variation of the corner frequency of the trap. However, examining the raw corner frequency data (Fig. S5 in File S1) reveals no correlation between the amount of experimental variation in the corner frequency and the percentage of BSA in the Stain, HUVEC, and BC media: 2%, 5%, and 10% weight by volume, respectively. As such, BSA adsorption likely does not significantly contribute to the measured trap stiffness.

For most cases, applying the measured viscosity data to the raw corner frequency data (see supporting information) results in trap stiffness values that follow the trend in media viscosity. However, the L15 media on the 475-nm array produce an anomalously high (low) trap stiffness for the 7.4 (8.0) pH samples. Similarly, the stain media do not produce significantly larger stiffness than the other media for both plasmonic and conventional optical traps, despite having the largest overall viscosities. Possible causes of these deviations include variations in material parameters such as the

media refractive index, which alters the plasmon resonance and modifies the optical forces, or the thermal conductivity of the media, which changes the heat dissipation in the system. We note that determination of the trap stiffness from raw corner frequency data is strongly dependent on the value of $\epsilon(a,h)$, however, we apply the same value to both plasmonic and conventional trapping experiments. Moreover, variation between trapping systems is minimized by precisely controlling the axial position of particles in the conventional case. Given that $\epsilon(a,h)$ applies to the drag coefficient, it does not alter the corner frequency of the trap [34], and thus the 3–4x higher corner frequencies measured in the plasmonic case implies that the stiffness is indeed higher for plasmonic traps.

Conclusion

We have shown that the trapping performance of plasmonic nanotweezers is largely unaltered when using biologically relevant media, producing trap stiffness values comparable to nanotweezers in a water-only environment and 3–4x higher than a conventional, high-NA optical tweezer. Our study confirms the Newtonian nature of several media commonly used in biological research via high-precision rheological measurements. In doing so, we validate the applicability of standard optical force-determination schemes (e.g., stiffness or drag-force efficiency) in biological media for both conventional and plasmonic optical tweezers. Variations in trap stiffness correspond reasonably well with trends in measured viscosity data, indicating that viscosity is the main factor contributing to the trap stiffness measured in a given medium.

Supporting Information

File S1   Supporting figures. Figure S1, Digital photograph of the buffer media. Figure S2, Temperature-dependent viscosities. Figure S3, Shear-rate dependence of Stain medium. Figure S4, Representative power-spectral data. Figure S5, Raw corner frequency data.

Acknowledgments

The authors thank Shaneen Braswell for preliminary studies.

Author Contributions

Conceived and designed the experiments: BJR MTJ RHE PII KCT. Performed the experiments: BJR MTJ. Analyzed the data: BJR MTJ. Contributed reagents/materials/analysis tools: FTLM PII. Wrote the paper: BJR MTJ FTLM RHE PII KCT.

References

1. Ashkin A, Dziedzic JM, Bjorkholm JE, Chu S (1986) Observation of a single-beam gradient optical trap for dielectric particles. Opt Lett 11: 288–290.
2. Smith SB, Cui Y, Bustamante C (1996) Overstretching b-dna: the elastic response of individual double-stranded and single-stranded DNA molecules. Science 271: 795–799.
3. Abbondanzieri EA, Greenleaf WJ, Shaevitz JW, Landick R, Block SM (2005) Direct observation of base-pair stepping by RNA polymerase. Nature 438: 460–465.
4. Shank EA, Cecconi C, Dill JW, Marqusee S, Bustamante C (2010) The folding cooperativity of a protein is controlled by its chain topology. Nature 465: 637–640.
5. Block SM, Blair DF, Berg HC (1989) Compliance of bacterial flagella measured with optical tweezers. Nature 338: 514–518.
6. Martinez IA, Campoy S, Tor M, Llagostera M, Petrov D (2013) A simple technique based on single optical trap for the determination of bacterial swimming pattern. PLoS one 8: e61630.

7. Veigel C, Coluccio LM, Jontes JD, Sparrow JC, Milligan RA, et al. (1999) The motor protein myosin-I produces its working stroke in two steps. Nature 398: 530–533.
8. Watanabe TM, Iwane AH, Tanaka H, Ikebe M, Yanagida T (2010) Mechanical characterization of one-headed myosin-V using optical tweezers. PLoS one 5: e12224.
9. Juan ML, Righini M, Quidant R (2011) Plasmon nano-optical tweezers. Nature Photon 5: 349–356.
10. Righini M, Ghenuche P, Cherukulappurath S, Myroshynchenko V, de Abajo FJG, et al. (2009) Nano-optical trapping of Rayleigh particles and Escherichia coli bacteria with resonant optical antennas. Nano Lett 9: 3387–3391.
11. Liu Y, Sonek GJ, Berns MW, Tromberg BJ (1996) Physiological monitoring of optically trapped cells: assessing the effects of confinement by 1064-nm laser tweezers using microfluorometry. Biophys J 71: 2158–2167.
12. Roxworthy BJ, Ko KD, Kumar A, Fung KH, Liu GL, et al. (2012) Application of plasmonic bowtie nanoantennna arrays for optical trapping, stacking, and sorting. Nano Lett 12: 796–801.

13. Grigorenko AN, Roberts NW, Dickinson MR, Zhang Y (2008) Nanometric optical tweezers based on nanostructured substrates. Nature Photon 2: 365–370.

14. Kang JH, Kim K, Ee HS, Lee YH, Yoon TY, et al. (2011) Low-power nano-optical vortex trapping via plasmonic diabolo nanoantennas. Nat Commun 2: 582.

15. Roxworthy BJ, Toussaint Jr KC (2012) Plasmonic nanotweezers: strong influence of adhesion layer and nanostructure orientation on trapping performance. Opt Express 20: 9591–9603.

16. Zhang W, Huang L, Santaschi C, Martin OJF (2010) Trapping and sensing 10 nm metal nanoparticles using plasmonic dipole antennas. Nano Lett 10: 1006–1011.

17. Shoji T, Saitoh J, Kitamura N, Nagasawa F, Murakoshi K, et al. (2013) Permanent fixing or reversible trapping and release of DNA micropatterns on gold nanostructures using continuous-wave or femtosecond-pulsed near-infrared laser light. J Am Chem Soc 135: 6643–6648.

18. Roxworthy BJ, Toussaint Jr KC (2012) Femtosecond-pulsed plasmonic nanotweezers. Sci Rep 2: 660.

19. Huang L, Maerkl J, Martin OJF (2009) Integration of plasmonic trapping in a microfluidic environment. Opt Express 17: 6018–6024.

20. Miao X, Lin LY (2007) Trapping and manipulation of biological particles through a plasmonic platform. IEEE J Sel Top Quant Electron 13: 1655–1662.

21. Brunner D, Frank J, Appl H, Schoffl H, Pfaller W, et al. (2010) Serum-free cell culture: the serum-free media interactive online database. Altex 27: 53–62.

22. Imoukhuede PI, Popel AS (2013) Quantitative fluorescent profiling of VEGFRS reveals tumor cell and endothelial cell heterogeneity in breast cancer xenografts. Cancer Med *In Press*.

23. Imoukhuede PI, Popel AS (2011) Quantification and cell-to-cell variation of vascular endothelial growth factor receptors. Exp Cell Res 317: 955–965.

24. Imoukhuede PI, Popel AS (2012) Expression of VEGF receptors on endothelial cells in mouse skeletal muscle. PLoS One 7: e44791.

25. Johnston MT, Ewoldt RH (2013) Precision rheometry: surface tension effects on low-torque measurements in rotational rheometers. J Rheol 57: 1515–1532.

26. Baffou G, Quidant R (2013) Thermo-plasmonics: using metallic nanostructures as nano-sources of heat. Laser Photonics Rev 7: 171–187.

27. Neuman KC, Block SM (2004) Optical trapping. Rev Sci Instrum 75: 2787–2809.

28. Marchington RF, Mazilu M, Kuriakose S, Garces-Chavez V, Reece PJ, et al. (2008) Optical deflection and sorting of microparticles in a near-field optical geometry. Opt Express 16: 3712–3726.

29. Krishnan GP, Leighton, Jr DT (1995) Inertial lift on a moving sphere in contact with a plane walll in a shear flow. Phys Fluids 7: 2538–2545.

30. Ploschner M, Mazilu M, Krauss TM, Dholakia K (2010) Optical forces near a nanoantenna. J. Nanophotonics 4: 041570.

31. Kellum JA (2000) Determinants of blood pH in health and disease. Crit. Care 4: 6–14.

32. Burckhardt P, Dawson-Hughes B, Weaver C (2010) Nutritional Influences on Bone Health. London: Springer, 167–171 pp.

33. Carter DC, Ho JX (1994) Structure of serum albumin. Adv. Protein. Chem. 45: 153–203.

34. Tolic-Norrelykke IM, Berg-Sorensen K, Flyvbjerg H (2004) Matlab program for precision calibration of optical tweezers. Comput. Phys. Commun. 159, 225–240.

9

Generic Delivery of Payload of Nanoparticles Intracellularly via Hybrid Polymer Capsules for Bioimaging Applications

Haider Sami[1], Auhin K. Maparu[2], Ashok Kumar[1]*, Sri Sivakumar[2]*

1 Department of Biological Sciences and Bioengineering, Indian Institute of Technology Kanpur, Kanpur, Uttar Pradesh, India, 2 Unit of Excellence on Soft Nanofabrication, Department of Chemical Engineering, Indian Institute of Technology Kanpur, Kanpur, Uttar Pradesh, India

Abstract

Towards the goal of development of a generic nanomaterial delivery system and delivery of the 'as prepared' nanoparticles without 'further surface modification' in a generic way, we have fabricated a hybrid polymer capsule as a delivery vehicle in which nanoparticles are loaded within their cavity. To this end, a generic approach to prepare nanomaterials-loaded polyelectrolyte multilayered (PEM) capsules has been reported, where polystyrene sulfonate (PSS)/polyallylamine hydrochloride (PAH) polymer capsules were employed as nano/microreactors to synthesize variety of nanomaterials (metal nanoparticles; lanthanide doped inorganic nanoparticles; gadolinium based nanoparticles, cadmium based nanoparticles; different shapes of nanoparticles; co-loading of two types of nanoparticles) in their hollow cavity. These nanoparticles-loaded capsules were employed to demonstrate generic delivery of payload of nanoparticles intracellularly (HeLa cells), without the need of individual nanoparticle surface modification. Validation of intracellular internalization of nanoparticles-loaded capsules by HeLa cells was ascertained by confocal laser scanning microscopy. The green emission from Tb^{3+} was observed after internalization of $LaF_3:Tb^{3+}$(5%) nanoparticles-loaded capsules by HeLa cells, which suggests that nanoparticles in hybrid capsules retain their functionality within the cells. *In vitro* cytotoxicity studies of these nanoparticles-loaded capsules showed less/no cytotoxicity in comparison to blank capsules or untreated cells, thus offering a way of evading direct contact of nanoparticles with cells because of the presence of biocompatible polymeric shell of capsules. The proposed hybrid delivery system can be potentially developed to avoid a series of biological barriers and deliver multiple cargoes (both simultaneous and individual delivery) without the need of individual cargo design/modification.

Editor: Sangaru Shiv Shankar, King Abdullah University of Science and Technology, Saudi Arabia

Funding: This work is funded by IIT Kanpur (IIT/CHE/20080352), Department of Biotechnology (DBT/CHE/20100304), Department of Science and Technology and India-UK Science Bridge (DST/BSBE/20090218). HS also acknowledges a senior research fellowship under the DBT Junior Research Fellowship (DBT-JRF) programme. The funders had no role in study design, data collection and analysis, decision to publish, or preparation of the manuscript.

Competing Interests: The authors have declared that no competing interests exist.

* E-mail: ashokkum@iitk.ac.in (AK); srisiva@iitk.ac.in (SS)

Introduction

Nanomaterials have attracted a great deal of interest in diverse fields, in particular bioimaging, drug delivery, and biosensing [1]. Different kinds of nanomaterials have been used in a variety of imaging applications such as X-ray computed tomography (noble metal nanoparticles-Au, Ag), near-IR optical imaging (noble metal nanoparticles-Au, Ag), fluorescence based imaging (semiconductor quantum dots, lanthanide-doped nanoparticles), magnetic resonance imaging (Iron and gadolinium based materials) etc [2]. For diagnostic and therapeutic applications in biomedicine, nanoparticles must overcome a series of biological barriers (degradation/aggregation in body fluids, phagocytic clearance by reticuloendothelial system, crossing of the plasma membrane etc) so as to ultimately perform their desired function [2]. Designing of nanomaterials to provide them with the ability of evading these hurdles and achieving desired localization, is key for exploiting their true potential for biomedical applications. For instance, intracellular delivery of nanoparticles can be achieved by engineering the physical (e.g. size and shape) and surface properties (charge, chemical and biomoties) of the nanoparticles [3–5]. However, this may require multiple step processes to achieve the required physical/surface properties of nanoparticles which can lead to aggregation and reduction in the desired properties (e.g. optical, magnetic, electrical, etc) of nanoparticles. Additionally, these multiple step processes can be different for different types of nanoparticles. To circumvent these issues, it is desired to have a generic delivery system to deliver the nanoparticle intracellularly without any individual nanoparticle surface modification. Moreover, the delivery system can serve the purpose of avoiding direct contact of the nanoparticles with the body fluids, and thus avoid degradation/aggregation of nanoparticles and their clearance by phagocytic cells. Fabrication of a generic nanomaterial delivery system encompasses different tasks namely, employment of a generic method to load different kinds of nanomaterials in a delivery vehicle, which in itself is capable/designed of avoiding physiological barriers and using this nanomaterial-loaded vehicle to finally deliver the cargo intracellularly. To this end, we have explored the prospect of loading variety of nanomaterials inside the cavity of polymer capsules and use them as generic nanomaterial delivery systems without individual nanoparticle design/surface modification.

The choice of polymer capsules as nanomaterial delivery vehicles is based on their increasingly potential applications such as drug delivery, biosensing, bioimaging, catalysis and biomedicine [6–12]. Additionally, these PEM capsules can be used as multifunctional biovehicle because, their physico-chemical properties (e.g. size, composition, porosity and surface functionality) can be easily tailored and controlled [13]. Integration of nanoparticles with polymer capsules is desirable for multifunctional applications (release of cargo, enhance mechanical properties, etc) [14,15], and can be achieved by incorporating nanoparticles within the two available compartments in PEM capsules, namely- shell and the cavity. To this end of loading nanoparticles in the shell of polymer capsules, few reports are available on the development of multifunctional PEM capsules which possess optical/magnetic nanoparticles sandwiched between the PEM layers along with the prospect of filling the cavity with desirable cargo [16–19]. Koo *et al.* have reported loading of gold nanorods on the surface of polymer capsules [20]. However, this approach of loading of 'pre-synthesized nanoparticles' within the shell has limitations such as lesser loading of nanoparticles [21], possible release of nanoparticles before reaching the targeted site due to disruption of few layers of polymers, and change in the property of nanoparticles [22]. Recently, Caruso and his co-workers have shown the synthesis of magnetic nanoparticles/QDs-loaded PEM capsules by using preformed nanoparticles in the emulsion template [23]. Even loading of pre-synthesized nanoparticles in template may also lead to change in property of nanoparticles as the nanoparticles will be exposed to the process of capsule synthesis (e.g. emulsification etc) and template removal. To circumvent these issues of using 'pre-synthesized nanoparticles' either in the shell or cavity, the nanoparticles can be loaded inside the cavity of polymer capsules, by synthesizing them within the capsule interior. Shchukin *et al.* have reported synthesis of YF_3 (for yttrium recovery from aqueous solutions) and rare earth phosphates nanoparticles in polyelectrolyte capsules [24,25]. However, to our knowledge, there is no report available on a general approach to 'synthesize' variety of nanomaterials inside the 'cavity' of polymer capsules and generic delivery of nanoparticles via polymer capsules.

The proposed work attempts to address the challenge of fabrication of generic nanomaterial delivery system by dealing with it at two different steps. Firstly, we have employed a general and versatile method to synthesize variety of nanomaterials inside the cavity of polymer capsules by templating polystyrene sulfonate (PSS)/polyallylamine hydrochloride (PAH) and cross-linked PAH polymer capsules as micro/nanoreactors. Secondly, nanoparticles-loaded capsules were interacted with human cervical cancer cells (HeLa) to deliver the 'as prepared' nanoparticles in a functional state intracellularly, without needing individual nanoparticle design/surface modification. In this report, the microvolume of polymer capsules was exploited to synthesize gold, silver, cadmium sulfide and lanthanide ion-doped nanoparticles (LaF_3:Tb^{3+}(5%), $LaVO_4$:Eu^{3+}(5%), GdF_3:Tb^{3+}(5%)), stabilized with citrate ligand inside the PSS/PAH capsule. In addition, different shapes of gold nanostructures (nanorods, nanoprisms, and multifaceted nanostructures) and co-loading of two types of nanoparticles within the same polymer capsule have been demonstrated. We have also demonstrated the co-loading of protein (RITC-BSA; Rhodamine B isothiocyanate-bovine serum albumin) along with gold nanoparticles for potential stimuli-responsive (e.g. laser) drug delivery applications. Furthermore, different sizes of polymer capsules (400 nm, 1 μm, and 5 μm) have been used as nano/microreactors to synthesize gold nanoparticles and to prove the generality and versatility of our method. Interaction of these nanoparticles-loaded PSS/PAH capsules with HeLa cells was examined for uptake kinetics and *in vitro* cytotoxicity studies, thereby demonstrating a generic platform for delivery of a variety of nanoparticles to cells. We have demonstrated synthesis of a) metal nanoparticles (Au and Ag) in capsules as candidates for potential micro-CT imaging applications and laser induced release of cargo, b) lanthanide ion doped inorganic nanoparticles (LaF_3:Tb^{3+}, $LaVO_4$:Eu^{3+} and GdF_3:Tb^{3+}) and CdS nanoparticles as candidates for fluorescence based imaging applications and c) GdF_3 nanoparticles as MRI contrast agents. One of the nanoparticles-loaded capsules was used to demonstrate fluorescence based imaging in HeLa cells *in vitro*.

This approach has several advantages from the standpoint of both generic synthesis and generic delivery of nanomaterials: (i) this method can be applied to load variety of nanoparticles (e.g. Au, Ag, CdS, LaF_3, GdF_3, $LaVO_4$, etc.) inside the polymer capsule along with loading of therapeutic molecules (e.g. RITC-BSA); (ii) different kinds of nanoparticles can be simultaneously loaded inside the capsule (e.g. Au and $LaVO_4$); (iii) different shapes of nanomaterials can be loaded (e.g. gold nanorods, nanoprisms, multifaceted nanostructures) (iv) size, composition, and morphology of capsules/nanoparticles can be easily tailored and controlled; (v) the number of nanoparticles inside the polymer capsules can be easily controlled by varying the concentration of nanoparticle precursors; (vi) variety of nanoparticles can be delivered intracellularly without individual nanoparticle surface modification (e.g. 'as prepared' Au and LaF_3:Tb^{3+} nanoparticles were delivered intracellularly to HeLa cells via polymer capsules); (viii) prevention of nanoparticle exposure to body fluids, avoidance of any nanoparticle release/degradation/change in property before reaching the target, evading direct contact of nanoparticles with cells and generic delivery of a payload of nanoparticles without the step of nanomaterial surface functionalization/design (of every single nanoparticle; required for all types of nanoparticles and is different for different types of nanoparticles) needed for delivery and avoidance of biological barriers.

These hybrid polymer capsules can be potentially used as active targeting (surface modification with ligands/antibodies) as well as passive targeting (based on their size, ~300–700 nm) [26,27]. Moreover, these nanomaterials-loaded PEM capsules can be potentially used as multimodal bioimaging agents [28,29] (magnetic resonance imaging (MRI), X-ray computed tomography (CT), fluorescence imaging, and nuclear imaging), drug delivery vehicles and biosensors. Loading of multiple cargoes (simultaneous loading of two types of nanoparticles/simultaneous loading of nanoparticles and therapeutics) within the capsule can be explored to design multifunctional vehicles for biomedical applications [30]. In addition, the residence time of these nanoparticles-loaded polymer capsules can potentially be longer in the affected tissues compared to bare nanoparticles (size <20 nm) which can facilitate better imaging of affected tissues. One has to bear in mind that the residence time of nanoparticles in the affected tissues depends on their size, shape, and surface [27].

Results and Discussion

Generic Method to Prepare Nanoparticles-Loaded Capsules

Scheme S1 shows the schematic representation of synthesis of nanomaterials-loaded polymer capsules. The PSS/PAH capsules were formed via layer-by-layer (LbL) assembly of polymers on monodisperse silica particles (size ~5 μm) followed by removal of the core by etching with HF. Further, these capsules were incubated in nanoparticle precursor salts solution along with citrate ions as ligand. The excess nanoparticles formed outside the polymer capsules were removed by washing thrice with water. All the nanoparticles-loaded polymer capsules were easily dispersible in PBS buffer (pH~7.2) as can be observed from the digital image (Scheme S1 inset). Figure 1a demonstrates the UV-Vis absorption spectra of Au and Ag nanoparticles-loaded PSS/PAH capsules. The appearance of absorbance peaks due to surface plasmon resonance at ~524 nm (green curve) and ~400 nm (red curve) suggest the formation of gold and silver nanoparticles, respectively. Furthermore, the absorbance of nanoparicles-loaded capsules matches with the absorbance of blank nanoparticles (Figure S1). Figure 1b shows the photoluminescence (PL) emission spectra of LaF$_3$:Tb^{3+}(5%), LaVO$_4$:Eu^{3+}(5%), and GdF$_3$:Tb^{3+}(5%) nanoparticles-loaded PSS/PAH capsules by excitation with laser. The emission bands (black and red curve, Figure 1b) around 544, 584 and 619 nm are assigned to $^5D_4$ to $^7F_5$, $^7F_4$, and $^7F_3$ transitions, respectively of Tb^{3+} ions. In addition, the average life time of Tb^{3+} ions in LaF$_3$ and GdF$_3$ are 1.7 ms and 2.8 ms, respectively, which clearly suggests that the Tb^{3+} ions are doped in an inorganic matrix.[31] The emission bands around 591 nm ($^5D_0$ to $^7F_1$), 615 nm ($^5D_0$ to $^7F_2$), and 696 nm ($^5D_0$ to $^7F_4$) are assigned to Eu^{3+} ions. Additionally, the average lifetime of Eu^{3+} is reported to be 5.3 ms. (Table S2 and Figure S2) [31]. We note that, the background emission in PL emission spectra arises from blank PSS/PAH capsules (Figure S3). Furthermore, the optical properties of the nanoparticles-loaded capsules match with the blank nanoparticles (Figure S4), which clearly suggested that properties of nanoparticles do not change when trapped inside the polymer capsule. Over all, the nanoparticles-loaded PSS/PAH capsules show excellent optical properties warranting their potential for various bioimaging applications.

The size and surface morphology of nanoparticles-loaded PSS/PAH capsules have been characterized by transmission electron microscopy (TEM) and scanning electron microscopy (SEM). Figure 2 demonstrates the TEM image of Au, Ag, LaVO$_4$:Eu^{3+}(5%), LaF$_3$:Tb^{3+}(5%), CdS, and GdF$_3$:Tb^{3+}(5%) nanoparticles-loaded PSS/PAH capsules. The contrast in the TEM image clearly shows that the interior of the polymer capsules are loaded with nanoparticles and the size/shape of the nanoparticles are uniform. As can be observed in the TEM images, the cavity has more contrast when compared to the walls of the capsule, which clearly suggests that the nanoparticles are inside the polymer capsules (if the particles were in the wall, the contrast of the wall of the polymer capsules would have been more as compared to cavity). High loading of nanoparticles can be observed in figure 2a and 2b, showing gold and silver nanoparticles in the size range of 5–10 nm and 5–15 nm respectively. We note that the aggregation observed in the TEM images is due to the drying effect during the sample preparation. To validate the formation of nanoparticles inside polymer capsules, the nanoparticles-loaded polymer capsules were subjected to SEM analysis. The surface morphology of blank capsules and gold nanoparticles-loaded polymer capsules are similar, which clearly suggests that nanoparticles are not on the surface of the PEM capsules (Figure 3). The synthesis of different kinds of nanoparticles inside the polymer capsule clearly proves the generality and versatility of our approach.

Figure 4a, 4b and 4c shows TEM images of PSS/PAH capsules loaded with different shapes of gold nanostructures, namely gold nanoprisms, gold nanorods and multifaceted gold nanostructures respectively. Figure 4b and 4c suggests that the formation of gold nanorods and multifaceted nanostructures inside the polymer capsule. It was also observed that the capsules possess seed gold particles along with gold nanoprisms. We note that Murphy *et al.* have reported that the synthesis of uniform size and shape of gold nanorods is a challenging process; however, they have improved the yield of nanorods by centrifugation processes [32]. Selective synthesis of different shapes of nanostructures by templating polymer capsules will be further investigated to evade the separation steps. This shows the versatility of the approach in terms of control on shape and adaptability of seed mediated growth of nanoparticles within the capsule. Figure S5a and S5b shows the UV-Vis absorption spectra of PSS/PAH capsules loaded with gold nanorods and multifaceted gold nanostructures respectively. The absorbance in the NIR region further suggests the formation of nanorods and multifaceted nanostructures. This clearly suggests that these capsules loaded with different shapes of nanomaterials can further widen their applications such as stimuli (near infra red (NIR)) responsive drug delivery, photothermal ablation therapy, etc [16,33].

Co-loading of two types of nanoparticles inside PSS/PAH capsules was done by loading the capsules with LaVO$_4$:Eu^{3+}(5%) nanoparticles first and then using these LaVO$_4$:Eu^{3+}(5%) nanoparticles-loaded capsules for synthesis of gold nanoparticles. As can be observed from TEM image (Figure 4d), the polymer capsules were successfully loaded with two types of particles having different electron densities. The lighter particles (marked with black arrow) are LaVO$_4$:Eu^{3+}(5%) nanoparticles and gold nanoparticles are the darker particles (marked with white colored

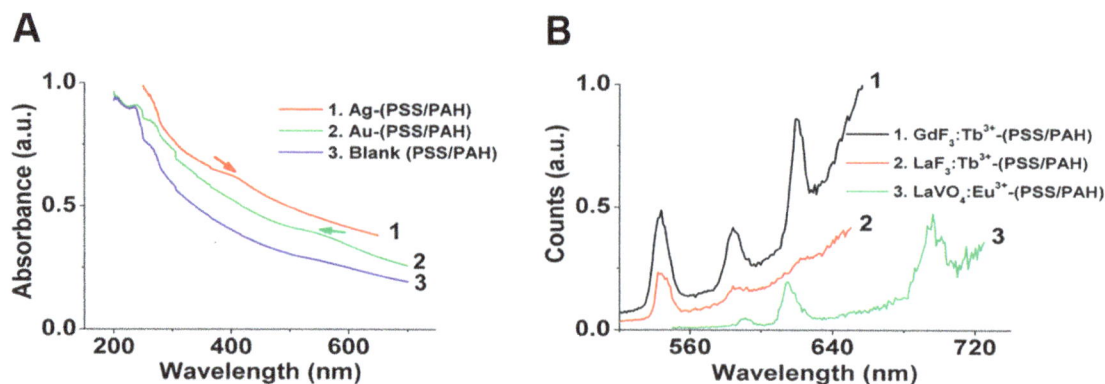

Figure 1. Optical properties of nanoparticles-loaded PSS/PAH capsules. (a) UV-Vis absorbance spectra of (1) Ag and (2) Au nanoparticles-loaded PSS/PAH capsules and (3) blank PSS/PAH capsules; (b) photoluminescence emission spectra of (1) GdF$_3$:Tb^{3+}(5%), (2) LaF$_3$:Tb^{3+}(5%) and (3) LaVO$_4$:Eu^{3+}(5%) nanoparticles-loaded PSS/PAH capsules [Excitation wavelength used was 488 nm for GdF$_3$:Tb^{3+}(5%) and LaF$_3$:Tb^{3+}(5%) and 465 nm for LaVO$_4$:Eu^{3+}(5%)].

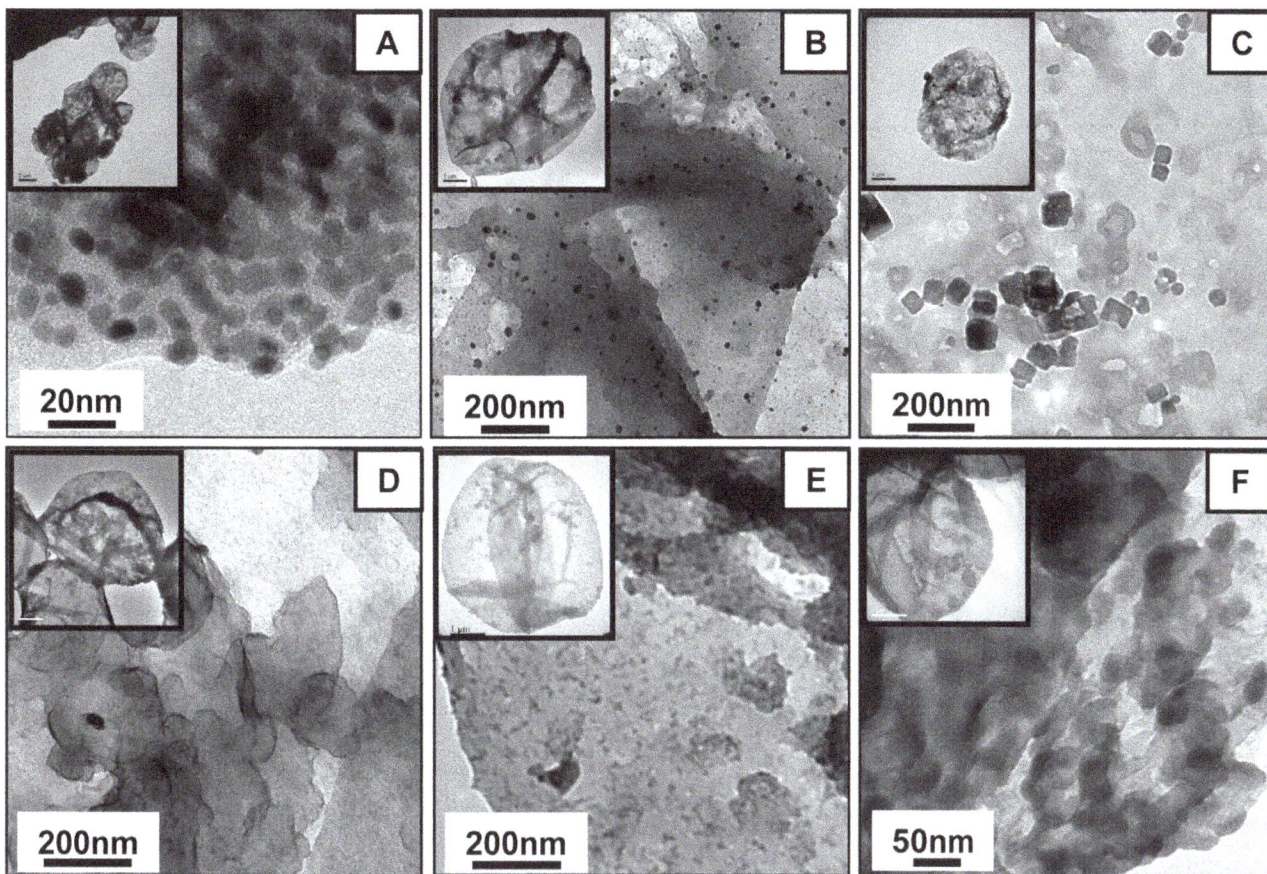

Figure 2. Validation of nanoparticle presence inside the PSS/PAH capsules. Representative TEM image of (a) Au, (b) Ag, (c) LaVO$_4$:Eu^{3+}(5%), (d) LaF$_3$:Tb^{3+}(5%), (e) CdS, and (f) GdF$_3$:Tb^{3+}(5%) nanoparticles-loaded PSS/PAH capsules. (Inset in each figure shows the respective nanoparticles-loaded PSS/PAH capsule at lower magnification.).

arrow). Figure 5c shows the energy-dispersive x-ray (EDX) spectroscopy of Au and LaVO$_4$:Eu^{3+}-loaded polymer capsules which indicates the presence of La and Au. This further confirms the co-loading of LaVO$_4$:Eu^{3+}(5%) and gold nanoparticles inside the PSS/PAH capsules. The PSS/PAH capsules co-loaded with LaVO$_4$:Eu^{3+}(5%) and gold nanoparticles were further investigated for their optical properties to ensure the functionality of the two types of nanoparticles within the capsule interior. Figure 5a shows

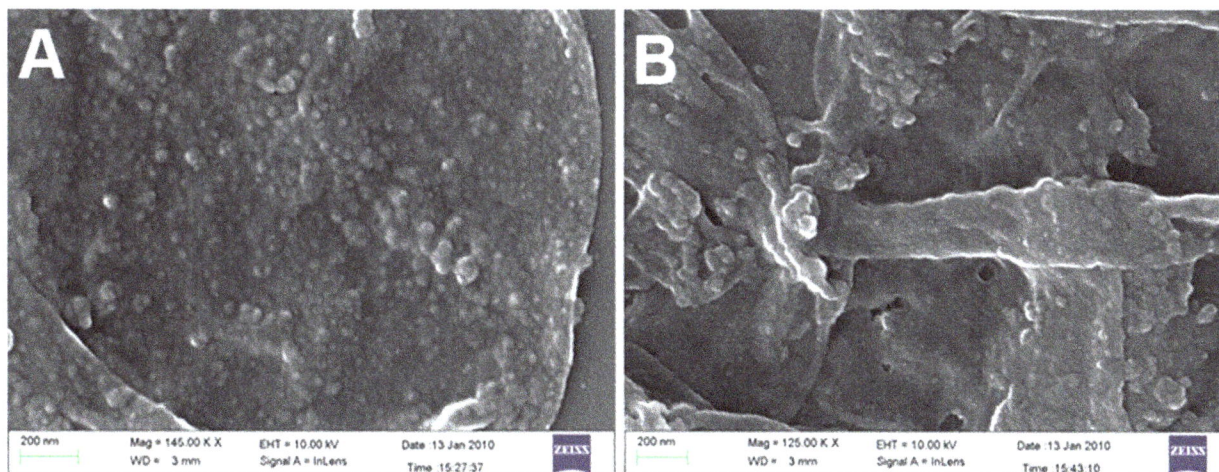

Figure 3. Surface characterization of nanoparticles-loaded PSS/PAH capsules. Scanning electron micrograph of (a) Au nanoparticles-loaded PSS/PAH capsules and (b) blank PSS/PAH capsules.

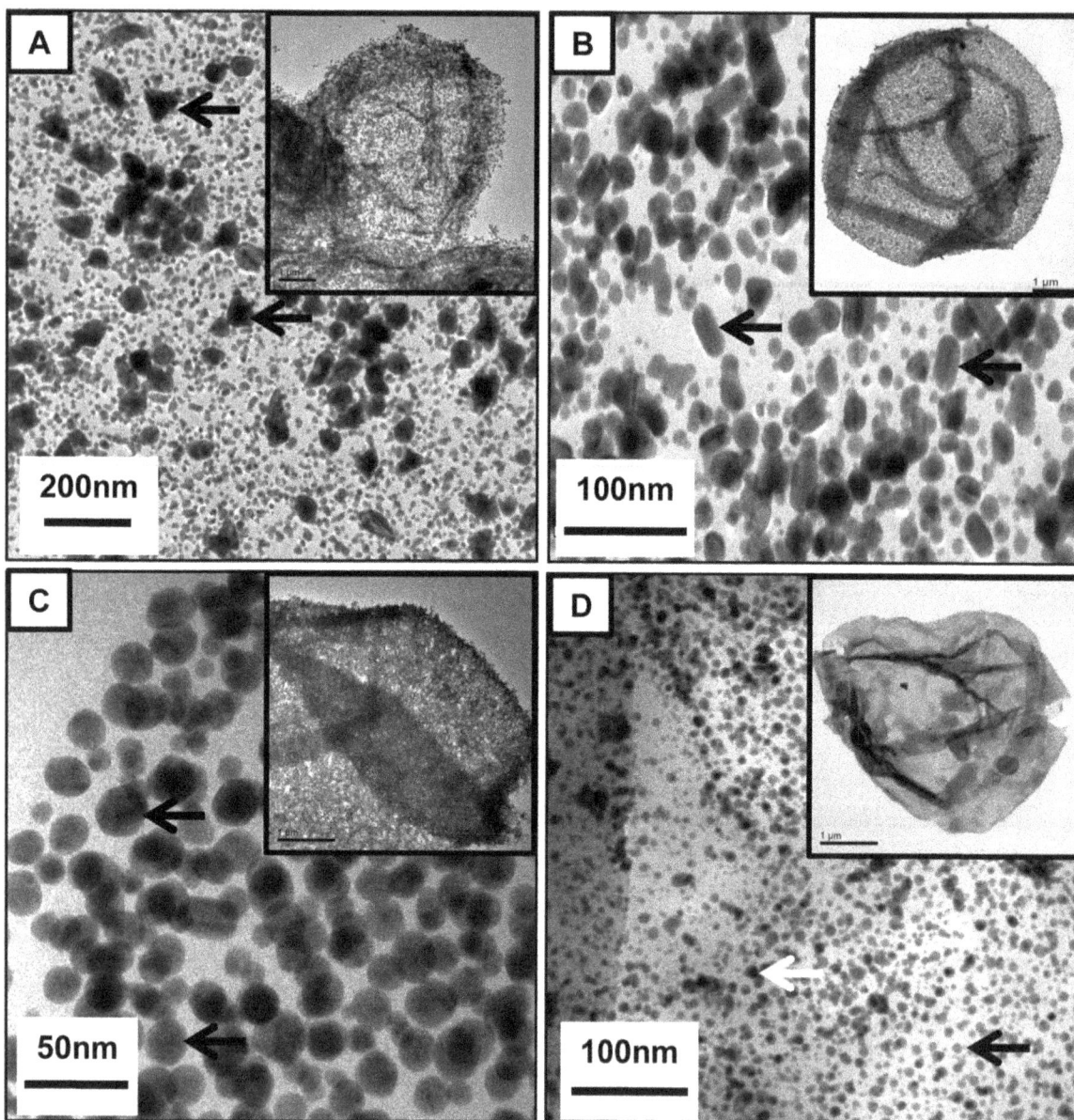

Figure 4. PSS/PAH capsules loaded with different shapes of nanoparticles and co-loaded with two types of nanoparticles. TEM image of PSS/PAH capsules loaded with (a) gold nanoprism, (b) gold nanorods, (c) multifaceted gold nanostructures and (d) Au and LaVO$_4$:Eu^{3+}(5%) nanoparticles (co-loaded, black and white arrow show LaVO$_4$:Eu^{3+}(5%) and Au nanoparticles respectively.).

the photoluminescence emission spectrum of PSS/PAH capsules co-loaded with LaVO$_4$:Eu^{3+}(5%) and gold nanoparticles, depicting the presence of emission bands around 591 nm ($^5D_0$ to $^7F_1$), 615 nm ($^5D_0$ to $^7F_2$), and 696 nm ($^5D_0$ to $^7F_4$), which are assigned to Eu^{3+} ions. The absorbance peak at 529 nm in Figure 5b indicates the formation of gold nanoparticles in the co-loaded capsules. We note that the life time of Eu^{3+} ions (Table S2) is less in co-loaded sample (LaVO$_4$:Eu^{3+}(5%) and Au; 1.6 ms) when compared to LaVO$_4$:Eu^{3+}(5%)(5.3 ms)-loaded polymer capsule. We depict that the reduction in life time may be due to reduction in sizes of LaVO$_4$ nanoparticles.

To demonstrate further the generality of our method three different sized polymer capsules were employed for synthesis of gold nanoparticles, namely 5 μm PSS/PAH capsules, 1 μm PSS/PAH capsules and cross-linked PAH nanocapsules ($\sim$ 400 nm).

Gold nanoparticles were successfully synthesized inside all the three types of capsules as can be observed in Figure 2a, 6a and 6b. The PSS/PAH capsules were co-loaded with RITC-BSA (model drug) and Au nanoparticles to exhibit the multifunctionality of the proposed hybrid materials. To confirm the loading of RITC-BSA, the loaded capsules were subjected to fluorescence microscopy (Figure S6), where the red emission corresponds to the presence of RITC-BSA in the interior of the polymer capsule. These RITC-BSA loaded capsules were then used as templates to synthesize Au nanoparticles by the same method as described above.

Versatility and generality of the approach was emphasized by synthesizing different shaped nanoparticles (nanorods, nanoprisms, and multifaceted nanostructures), co-loading of two types of nanoparticles and co-loading of protein along with Au nanoparticles inside PSS/PAH capsules. Nanoparticle loading inside these

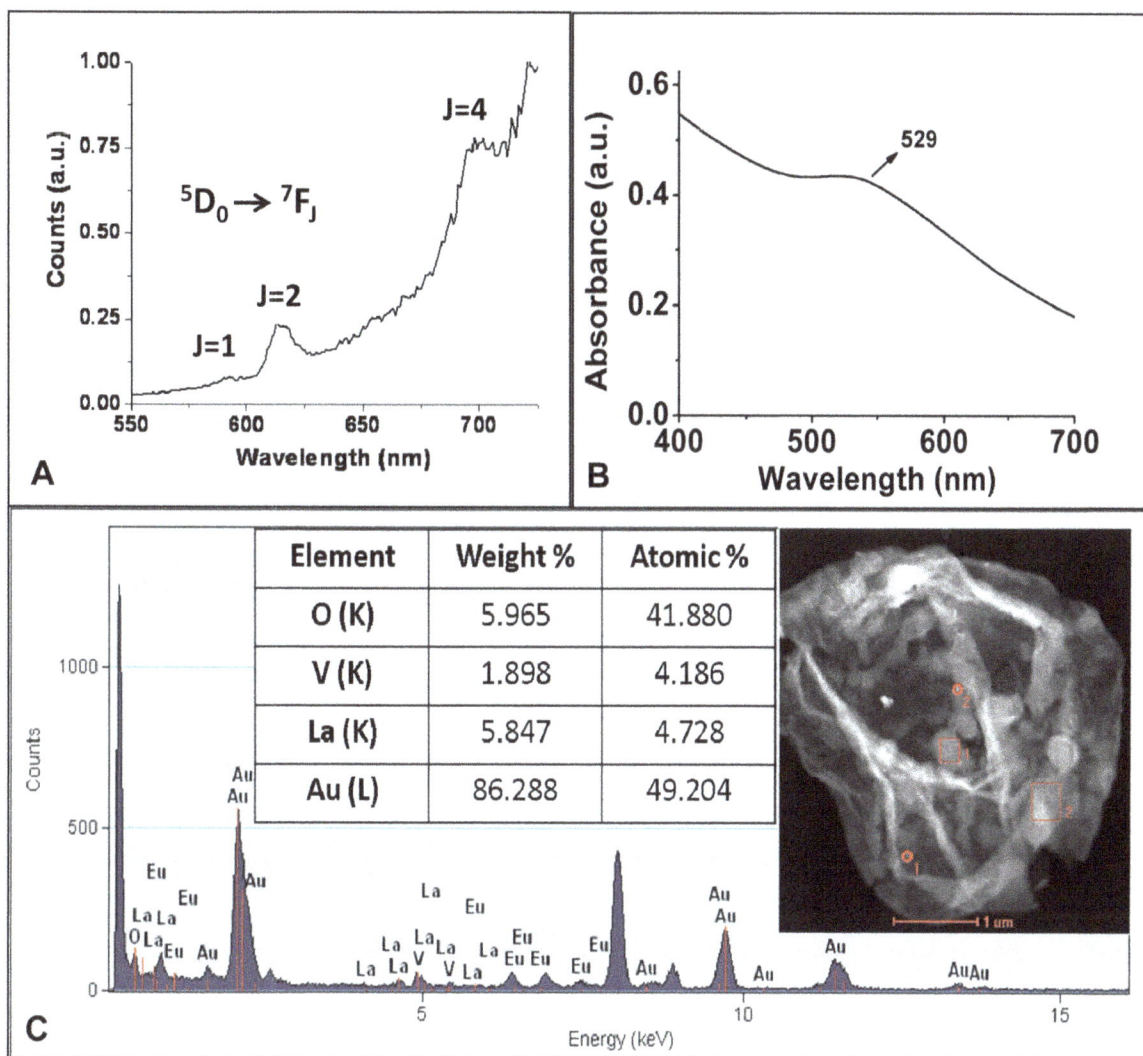

Element	Weight %	Atomic %
O (K)	5.965	41.880
V (K)	1.898	4.186
La (K)	5.847	4.728
Au (L)	86.288	49.204

Figure 5. Optical properties of PSS/PAH capsules co-loaded with two types of nanoparticles. (a) Photoluminescence emission spectrum, (b) UV-Vis absorption spectrum and (c) EDX analysis of PSS/PAH capsules co-loaded with Au and LaVO$_4$:Eu^{3+}(5%) nanoparticles.

capsules was shown by employment of a general procedure using citrate as ligand for all the nanoparticles loaded (Au, Ag, CdS, LaF$_3$, GdF$_3$, LaVO$_4$). Loading of gold nanoparticles inside different sized capsules (PSS/PAH capsules [5 μm, 1 μm] and cross-linked PAH nanocapsules [~400 nm]) further highlights the generality of our approach.

Delivery of Nanoparticles Payload to Cancer Cells

In order to investigate the delivery capabilities of these nanoparticles-loaded capsules, they were subjected to interaction with HeLa cells. Figure 7 shows the uptake kinetics of LaF$_3$:Tb^{3+}(5%) nanoparticles-loaded PSS/PAH capsules as a function of incubation time with HeLa cells, where the uptake was followed by fluorescence microscopy and scanning electron microscopy. To visualize the uptake by fluorescence microscopy, LaF$_3$:Tb^{3+}(5%) nanoparticles were loaded in RITC-labeled capsules (red emission) so as to locate the nanoparticles-loaded capsules during the internalization process. The uptake of nanoparticles-loaded capsules by HeLa cells has started around 2–5 hours as binding events [binding of nanoparticles-loaded RITC-labeled capsules (red) to cells (green)] can be observed

(Figure 7a, 7b, 7d and 7e) during that durations. Moreover, internalization events can be observed around 8 hours (Figure 7c and 7f) suggesting that the internalization process has initiated around 2–5 hours and is probably near completion around 8 hours of treatment. Further, uptake of LaF$_3$:Tb^{3+}(5%) and Au nanoparticles-loaded capsules by HeLa cells was visualized by confocal laser scanning microscopy (CLSM) after 16 hours of incubation, where it can be observed that most of the cells (green emission) have internalized the nanoparticles-loaded capsules (red emission from RITC of capsule) (Figure 8a). Moreover, it can also be seen that the internalized capsules are collapsed (Figure 8a inset) suggesting their uptake (also observed in other reports) [34]. CLSM imaging of various Z stacks of cells (Figure S7) after the incubation with nanoparticles-loaded capsules clearly suggests that the nanoparticles-loaded polymer capsules were efficiently internalized by the cells, thus delivering the payload of nanoparticles to the cells. Efficiency of uptake of Au nanoparticles-loaded capsules and blank PSS/PAH capsules by HeLa cells was investigated by flow cytometry to inspect differences in uptake. There was no significant difference in efficiency of uptake of nanoparticles-loaded capsules and blank capsules by HeLa

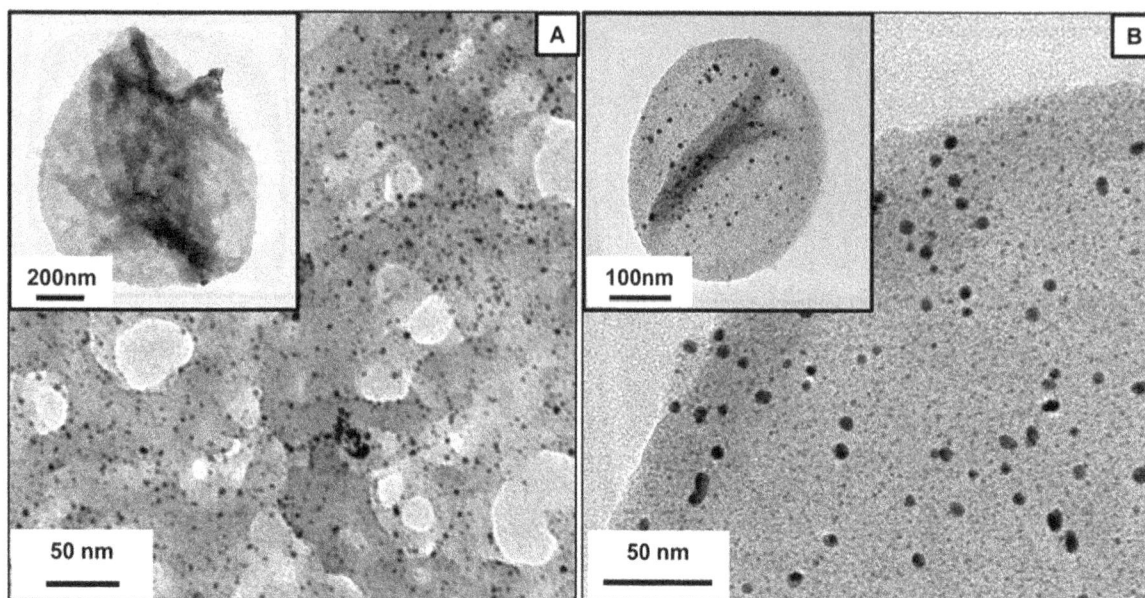

Figure 6. Nanoparticles in different sizes of capsules. TEM image of Au nanoparticles loaded (a) 1 μm sized PSS/PAH capsules and (b) cross-linked PAH nanocapsules (~400 nm).

cells as suggested by the FACS analysis (Figure 8b), where the average percentage positive cells was 72.85% and 79.25% for nanoparticles-loaded capsules and blank capsules respectively. This suggests that the process of loading of nanoparticles inside these PSS/PAH capsules does not affect the properties of the capsules required for uptake by the cells. The percentage positive cells were estimated from the 2D dot plots of the events corresponding to association/uptake of the nanoparticles-loaded polymer capsules by the HeLa cells (Figure S8). Events with high forward scattering and high fluorescence intensity were assigned to capsules/nanoparticles-loaded capsules bound/internalized with the cells. Events with low forward scattering and high fluorescence intensity correspond to free capsules/nanoparticles-loaded capsules. Only cells show high forward scattering with low fluorescence intensity.

Post uptake, examination of functionality of the nanoparticles is imperative from application point of view. Investigation of emission from Tb^{3+} from the internalized $LaF_3:Tb^{3+}(5\%)$ nanoparticles-loaded PSS/PAH capsules by HeLa cell was needed, to assess the optical properties of the nanoparticles post uptake process and while inside the cell. HeLa cells were imaged by CLSM post uptake of $LaF_3:Tb^{3+}(5\%)$ nanoparticles-loaded PSS/RITC-PAH capsules and the characteristic green emission from Tb^{3+} was observed (Figure 9) confirming the presence of Tb^{3+} doped nanoparticles in a functional state inside the cell. The green emission (from Tb^{3+}) localized with the red emission from RITC-PAH of the polymer wall of the capsules thus confirming the presence of nanoparticles inside the capsules in which they were loaded.

In vitro cytotoxicity of $LaF_3:Tb^{3+}$ (5%), $GdF_3:Tb^{3+}$ (5%), and Au nanoparticles-loaded PSS/PAH capsules was studied on HeLa cells by MTT assay, so as to investigate the effect of internalization of these nanoparticles-loaded capsules on cell viability. It was observed that cell proliferation percentages for Au, $LaF_3:Tb^{3+}(5\%)$ and $GdF_3:Tb^{3+}(5\%)$ nanoparticles-loaded polymer capsules were nearly similar to that of non-treated cell control and blank capsules (Figure 10). It clearly suggests that nanoparticles-loaded polymer

capsules show less/no cytotoxicity on HeLa cells, thus confirming their biocompatibility.

Thus a generic way of delivering Au nanoparticles and $LaF_3:Tb^{3+}(5\%)$ nanoparticles to HeLa cells was shown via PSS/PAH capsules and validated by CLSM and flow cytometry. We note that both the types of nanoparticles have been delivered intracellularly without the need of any surface modification for individual nanoparticles. Furthermore, this strategy can be extended for delivery of other nanoparticles also, as the method of loading nanoparticles inside the capsules employs a generic approach. Additionally, the fluorescence emission from Tb^{3+} was observed even after the uptake of Tb^{3+} doped nanoparticles-loaded PSS/PAH capsules by the cell, indicating the presence of nanoparticles in their functional state inside the cell and showing their potential as bioimaging agents. Further it was observed that these nanoparticles-loaded capsules show less or no cytotoxicity to the cells post uptake (in comparison to blank capsules or untreated cells), suggesting their *in vitro* compatibility with biological systems.

In conclusion, a general and facile delivery approach to deliver variety of nanoparticles to cells without individual nanoparticles surface modification has been demonstrated. Loading of different kinds of nanoparticles (Au, Ag, CdS, $LaF_3:Tb^{3+}$, $GdF_3:Tb^{3+}$, $LaVO_4:Eu^{3+}$) inside the polymer capsules supports the generality and versatility of the method. Additionally, different morphology of gold nanostructures (nanorods, nanoprisms and multifaceted nanostructures) and co-loading of Au and $LaVO_4:Eu^{3+}$ nanoparticles further demonstrates the versatility of our approach. Delivery of $LaF_3:Tb^{3+}$ and Au nanoparticles to HeLa cells thorough polymer capsules as a carrier have been shown. This clearly suggests that this approach can be used as a general method to deliver a payload of different types of nanoparticle without the need of individual nanoparticle surface modification. Furthermore, these nanoparticles-loaded polymer capsules show high biocompatibility and efficient internalization within the cells. Green fluorescence emission from internalized $LaF_3:Tb^{3+}$ nanoparticle-loaded polymer capsules in HeLa cells demonstrates their

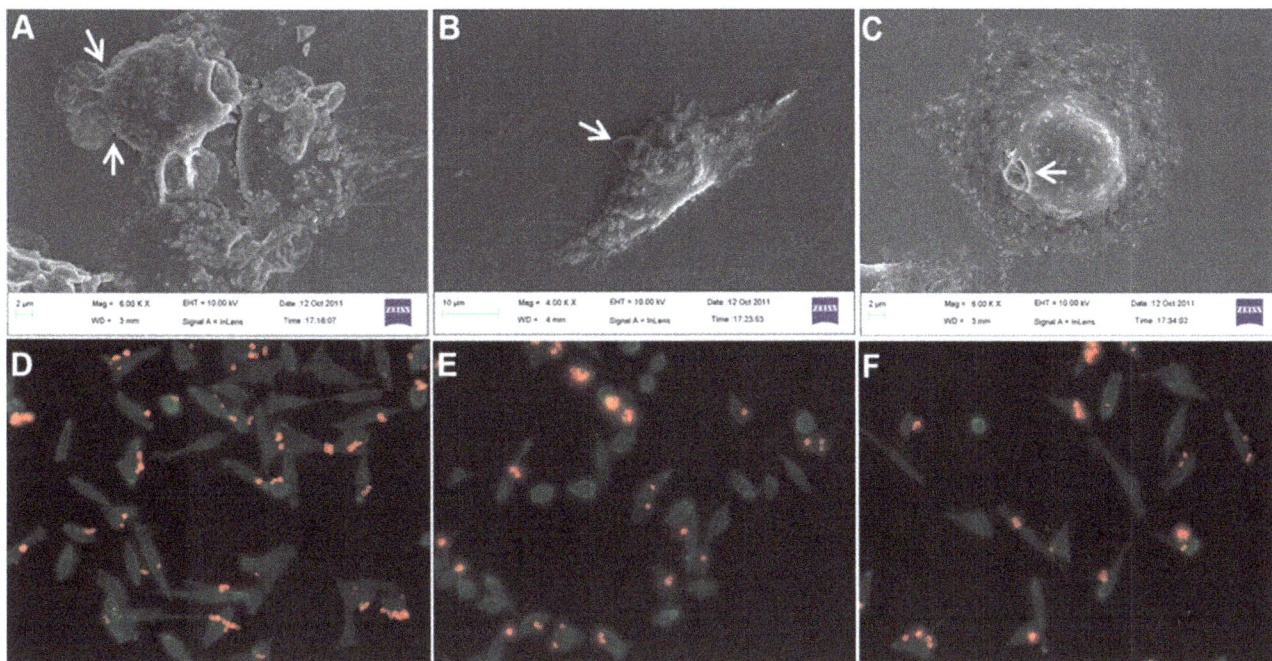

Figure 7. Kinetics of uptake of nanoparticles-loaded capsules by HeLa cells. Cell uptake kinetics of nanoparticles-loaded PSS/PAH capsules by HeLa cells followed by scanning electron microscopy (a–c) and fluorescence microscopy (d–f) as a function of time 2 h (a and d), 5 h (b and e) and 8 h (c and f). To visualize the uptake by fluorescence microscopy, LaF$_3$:Tb^{3+}(5%) nanoparticles were loaded in RITC-labeled PSS/PAH capsules (red emission) and actin cytoskeleton of cells was stained with FITC-Phalloidin (green). In SEM micrographs, nanoparticles-loaded capsules are indicated by white arrows.

potential use in bioimaging applications. These nanomaterials-loaded capsules can have potential applications in multimodal bioimaging and drug delivery.

Materials and Methods

Materials

All the lanthanide salts, poly-(sodium 4-styrene sulfonate) (PSS, M$_w$ 70 kDa), poly-(allylamine hydrochloride) (PAH, M$_w$ 70 kDa),

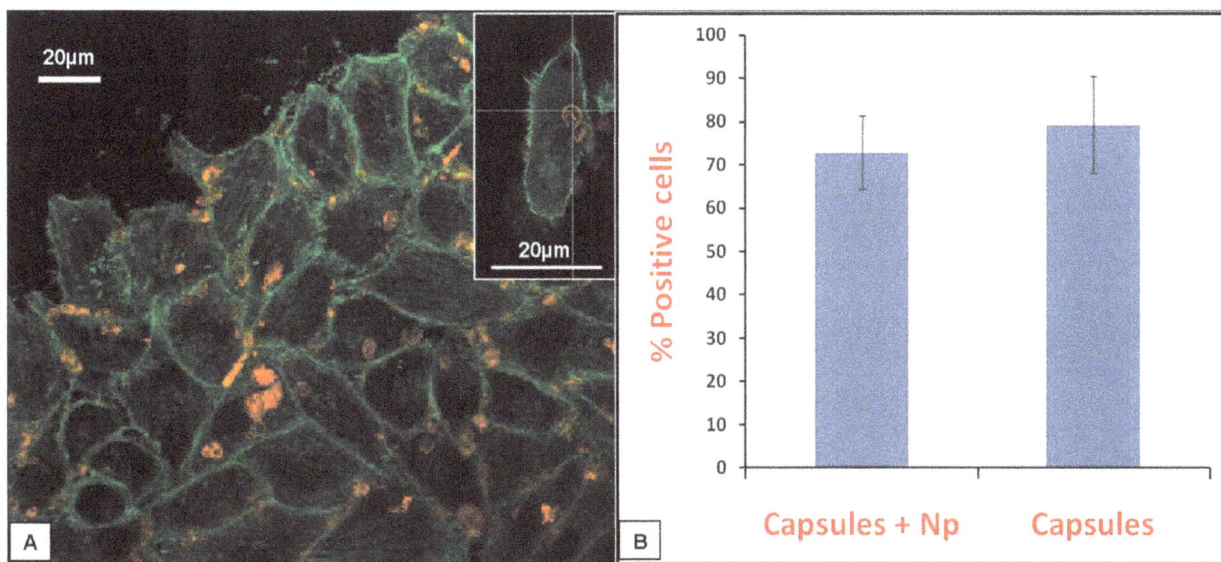

Figure 8. Uptake of nanoparticles-loaded PSS/PAH capsules by HeLa cells. (a) Confocal laser scanning microscopy (CLSM) image of HeLa cells after uptake of LaF$_3$:Tb^{3+}(5%) nanoparticles-loaded PSS/PAH capsule. Inset shows high magnification image of a HeLa cell, which has internalized Au nanoparticles-loaded capsule. Actin cytoskeleton of cells was stained with FITC-Phalloidin (green). The outer PAH layer of the capsules is labeled with RITC (red) and (b) efficiency of uptake of Au nanoparticles-loaded PSS/PAH capsules and blank PSS/PAH capsules by HeLa cells as investigated by flow cytometry. Efficiency of uptake is represented by percentage positive cells i.e. cells associated (both internalized and bound) with capsules.

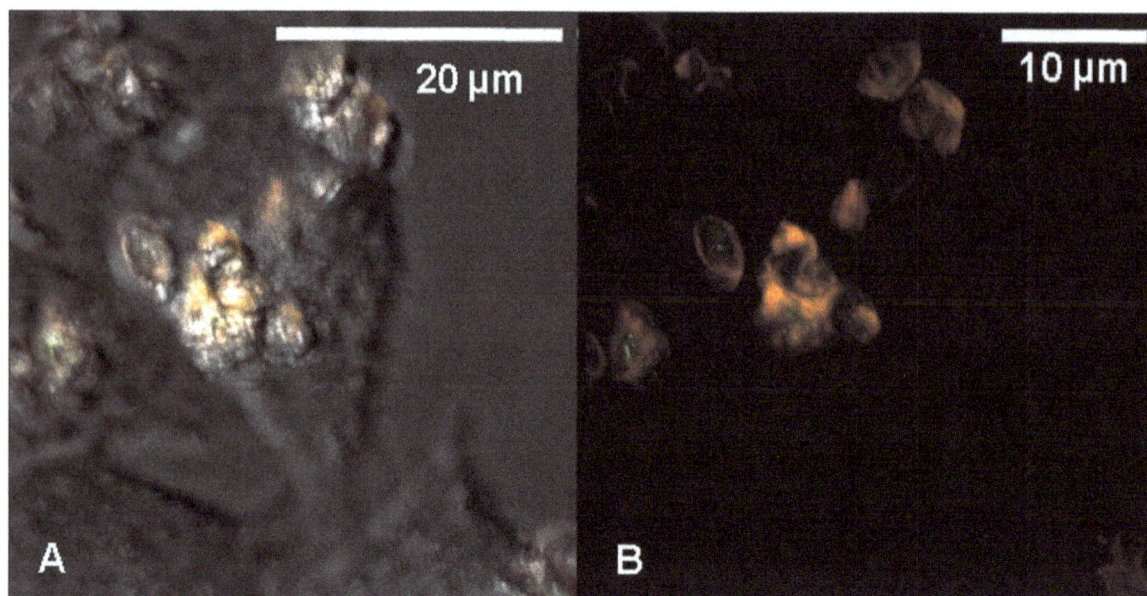

Figure 9. CLSM images showing the characteristic emission of Tb^{3+} from the HeLa cell internalized LaF$_3$:Tb^{3+}(5%) nanoparticles-loaded PSS/RITC-PAH capsule. (a) Composite image of bright field and fluorescent signals arising from LaF$_3$:Tb^{3+}(5%) nanoparticles (green emission from Tb^{3+}) and RITC-PAH (red); and (b) composite image of fluorescent signals arising from LaF$_3$:Tb^{3+}(5%) nanoparticles (characteristic green emission from Tb^{3+}) and RITC-PAH (red) showing internalized LaF$_3$:Tb^{3+}(5%) nanoparticles-loaded capsules.

FITC-phalloidin (Fluorescein isothiocyanate-phalloidin), trypsin-EDTA, dulbecco's modified eagle's medium (DMEM), penicillin-streptomycin antibiotic, (3-(4,5-Dimethylthiazol-2-yl)-2,5-diphenyltetrazolium bromide (MTT), and gelatin (from cold water fish skin) were purchased from Sigma Aldrich and used without further purification. Sodium borohydride and dimethyl sulphoxide were obtained from Merck's chemicals, India. Fetal bovine serum and chloroauric acid were bought from Hyclone (UT, USA) and LOBA Chemie, India respectively. Hydrofluoric acid, CTAB and cadmium nitrate were obtained from S.D. fine chem. Ltd. Silver nitrate, sodium sulfide, sodium chloride, liquor ammonia, triethylamine, and methanol were purchased from Fisher Scientific. 3-aminopropyltriethoxysilane (APTES) and tri sodium citrate dihydrate was obtained from Spectrochem Pvt. Ltd and RFCL Ltd. respectively. Citric acid and sodium fluoride

were purchased from Qualigens. Mesoporous silica particles (5 μm) were purchased from Tessek (Czech Republic). 1 μm sized silica particles were purchased from microparticles GmbH (Berlin, Germany). APTES modified mesoporous silica particles were prepared as described elsewhere [35]. FITC-PAH (Fluorescein isothiocyanate-PAH) and RITC-PAH (Rhodamine B isothiocyanate-PAH) was obtained by labeling PAH with FITC and RITC respectively [36]. HeLa cells were purchased from National Centre for Cell Science Pune, India which is a national repository of cell lines in India.[/LOOSER]

Preparation of PSS/PAH Capsules

PSS/PAH capsules were prepared as described in detail elsewhere [37]. Briefly, APTES-modified mesoporous silica (10 mg, 5 μm in size) were incubated with PSS and PAH (1 mg/ml of 0.5 M NaCl) solutions alternatively for 15 min to build-up the layers followed by washing thrice with water. Typically, five bilayers were developed and the silica core was removed by treating with 1 ml of 5 M HF for 2 minutes to obtain the PSS/PAH capsules (*Caution: HF is very toxic, so should be handled with all safety precautions*). The capsules were isolated via centrifugation/redispersion cycles (in water). Synthesis of 1 μm sized PSS/PAH capsules were done in a similar by templating 1 μm sized silica particles.

Synthesis of Cross-linked PAH Nanocapsules

For synthesis of cross-linked PAH nanocapsules, SC/MS (solid core mesoporous shell) silica template particles were synthesized as reported [38]. PAH nanocapsules were prepared as described by Wang *et al* [39]. In brief, the SC/MS template silica particles were taken and polymer was infiltrated in the mesoporous shell by incubating the template particles with PAH solution (5 mg/ml in 0.2 M NaCl; pH 8.5) for overnight with gentle mixing. Post infiltration, the unbound polymer was washed off thrice in water by centrifugation. The polymer was then cross-linked by incubating the polymer-infiltrated particles with glutaraldehyde for

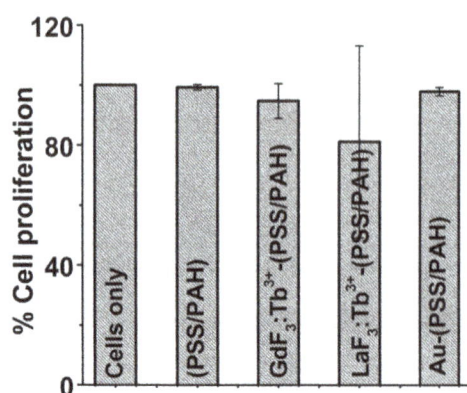

Figure 10. *In vitro* **biocompatibility of nanoparticles-loaded PSS/PAH capsules.** MTT assay results of HeLa cells, blank PSS/PAH capsules, GdF$_3$:Tb^{3+}(5%), LaF$_3$:Tb^{3+}(5%), and Au nanoparticles-loaded PSS/PAH capsules on incubation with HeLa cells.

20 min. Finally, the particles were washed with water and the template was etched out by incubating them with 2 M HF: 8 M NH_4F (pH 5) to get PAH capsules (*Caution: HF is very toxic, so should be handled with all safety precautions*).

Synthesis of Nanoparticles-loaded Polymer Capsules and Bare Nanoparticles

The PSS/PAH capsules were mixed with stock solution A for 15 min, followed by the incubation with the stock solution B (15 min). Stock solution C was added to the above and the mixture was kept under shaking for 1 h at ambient temperature. Excess nanoparticles formed outside the capsules were removed out by repeated centrifugation and washing steps with water. This procedure was followed to prepare Au, Ag, CdS, $LaF_3:Tb^{3+}$, $GdF_3:Tb^{3+}$, and $LaVO_4:Eu^{3+}$ nanoparticles-loaded PSS/PAH capsules by taking the respective stock solution (Table S1). Blank nanoparticles were synthesized by using the above procedure without templating the capsules. Similar steps were followed to synthesize gold nanoparticles in 1 μm sized PSS/PAH capsules and cross-linked PAH nanocapsules.

Synthesis of PSS/PAH Capsules-loaded with Gold Nanorods, Gold Nanoprisms and Multifaceted Gold Nanostructures

Gold nanostructures of various shapes were formed inside PSS/PAH capsules by following the seed mediated approach as described by Murphy *et al.*, with modifications in seed and ligand concentrations [32,40]. First seed was synthesized inside PSS/PAH capsules by dispersing the capsules in 1 ml of aqueous solution containing 0.25 mM tri sodium citrate and 0.25 mM chloroauric acid, followed by gentle mixing for 15 minutes. Freshly prepared ice cold 0.1 (M) sodium borohydride was added to the capsule solution drop wise while stirring and was left undisturbed for 2 hrs so as to form gold seeds inside capsules. Then, 900 μl of supernatant was removed by centrifuging at 5000 rpm for 4 min and the remaining solution containing Au nanoparticles loaded capsules was used as stock seed solution.

For synthesis of gold nanorods, these seed loaded PSS/PAH capsules (100 μl) were added in 40 ml of growth solution (containing 0.25 mM chloroauric acid, 0.01 (M) cetyltrimethylammonium bromide (CTAB) and 200 μl of 0.1 (M) freshly prepared ascorbic acid solution). The color of the solution changed to reddish brown after 5 min and the stirring was stopped at that point of time. After 20 min, the nanoparticles-loaded capsules were separated by centrifugation at 5000 rpm for 4 minutes followed by washing with water and redispersed in water.

For synthesis of gold nanoprisms, similar protocol as above was employed but the seed concentration was reduced 2.5 times. For synthesis of multifaceted gold nanostructures, same steps as in the above protocol were repeated but CTAB concentration used was 0.1(M) instead of 0.01(M).

Co-loading of Au and $LaVO_4:Eu^{3+}$(5%) Nanoparticles in PSS/PAH Capsules

$LaVO_4:Eu^{3+}$(5%) nanoparticles were first synthesized in PSS/PAH capsules as per the protocol given above. These $LaVO_4:Eu^{3+}$(5%) nanoparticles-loaded capsules were then washed thrice and Au nanoparticles were prepared in them by the same method as described for loading of gold nanoparticles in PSS/PAH capsules.

Co-loading of Au Nanoparticles and RITC-labeled Bovine Serum Albumin (BSA) in PSS/PAH Capsules

First PSS/PAH capsules were loaded with RITC-labeled BSA (BSA was labeled with RITC as per the same method as used for RITC-PAH above) as per the following protocol. 10 mg APTS-MS was incubated with 0.5 ml of RITC-BSA solution (1 mg/ml in DW) for ~30 h with gentle mixing. The above particles were then coated with five bilayers of PSS/PAH. The silica core was then etched by using 5 M HF and the resulting RITC-BSA loaded capsules were washed five times with distilled water. These RITC-BSA loaded capsules were then used for loading of gold nanoparticles by the same method as described for loading of gold nanoparticles in PSS/PAH capsules.

Cell Uptake Studies

HeLa cells were cultured in DMEM medium containing heat inactivated FBS (10% v/v) and penicillin/streptomycin (1% v/v), grown at 37°C in a humidified atmosphere containing 5% CO_2.

To investigate the uptake kinetics as a function of incubation time of nanoparticles-loaded capsules, HeLa cells were treated with $LaF_3:Tb^{3+}$ nanoparticles-loaded PSS/PAH capsules for different durations (2 h, 5 h, and 8 h; treatment ratio was 50 capsules (nanoparticles-loaded) per seeded cell; capsules were labeled with RITC-PAH). After the respective treatment, media was discarded and cells were washed thrice with PBS buffer (pH 7.4) to remove loosely bound capsules to the cells. Post washing, these cells were trypsinized by using 0.25% Trypsin-EDTA to further remove any capsules which have not yet internalized. The cells were then seeded on cover slips and left for few hours. After the cells adhered, they were fixed (and stained with FITC-Phalloidin) and imaged by fluorescence microscopy and scanning electron microscopy.

To determine cellular uptake of the nanoparticles-loaded capsules, HeLa cells (10^5) were seeded on glass cover slip (13 mm, 0.2% gelatin coated) for 20 hours. The cells were then treated with $LaF_3:Tb^{3+}$ and Au nanoparticles-loaded PSS/PAH capsules (capsules were labeled with RITC-PAH, treatment ratio was 50 capsules (nanoparticles-loaded) per seeded cell) for 16 h. Further, the cells were washed; actin cytoskeleton of the cells was stained with FITC-Phalloidin (green) and imaged by confocal laser scanning microscopy.

Flow Cytometry Studies

HeLa cells (10^5 cells per well) were grown as described above and the cells were incubated with blank PSS/PAH capsules and Au nanoparticles-loaded PSS/PAH capsules for 26 h (treatment ratio was kept same for both blank capsules and nanoparticles-loaded capsules i.e. 100 capsules per seeded cell). After the incubation the cells were washed with PBS buffer (pH 7.4) and trypsinized by using 0.25% Trypsin-EDTA. The cells were then resuspended in complete media and analyzed by flow cytometry. The experiments were done in triplicate.

MTT Assay

In vitro cytotoxicity studies of blank PSS/PAH capsules, $GdF_3:Tb^{3+}$(5%), $LaF_3:Tb^{3+}$(5%), and Au nanoparticles-loaded PSS/PAH capsules with HeLa cells were investigated by MTT assay [41]. 10^5 cells per well were seeded in a 24 well plate. The cells were grown for 18 h, followed by changing it with media containing respective capsules (100 capsules (nanoparticles-loaded) per seeded cell) and blank capsules (100 capsules per seeded cell). In the control wells (i.e. only cells), media was added without capsules. The cells were incubated with the capsules for ~22 h in a 37°C, 5% CO_2 humidified incubator followed by removal of media. 500 μL of basal media having MTT (0.5 mg/ml MTT) was added to the wells and allowed to incubate for 4 h. Further, media containing MTT was removed from the wells and 1 ml of dimethyl sulphoxide (DMSO) was added into each well. The blue color solution from the wells was transformed into a cuvette and absorbance at 570 nm was measured. All the assays were done in triplicate.

Characterization

UV-Vis spectra were recorded from UV-1800 Shimadzu UV spectrometer. The photoluminescence spectra and decay curves were recorded using Edinburgh instruments FLSP 920 fluorescence system. Emission and lifetime analyses of nanoparticles-loaded polymer capsules were done by exciting the samples with an Nd:YAG laser, attached with an optical parametric oscillator (OPO) with an optical range from 210–2400 nm. 450 W steady state Xe lamp was used to record the photoluminescence spectra of bare nanoparticles. The detector used was a red-sensitive Peltier element cooled Hamamatsu R928-P PMT. All the TEM images were obtained from FEI Technai G^2 U-Twin (200 KeV) instrument. The size and surface morphology of nanoparticles-loaded capsules and blank capsules were characterized by Scanning Electron Microscope (SUPRA 40 VP Gemini, Zeiss, Germany). The Confocal laser scanning microscopy images of cells were obtained from Leica PCS SP5 confocal microscope (40x, oil objective). Flow cytometry measurements were done by using Partec CyFlow® space cell scanner. Number of blank PSS/PAH capsules and PSS/PAH capsules (loaded with nanoparticles) were quantified by flow cytometry.

Supporting Information

Figure S1   UV-Vis absorption spectra of Au and Ag bare nanoparticles.

Figure S2   Lifetime of lanthanide-doped nanoparticles-loaded polymer capsules. Decay curve of a) LaVO$_4$:Eu^{3+}(5%), b) LaF$_3$:Tb^{3+}(5%), and c) GdF$_3$:Tb^{3+}(5%) nanoparticles-loaded PSS/PAH capsule. The emission was monitored at 541 nm for Tb^{3+} doped sample and 612 nm for Eu^{3+} doped sample.

Figure S3   Photoluminescence emission spectrum of blank PSS/PAH capsules. Blank PSS/PAH capsules were subjected to photoluminescence spectroscopy so as to investigate the background emission contribution from the blank capsules.

Figure S4   Photoluminescence emission spectra of bare nanoparticles. Photoluminescence emission spectra of LaF$_3$:Tb^{3+}(5%), GdF$_3$:Tb^{3+}(5%), and LaVO$_4$:Eu^{3+}(5%) bare nanoparticles. The emission bands (green and red curve, Figure 1c) around 544, 584 and 619 nm are assigned to $^5D_4$ to $^7F_5$, $^7F_4$, and $^7F_3$ transitions, respectively of Tb^{3+} ions. The emission bands (black curve) around 591 nm ($^5D_0$ to $^7F_1$), 615 nm ($^5D_0$ to $^7F_2$), and 696 nm ($^5D_0$ to $^7F_4$) are assigned to Eu^{3+} ions.

Figure S5   UV-Vis absorption spectra of 5 μm PSS/PAH capsules loaded with (a) gold nanorods and (b) multi-faceted gold nanoparticles.

Figure S6   Co-loading of drug and nanoparticles within PSS/PAH capsules. Fluorescence microscopy image of PSS/PAH capsules (~5 μm) co-loaded with RITC-labeled BSA and Au nanoparticles.

Figure S7   Validation of internalization of nanoparticles-loaded capsules by HeLa cells. Confocal laser scanning microscopy sections (XY) of HeLa cells after uptake of LaF$_3$:Tb^{3+} nanoparticles-loaded PSS/PAH capsules at different Z positions; (a) basal, (b) inside and (c) apical of the cell, and (d) XY,YZ and XZ section of the cell. Scale bar is 20 μm.

Figure S8   2D plots of the events recorded with FACS for uptake studies. (a) HeLa cells, (b) HeLa cells after incubation with blank PSS/PAH (FITC-PAH) capsules, and (c) HeLa cells after incubation with Au nanoparticles-loaded PSS/PAH capsules (FITC-PAH).

Table S1   Stock solutions for preparing the nanoparticles-loaded polymer capsules.

Table S2   Lifetime of lanthanide-doped nanoparticles-loaded polymer capsules. Average lifetime of Tb^{3+} or Eu^{3+} ions for different types of nanoparticles inside PSS/PAH capsules.

Scheme S1   Schematic representation for the synthesis of nanomaterials-loaded PEM capsules. Nanoparticles were synthesized inside the microvolume of capsules by incubating the capsules with nanoparticles precursors and citrate as ligand. (Inset is digital photograph of A) Au, B) Ag, C) LaVO$_4$:Eu^{3+}(5%), D) LaF$_3$:Tb^{3+}(5%), E) CdS, and F) GdF$_3$:Tb^{3+}(5%) nanoparticles-loaded PSS/PAH capsules and G) blank PSS/PAH capsules dispersed in PBS buffer (pH~7.2)).

Acknowledgments

Laser scanning confocal microscopy was done at BSBE department and help from Prof Pradip Sinha is duly acknowledged.

Author Contributions

Conceived and designed the experiments: HS AKM AK SS. Performed the experiments: HS AKM. Analyzed the data: HS AKM AK SS. Contributed reagents/materials/analysis tools: HS AKM AK SS. Wrote the paper: HS AKM AK SS.

References

1. Riehemann K, Schneider SW, Luger TA, Godin B, Ferrari M, et al. (2009) Nanomedicine–Challenge and Perspectives. Angewandte Chemie International Edition 48: 872–897.
2. Kievit FM, Zhang M (2011) Cancer Nanotheranostics: Improving Imaging and Therapy by Targeted Delivery Across Biological Barriers. Advanced Materials 23: H217–H247.
3. Zhao F, Zhao Y, Liu Y, Chang X, Chen C, et al. (2011) Cellular Uptake, Intracellular Trafficking, and Cytotoxicity of Nanomaterials. Small 7: 1322–1337.
4. Thurn K, Brown E, Wu A, Vogt S, Lai B, et al. (2007) Nanoparticles for Applications in Cellular Imaging. Nanoscale Research Letters 2: 430–441.
5. Chithrani BD, Ghazani AA, Chan WCW (2006) Determining the Size and Shape Dependence of Gold Nanoparticle Uptake into Mammalian Cells. Nano Letters 6: 662–668.

6. Pavlov AM, Saez V, Cobley A, Graves J, Sukhorukov GB, et al. (2011) Controlled protein release from microcapsules with composite shells using high frequency ultrasound-potential for in vivo medical use. Soft Matter 7: 4341–4347.

7. del Mercato LL, Rivera-Gil P, Abbasi AZ, Ochs M, Ganas C, et al. (2010) LbL multilayer capsules: recent progress and future outlook for their use in life sciences. Nanoscale 2: 458–467.

8. De Geest BG, De Koker S, Sukhorukov GB, Kreft O, Parak WJ, et al. (2009) Polyelectrolyte microcapsules for biomedical applications. Soft Matter 5: 282–291.

9. Johnston APR, Cortez C, Angelatos AS, Caruso F (2006) Layer-by-layer engineered capsules and their applications. Current Opinion in Colloid & Interface Science 11: 203–209.

10. Sivakumar S, Wark KL, Gupta JK, Abbott NL, Caruso F (2009) Liquid Crystal Emulsions as the Basis of Biological Sensors for the Optical Detection of Bacteria and Viruses. Advanced Functional Materials 19: 2260–2265.

11. Sivakumar S, Bansal V, Cortez C, Chong S-F, Zelikin AN, et al. (2009) Degradable, Surfactant-Free, Monodisperse Polymer-Encapsulated Emulsions as Anticancer Drug Carriers. Advanced Materials 21: 1820–1824.

12. Städler B, Price AD, Zelikin AN (2011) A Critical Look at Multilayered Polymer Capsules in Biomedicine: Drug Carriers, Artificial Organelles, and Cell Mimics. Advanced Functional Materials 21: 14–28.

13. Caruso F, Caruso RA, Mohwald H (1998) Nanoengineering of inorganic and hybrid hollow spheres by colloidal templating. Science 282: 1111–1114.

14. Skirtach AG, Muñoz Javier A, Kreft O, Köhler K, Piera Alberola A, et al. (2006) Laser-Induced Release of Encapsulated Materials inside Living Cells. Angewandte Chemie International Edition 45: 4612–4617.

15. Bedard MF, Munoz-Javier A, Mueller R, del Pino P, Fery A, et al. (2009) On the mechanical stability of polymeric microcontainers functionalized with nanoparticles. Soft Matter 5: 148–155.

16. Angelatos AS, Radt B, Caruso F (2005) Light-Responsive Polyelectrolyte/Gold Nanoparticle Microcapsules. The Journal of Physical Chemistry B 109: 3071–3076.

17. Skirtach AG, Antipov AA, Shchukin DG, Sukhorukov GB (2004) Remote activation of capsules containing Ag nanoparticles and IR dye by laser light. Langmuir 20: 6988–6992.

18. Gil PR, del Mercato LL, del_Pino P, Muñoz_Javier A, Parak WJ (2008) Nanoparticle-modified polyelectrolyte capsules. Nano Today 3: 12–21.

19. Choi WS, Koo HY, Park JH, Kim DY (2005) Synthesis of two types of nanoparticles in polyelectrolyte capsule nanoreactors and their dual functionality. Journal of the American Chemical Society 127: 16136–16142.

20. Koo HY, Choi WS, Kim D-Y (2008) Direct Growth of Optically Stable Gold Nanorods onto Polyelectrolyte Multilayered Capsules. Small 4: 742–745.

21. Xu X, Majetich SA, Asher SA (2002) Mesoscopic Monodisperse Ferromagnetic Colloids Enable Magnetically Controlled Photonic Crystals. Journal of the American Chemical Society 124: 13864–13868.

22. Rogach AL, Nagesha D, Ostrander JW, Giersig M, Kotov NA (2000) "Raisin bun"-type composite spheres of silica and semiconductor nanocrystals. Chemistry of Materials 12: 2676–2685.

23. Cui J, Wang Y, Postma A, Hao J, Hosta-Rigau L, et al. (2010) Monodisperse Polymer Capsules: Tailoring Size, Shell Thickness, and Hydrophobic Cargo Loading via Emulsion Templating. Advanced Functional Materials 20: 1625–1631.

24. Shchukin DG, Sukhorukov GB (2003) Selective YF3 nanoparticle formation in polyelectrolyte capsules as microcontainers for yttrium recovery from aqueous solutions. Langmuir 19: 4427–4431.

25. Shchukin DG, Sukhorukov GB, Mohwald H (2004) Fabrication of fluorescent rare earth phosphates in confined media of polyelectrolyte microcapsules. Journal of Physical Chemistry B 108: 19109–19113.

26. Cortez C, Tomaskovic-Crook E, Johnston APR, Radt B, Cody SH, et al. (2006) Targeting and Uptake of Multilayered Particles to Colorectal Cancer Cells. Advanced Materials 18: 1998–2003.

27. Yuan F, Dellian M, Fukumura D, Leunig M, Berk DA, et al. (1995) Vascular permeability in a human tumor xenograft: molecular size dependence and cutoff size. Cancer Res 55: 3752–3756.

28. Song Y, Xu X, MacRenaris KW, Zhang X-Q, Mirkin CA, et al. (2009) Multimodal Gadolinium-Enriched DNA-Gold Nanoparticle Conjugates for Cellular Imaging. Angewandte Chemie International Edition 48: 9143–9147.

29. Alric C, Taleb J, Duc GrL, Mandon Cl, Billotey C, et al. (2008) Gadolinium Chelate Coated Gold Nanoparticles As Contrast Agents for Both X-ray Computed Tomography and Magnetic Resonance Imaging. Journal of the American Chemical Society 130: 5908–5915.

30. Gorin DA, Portnov SA, Inozemtseva OA, Luklinska Z, Yashchenok AM, et al. (2008) Magnetic/gold nanoparticle functionalized biocompatible microcapsules with sensitivity to laser irradiation. Physical Chemistry Chemical Physics 10: 6899–6905.

31. Sivakumar S, Diamente PR, van Veggel FCJM (2006) Silica-Coated Ln3+- Doped LaF3 Nanoparticles as Robust Down- and Upconverting Biolabels. Chemistry – A European Journal 12: 5878–5884.

32. Jana NR, Gearheart L, Murphy CJ (2001) Wet Chemical Synthesis of High Aspect Ratio Cylindrical Gold Nanorods. The Journal of Physical Chemistry B 105: 4065–4067.

33. Melancon M, Lu W, Li C (2009) Gold-Based Magneto/Optical Nanostructures: Challenges for In Vivo Applications in Cancer Diagnostics and Therapy. Mater Res Bull 34: 415–421.

34. Muñoz Javier A, Kreft O, Semmling M, Kempter S, Skirtach AG, et al. (2008) Uptake of Colloidal Polyelectrolyte-Coated Particles and Polyelectrolyte Multilayer Capsules by Living Cells. Advanced Materials 20: 4281–4287.

35. Wang Y, Caruso F (2006) Template Synthesis of Stimuli-Responsive Nanoporous Polymer-Based Spheres via Sequential Assembly. Chemistry of Materials 18: 4089–4100.

36. Caruso F, Yang W, Trau D, Renneberg R (2000) Microencapsulation of Uncharged Low Molecular Weight Organic Materials by Polyelectrolyte Multilayer Self-Assembly. Langmuir 16: 8932–8936.

37. Sivakumar S, Gupta JK, Abbott NL, Caruso F (2008) Monodisperse Emulsions through Templating Polyelectrolyte Multilayer Capsules. Chemistry of Materials 20: 2063–2065.

38. Büchel G, Unger KK, Matsumoto A, Tsutsumi K (1998) A Novel Pathway for Synthesis of Submicrometer-Size Solid Core/Mesoporous Shell Silica Spheres. Advanced Materials 10: 1036–1038.

39. Wang Y, Bansal V, Zelikin AN, Caruso F (2008) Templated Synthesis of Single-Component Polymer Capsules and Their Application in Drug Delivery. Nano Letters 8: 1741–1745.

40. Sau TK, Murphy CJ (2004) Seeded high yield synthesis of short Au nanorods in aqueous solution. Langmuir 20: 6414–6420.

41. Mosmann T (1983) Rapid colorimetric assay for cellular growth and survival: Application to proliferation and cytotoxicity assays. Journal of Immunological Methods 65: 55–63.

10

Formation of Asymmetrical Structured Silica Controlled by a Phase Separation Process and Implication for Biosilicification

Jia-Yuan Shi[1], Qi-Zhi Yao[2], Xi-Ming Li[2], Gen-Tao Zhou[1]*, Sheng-Quan Fu[3]

1 Key Laboratory of Crust-Mantle Materials and Environments, Chinese Academy of Sciences, School of Earth and Space Sciences, University of Science and Technology of China, Hefei, People's Republic of China, **2** School of Chemistry and Materials, University of Science and Technology of China, Hefei, People's Republic of China, **3** Hefei National Laboratory for Physical Sciences at Microscale, University of Science and Technology of China, Hefei, People's Republic of China

Abstract

Biogenetic silica displays intricate patterns assembling from nano- to microsize level and interesting non-spherical structures differentiating in specific directions. Several model systems have been proposed to explain the formation of biosilica nanostructures. Of them, phase separation based on the physicochemical properties of organic amines was considered to be responsible for the pattern formation of biosilica. In this paper, using tetraethyl orthosilicate (TEOS, $Si(OCH_2CH_3)_4$) as silica precursor, phospholipid (PL) and dodecylamine (DA) were introduced to initiate phase separation of organic components and influence silica precipitation. Morphology, structure and composition of the mineralized products were characterized using a range of techniques including field emission scanning electron microscopy (FESEM), transmission electron microscope (TEM), X-ray diffraction (XRD), thermogravimetric and differential thermal analysis (TG-DTA), infrared spectra (IR), and nitrogen physisorption. The results demonstrate that the phase separation process of the organic components leads to the formation of asymmetrically non-spherical silica structures, and the aspect ratios of the asymmetrical structures can be well controlled by varying the concentration of PL and DA. On the basis of the time-dependent experiments, a tentative mechanism is also proposed to illustrate the asymmetrical morphogenesis. Therefore, our results imply that in addition to explaining the hierarchical porous nanopatterning of biosilica, the phase separation process may also be responsible for the growth differentiation of siliceous structures in specific directions. Because organic amine (e.g., long-chair polyamines), phospholipids (e.g., silicalemma) and the phase separation process are associated with the biosilicification of diatoms, our results may provide a new insight into the mechanism of biosilicification.

Editor: Vipul Bansal, RMIT University, Australia

Funding: This work was financially supported by the Chinese Ministry of Science and Technology (No. 2011CB808800), the Natural Science Foundation of China (No. 41172049) and the Knowledge Innovation Program of the Chinese Academy of Sciences, Grant No. KZCX2-YW-QN501. The funders had no role in study design, data collection and analysis, decision to publish, or preparation of the manuscript.

Competing Interests: The authors have declared that no competing interests exist.

* E-mail: gtzhou@ustc.edu.cn

Introduction

Biomineralization is the formation of hard tissues with complex structures and multifunctional properties, which occurs in almost all the living organisms from prokaryotes to humans [1,2]. Some of the morphologically gorgeous and structurally intricate biominerals are exemplified by the biosilica formed in the aquatic organisms including diatoms and sponges [3,4]. These biogenic minerals are structured in the nanometer to micrometer scale range, and composed of amorphous silica [5–7].

Diatom is well known for the spectacular design of its silica-based cell wall (termed frustules) [2,8,9]. More than 40 years ago, Nakajima and Volcani have noticed that diatom biosilica contained unusual amino acid derivatives such as N,N,N-trimethylhydroxylysine and dihydroxyproline [10,11]. This observation is the first to indicate that diatom silica is a composite material. In recent decades, a variety of organic and biological molecules have been successfully separated and identified from cell-wall extracts of diatoms[12,13]. An emerging consensus is that polysaccharides [14,15], proteins [16–20], and polyamines [21]

are general organic components of diatom cell walls. In such a context, many efforts have been made to explore how these components interact with silicic acid, silicate, or silicon-containing compound, and influence silica morphogenesis [2,16,22,23].

In terms of polyamines, all genera of diatoms investigated so far incorporate polyamines into their silica-based cell walls [24]. Most surprisingly, cell-wall extracts from *Coscinodiscus* diatoms exhibit predominately polyamines, whereas silaffin-related peptides appear to be absent [25]. These observations stimulate a polyamines-based phase separation model to be proposed for the pattern formation of the diatom cell-walls with hierarchically hexagonal porous structures [25]. In this model, polyamines are able to undergo a phase-separation process within a specialized membrane-bound compartment termed silica deposition vesicle to form an emulsion of microdroplets. These droplets form a hexagonally arranged monolayer within the silica deposition vesicle. Silica precipitation occurs at the interface between the solution and the organic microdroplets [26], which cause the formation of honeycomb-like framework. A defined fraction of the polyamine population is consumed by its co-precipitation with silica. As a result, smaller droplets separate from the surface of the original microdroplet. Silicification continues at these newly created water/polyamine interfaces of smaller droplets and a smaller hexagonal package of silica is thereupon developed. This mechanism would allow the creation of additional hexagonal frameworks at smaller and smaller scales. Finally, hierarchically porous structures and spectacular patterns are exhibited in the silica-based frustules. Polyamines in diatoms appear to be species-specific, which play an important role in the formation of frustules with species-specific patterns [21]. In other words, biosilicification in diatoms might be modulated by the specific structure of polyamines involved in the precipitation process [27].

Sponge spicules also possess highly hierarchical and organized siliceous nanostructures. The laminated spicule structure consists of alternating layers of silica and organic material [28]. Although the mechanism of biosilicification in sponges is distinct from that of silica formation in diatoms [29], organic amines have also been identified from the marine sponge *Axinyssa aculeata* [30]. These polyamines separated from sponge can deposit silica and the polyamine-derived macromolecules are chemical factors involved in silica deposition in sponges [30].

Phospholipids also play an important role in biosilicification [31]. Diatom silicification takes place in the silica deposition vesicle [32], whose membrane, called the silicalemma, consists of a typical lipid bilayer [33]. The overall outline of diatom's silica structure is determined by shaping of this kind of membrane-bound compartment [34]. Hildebrand et al. found that the silicalemma is tightly clung to siliceous structures in areas where silica is deposited [35]. This indicates that membrane components of silica deposition vesicle could become part of the silica structure [36,37]. Recently, X-ray photoelectron spectroscopy (XPS) [38] and solid state NMR (SSNMR) [39] studies were performed on diatom cells for analyzing the chemical composition of the diatom surface. The XPS analysis revealed a high concentration of lipids present as a structural part of the cell wall in the form of carboxylic esters. The SSNMR study also demonstrated that lipids are tightly associated with silica, even after harsh chemical treatment. All these imply that phospholipids may involve in the amines-mediated biosilica deposition in diatoms. [40,41].

Although the phase separation model successfully explain the important aspects of silica patterning in diatoms, biosilica in diatoms and sponges have other nanometer-scale details, and their nuanced structural and biological functions are well beyond the current ranges used in advanced materials [42]. Taking the centric diatom *Thalassiosira eccentrica* as an example, the ground-plan of its areolae is a two-dimensional system of hexagonal meshes [43]. Moreover, starting from this ground plan, the vertical growth of areolae walls and the horizontal extension on the distal side of areolae walls occur in sequence. It indicates that the asymmetrical development of silica deposition can be well achieved in diatom silicification [44]. However, it is still difficult to understand how the differentiation of solid siliceous structures would occur in different directions [34].

In this study, dodecylamine (DA) and phospholipid (PL) were selected as model organic additives to initiate phase separation and influence silica precipitation. Phospholipid, which has a hydrophilic head and two hydrophobic tails, is a major component of all the plasma membranes including the silicalemma in diatoms and sponges [45]. The goal of this study is to examine the effect of phase separation of biosilicification-associated model organic components on the development of silica morphology, and thus to reveal the contribution of the organic phase separation to growth differentiation of biogenic silica. As a consequence, asymmetrical discus-like silica particles with controlled aspect

ratios were indeed obtained during the phase separation of PL and DA, and the morphological evolution of the deposited silica from spherical through sunflower-looking to discus-like features were also exhibited at different conditions. Since the organic amines, membrane lipids, and the phase separation process are the important features of diatom silicification, our results may be useful for a deeper insight into biosilicification.

Materials and Methods

Materials

All starting chemicals were purchased from Sinopharm Chemical Reagent Co., Ltd, and used as received without further purification. Phospholipids (PL) are of biotech grade while all other reagents, such as ethanol, dodecylamine and tetraethyl orthosilicate, are of analytical grade. Deionized water was also used in these syntheses. For all experiments, glassware was cleaned with aqua regia (3:1 HCl/HNO_3), rinsed thoroughly with ultrapure water, and oven-dried overnight before use.

Preparation

In a typical biomimetic synthesis, 0.10 g of PL and 0.16 g of DA (0.863 mmol) were dissolved in 30 mL of ethanol through ultrasonification, and then stirring for about 5 min in a closed 100 mL flask until the solution became clear (Fig. S1a in Supplementary Information). Afterwards 30 μL of TEOS (0.134 mmol, 2.2 mM) was injected into the solution using a 50 μL syringe with stirring. In succession, 30 mL of H_2O was added to the above solution to obtain a turbid suspension (Fig. S1b). This suspension was then heated in a 80°C thermostated water bath, and became clear again with the increase of temperature (Fig. S1c). After 24 h of thermostated reaction, the solution was moved out of the water bath, and cooled down to room temperature naturally. As the temperature of the solution lowered, a white turbidity gradually appeared. Notably, the turbidness could be explicitly distinguished after the flask was cooled down for an hour at room temperature (Fig. S1d). Nevertheless, the centrifuged precipitate could dissolve in ethanol, and thus no silica could be obtained in this case, indicating that the isolated precipitate should be organics, i.e., an undissolvable organic phase was first formed at room temperature. After the flask was continuously stationed for another 1 day (Fig. S1e), the resultant particles were isolated by centrifugation, cleaned by three cycles of centrifugation/washing/redispersion in ethanol, and dried at room temperature for 1 day in vacuum. The obtained sample was named as sample L5. For other morphogenesis of silica structures, the same synthetic procedures were deployed except that some experimental parameters were varied. The detailed experimental conditions and the corresponding aspect ratios of the silica particles are listed in Table S1. Moreover, in order to understand the detailed microstructures, some samples were also calcined at 550°C in air for 6 h to remove the occluded organic components, and XRD and nitrogen physisorption analyses were performed.

Characterization

Several analytical techniques were used to characterize the products. Field emission scanning electron microscopy (FESEM) (JEOL JSM-6700 F) was applied to investigate the size and morphology. Transmission electron microscope (TEM) images were obtained on a JEM 2010 transmission electron microscope with an accelerating voltage of 200 kV. The samples for the TEM measurements were prepared by dropping a few drops of sample suspension with ethanol as the solvent on a copper grid, and the solvent was allowed to evaporate to dry state before analysis. The powder X-ray diffraction (XRD) patterns of the samples were recorded with a Japan Rigaku TTR-III X-ray diffractometer 0.154056 nm), employing a scanning rate of $0.02°s^{-1}$ in the 2θ range 0.8–10°. Infrared spectra were collected using a Nicolet 8700 FT-IR spectrometer on KBr pellets. Thermogravimetric and differential thermal analysis (TG-DTA) was carried out using a SDTQ 600 TG/DTA thermal analyzer (TA, USA) with a heating rate of 10°C/min from room temperature to 800°C in a flow air atmosphere. N_2-sorption isotherms of the samples were measured by using a Micromeritics Tristar II 3020 M instrument at liquid-nitrogen temperature. From the adsorption isotherm, the Barrett-Joyner-Halenda theory (BJH) was used to calculate the mesopore volume and its size distribution. Specific surface areas were calculated by using the Brunauer-Emmett-Teller (BET) method in the relative pressure range of $P/P_0 = 0.05$–0.3. Pore volumes were obtained from the volumes of N_2 adsorbed at $P/P_0 = 0.95$ or in the vicinity. The dispersibility of suspensions was estimated by dynamic light scattering (DLS, DYNAPRO-99).

Results and Discussion

Figure 1a depicts the low-magnification FESEM image of sample L5. The product is solely composed of the discus-like particles with a diameter of 2.0–3.0 μm, and no aggregation among the particles occurs. Figure 1b and c present the side and front view of an individual particle, respectively. The discus-like morphology is further confirmed and a ridge between the two halves is visible (indicated by black arrowheads in Fig. 1b). The ratio (D/T, *i.e.* aspect ratio) of particle diameter ("D" in Fig. 1c) to thickness ("T" in Fig. 1b) is 1.60±0.06. It should be pointed out that the two halves are not completely symmetric (e.g., Fig. 1b),

which is also observed in the corresponding TEM image (Fig. 2a). The TEM analyses (e.g., Fig. 2b) also show that the discus-like particles are not hollow, but solid. Fig. 2c and d depicts the local-magnification TEM images of the areas framed in Fig. 2a and b, respectively. The disordered pores are obviously discernable, and no resolved diffraction peaks can be observed in the XRD patterns including calcined sample L5 (Fig. 3a), indicating that the arrangement of the pore channels may be random [46]. Fig. 3b presents the N_2 adsorption–desorption isotherm with the inset of the BJH pore size distribution of the calcined sample L5. One can see a typical type IV isotherm with a N_2 hysteresis loop in the calcined sample, indicating the mesoporous property [47]. The adsorption isotherm shows a well-defined capillary condensation step at relative pressure (P/P_0) of 0.40–0.50, corresponding to the pore size of 3.3 nm. The Brunauer-Emmett-Teller (BET) surface area is calculated at 730 $m^2 \cdot g^{-1}$ and the pore volume is 0.62 $cm^3 \cdot g^{-1}$. Therefore, the silicified product is an asymmetrical discus-like structure possessing disordered mesoporous character.

The FT-IR spectrum (Fig. 4) of sample L5 displays three characteristic peaks of silica: Si-O-Si asymmetric stretching at 1081 cm^{-1}, symmetric stretching at 800 cm^{-1}, and Si-OH stretching at 965 cm^{-1} [48–51]. The H-bonded OH groups with various OH⋯H distances are responsible for the intense absorption at 3428 cm^{-1}, and the band at 1633 cm^{-1} is due to the $\delta(HOH)$ of physisorbed water [52]. Bands detected at 2926 and 2855 cm^{-1} are assigned to the stretching vibrations of the CH groups, which indicate the existence of organic components [50]. The characteristic vibration of C-C bonds at 1468 cm^{-1} is also observed. Moreover, the bands at 553 and 1722 cm^{-1} can be assigned to the O-P-O bond and the carbonyl group, respectively, both of which should originate from phospholipid molecules [53].

Figure 5 presents the TG-DTA curves of the original silica particles (sample L5). The TG curve reveals ~25.4% total weight loss from room temperature to 800°C. A ~5.2% of weight loss from room temperature to 120°C and the corresponding endothermal peak at 50°C in DTA curve indicate the evaporation of the surface-adsorbed water and ethanol. The small endothermic peak at 218°C in the DTA curve is believed to originate from the organic component decomposition and/or the polycondensation

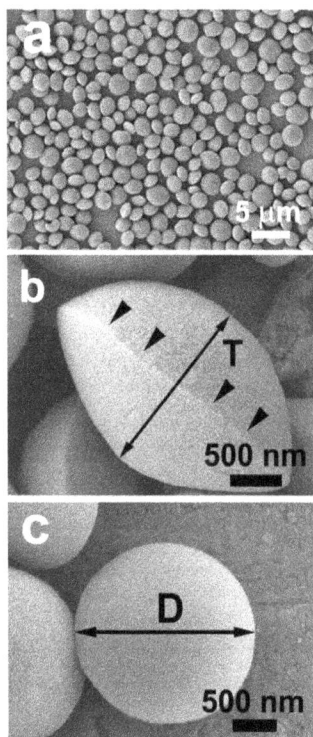

Figure 1. SEM images of discus-like silica particles (sample L5): low magnification (a), the side- (b) and front-view (c) observations of individual particles.

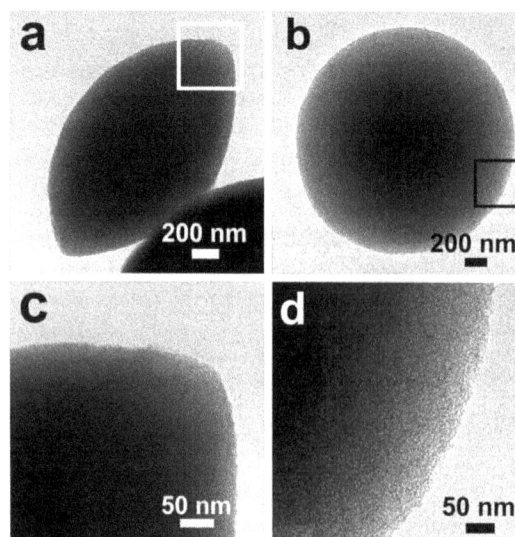

Figure 2. TEM images of individual particles in sample L5 by a side (a) and front (b) view, and their local high-magnification images (c, d).

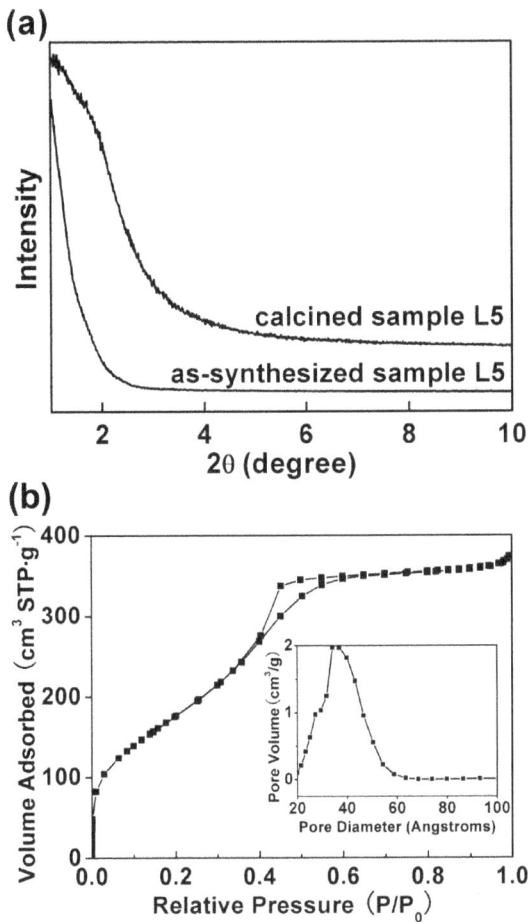

Figure 3. XRD patterns (a) of sample L5 before and after calcined and N₂ sorption isotherms (b) of the calcined sample L5.

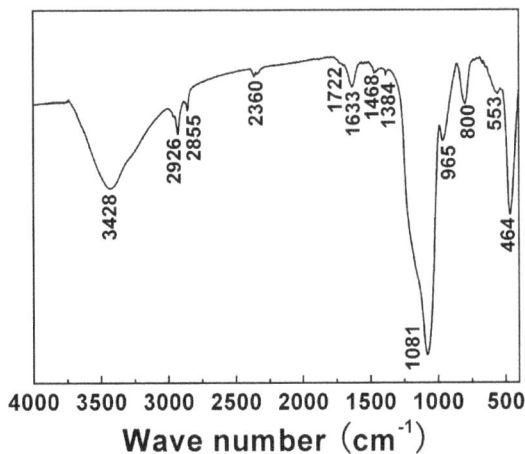

Figure 4. FTIR spectrum of the discus-like particles (sample L5).

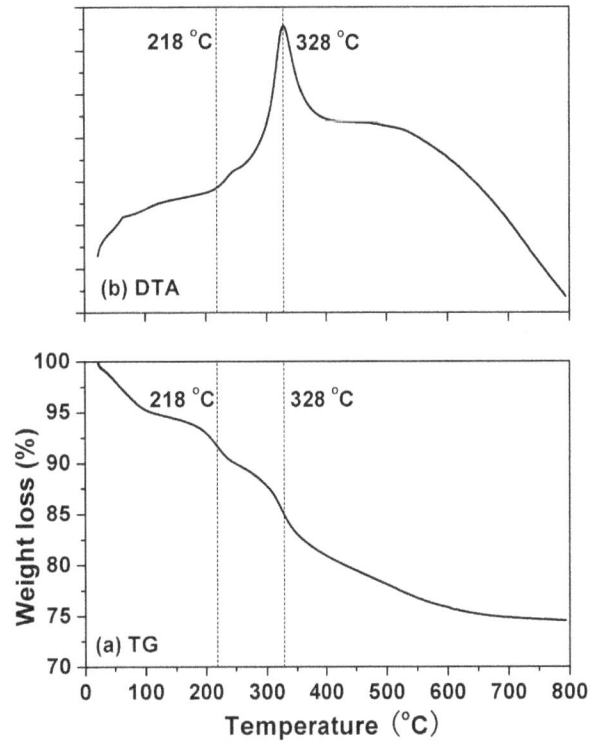

Figure 5. TG (a) and DTA (b) curves of sample L5.

of the silica network [54]. The exothermic peak at 328°C can be ascribed to the combustion of the incorporated organic substances [55]. The weight loss at temperatures above 400°C (6.3%) is generally due to the further condensation and dehydration of silanol groups [56]. FT-IR and TG-DTA analyses of the as-synthesized product confirm the co-existence of silica and organic components, indicating the formation of an organic-inorganic composite.

Moreover, our results also show that the asymmetrical structured silica including semispherical or discus-like particles can be obtained in a relatively broad range of ethanol/water volume ratio (E/W), as shown in Figs. 6. When the E/W is varied between 15/45 and 25/35, the interconnected semispherical particles are always obtained (Fig. 6 and 7b). This is due to the fact that the lower the E/W ratio, the more the precipitated turbidness. As a result, more and more oil droplets are formed. Therefore, the silicified particles are inclined to connect each other. When the E/W is under 15/45, however, both PL and DA can not be well dissolved, and an irregular aggregate is formed (Fig. 6 and Fig. 7a). Conversely, when the E/W is over 35/25, the morphologies of the products change from discus (Fig. 1) to microsphere (Fig. 6 and Fig. 7c). This is possible because higher ethanol concentrations facilitate the dissolution of organic components, and do not favor the formation of the oil-water interface [57,58]. Further increasing E/W to 40/20 leads to the formation of ca. 60-nm-diameter spherical nanoparticles (Fig. 6 and Fig. 7d), which is probably because of the shrinking effect of ethanol at such a high E/W [59,60].

The concentration of silica precursor (TEOS) is also another important factor for the morphogenesis of the asymmetrical silica (Fig. 8). Monodisperse discus-like particles can be obtained with a TEOS concentration of 2.2 mM (sample L5, Fig. 1). However, the interconnection among the particles becomes more significant in the case of both higher and lower concentrations of TEOS. With the decrease of TEOS concentration to 1.5 mM, the development of the two halves is insufficient and the deposition region of silica is predominantly confined to the water/organics interfaces. Therefore, the obtained particles become thinner (as indicated in Fig. 8b)

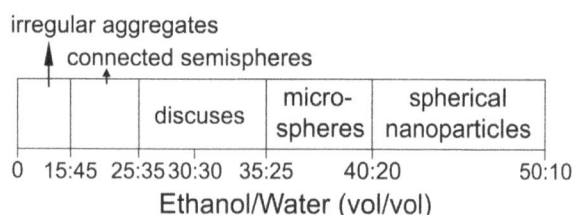

Figure 6. Schematic illustration of the product morphology dependence on ethanol/water volume ratio. The amounts of PL, DA, and TEOS, and the total solution volume were fixed at 0.10 g, 0.16 g, 30 μL, and 60 mL, respectively.

and the siliceous extension at the interfaces makes some particles interconnect each other. Further interconnection among the silica wafers occurs with decreasing the TEOS concentration to 0.7 mM (Fig. 8a). Nevertheless, it is almost impossible to collect any precipitate as the concentration is lower than 0.7 mM. Conversely, as the concentration of TEOS is over 2.2 mM, the particles show better growth with an increase in thickness and diameter (Fig. 8c-f): while the TEOS concentration is 3.0 mM, particles with obvious ridges exhibit asymmetrical discus-like shapes, and some asymmetrical particles fuse together along their ridges (as indicated by arrows in Fig. 8c). Further increasing the concentration to 3.7 mM results in the extra formation of spherical particles together with the asymmetrical aggregates of silica (as indicated by arrows in Fig. 8d). More spherical particles with lager diameters can be observed as 4.5 mM or 5.2 mM of TEOS is used (Fig. 8e and f). The emergence of extra spherical silica at the higher concentrations of TEOS can be ascribed to the independent nucleation and growth of silica in the reaction solutions. We have noted that the reaction solution with 3.7, 4.5 or 5.2 mM of TEOS cannot become clear under the same heating conditions. In other words, silica precipitation has occurred before the cooling-down, which may result in the formation of the extra silica spheres at the higher TEOS concentrations. In summary, the morphology of silica is sensitive to the concentrations of TEOS over the range of 0.7 to 5.2 mM. Thicker and more robust siliceous structures are formed with increasing the concentrations of silica precursor. Similar phenomenon has been found by Finkel et al [61] when they tried to quantify silicification in marine diatoms. The frustules

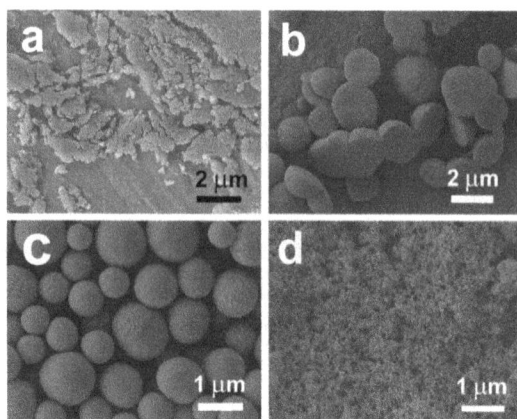

Figure 8. SEM images of the samples prepared with different concentrations of TEOS: (a) 0.7 mM; (b) 1.5 mM; (c) 3.0 mM; (d) 3.7 mM; (e) 4.5 mM and (f) 5.2 mM.

became more heavily silicified with increasing silicate concentrations over the range of 0.02–1.1 mM. Therefore, changes in the frustules thickness of diatoms may provide a paleoproxy for surface silicate concentrations under conditions where they lived [61].

For a better understanding of the morphogenesis details of the asymmetrical siliceous structures, a series of experiments with different concentrations of PL or DA were carried out. The experimental details are depicted in Table S1. Increasing PL concentration from 0 to 1.70 g/L (samples L1-L5; Fig. 9a-e and Fig. 1) leads to an increase in the aspect ratio of the obtained particles (see the line symbolized with '●' in Fig. 10). Many connected particles appear with the further increase of PL concentration to 2.0 g/L (sample L6; Fig. 9f), so their aspect ratios are not calculated. It should also be pointed out that the asymmetry between the two halves is much more significant in Fig. 9c-e relative to Fig. 1. Nevertheless, cracked spheres are prepared without the addition of PL (Fig. 9a), which can be determined by the shape of the initial DA micelle [62]. On the other hand, when the concentration of PL and the pH of initial reaction mixture are fixed at 1.70 g/L and 11.6, respectively, the particles become thinner with decreasing DA concentration (see the line symbolized with '○' in Fig. 10). In the absence of DA, silica films are finally produced (data not shown). It is probably due to the fact that the property of organic aggregates and the DA concentration in the system pose an important influence on the silica morphogenesis. DA can interact with PL in solution [63,64]. The incorporation of DA molecules can introduce more amino groups into the organic aggregates. These amino groups further interact with silanol groups of silicates, and induce the preferential deposition of silica at the organic interface. Meanwhile, increasing DA concentration inevitably leads to more DA molecules

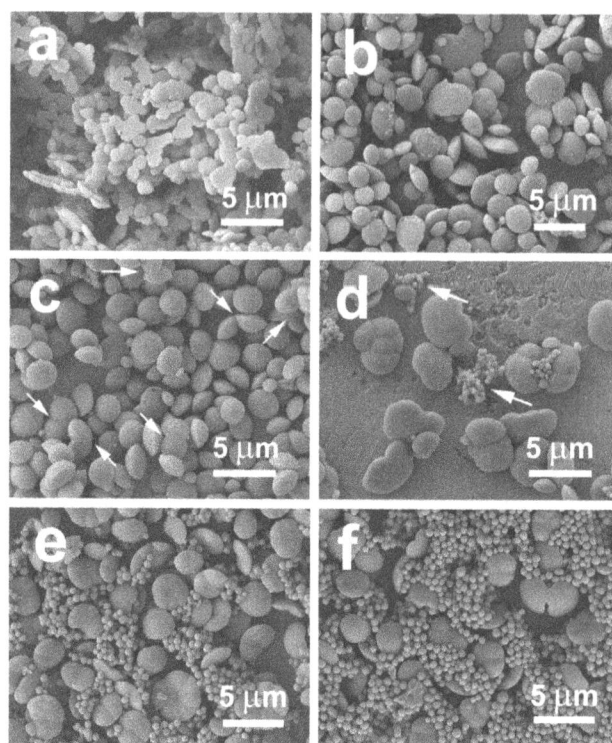

Figure 7. SEM images of the samples prepared at different volume ratios of ethanol to water: (a) 15/45; (b) 25/35; (c) 35/25 and (d) 40/20.

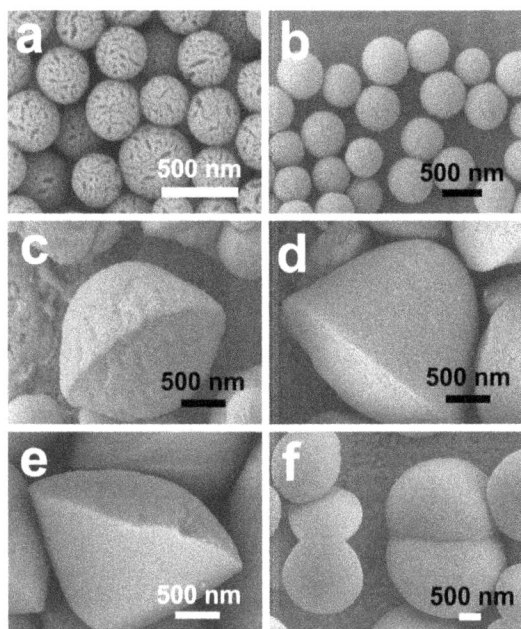

Figure 9. SEM images of the samples obtained at different concentrations of PL: (a) 0.00 g/L; (b) 0.35 g/L; (c) 0.70 g/L; (d) 1.00 g/L; (e) 1.35 g/L; (f) 2.00 g/L.

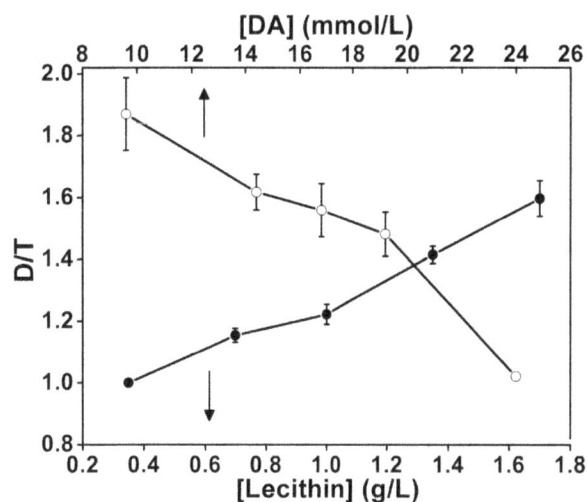

Figure 10. Relationship between the aspect ratio of particles and the concentration of PL or DA in the mixed solvent of 30 mL ethanol and 30 mL water. The amount of DA was fixed at 0.16 g for the solid circular symbols. The amount of PL was fixed at 0.10 g for the hollow circular symbols. The aspect ratio of particles was the average value obtained in the SEM images, and at least 50 particles were measured in each case.

coprecipitation into silica and/or anchoring to the surfaces of siliceous structures. These also favor the growth and thickening of the siliceous structures. As a result, the thicker silica structures can be formed at the higher concentration of DA. In contrast, thinner silica particles with higher diameter/thickness (D/T) ratios are obtained at the lower concentration of DA. Meanwhile, the excessive extension of silica at the interface causes the connection of the neighboring particles to be easier. Especially, in the absence of DA, silica deposition predominately occurs at the PL interface owe to the electrostatical interaction between Si-O$^-$ groups from silicates and ammonium head groups from PL molecules [65]. Silica deposition is confined to the extension of silica at the interface. The connection of the neighboring particles occurs commonly and the film-like siliceous structures are finally obtained without DA. On the basis of the above results, it can be concluded that both PL and DA are indispensable factors during the formation of asymmetrical silica particles. Moreover, it can be seen from Fig. 10 that these two additives display opposite effects on the aspect ratio of the resultant particles. The change of aspect ratio can be considered as an indication of silica asymmetrical growth in different directions. Therefore, the asymmetrical growth of siliceous structures can be well controlled by changing the proportion of organic components PL and DA in our experiments. In the past few years, the fabrications of asymmetrically structured silica have been reported. Non-spherical silica Janus colloids, for instance, were produced by asymmetric wet-etching at the wax/water interface [66]. However, it is not achieved directly by the asymmetrical deposition of silica. Wang et al. [67] used a single-step emulsion templating method creating budded mesoporous silica capsules with the protruding stumps formed in particular orientations, and the radiolaria-like morphology of silica with multicellular structured spines has also been obtained [68]. However, to the best of our knowledge, no report on the preparation of asymmetrical silica structures in the presence of phospholipid and organic amine can be found, and the aspect

ratio (diameter-to-thickness ratio) of the obtained particles can be finely controlled by tuning the feeding amount of organic components (Fig. 9 and 10).

In our biomimetic experiments, PL and DA are used as the biosilicification-associated model organic components to form PL-DA composite emulsion by a deliberate heating-cooling process (see the experiment details and Fig. S1) and create oil-water interface at room temperature for the deposition of silica. Specifically, PL can dissolve in the ethanol/water mixture at 80°C [69]. Therefore, a 80°C pretreatment temperature was selected to promote PL dissolution and reinforce PL-DA interaction. In fact, the solution became clear during continuous heating process, which suggests that neither organic turbidness nor silica precipitation formed in this process (Fig. S1c). After the flask is removed out of the water bath and cooled down naturally, however, white organic turbidness appears with the gradual decrease in temperature, and the phase separation can be directly observed at room temperature (Fig. S1d) [57]. Furthermore, our DLS results also reveal that the larger micelles (1781.5±712.4 nm in diameter) indeed occur in the suspension at room temperature, confirming the phase separation process present. It has been well known that the dodecyl chains of DA molecules can interact with the PL hydrophobic chains by van der waals force, while their NH$_2$ or NH$_3^+$ heads interact with P-O$^-$ groups of PL by hydrogen bonding and electrostatic interaction [63]. Therefore, in such physico-chemical environment, the organic emulsion is formed, and subsequently the hydrolysis of TEOS occurs near the oil/water interfacial region owe to the electrostatical interaction of Si-O$^-$ from silicates and the ammonium groups from PL and DA molecules [65]. As previously reported, asymmetrical polystyrene particles with flattened shapes were produced at an oil-water interface [70]. Driven by surface tension [57,70], the particles appear to be spreading at the fluid interface, which leads to the appearance of ridge and subsequent formation of discus-like particles. It should be pointed out that although the preheating process was carried out first, the formation of organic turbidness and silica precipitation did occur at room temperature. These results suggest that the precipitation of asymmetrical silica structures can be achieved by phase separation of the organic components (e.g., Fig. 1, Fig. 9c-f and Fig. 10). It appears that the interaction between the different organic molecules and their phase separation can significantly affect physico-chemical growth environment of the siliceous structures, and finally control the silica morphologies [71].

To further understand the formation details of the discus-like silica particles, some time-dependent silicification experiments are also carried out. It is found that the reaction solution turns gradually turbid during the cooling process at room temperature, and the precipitate obtained by centrifugation after 1 h of standing is organic components because the precipitate can completely dissolve in ethanol. However, the precipitate obtained after 1.25 h of standing can incompletely dissolve in ethanol, indicating that the silicified structures have formed. SEM analysis reveals that the silicified structures consist of small silica particles of ca. 250 nm in diameter (Fig. 11a), and the interconnection of the particles leads to the appearance of some larger aggregates with diameter above 400 nm, as arrowed in Fig. 11b. After 1.5 h of reaction, however, some particles with thin margin can be found (arrowed in Fig. 11c), indicating that the morphological development of the particles may occur at the oil/water interface. When the silicification system continues standing for 1.75 h, the particles have developed into discus-like embryos with diameter up to 1 μm (Fig. 11d). Moreover, a few discus-like particles with the expanding margin can be clearly observed (typically arrowed in Fig. 11d), further supporting that the formation of the discus-like structures occurs at the oil/water interface. Further prolonging the standing time to 2 or 2.25 h leads to the appearance of the well-developed asymmetric discuses of ca. 2 μm in diameter, and many of them exhibit conjoined structures (Fig. 11e-h). Combined with the results depicted in Fig.1, it is not difficult to find that at the oil/water interface, aggregation, fusion and margin expansion of the small siliceous particles, as well as further growth lead to the monodisperse perfect discus-like asymmetric structures.

On the basis of our time-dependent experiments, a tentative mechanism is proposed and illustrated in Fig. 12 for the formation of discus-like asymmetric silica. Namely, when the bulk solution is cooled down naturally, the hydrolysis of TEOS and the precipitation of silica occur slowly near the oil/water interfacial region with the phase separation of organic components. The silica formation begins with the appearance of small particles (Fig. 12a, Fig. 11a). With the growth and aggregation of them, larger aggregates of silica particles can be formed (Fig. 12b, Fig. 11b). Further growth of these aggregates get their surfaces smoother, and the growth environment (oil/water interfacial region) facilitates their expansion at the oil/water interfaces. Therefore, flake-like silica structures appear (Fig. 12c, Fig. 11c), and further develop into discus-like particles with a diameter of ca. 1 μm, which is much smaller than the final product (2–3 μm) (Fig. 12d, Fig. 11d). As the margin expansion process continues, several neighboring particles (e.g., two particles) are joined together to form the "conjoined structures" (Fig. 12e, Fig. 11e-f). The further fusion and growth of the conjoined structures lead to discus-like particles with diameter above 2 μm (Fig. 12f, Fig. 11g-h). Finally, the fully development of their two halves results in the formation of well-defined asymmetric discus-like structures of 2–3 μm in diameter (Fig. 12g, Fig. 1).

Implication for biosilicification

Silicification in diatoms is a complicated process involving architecture design from nano- to microsize level [72]. The siliceous structures formed in different scales and stages can be

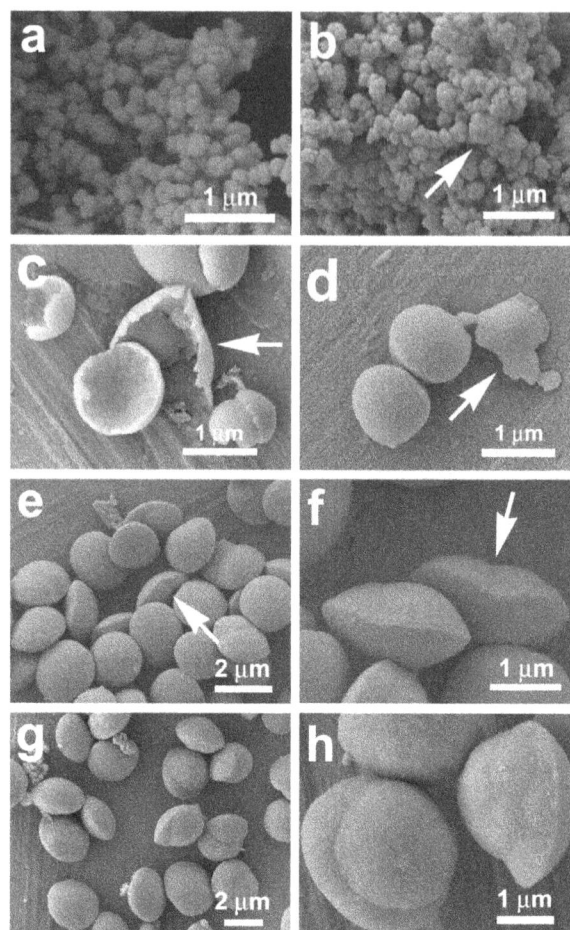

Figure 11. FESEM images of silica particles after the reaction mixtures were first heated at 80°C for 1 day and then cooled down at room temperature for (a,b) 1.25 h; (c) 1.5 h; (d) 1.75 h; (e, f) 2 h and (g, h) 2.25 h.

unified in the mineralization system of diatoms, and finally assemble into hierarchical and multifunctional frustules. The valve development of *Thalassiosira eccentrica* can be divided into three stages. Formation of base layer (areolae) defines the structure in the x, y plane (Stage 1), and subsequent deposition (Stage 2) involves expansion in the z axis but only in one direction [34,43,44]. During the development of the outer layer (Stage 3), however, the differentiation of the plane occurs again, forming a

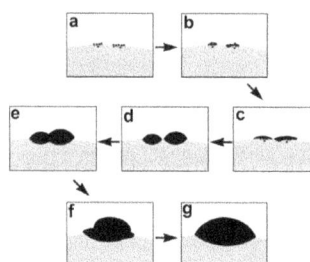

Figure 12. Schematic illustration of the formation of discus-like silica particles. The organic precipitates, silica particles and reaction solution are stained in gray, black and white, respectively.

right angle to the previous plane (Stage 2) and lying parallel to the base layer (Stage 1) [43]. The formation of the two-dimensional system of hexagonal meshes (areolae) in stage 1 can be well explained by the phase separation model [25]. However, it is not clear whether this model is also suitable for the asymmetrical precipitation of silica including vertical expansion in stage 2 and horizontal growth in stage 3.

Space-limited by the membrane-bound compartment and promoted by organic amines, siliceous base layer with pores in a hexagonal arrangement formed during the phase separation of organic droplets [40,41]. However, the role of organic amines and phospholipids on biosilicification may be not restricted to influencing the development of base layer. Our experiments exhibit the controlled deposition of asymmetrical silica particles during the phase separation. The asymmetrical particles emerge as the concentration of PL is over 0.70 g/L. The addition of PL favors the morphology transition from spherical to discus-like particles and the aspect ratio regularly increases with increasing the concentration of PL (e.g., Fig. 9). These results show that phospholipids can provide distinct chemical influences in organic-amine-induced silica precipitation [34,43,64]. That is, their aspect ratios can be easily adjusted through varying the stoichiometric compositions of the mineralization system (including DA and PL, Fig. 9 and 10). And the degree of fusion among the neighboring siliceous structures is drastically affected by the concentration of silica precursor (Fig. 8). Therefore, it can be presumed that the phase separation of organic droplets is still an important process for the oriented differentiation of silica. In other words, the phase separation model may be broadened to explain the formation of siliceous structures in the last two stages.

Conclusions

A series of experiments were accomplished by introducing PL and DA into the reaction system to initiate phase separation of organic components and influence the morphogenesis of silica. The results show that this phase separation process leads to the formation of asymmetrically non-spherical silica structures, and the aspect ratios of the asymmetrical structures can be well controlled by varying the concentrations of PL and DA. A tentative mechanism is also proposed based on the time-dependent experiments. Moreover, controlling the degree of fusion among the neighboring siliceous structures can be achieved via modulating the concentration of silica precursor (TEOS) in the silicified region. Based on the special importance of phospholipids (e.g., silicalemma), organic-amines and the phase separation process for biosilicification, our results suggest that in addition to explaining the biosilica nanopatterning, the phase separation process may be also involved in the growth differentiation of siliceous structures in specific directions. This provides a new insight into the mechanism of biosilicification.

Acknowledgments

We thank Mr. Jianliu Huang and Mr. Ming Li for the help of FESEM analysis.

Author Contributions

Conceived and designed the experiments: J-YS Q-ZY. Performed the experiments: J-YS X-ML. Analyzed the data: G-TZ J-YS Q-ZY. Contributed reagents/materials/analysis tools: S-QF. Wrote the paper: J-YS G-TZ.

References

1. Gower LB (2008) Biomimetic model systems for investigating the amorphous precursor pathway and its role in biomineralization. Chem Rev 108: 4551–4627.
2. Kröger N, Poulsen N (2008) Diatoms—From cell wall biogenesis to nanotechnology. Annu Rev Genet 42: 83–107.
3. Mann S (2001) Biomineralization: Principles and concepts in bioinorganic materials chemistry (Oxford Chemistry Masters). Oxford: Oxford University Press. 210 p.
4. Bäuerlein E, ed (2004) Biomineralization: Progress in Biology, Molecular Biology and Application, 2nd, Completely Revised and Extended Edition. Weinheim :Wiley-VCH. 361 p.
5. Gordon R, Drum RW (1994) The chemical basis of diatom morphogenesis. Int Rev Cytol 150: 243–372.
6. Mann S (1993) Molecular tectonics in biomineralization and biomimetic materials chemistry. Nature 365: 499–505.
7. Oliver S, Kupermann A, Coombs N, Lough A, Ozin GA (1995) Lamellar aluminophosphates with surface patterns that mimic diatom and radiolarian microskeletons. Nature 378: 47–50.
8. Round F, Crawford R, Mann D (1990) The Diatoms: Biology. & Morphology of the Genera. Cambridge :Cambridge University Press. 745 p.
9. Livage J, Coradin T (2006) Living Cells in Oxide Glasses. Rev Mineral Geochem 64: 315–332.
10. Nakajima T, Volcani BE (1969) 3,4-Dihydroxyproline: A new amino acid in diatom cell walls. Science 164: 1400–1401.
11. Nakajima T, Volcani BE (1970) ε-N-trimethyl-L-δ-hydroxylysine phosphate and its nonphosphorylated compound in diatom cell walls. Biochem Biophys Res Commun 39: 28–33.
12. Patwardhan SV, Clarson SJ, Perry CC (2005) On the role(s) of additives in bioinspired silicification. Chem Commun: 1113–1121.
13. Matsukizono H, Jin RH (2012) High-Temperature-Resistant Chiral Silica Generated on Chiral Crystalline Templates at Neutral pH and Ambient Conditions. Angew Chem Int Ed 51: 5862–5865.
14. Hoagland KD, Rosowski JR, Gretz MR, Roemer SC (1993) Diatom extracellular polymeric substances-function, fine-structure, chemistry, and physiology. J Phycol 29: 537–566.
15. Kinrade SD, Gillson AME, Knight CTG (2002) Silicon-^{29}NMR evidence of a transient hexavalent silicon complex in the diatom *Navicula pelliculosa*. J Chem Soc Dalton Trans: 307–309.
16. Poulsen N, Sumper M, Kröger N (2003) Biosilica formation in diatoms: characterization of native silaffin-2 and its role in silica morphogenesis. PNAS 100: 12075–12080.
17. Poulsen N, Kröger N (2004) Silica morphogenesis by alternative processing of silaffins in the diatom *Thalassiosira pseudonana*. J Biol Chem 279: 42993–42999.
18. Davis AK, Hildebrand M, Palenik B (2005) A stress-induced protein associated with the girdle band region of the diatom *Thalassiosira pseudonana* (Bacillariophyta). J Phycol 41: 577–589.
19. Dickerson MB, Sandhage KH, Naik RR (2008) Protein- and peptide-directed syntheses of inorganic materials. Chem Rev 279: 4935–4978.
20. Ehrlich H, Deutzmann R, Brunner E, Cappellini E, Koon H, et al. (2010) Mineralization of the metre-long biosilica structures of glass sponges is templated on hydroxylated collagen. Nat Chem 2: 1084–1088.
21. Kröger N, Deutzmann R, Bergsdorf C, Sumper M (2000) Species-specific polyamines from diatoms control silica morphology. PNAS 97: 14133–14138.
22. Patwardhan SV (2011) Biomimetic and bioinspired silica: recent developments and applications. Chem Commun 47: 7567–7582.
23. Nassif N, Livage J (2011) From diatoms to silica-based biohybrids. Chem Soc Rev 40: 849–859.

24. Sumper M, Brunner E (2006) Learning from diatoms: Nature's tools for the production of nanostructured silica. Adv Funct Mater 16: 217–228.

25. Sumper M (2002) A phase separation model for the nanopatterning of diatom biosilica. Science 295: 2430–2433.

26. Meldrum FC, Cölfen H (2008) Controlling mineral morphologies and structures in biological and synthetic systems. Chem Rev 108: 4332–4432.

27. Sumper M, Lehmann G (2006) Silica pattern formation in diatoms: Species-specific polyamine biosynthesis. ChemBioChem 7: 1419–1427.

28. Coradin T, Lopez PJ (2003) Biogenic silica patterning: simple chemistry or subtle biology? ChemBioChem 4: 251–259.

29. Schröder HC, Wang X, Tremel W, Ushijimad H, Müller WEG (2008) Biofabrication of biosilica-glass by living organisms. Nat Prod Rep 25: 455–474.

30. Matsunaga S, Sakai R, Jimbo M, Kamiya H (2007) Long-chain polyamines (LCPAs) from marine sponge: Possible implication in spicule formation. ChemBioChem 8: 1729–1735.

31. Müller WE, Rothenberger M, Boreiko A, Tremel W, Reiber A, et al. (2005) Formation of siliceous spicules in the marine demosponge *Suberites domuncula*. Cell Tissue Res 321: 285–297.

32. Drum RW, Pankratz HS (1964) Post mitotic fine structure of *Gomphonema parvulum*. J Ultrastruct Res 10: 217–223.

33. Bäuerlein E (2003) Biomineralization of Unicellular Organisms: An Unusual Membrane Biochemistry for the Production of Inorganic Nano- and Microstructures. Angew Chem Int Ed 42: 614–641.

34. Hildebrand M (2008) Diatoms, biomineralization processes, and genomics. Chem Rev 108: 4855–4874.

35. Hildebrand M, Kim S, Shi D, Scott K, Subramaniam S (2009) 3D imaging of diatoms with ion-abrasion scanning electron microscopy. J Struct Biol 166: 316–328.

36. Ji Q, Iwaura R, Kogiso M, Jung JH, Yoshida K, et al. (2004) Direct sol-gel replication without catalyst in an aqueous gel system: From a lipid nanotube with a single bilayer wall to a uniform silica hollow cylinder with an ultrathin wall. Chem Mater 16: 250–254.

37. Ji Q, Iwaura R, Shimizu T (2007) Regulation of silica nanotube diameters: Sol-gel transcription using solvent-sensitive morphological change of peptidic lipid nanotubes as templates. Chem Mater 19: 1329–1334.

38. Tesson B, Masse S, Laurent G, Maquet J, Livage J, et al. (2008) Contribution of multi-nuclear solid state NMR to the characterization of the *Thalassiosira pseudonana* diatom cell wall. Anal Bioanal Chem 390: 1889–1898.

39. Tesson B, Genet MJ, Fernandez V, Degand S, Rouxhet PG, et al. (2009) Surface chemical composition of diatoms. ChemBioChem 10: 2011–2024.

40. Noll F, Sumper M, Hampp N (2002) Nanostructure of Diatom Silica Surfaces and of Biomimetic Analogues. Nano Lett 2: 91–95.

41. Zurzolo C, Bowler C (1999) Exploring Bioinorganic Pattern Formation in Diatoms. A Story of Polarized Trafficking. Plant Physiol 127: 1339–1345.

42. Losic D, Mitchell JG, Voelcker NH (2009) Diatomaceous lessons in nanotechnology and advanced materials. Adv Mater 21: 2947–2958.

43. Schmid AMM, Schulz D (1979) Wall Morphogenesis in Diatoms: Deposition of Silica by Cytoplasmic Vesicles. Protoplasma 100: 267–288.

44. Hildebrand M, York E, Kelz JI, Davis AK, Frigeri LG, et al. (2006) Nanoscale control of silica morphology and three-dimensional structure during diatom cell wall formation. J Mater Res 21: 2689–2698.

45. Palsdottir H, Hunte C (2004) Lipids in membrane protein structures. Biochimica et Biophysica Acta 1666: 2–18.

46. Yan Z, Li Y, Wang S, Xu Z, Chen Y, et al. (2010) Artificial frustule prepared through a single-templating approach. Chem Commun 46: 8410–8412.

47. Qu XF, Yao QZ, Zhou GT, Fu SQ, Huang JL (2010) Formation of hollow magnetite microspheres and their evolution into durian-like architectures. J Phys Chem C 114: 8734–8740.

48. Ji Q, Kamiya S, Jung JH, Shimizu T (2005) Self-assembly of glycolipids on silica nanotube templates yielding hybrid nanotubes with concentric organic and inorganic layers. J Mater Chem 15: 743–748.

49. Michaux F, Carteret C, Stébé MJ, Blin JL (2008) Hydrothermal stability of mesostructured silica prepared using a nonionic fluorinated surfactant. Micropor Mesopor Mat 116: 308–317.

50. Venkatathri N, Srivastava R, Yun DS, Yoo JW (2008) Synthesis of a novel class of mesoporous hollow silica from organic templates. Micropor Mesopor Mat 112: 147–152.

51. Patwardhan SV, Maheshwari R, Mukherjee N, Kiick KL, Clarson SJ (2006) Conformation and Assembly of Polypeptide Scaffolds in Templating the Synthesis of Silica: An Example of a Polylysine Macromolecular "Switch". Biomacromolecules 7: 491–497.

52. Zhao Y, Qi Y, Wei Y, Zhang Y, Zhang S, et al. (2008) Incorporation of Ag nanostructures into channels of nitrided mesoporous silica. Micropor Mesopor Mat 111: 300–306.

53. Sadasivan S, Khushalani D, Mann S (2005) Synthesis of Calcium Phosphate Nanofilaments in Reverse Micelles. Chem Mater 17: 2765–2770.

54. Lin HY, Chen YW (2005) Preparation of spherical hexagonal mesoporous silica. J Porous Mat 12: 95–105.

55. Dimos K, Stathi P, Karakassides MA, Deligiannakis Y (2009) Synthesis and characterization of hybrid MCM-41 materials for heavy metal adsorption. Micropor Mesopor Mat 126: 65–71.

56. Hukkamaki J, Pakkanen TT (2003) Amorphous silica materials prepared by neutral templating route using amine-terminated templates. Micropor Mesopor Mat 65: 189–196.

57. Zhang HA, Bandosz TJ, Akins DL (2011) Template-free synthesis of silica ellipsoids. Chem Commun 47: 7791–7793.

58. Jiang S, Granick S (2008) Controlling the geometry (Janus balance) of amphiphilic colloidal particles. Langmuir 24: 2438–2445.

59. Di Renzo F, Testa F, Chen JD, Cambon H, Galarneau A, et al. (1999) Textural control of micelle-templated mesoporous silicates: the effects of co-surfactants and alkalinity. Micropor Mesopor Mat 28: 437–446.

60. Shan W, Wang B, Zhang Y, Tang Y (2005) Fabrication of lotus-leaf-like nanoporous silica flakes with controlled thickness. Chem Commun 1877–1879.

61. Finkel ZV, Matheson KA, Regan KS, Irwin AJ (2010) Genotypic and phenotypic variation in diatom silicification under paleo-oceanographic conditions. Geobiology 8: 433–445.

62. Hu J, Shan W, Zhang W, Zhang Y, Tang Y (2010) Morphological diversity of dual meso-structured HMS and their transformation process. Micropor Mesopor Mat 129: 210–219.

63. Galarneau A, Sartori F, Cangiotti M, Mineva T, Di Renzo F, et al. (2010) Sponge mesoporous silica formation using disordered phospholipid bilayers as template. J Phys Chem B 114: 2140–2152.

64. Shi JY, Yao QZ, Li XM, Zhou GT, Fu SQ (2012) Controlled morphogenesis of amorphous silica and its relevance to biosilicification. Am Mineral 97: 1381–1393.

65. Baral S, Schoen P (1993) Silica-deposited phospholipid tubules as a precurosor to hollow submicron-diameter silica cylinders. Chem Mater 5: 145–147.

66. Liu B, Zhang CL, Liu JG, Qu XZ, Yang ZZ (2009) Janus non-spherical colloids by asymmetric wet-etching. Chem Commun 3871–3873.

67. Wang J, Xiao Q, Zhou H, Sun P, Yuan Z, et al. (2006) Mesoporous silica hollow spheres: Hierarchical structure controlled by kinetic self-assembly. Adv Mater 18: 3284–3288.

68. Wang J, Xiao Q, Zhou H, Sun P, Li B, et al. (2007) Radiolaria-like silica with radial spines fabricated by a dynamic self-organization. J Phys Chem C 111: 16544–16548.

69. Konno Y, Naito N, Yoshimura A, Aramaki K (2010) Phase behavior and hydrated solid structure in lysophospholipid/long-chain alcohol/water system and effect of cholesterol addition. J Oleo Sci 59: 581–587.

70. Park BJ, Furst EM (2010) Fabrication of unusual asymmetric colloids at an oil-water interface. Langmuir 26: 10406–10410.

71. Ramanathan R, Campbell JL, Soni SK, Bhargava SK, Bansal V (2011) Cationic amino acids specific biomimetic silicification in ionic liquid: a quest to understand the formation of 3-D structures in diatoms. PLoS One 6: e17707.

72. Davis A, Hildebrand M (2007) Molecular processes of biosilicification in diatoms. In: Sigel H, Sigel A (Eds.), Metal Ions in Life Sciences. Biomineralization. From Nature to Application, vol. 4. London : Wiley. 255–294.

11

Unsaturated Fatty Acid, *cis*-2-Decenoic Acid, in Combination with Disinfectants or Antibiotics Removes Pre-Established Biofilms Formed by Food-Related Bacteria

Shayesteh Sepehr, Azadeh Rahmani-Badi*, Hamta Babaie-Naiej, Mohammad Reza Soudi

Department of Biology, Alzahra University, Tehran, Iran

Abstract

Biofilm formation by food-related bacteria and food-related pathogenesis are significant problems in the food industry. Even though much disinfection and mechanical procedure exist for removal of biofilms, they may fail to eliminate pre-established biofilms. *cis*-2 decenoic acid (CDA), an unsaturated fatty acid messenger produced by *Pseudomonas aeruginosa*, is reportedly capable of inducing the dispersion of established biofilms by multiple types of microorganisms. However, whether CDA has potential to boost the actions of certain antimicrobials is unknown. Here, the activity of CDA as an inducer of pre-established biofilms dispersal, formed by four main food pathogens; *Staphylococcus aureus*, *Bacillus cereus*, *Salmonella enterica* and *E. coli*, was measured using both semi-batch and continuous cultures bioassays. To assess the ability of CDA combined biocides treatments to remove pre-established biofilms formed on stainless steel discs, CFU counts were performed for both treated and untreated cultures. Eradication of the biofilms by CDA combined antibiotics was evaluated using crystal violet staining. The effect of CDA combined treatments (antibiotics and disinfectants) on biofilm surface area and bacteria viability was evaluated using fluorescence microscopy, digital image analysis and LIVE/DEAD staining. MICs were also determined to assess the probable inhibitory effects of CDA combined treatments on the growth of tested microorganisms' planktonic cells. Treatment of pre-established biofilms with only 310 nM CDA resulted in at least two-fold increase in the number of planktonic cells in all cultures. While antibiotics or disinfectants alone exerted a trivial effect on CFU counts and percentage of surface area covered by the biofilms, combinational treatments with both 310 nM CDA and antibiotics or disinfectants led to approximate 80% reduction in biofilm biomass. These data suggests that combined treatments with CDA would pave the way toward developing new strategies to control biofilms with widespread applications in industry as well as medicine.

Editor: Mark Alexander Webber, University of Birmingham, United Kingdom

Funding: The authors have no support or funding to report.

Competing Interests: The authors have declared that no competing interests exist.

* Email: a.rahmanibadi@gmail.com

Introduction

The biofilm mode of growth is a basic survival strategy deployed by microorganisms in a wide range of environmental, industrial and clinical settings [1]. Biofilms are defined as sessile communities of cells attached to each other and/or to surfaces or interfaces which are embedded in a self-produced matrix of extracellular polymeric substances (EPS) [2,3]. A function frequently attributed to EPS is their general protective effect on sessile microorganisms against adverse conditions including presence of most antimicrobial agents [3]. This is supposed to be due mainly to physiological characteristics of biofilm bacteria, but also to a barrier function of EPS [4]. According to Körstgens *et al.* [5] the EPS matrix also provides biofilm mechanical stability by filling and forming the space between the bacterial cells, keeping them together.

Biofilm formation by food-related bacteria and food-related pathogenesis are significant problems in the food industry. The attachment of the bacteria to the food product or the product contact surfaces leads to serious hygienic problems and economic losses due to food spoilage [6,7].

For the sanitation and removal of biofilms in food industry, chemical agents and mechanical forces (sonication, flushing, etc.) are parameters often involved simultaneously. Mechanical actions only allow the removal of the biofilms from the surfaces and once established, biofilms are harder to be removed completely [8]. They also cannot kill biofilms and biofilm cells might later re-attach to other surfaces and form a biofilm [8,9]. Thus, disinfection procedure is indispensible with the intention of killing them. However, it is important to note that most of the disinfection processes that are implemented are based upon the results of planktonic tests [10]. Therefore, such tests do not mimic the behavior of sessile cells and can be highly ineffective when applied to control biofilms. Biofilms have been reported as possessing susceptibilities towards antimicrobials that are 100–1000 times less than equivalent populations of planktonic counterparts [11]. If a microbial population faces high concentrations of an antimicrobial product, susceptible cells will be inactivated. Although some cells may possess a degree of natural resistance and physiological plasticity or they may acquire it later through mutation or genetic exchange. These processes allow the microorganisms to survive and grow [4].

To address the need for novel and improved measures against biofilms especially pre-established biofilms, a clear strategy is to study the biofilm life cycle and identify key trigger points that regulate biofilm development. To control biofilm, the last stage of biofilm development presents several advantages, where a coordinated dispersal of biofilm cells is possible. Induction of biofilm dispersal could potentially use the microorganisms' own energy to remove established biofilms, revert cells to a planktonic phenotype and restore their susceptibility to disinfectants and antibiotics.

It has been recently reported that *P. aeruginosa* produces an unsaturated fatty acid, *cis*-2-decenoic acid (C_{10}: Δ^2, CDA), which is capable of inducing the dispersion of pre-established biofilms by multiple types of bacteria [12]. Furthermore, CDA is also capable of inducing dispersion in biofilms of *Candida albicans*, indicating that this signalling molecule is involved in inter-species and inter-kingdom signalling where it can modulate the behavior of other microorganisms that do not produce the signal [12]. CDA is a promising candidate for control of biofilms in different industrial and clinical settings as it has a broad-spectrum of activity in addition to the fact that it has no cytotoxic effects to human cells at nano-molar ranges [13]. However, whether CDA has potential to boost the actions of certain disinfectants and antibiotics is unknown.

Therefore, in the current work, the ability of nano-molar concentrations of CDA to induce dispersal in pre-established biofilms, formed by four main food-borne biofilm producer bacteria (*Bacillus cereus*, *Staphylococcus aureus*, *Salmonella enterica* and *E. coli*) as well as to remove and kill their biofilms when combined with biocides or antibiotics were studied? Besides, the ability of CDA to increase the inhibitory effects of antimicrobials on the growth of tested microorganisms' planktonic cells was investigated.

Materials and Methods

Bacterial strains, media and growth conditions

The microorganisms used in the present study included *E. coli* (ATCC 25922), *Staphylococcus aureus* (ATCC 25923), *Bacillus cereus* (ATCC 11778) and *Salmonella enterica* (ATCC 14028). Overnight cultures were grown at optimum temperature for each microorganism in Luria Bertani (LB) medium (Merck, Germany) for *E. coli*, *B. cereus* and *S. enterica* and in Tryptic Soy Broth (TSB) medium (Merck, Germany) for *S. aureus*. Biofilm experiments were performed in 1/5 strength LB for *E. coli*, *B. cereus* and *S. enterica*, and in 1/5 strength TSB for *S. aureus*.

Chemicals and antimicrobial compounds

Three different concentrations of CDA (U-Chemo, China) (100, 310 or 620 nM) were used. These concentrations were previously observed to have the most effect on inducing the dispersion of pre-established biofilms [12] with no cytotoxic effects on human cells [13]. Ethanol (10%) (Merck, Germany) was used as a carrier for CDA. Two commercial disinfectants, Epimax S (Epimax, Iran) and Percidine (Behban chemistry, Iran), were used for their widespread applications in food industry in Iran. Their active ingredients were hydrogen peroxide (45–50%) and peracetic acid (15%), respectively. Final concentration of 120 ppm hydrogen peroxide for Epimax S and 70 ppm peracetic acid for Percidine was used. These concentrations were respectively 3 and 4 times

lower than the manufacturer's recommended concentration for disinfection purposes. This study also examined three antibiotics commonly used in medical and veterinary practice; ciprofloxacin (Sigma, USA) for both gram positive and gram negative tested microorganisms, vancomycin (Sigma, USA) for only gram positive bacteria, and ampicillin (Sigma, USA) for gram negative strains. Ciprofloxacin (Sigma) was used at a final concentration of 1 µg.ml^{-1}, vancomycin at (4 µg.ml^{-1} and 256 µg.ml^{-1} for *S. aureus* and *B. cereus*, respectively) and ampicillin at 256 µg.ml^{-1}. The concentrations of antibiotic selected for use were established in our laboratory to be effective against planktonic cells but have no inhibitory effect on the tested pathogens' biofilm cells.

Biofilm dispersal bioassays in petri dishes

Biofilms were grown on the inside surface of petri dishes by using a semi-batch culture method in which the medium was replaced every 24 h. This was done to reduce the accumulation of native dispersion inducing factors and to allow mature biofilms form. Biofilms grown in this manner were then treated with three different concentrations of CDA (100, 310 or 620 nM) as dispersion inducer or just the carrier (10% ethanol) as a control to release cells into the bulk liquid and evaluate dispersed cell number by measuring the optical density (OD). To cultivate biofilms, overnight cultures of tested microorganisms were diluted 1:1,000 into fifteen ml of growth medium, (except for *B. cereus* that was diluted 200 times), inoculated in sterile petri dishes and incubated at room temperature with 30 rpm shaking. Medium in the plates was replaced every 24 h for 5 days. After the last exchange of medium, the cells were allowed to grow for about 1 h and then dispersion induction was tested by replacing the growth medium with fresh medium containing one of the indicated concentrations of CDA or just the carrier as a control and the cells were incubated for a further 1 h. Afterward, Medium containing dispersed cells was transferred by pipette to a 50 ml Erlenmeyer and was homogenized for 30 s at 5,000 rpm with a WiseTis-Homogenizer model HG-150 (Daihan Scientific Co., Ltd., Korea) to ensure the separation of cells. The cell density was then determined based on the OD$_{600}$ with an UV/VIS spectrophotometer model T80$^+$ (PG Instruments, Ltd., China). Biofilm dispersal bioassays were performed in triplicates in at least three individual experiments for each concentration.

Dispersion bioassays of biofilms in biofilm tube reactors

Biofilms were also grown on the interior surfaces of tubing reactors. A continuous once-through tube reactor system was configured by using eight silicone reactor tubes (40-cm length by 3-mm inner diameter), connected to an eight-roller head peristaltic pump (Baoding Longer Precision Pump Co., Ltd., China) and medium reservoir, via an additional silicone tubing. Medium was pumped through the tubing to a closed effluent medium reservoir. The entire system was closed to the outside environment but maintained in equilibrium with atmospheric pressure by a 0.2-µm-pore-size gas-permeable filter fitted to medium reservoir. The assembled system was sterilized by autoclaving prior to inoculation. The silicone tubes were inoculated by syringe injection through a septum 1 cm upstream from each reactor tube, with 3 ml of overnight cultures of each microorganism. Bacteria cells were allowed to attach (static incubation) to the tubing for 1 h, after which the flow was started at an elution rate of 280 µl.min^{-1}. After 5 days of biofilm cultures, the influent medium was switched from fresh medium in the test lines to one of the three concentrations of CDA. Control lines were switched to new lines containing just the carrier (ethanol 10%). Samples were collected in test tubes on ice and were subsequently homogenized and cell density was determined as mentioned above. All experiments were repeated three times.

The concentration of CDA that induced the most dispersal in the examined biofilms in both petri dish and tube reactor cultures was used for further studies.

Combined CDA and biocide treatment of pre-established biofilms, formed on stainless steel discs

For disinfectants alone and combined CDA susceptibility testing, biofilms were formed on stainless steel (SS) type 316 discs with a surface area of 2.7 cm^2, placed at the bottom of wells in 24-well plates. To grow biofilms, 2.5 ml of overnight cultures of each microorganism, previously diluted 1:1,000 in biofilm medium (except for *B. cereus* as indicated above), was added to each well and incubated at room temperature with gentle shaking. Medium in the wells was replaced every 24 h for 5 days to allow mature biofilms form. Biofilms were then treated for 1 h with indicated concentrations of disinfectants alone or combined with 310 nM CDA as CDA at this concentration induced the most dispersal in the tested biofilms in both petri dish and tube reactor cultures. At the end of the experimental period, the SS discs were washed with PBS to remove non-adherent bacteria, carefully transferred to sterile glass tubes containing 1 ml of sterile 0.89% NaCl and washed with another 1 ml of 0.89% NaCl. To remove the biofilm from the SS discs, the glass tubes with the biofilms were placed in an ultrasonic bath for 10 min at room temperature. CFU were enumerated after plating on LB agar to assess bacterial viability. All experiments were repeated three times.

Antibiotics combined CDA biofilm microtiter plate assays

To assess the effect of antibiotics alone and in combination with CDA, biofilms were grown on the inside surface of sterile polystyrene 96-well plates. For biofilm cultures, plates were inoculated with 150 μl/well of overnight culture containing the tested organism, previously diluted in growth medium (as indicated above) and incubated at 37 °C with shaking at 120 rpm. Medium within each well was replaced every 24 h for 5 days. Biofilms were then treated for 1 h with indicated concentrations of antibiotics alone or combined with 310 nM CDA. The plates were gently rinsed twice with PBS to remove planktonic and loosely adherent organisms. After rinsing, the plates were shaken dry and each well of each plate stained with 160 μl of an aqueous 0.1% crystal violet solution in distilled water. After allowing the stain to adhere to the biofilms for 15 min, each plate was again rinsed with PBS until no more stain could be rinsed from the plate. Each plate was again shaken dry, inverted and allowed to dry thoroughly for 30 min. Finally, 170 μl of a 30% acetic acid solution was pipetted into each well to desorb the adhered stain back into solution. After allowing 30 min for the adhered stain to dissolve into the destaining solution, the biofilm in each well was quantified via absorbance at OD_{590} using a ELx808 Absorbance Microplate Reader (BioTek Instruments, Inc., Winooski, VT) [14]. All experiments were repeated at least three times.

Combined CDA and antimicrobial treatment of planktonic cells

We have evaluated the probable inhibitory effects on the growth of tested microorganisms' planktonic cells by biocides or antibiotics alone and in combination with three different concentrations of CDA (100, 310 or 620 nM). The MICs were determined in triplicate in Mueller-Hinton broth by using microdilution assay with bacteria at a density of 10^5 CFU/ml. Plates were incubated for 24 h at optimum temperature for each bacterium. The lowest concentration of antibiotics or biocides where there was no growth after 24 h was taken as the MIC [15,16].

Flow cell (continuous-culture) biofilm experiments; disinfectants and antibiotics sensitivity assays and surface area coverage

To observe the effect of CDA combined antimicrobial treatments on biofilm surface area and bacteria viability, biofilms were also grown in continuous culture flow cells (channel dimensions, 1×4×40 mm). Appropriate sterile biofilm medium was pumped from a 5-Liter vessel through silicone tubing to the flow cell using an eight-roller-head peristaltic pump (Baoding Longer Precision Pump Co., Ltd., China) at a flow rate of 280 μl.min^{-1}. Medium leaving the flow cell was discharged to an effluent reservoir via silicone tubing. The entire system was closed to the outside environment but maintained in equilibrium with atmospheric pressure by a 0.2-μm-pore-size gas-permeable filter fitted to each vessel. Channels were inoculated with overnight cultures of tested organism and incubated without flow for 1 h, at room temperature. After 48 h of biofilm cultures, the influent medium was switched from fresh medium in the test lines to the antimicrobials in combination with 310 nM CDA. Control lines were switched to new lines containing only examined antimicrobial agents. After 1 h treatment, biofilms were stained with a LIVE/DEAD *Bac*Light bacterial viability kit (Molecular Probes). The two stock solutions of the stain (SYTO 9 and propidium iodide) were diluted to 3 μl.ml^{-1} in biofilm medium and injected into the flow channels. Live SYTO 9-stained cells and dead propidium iodide-stained cells were visualized using epifluorescence microscopy (CETI, Belgium). 15 selected fields of view per flow cell were imaged in the XY plane, at regular intervals and across the entire channels. Image analysis (ImageJ Software, NIH) was performed and results were presented as the percentage of total biofilm surface reduction in cultures treated with combined CDA and antimicrobial treatments relative to the total biofilm surface in control cultures that were not exposed to CDA. Three replicates per experiment were used and at least 2 independent repetitions of experiments were performed.

Statistical Analysis

All data were analyzed using analysis of variance (ANOVA) by the general linear model procedure of Minitab data analysis software (release 16, Minitab Inc., PA. USA). Pairwise comparisons were then made between all of the groups using Tukey's method. P values <0.05 were regarded as significant. All measurements were carried out in triplicate.

Results

Very low concentrations of CDA induce biofilm dispersal

We investigated the effect of exposure to nano-molar concentrations of CDA on pre-established biofilms in the petri dish cultures. In all cultures tested, CDA treatments resulted in a significant increase in the populations of planktonic cells released into the bulk liquid compared to untreated control samples (Figure 1A). The greatest effect was repeatedly observed with 310 nM CDA with at least two-fold increase in the number of planktonic cells. However, no significant differences were detected in the number of planktonic cells after exposure of *S. enterica* biofilms to 310 and 620 nM CDA (*P*-value <0.05) (Figure 1A). Following exposure to 310 nM CDA, the most significant increase in planktonic population was observed in the case of *E. coli* biofilms ($OD_{600} = 0.9 \pm 0.02$, SE, *P*-value <0.05) versus untreated controls

A

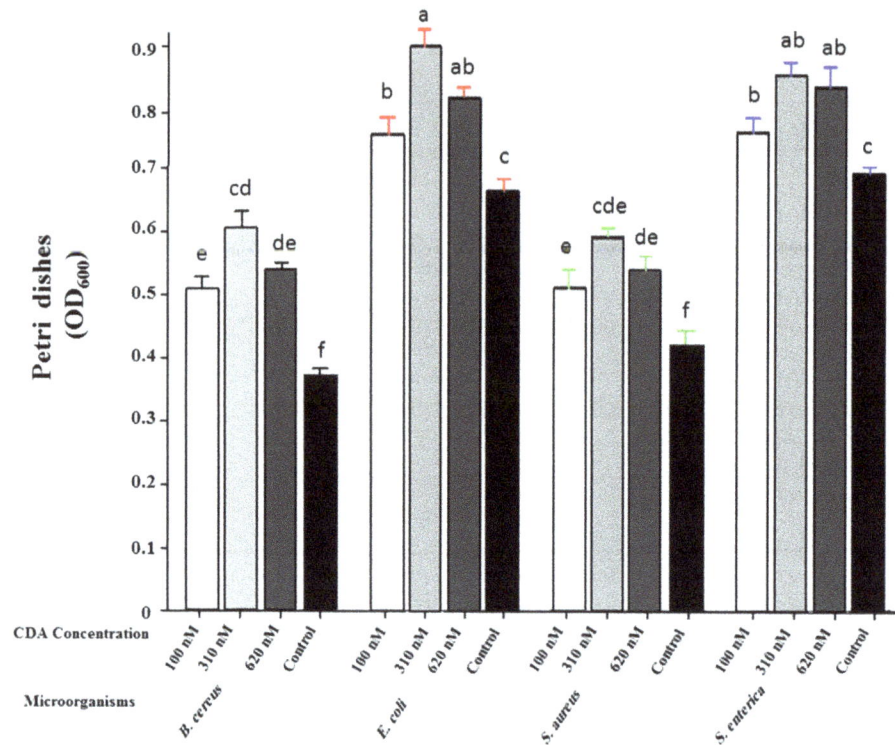

B

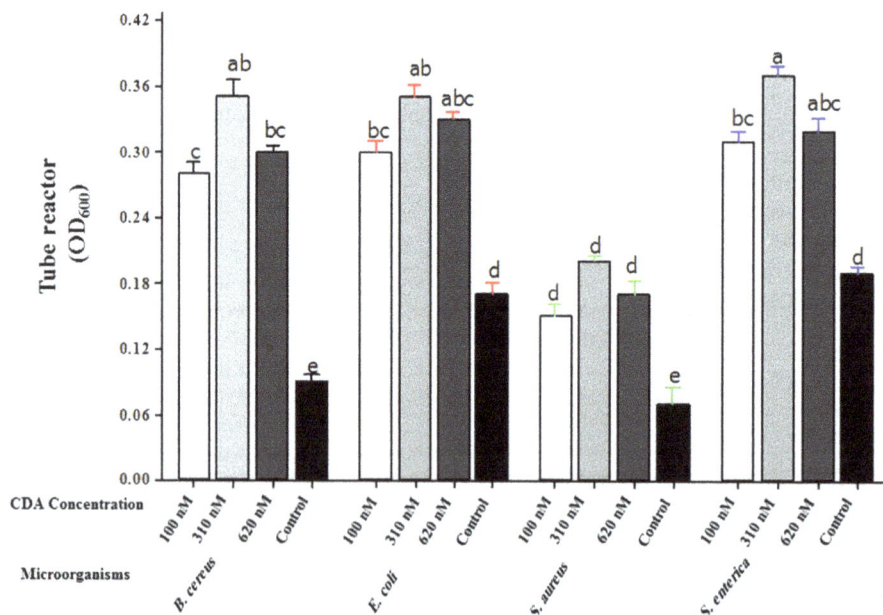

Figure 1. Induction of planktonic mode of growth in pre-established biofilms formed by food pathogens using CDA. (A) Biofilms were grown for 5 days in petri dishes in which the medium was replaced every 24 h. Dispersion induction was tested by replacing the growth medium with fresh medium containing three different concentrations of CDA (100, 310 or 620 nM) or just the carrier as a control and the cells were incubated for a further 1 h. Medium containing dispersed cells was then homogenized and cell density was determined by measuring the optical density. (B) After 5 days of biofilm growth in flow cell continuous cultures, the influent medium was switched from fresh medium in the test lines to three indicated concentrations of CDA and control lines were switched to new lines containing just the carrier. Effluent runoffs were then collected and cell density was determined by measuring the OD. Error bars indicate standard errors (n = 3) and mean values sharing at least one common lowercase letter shown above the bars are not significantly different (*P-value* <0.05).

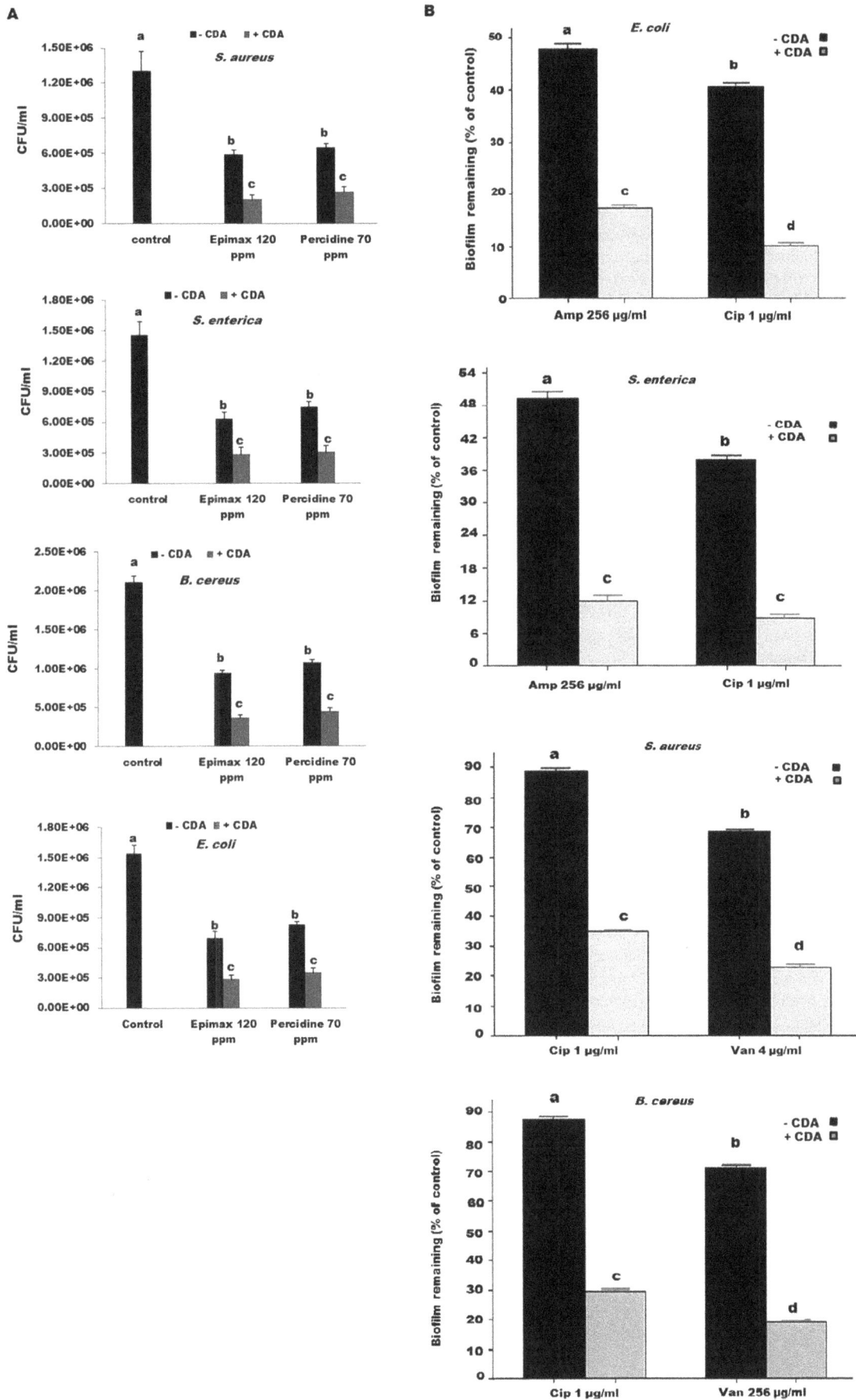

A

B

Figure 2. Effect of CDA combined antimicrobial treatments on eradication and killing of pre-established biofilms. (A) After 120 h of growth on the surface of SS discs, biofilms were treated for 1 h with biocides alone or combined with 310 nM CDA; CFU plate counts were then performed to assess the viability of the bacteria. (B) The amount of biofilm remaining was determined by the absorbance at 590 nm of crystal violet after staining the 120 h different biofilms in a microtiter plate assay after treatment with tested concentrations of antibiotics alone (- CDA) or in combination with 310 nM CDA (+CDA) for 1 h. All readings are corrected to reflect 0% and 100% controls (blank well, 0%; biofilms without any treatments, 100%). Error bars indicate standard errors (n = 3) and mean values sharing at least one common lowercase letter shown above the bars are not significantly different (*P-value* <0.05).

($OD_{600} = 0.66 \pm 0.01$, SE, *P-value* <0.05) (Figure 1A). The results from these experiments are summarized in Figure 1A. We also examined the effect of exposure to very low concentrations of CDA on pre-established biofilms grown in continuous cultures on the inner surface of silicone tubing. We again observed an increase in population of planktonic cells after treatment with CDA, indicating the release of biofilm bacteria into the effluent of cultures treated with CDA. As for semi-batch biofilm cultures, the most increase in population of planktonic cells in the effluents, with more than two-fold increase in the number of planktonic cells in comparison with control biofilms were observed when cultures were treated with 310 nM CDA (Figure 1B).

At this concentration, the most significant increase in population of planktonic cells was observed in *S. enterica* biofilms ($OD_{600} = 0.37 \pm 0.01$, SE, *P-value* <0.05) compared to results for untreated controls ($OD_{600} = 0.19 \pm 0.005$, SE, *P-value* <0.05) and no significant differences were detected between *B. cereus* and *E. coli* biofilms.

The results from these two different dispersal bioassays demonstrated the ability of nano-molar ranges of CDA to stimulate the release of cells from pre-established biofilms formed by different species of food related- bacteria.

Antimicrobial combined CDA survival assays of pre-established biofilms on stainless steel and polystyrene surfaces

To examine the effect of CDA combined antimicrobial agents on removal of biofilms; we tested Epimax S (hydrogen peroxide) and Percidine (peracetic acid) against pre-established biofilms grown on the surface of SS discs, in the presence and absence of 310 nM CDA. When 120 h biofilms were treated in the absence of CDA, both disinfectants caused approximate two-fold decrease in CFU counts compared to the untreated controls, while combined exposure of cultures to 310 nM CDA and 70 ppm Percidine or 120 ppm Epimax S, resulted in approximate five-fold decrease in CFU counts. No significant differences were observed between these two different combinational treatments in reduction of CFU counts (*P-value* <0.05). The results from these experiments are illustrated in Figure 2A.

We have also tested effectiveness of CDA combined with three antibiotics (ciprofloxacin, vancomycin and ampicillin). We observed that combined treatments with both CDA and antibiotics had a significant effect on removing pre-established biofilms

formed by examined microorganisms on polystyrene surfaces. For example, ciprofloxacin treatment of biofilms formed by *S. aureus* and *B. cereus* caused approximately 11% and 13% reductions in their biofilms, respectively (compared to biofilms without any treatments) while combined treatment of their biofilms with 1 μg of ciprofloxacin and 310 nM CDA resulted in 87% and 89% removal of their biofilms, respectively.

Significant differences were detected between two different combinational treatments applied for gram positive and gram negative bacteria; since the combination of CDA and ciprofloxacin was more effective than CDA combined ampicillin to eradicate biofilms formed by gram negative organisms. Similarly, combined treatments with both CDA and vancomycin were more effective to eliminate biofilms formed by gram positive bacteria. Results from these experiments are summarized in Figure 2B.

Thus, combined treatments using only low concentrations of CDA together with biocides or antibiotics were highly effective in removal and killing of pre-established biofilms formed by food pathogens.

Combined CDA and antimicrobial treatment of planktonic cells

To further investigate the effect of CDA on the sensitivity of tested microorganisms towards antimicrobial agents, we also evaluated very low concentrations of CDA for any inhibitory effects on growth of their planktonic cells. Compared to antibiotics or biocides alone, combination of antimicrobial treatment with nano-molar concentrations of CDA had no additional inhibitory effects on the growth of planktonic cells; for that reason only Minimum Inhibitory Concentrations (MICs) for antibiotics and disinfectants alone are presented in Table 1.

Biofilm surface coverage reduction by CDA combined biocides or antibiotics

To further examine the effect of CDA on biofilm surface area and bacteria viability, we also tested various disinfectants and antibiotics alone or combined with CDA against pre-established biofilms grown in continuous culture flow cells. When 48-h biofilms were treated in the absence of CDA, none of the disinfectants or antibiotics reduced biofilm biomass effectively (Figure 3). In contrast, after combined treatment, the biofilm cells remaining on the surface were easily removed and killed by antimicrobial compounds when examined by using the LIVE/

Table 1. MICs of tested microorganisms' planktonic cells to examined disinfectants and antibiotics.

Bacteria	Epimax S (ppm)	Percidine (ppm)	Ampicillin (μg.ml^{-1})	Vancomycin (μg.ml^{-1})	Ciprofloxacin (μg.ml^{-1})
E. coli	20	10	128	-	0.125
S. aureus	10	10	-	1	0.25
B. cereus	10	10	-	64	0.125
S. enterica	20	10	128	-	0.25

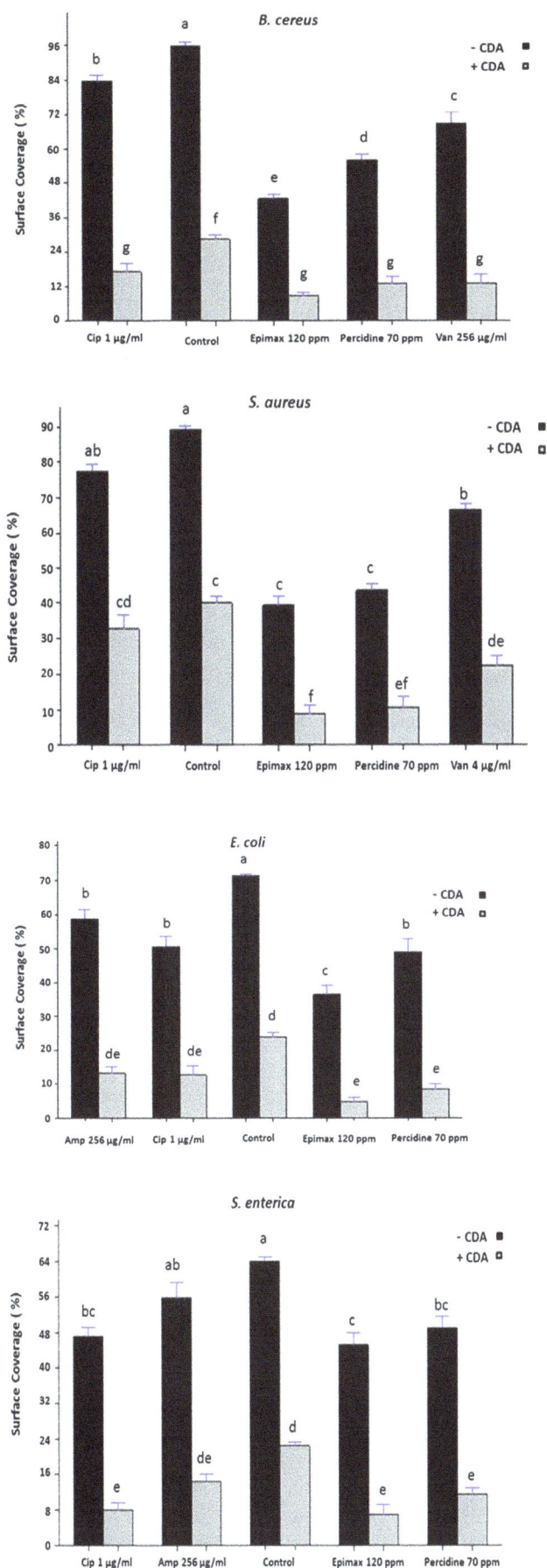

Figure 3. Effect of CDA combined disinfectant or antibiotic treatments on biofilms surface area. Following dispersion of biofilms by CDA, cells remaining on the surface are easily killed and removed by various disinfectants (Epimax S and Percidine) or antibiotics (vancomycin; Van, ampicillin; Amp, ciprofloxacin; Cip) in biofilms grown in continuous culture flow cells. Pre-established biofilms were grown for 48 h without any treatment and then were treated with indicated concentrations of antimicrobials alone (- CDA) or combined with 310 nM CDA (+ CDA) for 1 h, stained with LIVE/DEAD staining and quantified (percent surface coverage) using digital image analysis. The bars show the levels of biofilm biomass after treatment with antimicrobials alone or combined with 310 nM CDA. Error bars indicate standard errors (n = 3) and mean values sharing at least one common lowercase letter shown above the bars are not significantly different (*P-value* <0.05).

DEAD staining kit (Figure 4). The most significant reduction in biofilm surface area (*P-value* <0.05) was observed when biofilms were treated with combination of Epimax S and 310 nM CDA. For example, this combination resulted in eradication of more than 90% of the *E. coli* biofilms from the surface (Figure 3).

Combined treatments with both CDA and antibiotics or biocides caused almost-complete eradication of pre-established biofilms.

Discussion

The EPS matrix acts as a barrier in which diffusive transport prevails over convective transport [3]. EPS delay or prevent antimicrobials from reaching target microorganisms within the biofilm by diffusion limitation (like ciprofloxacin and ampicillin) [17,18] and/or chemical interaction with the matrix material (like peroxides such as peracetic acid and hydrogen peroxide) [19]. Against such a drawback still oxidizing compounds (like peroxides) for their nonspecific mode of actions and because of variation in the chemical composition of biofilms are among widely used disinfectants in food industry in most countries including Iran. Several studies have shown that strategies to induce biofilm dispersal could potentially use the microorganisms' own energy to disrupt EPS and remove pre-established biofilms [20]. In a previous study Davies and Marques [12] showed that a synthesized signalling molecule by *P. aeruginosa* induces dispersion of pre-established biofilms in *P. aeruginosa* as well as many other strains of microorganisms. They concluded that CDA most likely induce the production of degradative enzymes of EPS by these microorganisms. Differential microarray analysis, by Rahmani *et al.* (under preparation) indicated that 100 nM CDA (added exogenously to *P. aeruginosa* pre-established biofilms) significantly up regulates the expression of *P. aeruginosa* genes including EPS, alginate, degradative enzyme (alginate lyiase; *algL*) and negative regulator for this EPS biosynthesis (*mucB*). Their results also showed that CDA down regulates the expression of genes involved in *P. aeruginosa* attachment to the surfaces (*cupA* and *cupB*), which results in reversion of biofilms to a population of planktonic cells with increased susceptibility to antimicrobial agents compared to their sessile counterparts (Rahmani *et al.*, under preparation). Therefore, in this investigation we first examined the action of nano-molar concentrations of CDA (as an inducer of biofilm dispersal) on dispersion of pre-established biofilms, formed by four main food-borne pathogenic or spoilage microorganisms. Our results interestingly showed that only 310 nM of the signal was enough to reverse pre-established biofilms, formed by distant genera of bacteria, to their planktonic mode of growths. Since disinfectants and antibiotics have greater bactericidal efficacy against planktonic bacteria than their sessile counterparts, the

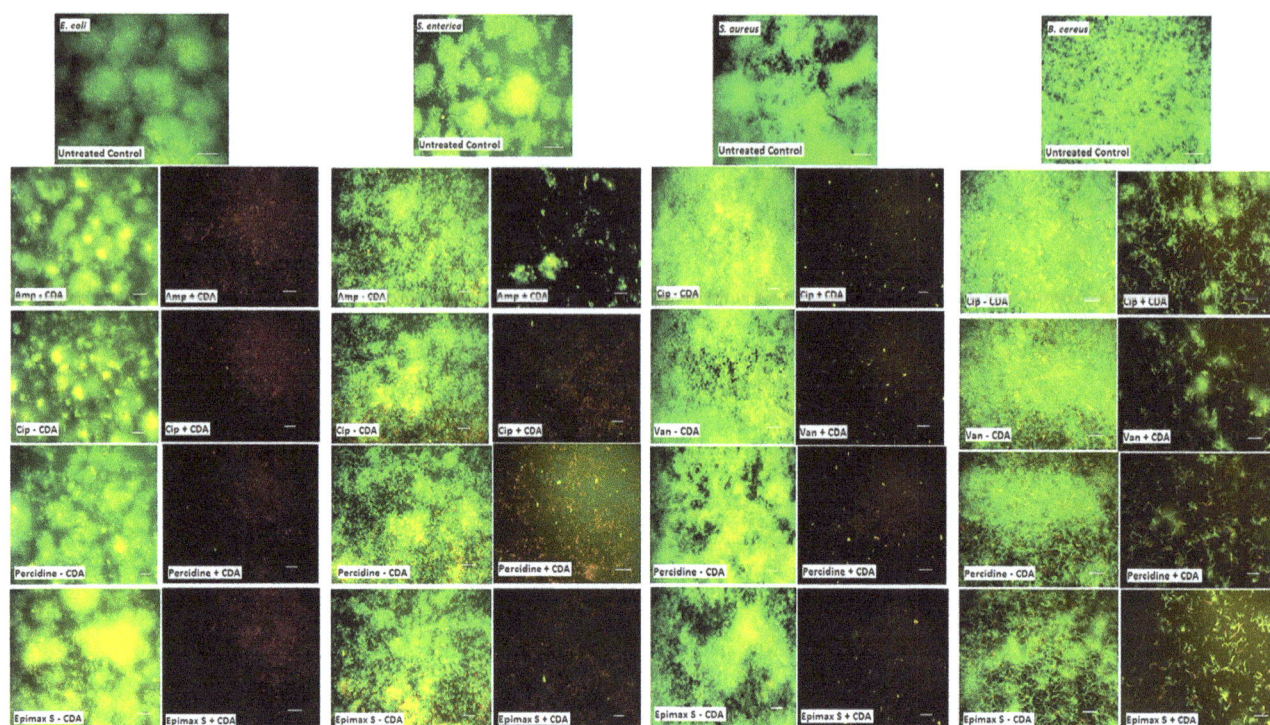

Figure 4. Effect of CDA combined antimicrobial treatments on killing of pre-established biofilms. CDA treatment reverses biofilm formation in pre-established biofilms and cells remaining on the surface are easily removed and killed various disinfectants (Epimax S and Percidine) or antibiotics (vancomycin; Van, ampicillin; Amp, ciprofloxacin; Cip) in biofilms grown in continuous culture flow cells. Pre-established biofilms were grown for 48 h without any treatment, then were treated with indicated concentrations of antimicrobials alone (- CDA) or combined with 310 nM CDA (+CDA) for 1 h and stained with LIVE/DEAD staining to allow analysis using fluorescence microscopy. The images show microscopic pictures of the biofilms on the surface of cover slip after combinatorial treatments. Images are top-down views (x-y plane); scale bars: 50 µm. Results are representative of 3 separate experiments.

combination of CDA with common antimicrobial agents could have improved bactericidal efficacy. Thus, we then tried to remove and kill pre-established biofilms by using the combination of CDA and traditional disinfectants or antibiotics which are broadly used in food processing environments and their related medical issues, at concentrations that had no significant effects against biofilms, to reach a novel mechanism for enhancing the activity of these treatments through the disruption of biofilms. The results presented here demonstrated that following exposure to low concentrations of CDA, biofilm cells on the surface were easily detached and then killed by antimicrobial agents where the combination of 310 nM CDA with examined disinfectants (Percidine and Epimax S) or antibiotics (ciprofloxacin, vancomycin and ampicillin), when added to their solutions, resulted in approximate 80% reduction in biofilm biomass in all cultures.

Numerous strategies to control microbial biofilms have been proposed, with different degrees of success. In various industrial settings, a range of biocides and toxic metals (e.g., tin and copper) has been used for antifouling coatings and sanitizing purposes [21,22]; however, these substances are not appropriate for use in food industries and clinical settings. In this work, we showed that CDA-based strategies to induce biofilm dispersal involve only nano-molar concentrations of CDA that should be safe to humans and to the environment. Besides, previous findings showed that CDA has no cytotoxic or stimulatory effect on human cells even at high concentrations (up to $250\ \mu g.ml^{-1}$) [13]. Because CDA mediates the transition from a biofilm to a planktonic phenotype via a signalling mechanism (because acts at nano-molar concen-

trations which are consistent with all known cell-to-cell signalling molecules) rather than toxic effect, CDA-based biofilm control strategies would not be expected to select for resistant strains as seen with antibiotics. Therefore, in this study we examined two different combination of CDA; CDA combined disinfectants and CDA combined antibiotics, to introduce a promising strategy which is appropriate to control biofilms both in food industry and clinical settings.

While some free fatty acids have antimicrobial properties [23,24] and play a vital role in maintaining the microbial flora of the skin [25,26], we demonstrated that CDA does not inhibit bacterial growth at nano-molar ranges that induce biofilm dispersal. These results were highly in consistent with the results from Jennings *et al.* study [13] where they showed that CDA inhibited bacterial growth only at high (micro-molar to milli-molar) concentrations. This lack of growth inhibition at lower concentrations was not surprising since bacteria produce this unsaturated fatty acid and use it as a signalling molecule [12].

Conclusions

Data from this study suggest that application of CDA prior to or in combination with disinfectants or antibiotics may allow for novel and improved strategies to control biofilms in industrial as well as clinical settings, with clear benefits such as reduced ecological impact and reduced treatment costs.

Acknowledgments

We sincerely thank Dr. David G. Davies (Department of Biological Sciences, State University of New York at Binghamton, USA) for valuable information about *cis*-2-decenoic acid and also providing protocols for biofilm dispersal bioassays. We also thank Dr. Sadegh Mousavi-Fard (Shahrekord University, Shahrekord, Iran) for statistical analysis and Dr. Hossein Fallahi (Department of Biology, Razi University, Kermanshah, Iran) for the comments on the manuscript.

Author Contributions

Conceived and designed the experiments: AR-B SS HB-N MRS. Performed the experiments: AR-B HB-N SS MRS. Analyzed the data: AR-B HB-N. Contributed reagents/materials/analysis tools: AR-B SS HB-N MRS. Wrote the paper: AR-B HB-N SS MRS.

References

1. Stoodley P, Sauer K, Davies DG, Costerton JW (2002) Biofilms as complex differentiated communities. Annu Rev Microbiol 56: 187–209.
2. Palmer J, Flint S, Brooks J (2007) Bacterial cell attachment, the beginning of a biofilm. J Ind Microbiol Biotechnol 34: 577–588.
3. Flemming HC, Wingender J (2010) The biofilm matrix. Nat Rev Microbiol 8: 623–632.
4. Mah TF, O'Toole GA (2001) Mechanisms of biofilm resistance to antimicrobial agents. Trends Microbiol 9: 34–39.
5. Körstgens V, Flemming HC, Wingender J, Borchard W (2001) Uniaxial compression measurement device for investigation of the mechanical stability of biofilms. J Microbiol Meth 46: 9–17.
6. Verran J, Airey P, Packer A, Whitehead KA (2008) Microbial retention on open food contact surfaces and implications for food contamination. Adv Appl Microbiol 64: 223–246.
7. Shi X, Zhu X (2009) Biofilm formation and food safety in food industries. Trends Food Sci Technol 20: 407–413.
8. Simo~es M, Pereira MO, Vieira MJ (2005) Effect of mechanical stress on biofilms challenged by different chemicals. Wat Res 39: 5142–5152.
9. Poppele EH, Hozalski RM (2003) Micro-cantilever method for measuring the tensile strength of biofilms and microbial flocs. J Microbiol Meth 55: 607–615.
10. European Standard – EN 1276 (1997) Chemical disinfectants and antiseptics – quantitative suspension test for the evaluation of bactericidal activity of chemical disinfectants and antiseptics used in food, industrial, domestic, and institutional areas – test method and requirements (phase 2, step 1).
11. Gilbert P, Allison DG, McBain AJ (2002) Biofilms in vitro and in vivo: do singular mechanisms influx cross-resistance? J Appl Microbiol 92: 98–110.
12. Davies DG, Marques CN (2009) A fatty acid messenger is responsible for inducing dispersion in microbial biofilms. J Bacteriol 191: 1393–1403.
13. Jennings JA, Courtney SH, Haggard OW (2012) *Cis*-2-decenoic acid inhibits *S. aureus* growth and biofilm in Vitro: a pilot study. Clin Orthop Relat Res 470: 2663–2670.
14. Musk Jr DJ, Hergenrother PJ (2008) Chelated iron sources are inhibitors of *Pseudomonas aeruginosa* biofilms and distribute efficiently in an *in vitro* model of drug delivery to the human lung. J Appl Microbiol 105: 380–388.
15. Clinical and Laboratory Standards Institute (2006) Methods for dilution antimicrobial susceptibility tests for bacteria that grow aerobically, 7th ed. Approved standard M7-A7. Clinical and Laboratory Standards Institute, Wayne, PA.
16. Clinical and Laboratory Standards Institute (2006) Performance standards for antimicrobial susceptibility testing; 16th informational supplement. M100-S16. Clinical and Laboratory Standards Institute, Wayne, PA.
17. Walters MC 3rd, Roe F, Bugnicourt A, Franklin MJ, Stewart PS (2003) Contributions of antibiotic penetration, oxygen limitation, and low metabolic activity to tolerance of *Pseudomonas aeruginosa* biofilms to ciprofloxacin and tobramycin. Antimicrob Agents Chemother 47: 317–323.
18. Anderl JN, Franklin MJ, Stewart PS (2000) Role of antibiotic penetration limitation in *Klebsiella pneumoniae* biofilm resistance to ampicillin and ciprofloxacin. J Antimicrob Chemother 44(7): 1818–1824.
19. Campanac C, Pineau L, Payard A, Baziard-Mouysset G, Roques C (2002) Interactions between biocide cationic agents and bacterial biofilms. Antimicrob Agents Chemother 46: 1469–1474
20. Yang L, Liu Y, Wu H, Song Z, Høiby N, et al. (2012) Combating biofilms. FEMS Immunol. Med Microbiol 65: 146–157
21. Cloete TE, Jacobs L, Brözel VS (1998) The chemical control of biofouling in industrial water systems. Biodegradation 9: 23–37.
22. Chambers LD, Stokes KR, Walsh FC, Wood RJK (2006) Modern approaches to marine antifouling coatings. Surf Coatings Technol 201: 3642–3652.
23. Desbois AP, Lebl T, Yan L, Smith VJ (2008) Isolation and structural characterization of two antibacterial free fatty acids from the marine diatom, *Phaeodactylum tricornutum*. Appl Microbiol Biotechnol 81: 755–764.
24. Desbois AP, Smith VJ (2010) Antibacterial free fatty acids: activities, mechanisms of action and biotechnological potential. Appl Microbiol Biotechnol 85: 1629–1642.
25. Kenny JG, Ward D, Josefsson E, Jonsson IM, Hinds J, et al. (2009) The *Staphylococcus aureus* response to unsaturated long chain free fatty acids: survival mechanisms and virulence implications. PLoS One 4: e4344
26. Takigawa H, Nakagawa H, Kuzukawa M, Mori H, Imokawa G (2005) Deficient production of hexadecenoic acid in the skin is associated in part with the vulnerability of atopic dermatitis patients to colonization by *Staphylococcus aureus*. Dermatology 211: 240–248.

12

Multicellular Tumor Spheroids for Evaluation of Cytotoxicity and Tumor Growth Inhibitory Effects of Nanomedicines *In Vitro*: A Comparison of Docetaxel-Loaded Block Copolymer Micelles and Taxotere®

Andrew S. Mikhail[1,2ⓨ], Sina Eetezadi[1ⓨ], Christine Allen[1,2,3]*

1 Leslie Dan Faculty of Pharmacy, University of Toronto, Toronto, Ontario, Canada, 2 Institute of Biomaterials and Biomedical Engineering, University of Toronto, Toronto, Ontario, Canada, 3 Spatio-Temporal Targeting and Amplification Radiation Response (STTARR) Innovation Centre, Toronto, Ontario, Canada

Abstract

While 3-D tissue models have received increasing attention over the past several decades in the development of traditional anti-cancer therapies, their potential application for the evaluation of advanced drug delivery systems such as nanomedicines has been largely overlooked. In particular, new insight into drug resistance associated with the 3-D tumor microenvironment has called into question the validity of 2-D models for prediction of *in vivo* anti-tumor activity. In this work, a series of complementary assays was established for evaluating the *in vitro* efficacy of docetaxel (DTX) -loaded block copolymer micelles (BCM+DTX) and Taxotere® in 3-D multicellular tumor spheroid (MCTS) cultures. Spheroids were found to be significantly more resistant to treatment than monolayer cultures in a cell line dependent manner. Limitations in treatment efficacy were attributed to mechanisms of resistance associated with properties of the spheroid microenvironment. DTX-loaded micelles demonstrated greater therapeutic effect in both monolayer and spheroid cultures in comparison to Taxotere®. Overall, this work demonstrates the use of spheroids as a viable platform for the evaluation of nanomedicines in conditions which more closely reflect the *in vivo* tumor microenvironment relative to traditional monolayer cultures. By adaptation of traditional cell-based assays, spheroids have the potential to serve as intermediaries between traditional *in vitro* and *in vivo* models for high-throughput assessment of therapeutic candidates.

Editor: Xiaoming He, The Ohio State University, United States of America

Funding: A. Mikhail is the recipient of post-graduate scholarships from NSERC and the Government of Ontario. S. Eetezadi is funded by the NSERC CREATE Biointerfaces training program and holds an Ontario Trillium scholarship. The funders had no role in study design, data collection and analysis, decision to publish, or preparation of the manuscript.

Competing Interests: The authors have declared that no competing interests exist.

* E-mail: cj.allen@utoronto.ca

ⓨ These authors contributed equally to this work.

Introduction

It has become increasingly clear that resistance to chemotherapy is not only facilitated by processes at the cellular level, but also by mechanisms associated with the tumor microenvironment [1,2]. In growing tumors, the heterogeneous architecture of the vasculature, irregular blood flow, large intervascular distances and nature of the extracellular matrix limit the access of cells to oxygen, nutrients, and systemically administered therapies [3,4]. Within the tumor interstitium, gradients in the rate of cell proliferation are established wherein rapidly dividing cells reside close to the tumor vasculature and quiescent cells are situated deep within the extravascular space. However, many anti-neoplastic agents exert limited toxicity against slowly- or non-proliferating cells and are less effective in the hypoxic and acidic microenvironments of poorly perfused tissues [5,6]. These therapeutic limitations are exacerbated by high interstitial fluid pressure which inhibits the penetration of chemotherapeutic agents through the tumor interstitium by limiting convective transport [7]. As a result cells located distant from blood vessels may be less sensitive to treatment and also be exposed to sub-therapeutic drug concentrations.

The use of *in vitro* cell culture is critical in drug discovery and formulation development for rapid identification of lead candidates and for investigating mechanisms of drug efficacy at the cellular and molecular levels. In contrast to *in vivo* tumor models, *in vitro* cultures are better suited for systematic studies of formulation parameters in a highly controlled environment. However, cytotoxic effects observed in conventional monolayer cultures often fail to translate into similar effects *in vivo* [8,9]. This is due to the inherent inability of 2-D cultures to account for mechanisms of drug resistance and transport restrictions associated with the 3-D tumor microenvironment. As such, there is increasing interest in applying 3-D *in vitro* models that enable rapid, high throughput screening of drug formulations for selection of lead candidates to move forward to *in vivo* evaluation [10–12].

As depicted in Figure 1, 3-D tissue cultures such as MCTS serve as an intermediary between the oversimplified structure of monolayer cultures and the highly complex nature of *in vivo* tumors. Spheroid cultures possess a complex network of cell-cell contacts and advanced extracellular matrix development, as well as pH, oxygen, metabolic and proliferative gradients analogous to the conditions in poorly vascularized and avascular regions of solid tumors [13–15]. In general, a spheroid is comprised of an outer region of proliferating cells which surrounds intermediate layers of quiescent cells and, if the spheroid is large enough, a necrotic core. This arrangement parallels the radial organization of tissues surrounding tumor blood vessels. To date, a variety of 3-D *in vitro* tissue models have been applied for the study of anticancer therapies including natural and synthetic tissue scaffolds [16,17], multicellular layers [18–22], and multicellular tumor spheroids [16,23,24]. MCTS are particularly relevant in the development of nanomedicines since the penetration of the encapsulated drug in tumor tissues may be significantly altered by properties of the delivery vehicle. To date, however, there remain limited examples of the use of MCTS for the evaluation of nanomedicines [25–29].

DTX is a potent chemotherapeutic agent that is administered as Taxotere® (Sanofi-Aventis) and used for treatment of cancers of the breast, prostate, lung, head and neck, and stomach [30]. DTX is also being investigated in a phase II clinical trial for treatment of metastatic colorectal adenocarcinoma in combination with gemcitabine and has been investigated as a single agent for treatment of cervical cancer [31,32]. However, Taxotere® is known to be associated with significant side effects that can require reduction of the administered dose [33]. Encapsulation of chemotherapeutic agents within biocompatible nanosystems such as block copolymer micelles (BCMs) has proven to be a promising approach for mitigating the burden of toxicity on normal tissues and increasing tumor-specific drug accumulation [34]. The primary objective of this study was to adapt and apply traditional cell-based assays in a systematic and complementary manner for the evaluation of Taxotere® and a DTX-containing nanomedicine in both monolayer and MCTS cultures (Figure 2).

Materials and Methods

Materials

Methoxy poly(ethylene glycol) (CH$_3$O-PEG-OH; Mn = 5000, Mw/Mn = 1.06) was obtained from Sigma-Aldrich (Oakville, ON, Canada). ε-Caprolactone and dichloromethane (Sigma-Aldrich) were dried using calcium hydride prior to use. Hydrogen chloride (HCl) (1.0 M in diethyl ether), N,N-dimethylformamide (DMF), diethyl ether, hexane and acetonitrile (Sigma-Aldrich) were used without further purification. Alexa Fluor 488 (AF488) carboxylic acid succinimidyl ester was purchased from Molecular Probes (Eugene, OR). The hypoxia marker, EF5, and Cy5-conjugated anti-EF5 antibody were purchased from the Department of Radiation Oncology, University of Pennsylvania, (Philadelphia,

PA). DTX was purchased from Jari Pharmaceutical Co. (Jiangsu, China).

Synthesis of CH$_3$O-PEG-*b*-PCL (PEG-*b*-PCL) Copolymers

PEG-*b*-PCL copolymer was prepared as previously described [35]. Briefly, CH$_3$O-PEG-OH was used to initiate the ring-opening polymerization of ε-CL in the presence of HCl. The reaction was carried out for 24 h at room temperature prior to termination by addition of triethylamine (TEA) and precipitation in diethyl ether and hexane (50:50, v/v%). The product was dried under vacuum at room temperature.

Preparation and Characterization of BCM+DTX

PEG-*b*-PCL copolymers and DTX were dissolved at a copolymer:drug weight ratio of 20:1 in DMF and stirred for 30 min. DMF was evaporated under N$_2$ at 30°C and residual solvent was removed under vacuum. Dry copolymer-drug films were then heated to 60°C in a water bath prior to the addition of PBS buffer (pH 7.4) at the same temperature. Resultant micelle solutions were vortexed, stirred for 24 h at room temperature and finally sonicated (Laboratory Supplies Co., NY) for 1 h. Undissolved drug crystals were removed by centrifugation at 4400 g for 12 min (Eppendorf 5804R). The final copolymer concentration was 10 mg/mL. The amount of physically entrapped DTX in BCM samples was determined by HPLC analysis (Agilent series 1200) with UV detection (Waters 2487) at a wavelength of 227 nm. An XTerra C18 reverse phase column was employed with ACN/water (60/40, v/v%) as the mobile phase. Drug loading was quantified using a calibration curve generated from a series of DTX standards.

Sizing of BCM+DTX

The average hydrodynamic diameter of the BCMs was determined by dynamic light scattering (DLS) using a 90Plus Particle Size Analyzer (Brookhaven Instruments Corp., Holtsville, NY) at an angle of 90° and temperature of 37°C. The samples were diluted to a copolymer concentration of 0.5 mg/mL prior to measurement. Analysis was performed using the 90Plus Particle Sizing Software.

Transmission Electron Microscopy (TEM)

BCMs were observed by TEM using a Hitachi 7000 microscope operating at an acceleration voltage of 75 kV (Schaumburg, IL). Samples were diluted in double distilled water immediately prior to analysis and negatively stained with a 1% uranyl acetate (UA) solution. The final copolymer concentration was 0.5 mg/mL. The samples were then deposited on copper grids that had been pre-

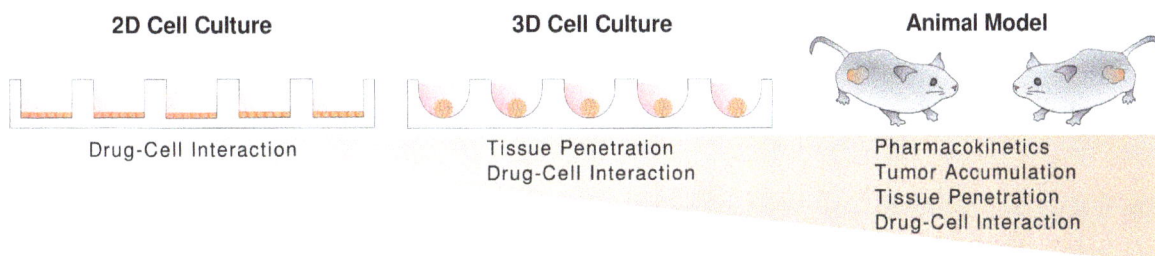

Figure 1. 3-D cultures as intermediary between 2-D cultures and animal models. Intermediate in complexity, 3-D cultures permit the systematic, high-throughput assessment of formulation properties in a controlled environment that approximates important properties of *in vivo* tumors in the absence of complex parameters which may confound data interpretation.

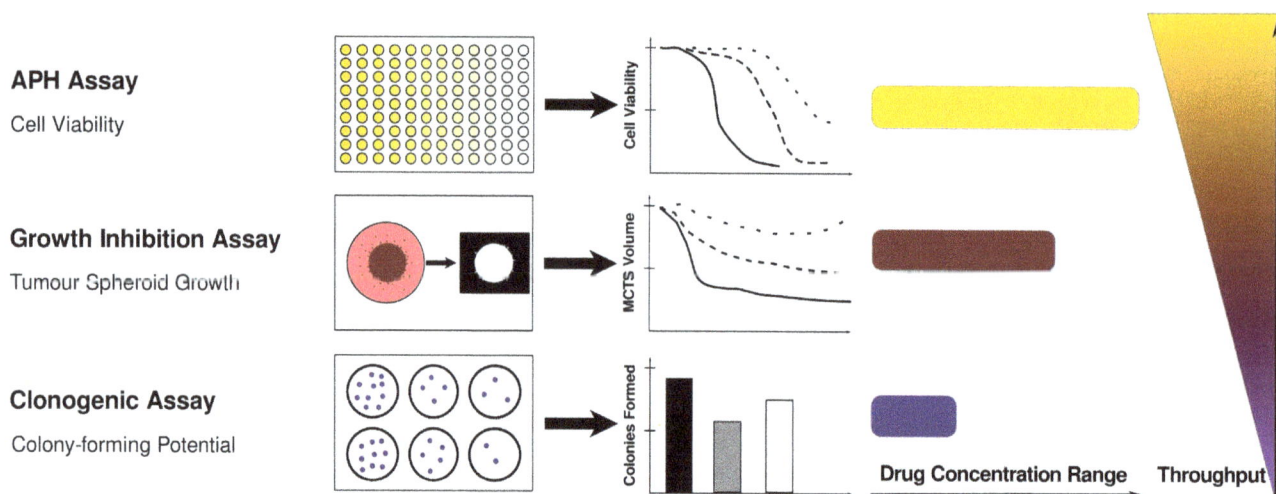

Figure 2. *In vitro* assays used in this study for analysis of formulation efficacy in spheroids.

coated with carbon and negatively charged (Ted Pella Inc., Redding, CA) and briefly air-dried prior to analysis.

Drug Release

The release of DTX from BCMs and Taxotere® was analyzed using a dialysis method. Aliquots (1 mL) of BCM+DTX, DTX in DMSO, and Taxotere® were placed in individual dialysis bags (MWCO 2 kDa, Spectra/Por, Rancho Dominguez, CA) and dialyzed separately against 100 mL of PBS at pH 7.4 in an incubator at 37°C ensuring that sink conditions were maintained. At selected timepoints, 50 μL samples were withdrawn from the dialysis bags and DTX content was measured by HPLC as described above.

Tissue Culture and Growth of MCTS

Human cervical (HeLa) and colon (HT29) (ATCC, Manassas, VA) cancer cells were incubated at 37°C and 5% CO_2 in DMEM containing 1% penicillin-streptomycin and supplemented with 10% FBS. For growth of MCTS, cells were suspended using trypsin-EDTA and 2000 and 5000 HT29 and HeLa cells were seeded onto non-adherent 96-well round-bottomed Sumilon PrimeSurface™ plates (Sumitomo Bakelite, Tokyo, Japan), respectively, in 200 μL of media per well. During growth, 50% of the media was exchanged every other day. MCTS were grown for 7 days until they reached ~ 500 μm in diameter before use.

Immunohistochemical Analysis of MCTS

MCTS were washed in PBS and transferred onto a vinyl specimen mold (Cryomold®, Tissue-Tek, Sakura Finetek, CA) prior to addition of Tissue-Tek® O.C.T. compound (Sakura Finetek, Torrance, CA). MCTS were then submersed in an isopentane bath cooled by liquid nitrogen, cut into 5 μm thick sections using a microtome and mounted on glass slides. Histological staining was conducted for the identification of cellular proliferation (Ki67) and stained with hematoxylin and eosin (H&E). For identification of hypoxic regions, MCTS were incubated with 0.5 mM EF5 and soaked in PBS prior to cryosectioning. EF5 in the MCTS sections was identified by binding with cyanine-5-conjugated mouse anti-EF5 (1/50) antibody. The positive signal distribution for Ki67 was analyzed using a customized MATLAB® algorithm, as described previously [36]. Briefly, images containing Ki67-stained MCTS sections were

thresholded for positive color intensity. Using a distance map, signal intensities were summed within three concentric regions of equidistant thickness (periphery, intermediate and core), each equivalent to 1/3 of the MCTS radius. The distribution of Ki67 positive signal is expressed as a percentage of total positive signal in the MCTS section.

Measurement of MCTS Growth

Spheroids were imaged using a light microscope with a 10× objective lens (VWR VistaVision ™) connected to a digital camera (VWR DV-2B). The diameter and volume of MCTS were determined by measuring their cross-sectional area using an automated image analysis macro developed for use with the ImageJ software package (NIH, Bethesda, MD, Version 1.44 m). The automated method was validated by comparison to manual determination of spheroid diameter and volume (Figure S1). For the automated method, images were converted into 8-bit greyscale and the perimeter of an individual MCTS was recognized by an automated threshold function and the image converted to a 2-D mask. The area of the spheroid mask was recorded, applying an image of known scale as calibration. Finally, the volume of the MCTS was calculated by assuming a spherical shape as follows: $V = 4/3 * \pi * (d/2)^3$. Data was fit using the Gompertz equation for tumor growth as follows: $V(t) = V(0)\exp(\alpha/\beta(1 - \exp(-\beta * t)))$ where $V(t)$ is volume at time t, $V(0)$ the initial volume and α and β are constants [37].

Cytotoxicity in Monolayer and Spheroid Tissue Cultures

The cytotoxicity of BCM+DTX and Taxotere® in monolayer and spheroid cell cultures was determined using the established acid phosphatase (APH) assay which is based on quantification of cytosolic acid phosphatase activity [38]. For this assay, *p*-nitrophenyl phosphate is added in cell culture and hydrolyzed in viable cells to *p*-nitrophenol via intracellular acid phosphatase. Briefly, MCTS (one spheroid per well) and monolayer cultures (4000 cells per well) were treated with Taxotere® or BCM+DTX for 24 h over a range of drug concentrations. Following treatment, monolayers and spheroids were washed three times with fresh media and cultured for an additional 48 h. Monolayers and spheroids were then washed with PBS buffer prior to the addition of 100 μL of freshly prepared reaction buffer (2 mg/ml *p*-nitrophenyl phosphate (Sigma) and 0.1% v/v Triton-X-100 in

0.1 M sodium acetate buffer at pH 5.5). Following incubation for 2 h in the cell incubator, 10 μL 1 M sodium hydroxide was added to each well and cell viability was determined by measuring the UV absorbance at 405 nm using an automated 96-well plate reader (SpectraMax Plus 384, Molecular Devices, Sunnyvale, CA). Results were normalized to controls as follows: % viability = (A$_{treatment}$ − A$_{media}$)/(A$_{control}$ − A$_{media}$), where A = mean absorbance. All experiments were performed in triplicate.

Growth Inhibition of MCTS

BCM+DTX or Taxotere® was administered to spheroids for 24 h at a DTX equivalent concentration of 2, 20 or 200 ng/mL. The culture media was replaced following the incubation period. Subsequently, half of the culture media was replaced by pipette every other day. Images of spheroids were captured using a light microscope with a 10× objective lens (VWR VistaVision™) connected to a digital camera (VWR DV-2B). Spheroid size was determined by measuring their 2-D cross-sectional area using the automated image analysis method described previously. The data are reported as the mean volume of six spheroids ± SD.

Clonogenic Survival Assay

The clonogenic assay was used to determine the ability of single cells to replicate and form colonies (>50 cells) following exposure to BCM+DTX and Taxotere®. Single cell suspensions derived from monolayer and disaggregated spheroids were diluted in culture media and cells were plated in 6-well plates in desired numbers. MCTS were disaggregated by incubation in trypsin-EDTA for 10 min, followed by gentle agitation. Drug formulations were added immediately at a DTX equivalent concentration of 20 ng/mL. After treatment for 24 h, cells were washed with PBS and 2 mL of fresh media was added to each well. For treatment of intact spheroids, drug formulations were added directly into wells containing individual MCTS. After 24 h, MCTS were collected and rinsed in PBS, suspended as single-cell suspensions in fresh media following trypsinization, and seeded onto 6-well plates. Cells were incubated for 14–16 days prior to fixation with methanol and staining with 1% crystal violet solution. Colonies consisting of at least 50 cells were counted. The surviving fraction (SF) was expressed as the number of colonies divided by the product of the number of cells plated and the plating efficiency. The plating efficiency was determined by dividing the number of colonies formed by the number of cells plated for untreated controls.

Results

Characterization of BCM+DTX

PEG-b-PCL copolymer micelles containing physically encapsulated DTX were formulated with a spherical morphology (Figure 3a). The size distribution of the micelles was monomodal with an average hydrodynamic diameter of 49.2±2.3 nm (Figure 3b). Drug loading resulted in a final DTX equivalent concentration of 258.7±35.5 μg/mL at a loading efficiency of 52.7±7.1%. Release of DTX from BCMs occurred over the course of 24 h wherein 74% of the drug was released by 12 h. In contrast, the release of docetaxel from Taxotere® was complete by 12 h (Figure 4).

Growth of MCTS

Spheroids were grown using a modified liquid overlay technique by seeding HT29 or HeLa cells onto non-adherent U-bottom tissue culture wells without the use of an agarose surface coating.

MCTS were spherical, followed a sigmoidal growth profile, and were grown until a diameter of ~500 μm was reached prior to use (Figure 5).

Cytotoxicity in Monolayer and MCTS Culture

Cell viability following exposure to BCM+DTX or Taxotere® was assessed using the APH assay (Figure 6). This assay was validated by assessment of the relationship between UV absorbance and cell number in both monolayer and spheroid cultures. As shown in Figure S2, a linear relationship was obtained. A well-established tetrazolium salt-based assay (WST-8) was also evaluated and did not yield a similar correlation (Figure S3). Spheroid cultures were substantially less sensitive to BCM+DTX and Taxotere® relative to their monolayer counterparts. HeLa cells were less responsive to treatment with either BCM+DTX or Taxotere® than HT29 cells in monolayer culture. However, in spheroid culture, HT29 cells were less sensitive to treatment. The IC$_{50}$ of HeLa and HT29 monolayer cultures treated with BCM+DTX were 0.37+/−0.01 and 0.01+/−0.004 ng/mL, respectively. When treated with Taxotere®, the IC$_{50}$ of HeLa and HT29 monolayer cultures were 2.2+/−0.5 and 0.09+/−0.01 ng/mL, respectively. The IC$_{50}$ of HeLa cells cultured as MCTS was 1396±198 ng/mL for BCM+DTX and 1558±103 ng/mL for Taxotere® whereas HT29 MCTS maintained a viability above 80% at all drug concentrations.

Inhibition of MCTS Growth

MCTS volume was plotted over a 30 day period following a 24 h incubation with 2, 20, and 200 ng/mL BCM+DTX or Taxotere® (Figure 7). The growth of HeLa MCTS was completely impeded following incubation with DTX concentrations of 20 and 200 ng/mL. No significant difference in growth was observed following exposure to 2 ng/mL of DTX relative to untreated controls. In the case of HT29 MCTS, incubation with 20 ng/mL of BCM+DTX and Taxotere® only resulted in a partial reduction in MCTS volume. Similarly to HeLa MCTS, complete inhibition of growth was observed following incubation with 200 ng/mL of drug. Unlike HeLa MCTS, however, a slight growth delay was also observed at 2 ng/mL. Interestingly, following re-treatment on day 14 at a DTX concentration of 20 ng/mL, BCM+DTX demonstrated greater inhibition of spheroid growth in HT29 cultures than Taxotere®.

Immunohistochemistry

Immunohistochemical analysis of MCTS cross-sections was performed in order to identify regions of necrosis, cellular proliferation and hypoxia (Figure 8). Staining with the proliferation marker Ki67 revealed a greater proportion of proliferative cells in HeLa MCTS relative to HT29. Quantitative image analysis revealed that 88.6% of proliferating cells were located within the periphery of HT29 MCTS (Figure 9). In contrast, only 51% of the total proliferating cells were located in the periphery of HeLa MCTS and 25% and 24% were located in the intermediate region and core, respectively. Signs of necrosis were visible following staining with H&E in HT29 MCTS. Incubation of MCTS with EF5 allowed for identification of regions of hypoxia following exposure to Cy5-conjugated anti-EF5 antibody. Hypoxic conditions were observed primarily in the core and intermediate regions of HT29 MCTS. In contrast, HeLa MCTS did not demonstrate any regional hypoxia. The relative distributions of cellular proliferation, hypoxia and necrosis in the MCTS are summarized in Figure 10.

a)

b)

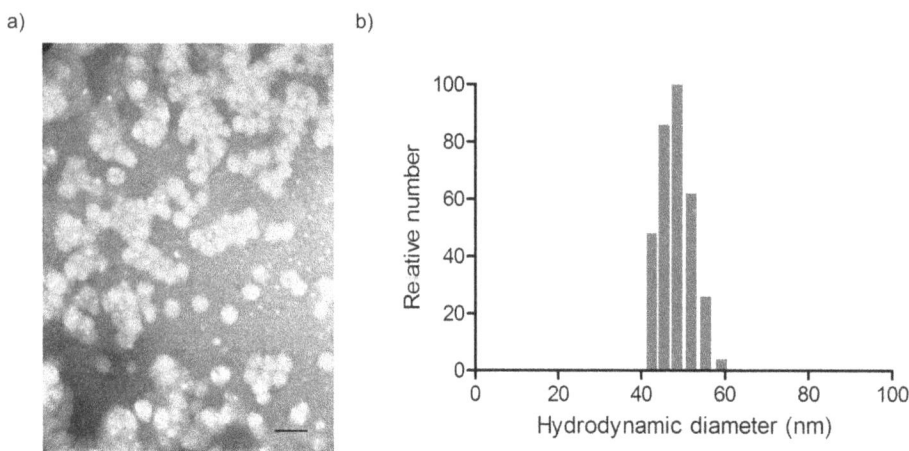

Figure 3. Characterization of micelle morphology and size. a) Transmission electron micrograph (Scale bar in represents 100 nm) and b) size distribution of BCM+DTX as determined by dynamic light scattering at 37°C.

Clonogenic Survival

The surviving fractions (SF) of HeLa and HT29 cells were determined following treatment with BCM+DTX or Taxotere® as monolayer and MCTS cultures (Figure 10).

The SF was higher for all intact MCTS cultures relative to monolayers. HeLa cells were less sensitive to treatment than HT29 when cultured as monolayers, but more sensitive than HT29 cells when the cells were exposed to treatment as MCTS. In all cases, the SF was lower when treated with BCM+DTX compared to Taxotere®. Furthermore, cells exposed to treatment immediately following MCTS disaggregation demonstrated residual resistance to both BCM+DTX and Taxotere®.

Discussion

In recent years, the tumor microenvironment has been implicated in the coordination of tumor growth, metastasis and resistance to anti-cancer therapies [39,40]. As such, effective evaluation of novel therapeutic agents requires the use of tissue models which closely mimic native conditions within the intratumoral space. Yet, the vast majority of chemotherapeutic agents are screened for cytotoxic effects in monolayer cultures which do not account for critical mechanisms of drug resistance associated with the tumor microenvironment. Consequently, these

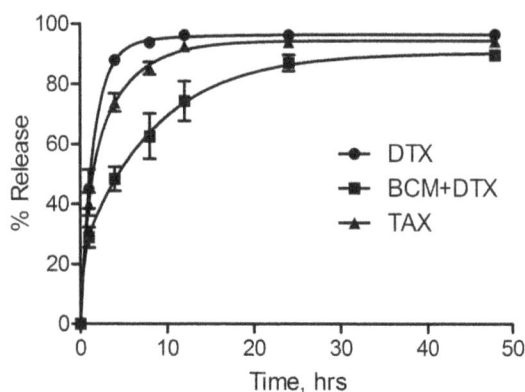

Figure 4. Drug release. Release of docetaxel from dialysis bags containing BCM+DTX, Taxotere®, and DTX in DMSO, n = 3.

models poorly predict a drug's therapeutic efficacy *in vivo* [8]. In contrast, 3-D MCTS better approximate the state of cancer cells in their native environment and thus can be used to more accurately estimate a drug's therapeutic potential. A variety of methods have been used to grow MCTS for use in cancer research including spinning culture flasks [41], hanging drops [42], liquid overlay on agarose [43], micropatterned plates [44], and recently, using intercellular linkers [45]. However, many of these techniques are impractical, time-consuming, and involve delicate handling procedures, limiting the use of the MCTS model in drug screening and development. In addition, practical application of traditional cell-based assays in MCTS cultures remains poorly established. In the current study, the performance of BCM+DTX and Taxotere® was evaluated by adaptation of conventional cytotoxicity and survival assays in monolayer and MCTS cultures using a robust MCTS culture technique.

MCTS grew according to sigmoidal growth patterns reflective of tumor growth *in vivo* (Figure 5) and possessed histological features similar to those of the native tumor microenvironment including gradients in cell proliferation and regions of hypoxia and necrosis (Figure 8, Figure S4). Cells grown in spheroid cultures demonstrated considerably greater resistance to treatment with BCM+DTX or Taxotere® relative to cells grown in monolayer cultures. This may be a result of the limited exposure of cells within MCTS to treatment due to poor penetration of DTX or BCMs, the limited sensitivity of cells within MCTS to DTX due to a reduction in cellular proliferation and/or resistance associated with 3-D cell adhesion (i.e. contact effect). In a study by Kyle et al., the penetration half-depth (the depth from the surface at which the amount of drug falls to half of its maximum concentration) of DTX in multicellular layers was found to be <25 μm following a 2 h incubation at a concentration of 0.3 μM [46]. Peak tissue levels did not increase proportionally following a 10-fold increase in drug concentration although the depth of penetration was improved indicating partial saturation of tissue binding. Therefore, it is likely that high intracellular binding and consumption of DTX by peripheral cells in the MCTS limits the toxicity to cells distant from the surface. For drugs which are rapidly consumed by cells, encapsulation in BCMs which minimize interactions and uptake by cells may improve drug penetration [47]. For example, Pun et al. reported ameliorated penetration of doxorubicin into MCTS when encapsulated in triblock copolymer micelles [25]. However, BCMs which pene-

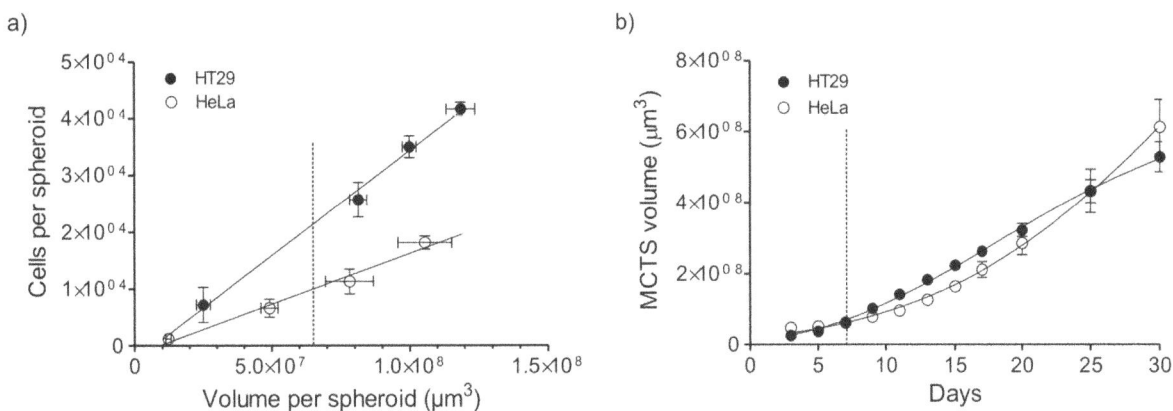

Figure 5. Spheroid packing density and growth. a) Cells per HeLa and HT29 spheroid of given volume, n = 12. b) Growth of HeLa and HT29 spheroids, n = 6. Data was fit using the Gompertz equation for tumor growth. The dashed lines indicate spheroid properties used in the studies.

trate poorly through tissues may limit the penetration of the encapsulated drug. Overall, the extent to which the BCMs influence drug penetration will depend on the relative rates of drug release and BCM penetration in the MCTS. We have previously found that PEG-*b*-PCL BCMs of 55 nm diameter can achieve a homogeneous distribution in MCTS following a 24 h incubation (unpublished data).

In addition to potential limitations in MCTS penetration associated with the drug and BCMs, the discrepancy between MCTS and monolayer cytotoxicity may also be a result of drug resistance imparted by the MCTS microenvironment. A marked decrease in the proportion of proliferating cells was observed in MCTS with increasing depth from the surface (Figure 9). Since DTX exerts its therapeutic effect on cycling cells, cells located near

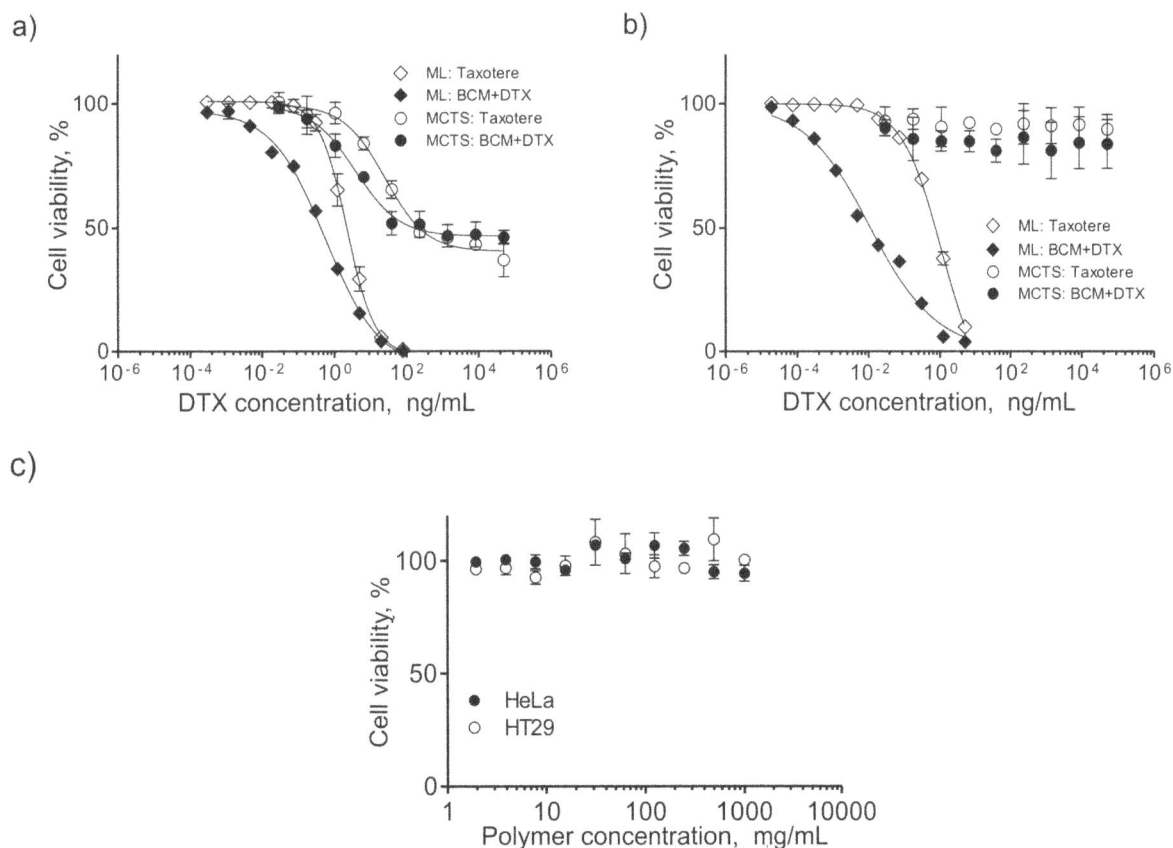

Figure 6. Cytotoxicity of Taxotere® and BCM+DTX in spheroid and monolayer cultures. Viability of a) HeLa and b) HT29 cells cultured as monolayers and spheroids as measured using the APH assay. Data is expressed as the percent viability relative to untreated controls and fit to the Hill equation. c) Cytotoxicity of blank PEG-*b*-PCL micelles as a function of copolymer concentration. Each plot represents the mean of three independent experiments ± SD (n = 3).

a)

b) c) d) e)

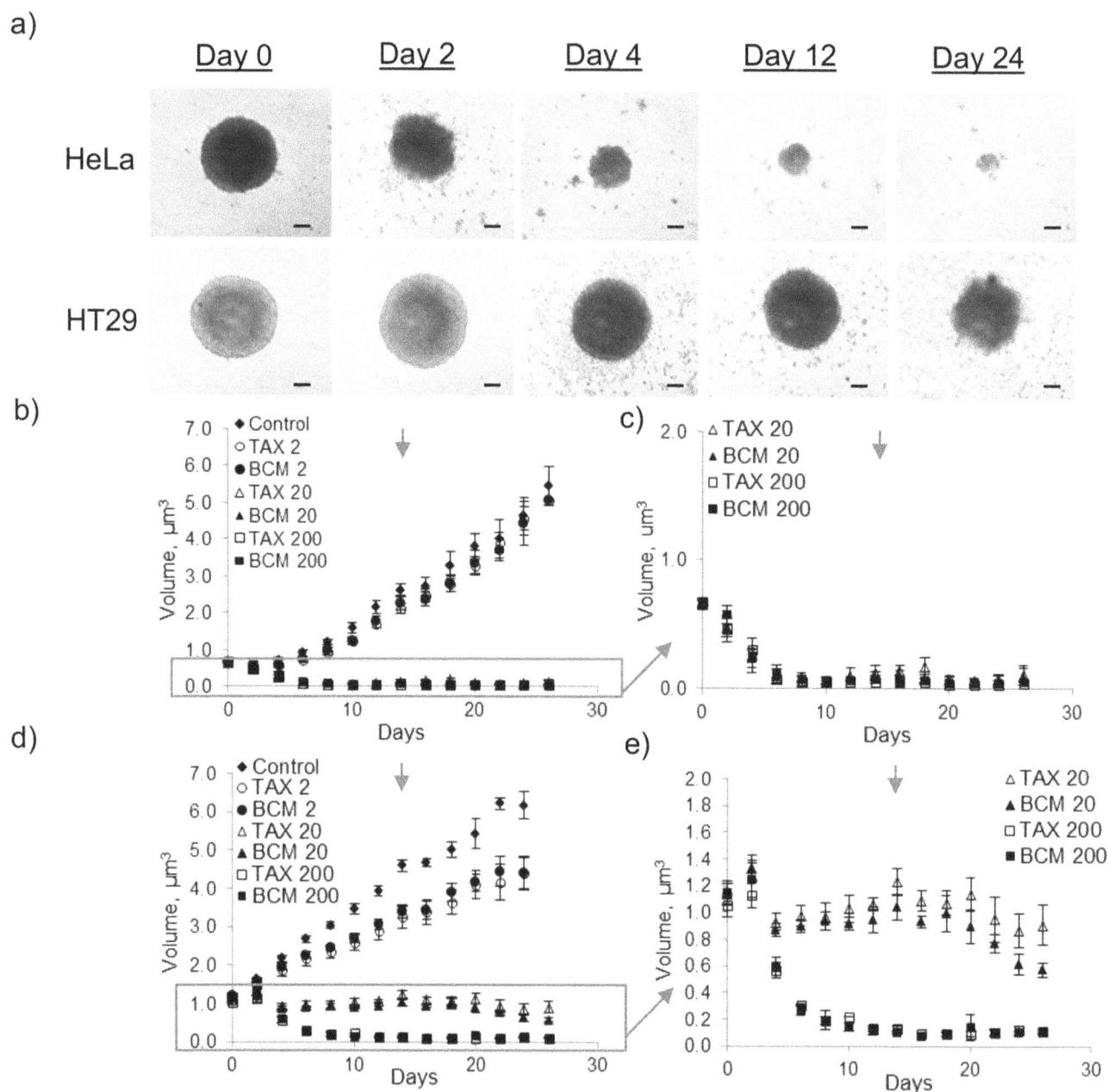

Figure 7. Inhibition of spheroid growth. a) Sequential images of the same HeLa and HT29 spheroids following treatment with BCM+DTX at a concentration of 20 ng/mL. Bars represent 100 μm. Growth inhibition of HeLa (b,c) and HT29 (d,e) MCTS by BCM+DTX and **Taxotere**® at concentrations of 2, 20 and 200 ng/mL. Cells were re-treated after two weeks (arrow). Box represents expanded region of plots b) and d). Data is expressed as the mean volume of six spheroids (n = 6) ± SD. "*" represents a significant difference between BCM 20 and TAX 20, p<0.05.

the MCTS surface will respond to treatment similarly to cells cultured as monolayers. By contrast, quiescent cells that are located in the intermediate and core regions of the MCTS will be less sensitive to treatment. This notion is supported by the observation that cells exposed to treatment immediately following disaggregation of MCTS demonstrated greater clonogenic survival than monolayer cells, but less than cells treated as intact MCTS. Therefore, there exists a population of cells within the MCTS that is more resistant to treatment than cells cultured as monolayers even in the absence of any physical barrier to drug penetration. As such, the limited sensitivity of MCTS to treatment is likely a result

of both restricted transport and mechanisms of drug resistance associated with the MCTS microenvironment.

The extent to which culturing cells as MCTS influenced the therapeutic effect of BCM+DTX and Taxotere® relative to monolayers was found to be cell-line specific. In monolayer cultures, BCM+DTX and Taxotere® demonstrated greater cytotoxicity against HT29 cells relative to HeLa cells. In contrast, culturing cells as MCTS imparted a greater enhancement in therapeutic resistance (i.e. greater increase in IC$_{50}$) to HT29 cells than to HeLa cells. We have previously shown significantly greater penetration of BCMs into HeLa MCTS than HT29 MCTS due to the former's lower cell packing density and large intercellular

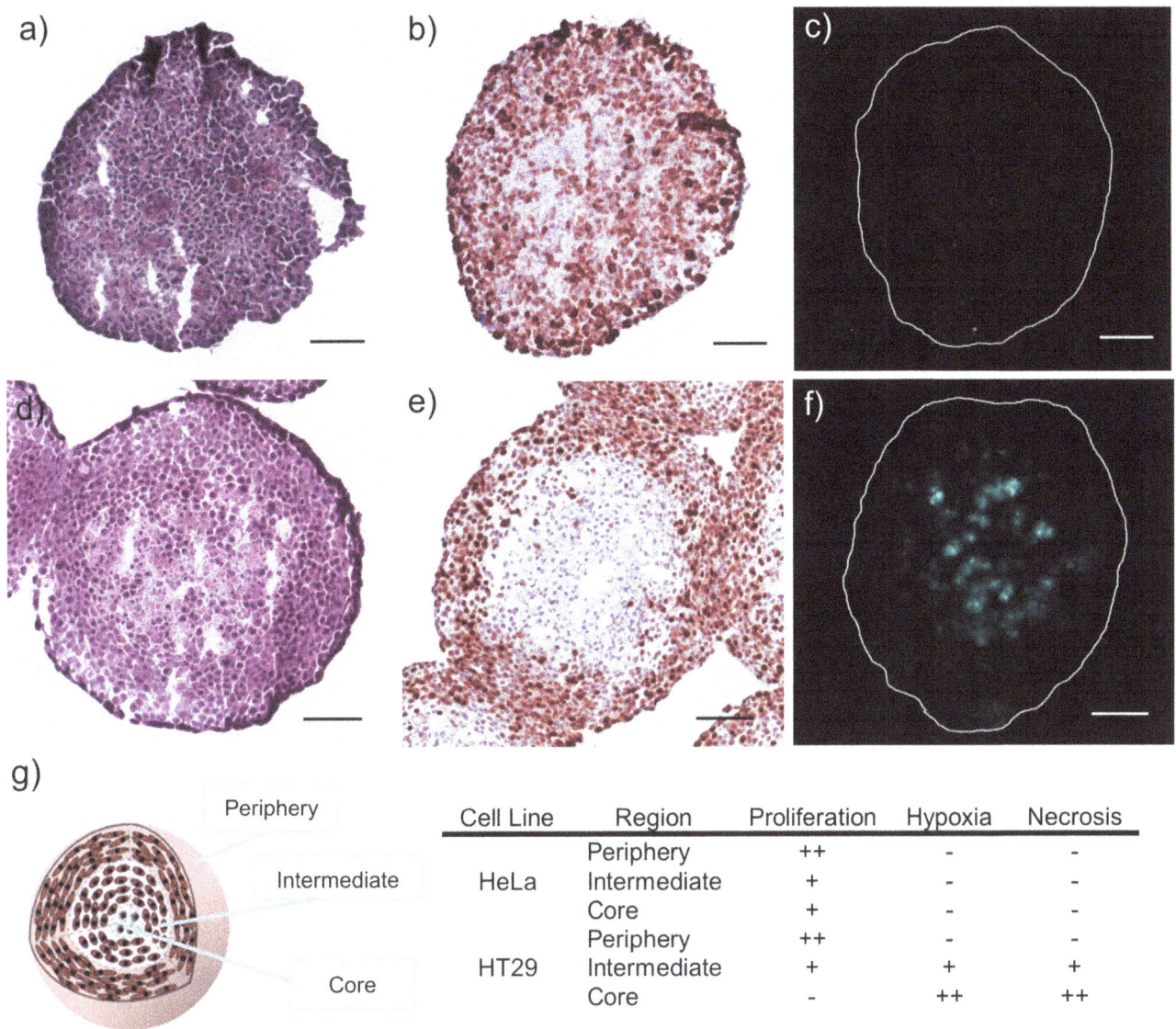

Figure 8. Histological assessment of spheroid microenvironment. HeLa (a–c) and HT29 (d–f) MCTS cross-sections stained with H&E (a, d), Ki67 proliferation marker (b, e) and EF5 (c, f), a marker of hypoxia. Scale bars represent 100 μm. g) Properties of the spheroid microenvironment and their spatial distribution. "++", "+", and "−", indicate high, intermediate and low levels of the corresponding feature, respectively.

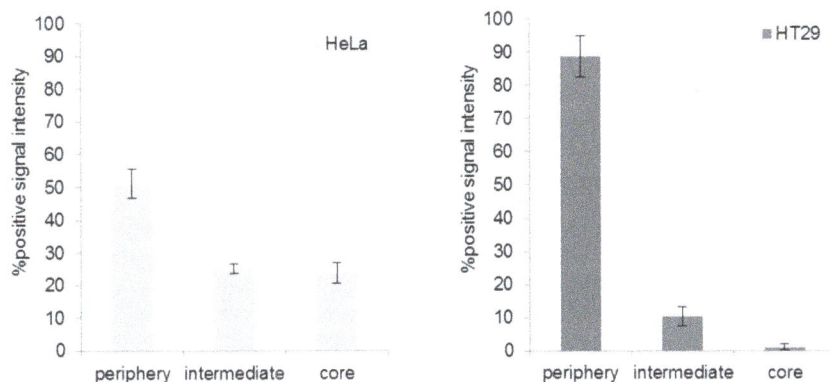

Figure 9. Spatial distribution of proliferating cells in spheroids. Ki67 positive signal distribution relative to radial position in a) HeLa and b) HT29 MCTS as a percent of total positive stain, n = 6.

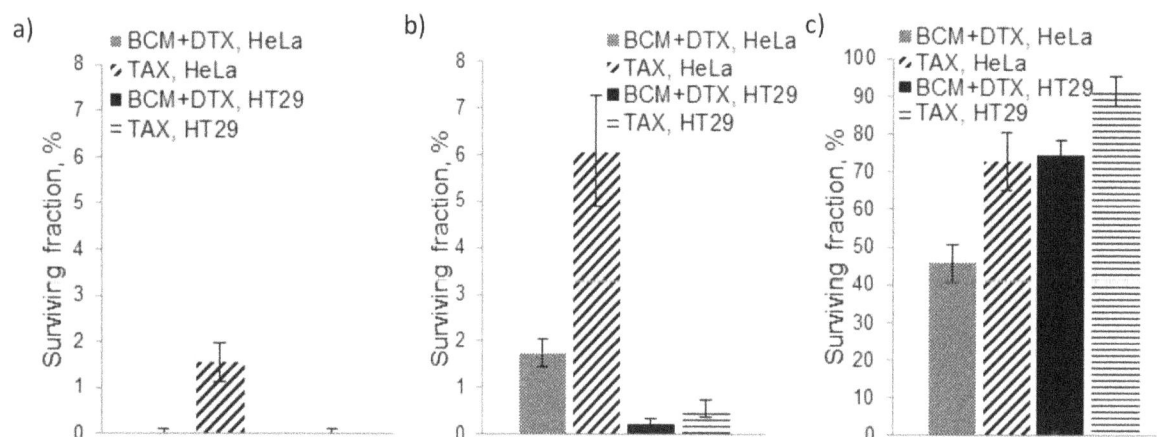

Figure 10. Clonogenic potential of cells following treatment. Clonogenic survival of HeLa and HT29 cells following 24 h treatment with 20 ng/mL of BCM+DTX or Taxotere® as a) monolayers, b) disaggregated spheroids and c) intact spheroids.

channels (unpublished data). In the current study, significant cell line-dependent differences in MCTS microenvironment were observed. Limited permeability of HT29 MCTS and/or high consumption of oxygen by peripheral cells was reflected by the presence of central hypoxia and necrosis. Importantly, HT29 MCTS contained a greater proportion of non-proliferating cells relative to HeLa MCTS. It is likely that some quiescent cells within the MCTS retained their clonogenic potential following exposure to sub-therapeutic amounts of DTX and were capable of recommencing proliferative activity when re-plated as monolayers. The greater clonogenic potential of HeLa cells following disaggregation of MCTS relative to HT29 cells likely reflects the greater sensitivity of HT29 monolayer cells to DTX rather than greater residual resistance of MCTS-derived HeLa cells.

One of the important advantages of the MCTS model is that it allows for treatment efficacy to be observed over an extended period of time. In order to evaluate the potential of surviving cells to repopulate MCTS, the growth of MCTS following treatment with BCM+DTX and Taxotere® was evaluated for 28 days with treatment re-applied after 14 days. The results of this study demonstrate both dose- and time-dependent changes in MCTS growth following incubation with the drug formulations. Near complete elimination of HeLa MCTS was observed following treatment at 20 ng/mL or greater with either BCM+DTX or Taxotere®. In contrast, only partial growth inhibition was observed in HT29 MCTS when exposed to the same concentration. This observation is consistent with the results obtained from the cytotoxicity and clonogenic assays in which HT29 MCTS demonstrated greater resistance to treatment relative to HeLa MCTS. A slight inhibitory effect in HT29 MCTS following administration of DTX formulations at 2 ng/mL was likely due to the cytotoxicity and shedding of surface cells, consistent with the response of HT29 cells to treatment in monolayer cultures. In addition, the apparent discrepancy between the limited cytotoxicity in HT29 spheroids revealed using the APH assay (measured 2 days post drug incubation) and the marked growth inhibition at 20 ng/mL is consistent with the observed 4 day delay in growth inhibitory effect. Interestingly, little difference in spheroid growth inhibition was observed between BCM+DTX and Taxotere® following initial treatment. It should be noted, however, that following retreatment after 14 days of culture, BCM+DTX demonstrated a greater growth inhibitory effect relative to Taxotere®.

Several factors may have contributed to the greater cytotoxicity of BCM+DTX relative to Taxotere® in monolayer and MCTS cultures. It has been hypothesized that DTX is taken up more rapidly by cells following release from BCMs in close proximity to the cell membrane due to an increase in the local transmembrane concentration gradient [48–50]. Slower efflux of BCM-encapsulated DTX relative to free DTX, by avoidance of membrane efflux pumps, may also contribute to the greater therapeutic effect of the DTX-loaded BCMs [51–53]. While these results are promising, further investigation is required to fully elucidate the mechanism of cytotoxicity that lead to enhanced therapeutic effects of BCM+DTX relative to Taxotere® *in vitro*.

Overall, as outlined in Figure 2, each of the three assays employed in this study is unique and together they provide complementary information on the therapeutic potential of drug formulations. Importantly, comparison of results obtained in monolayer and spheroid cultures demonstrated the important influence of the microenvironment and 3-D tissue structure on formulation efficacy. Therefore, 3-D cultures such as MCTS may serve as important tools for investigating the performance of nanomedicines in environments that more closely mimic intratumoral conditions *in vivo*. However, while spheroids share several important structural and microenvironmental properties with native tumors, there are important differences which may limit the extent to which this *in vitro* model can be used to predict drug efficacy *in vivo*. Notably, the MCTS model does not account for the potential influence of convective flow or presence of stromal cells on drug and nanoparticle transport. Despite these limitations, evaluation of formulation efficacy in spheroids rather than monolayer cultures is expected to more accurately reflect therapeutic performance *in vivo*.

Supporting Information

Figure S1 Measurement of spheroid volume. a) Schematic representation of the analysis process using a macro developed for ImageJ (version 1.44 m). b) Correlation between manual and automated volume measurements of HeLa MCTS. MCTS were imaged at selected intervals of growth. Manual measurement of MCTS volume was performed by determining the average of the largest and smallest diameters using the captured images and assuming a spherical MCTS morphology. Automated volume measurement was achieved using an image recognition technique in ImageJ. Firstly, MCTS images were converted into 8-bit greyscale and the perimeter of the MCTS was recognized by an automated threshold function. The area of the 2-D MCTS mask was recorded and converted to μm^2 by calibration using an image of known scale and subsequently used to calculate the volume.

Figure S2　Validation of the acid phosphatase (APH) assay. Results from the APH assay using HeLa (left column) and HT29 cells (right column) grown as spheroids (top row) and monolayers (bottom row) demonstrate a linear relationship between cell number and UV absorption at 405 nm. Each data point represents the mean of three independent experiments $\pm$ SD (n = 3).

Figure S3　Failure of WST-8 assay. Results from the WST-8 assay demonstrate a non-linear correlation between the number of cells and OD_{450} in spheroid culture.

Figure S4　Fluorescence images of HT29 (a) and HeLa (b) tumor xenografts displaying markers of hypoxia (EF5 - blue) and blood vessels (CD31 - red). Scale bars represent 100 μm.

Author Contributions

Conceived and designed the experiments: ASM SE CA. Performed the experiments: ASM SE. Analyzed the data: ASM SE CA. Contributed reagents/materials/analysis tools: CA. Wrote the paper: ASM CA.

References

1. Trédan O, Galmarini CM, Patel K, Tannock IF (2007) Drug resistance and the solid tumor microenvironment. J Natl Cancer Inst 99: 1441–1454.
2. Swartz MA, Iida N, Roberts EW (2012) Tumor Microenvironment Complexity: Emerging Roles in Cancer Therapy. Cancer Res 72: 2473–2480.
3. Jain RK, Stylianopoulos T (2010) Delivering nanomedicine to solid tumors. Nat Rev Clin Oncol 7: 653–664.
4. Minchinton AI, Tannock IF (2006) Drug penetration in solid tumours. Nat Rev Cancer 6: 583–592.
5. Vaupel P, Mayer A (2007) Hypoxia in cancer: significance and impact on clinical outcome. Cancer Metastasis Rev 26: 225–239.
6. Kim JW, Ho WJ, Wu BM (2011) The role of the 3D environment in hypoxia-induced drug and apoptosis resistance. Anticancer Res 31: 3237–3245.
7. Chauhan VP, Stylianopoulos T, Boucher Y, Jain RK (2011) Delivery of Molecular and Nanoscale Medicine to Tumors: Transport Barriers and Strategies. Annu Rev Chem Biomol Eng 2: 281–298.
8. Johnson JI, Decker S, Zaharevitz D, Rubinstein L V, Venditti JM, et al. (2001) Relationships between drug activity in NCI preclinical *in vitro* and *in vivo* models and early clinical trials. Br J Cancer 84: 1424–1431.
9. Phillips R, Bibby M, Double J (1990) A Critical Appraisal of the Predictive Value of in Vitro Chemosensitivity Assays. J Natl Cancer Inst 82: 1457–1463.
10. Elliott N (2010) A review of three-dimensional in vitro tissue models for drug discovery and transport studies. J Pharm Sci 100: 2–8.
11. Nyga A, Cheema U, Loizidou M (2011) 3D tumour models: novel in vitro approaches to cancer studies. J Cell Commun Signal 5: 239–248.
12. Vinci M, Gowan S, Boxall F, Patterson L, Zimmermann M, et al. (2012) Advances in establishment and analysis of 3D tumour spheroid-based functional assays for target validation and drug evaluation. BMC Biol 10: 1–20.
13. Rotin D, Robinson B, Tannock IF (1986) Influence of hypoxia and an acidic environment on the metabolism and viability of cultured cells: potential implications for cell death in tumors. Cancer Res 46: 2821–2826.
14. Nederman T, Norling B, Glimelius B, Carlsson J, Brunk U (1984) Demonstration of an Extracellular Matrix in Multicellular Tumor Spheroids. Cancer Res 44: 3090–3097.
15. Acker H, Carlsson J, Mueller-Klieser W, Sutherland RM (1987) Comparative pO2 measurements in cell spheroids cultured with different techniques. Br J Cancer 56: 325–327.
16. Fischbach C, Chen R, Matsumoto T, Schmelzle T, Brugge JS, et al. (2007) Engineering tumors with 3D scaffolds. Nat Methods 4: 6–11.
17. Horning JL, Sahoo SK, Vijayaraghavalu S, Dimitrijevic S, Vasir JK, et al. (2008) 3-D tumor model for in vitro evaluation of anticancer drugs. Mol Pharm 5: 849–862.
18. Hicks KO, Ohms SJ, Van Zijl PL, Denny WA, Hunter PJ, et al. (1997) An experimental and mathematical model for the extravascular transport of a DNA intercalator in tumours. Br J Cancer 76: 894–903.
19. Tannock IF, Lee CM, Tunggal JK, Cowan DSM, Egorin MJ (2002) Limited penetration of anticancer drugs through tumor tissue: a potential cause of resistance of solid tumors to chemotherapy. Clinical cancer research: an official journal of the American Association for Cancer Res 8: 878–884.
20. Cowan DS, Tannock IF (2001) Factors that influence the penetration of methotrexate through solid tissue. Int J Cancer 91: 120–125.
21. Chetprayoon P, Kadowaki K, Matsusaki M, Akashi M (2012) Survival and structural evaluations of three-dimensional tissues fabricated by the hierarchical cell manipulation technique. Acta Biomater *in press*.
22. Hosoya H, Kadowaki K, Matsusaki M, Cabral H, Nishihara H, et al. (2012) Engineering fibrotic tissue in pancreatic cancer: A novel three-dimensional model to investigate nanoparticle delivery. Biochem Biophys Res Commun 419: 32–37.
23. Grantab R, Sivananthan S, Tannock IF (2006) The penetration of anticancer drugs through tumor tissue as a function of cellular adhesion and packing density of tumor cells. Cancer Res 66: 1033–1039.
24. Hirschhaeuser F, Menne H, Dittfeld C, West J, Mueller-Klieser W, et al. (2010) Multicellular tumor spheroids: an underestimated tool is catching up again. J Biotechnol 148: 3–15.
25. Kim T-H, Mount CW, Gombotz WR, Pun SH (2010) The delivery of doxorubicin to 3-D multicellular spheroids and tumors in a murine xenograft model using tumor-penetrating triblock polymeric micelles. Biomaterials 31: 7386–7397.
26. Perche F, Patel NR, Torchilin VP (2012) Accumulation and toxicity of antibody-targeted doxorubicin-loaded PEG-PE micelles in ovarian cancer cell spheroid model. J Control Release *in press*.
27. Oishi M, Nagasaki Y, Nishiyama N, Itaka K, Takagi M, et al. (2007) Enhanced growth inhibition of hepatic multicellular tumor spheroids by lactosylated poly(ethylene glycol)-siRNA conjugate formulated in PEGylated polyplexes. ChemMedChem 2: 1290–1297.
28. Tsukioka Y, Matsumura Y, Hamaguchi T, Koike H, Moriyasu F, et al. (2002) Pharmaceutical and biomedical differences between micellar doxorubicin (NK911) and liposomal doxorubicin (Doxil). Jpn J Cancer Res 93: 1145–1153.
29. Bae Y, Nishiyama N, Fukushima S, Koyama H, Yasuhiro M, et al. (2005) Preparation and biological characterization of polymeric micelle drug carriers with intracellular pH-triggered drug release property: tumor permeability, controlled subcellular drug distribution, and enhanced in vivo antitumor efficacy. Bioconjug Chem 16: 122–130.
30. Prescribing Information, Taxotere (2010). Sanofi-Aventis. Available: http://products.sanofi.us/Taxotere/taxotere.html.
31. ClinicalTrials.gov. Available: http://clinicaltrials.gov/ct2/show/NCT01639131?term = colon+and+docetaxel&rank = 15. Accessed 2013 March 18.
32. ClinicalTrials.gov. Available: http://clinicaltrials.gov/ct2/show/NCT00003445?term = cervix+and+docetaxel&rank = 3. Accessed 2013 March 18.
33. Baur M, Van Oosterom AT, Diéras V, Tubiana-Hulin M, Coombes RC, et al. (2008) A phase II trial of docetaxel (Taxotere) as second-line chemotherapy in patients with metastatic breast cancer. J Cancer Res Clin Oncol 134: 125–135.
34. Mikhail A, Allen C (2009) Block copolymer micelles for delivery of cancer therapy: transport at the whole body, tissue and cellular levels. J Control Release 138: 214–223.
35. Mikhail A, Allen C (2010) Poly(ethylene glycol)-b-poly(epsilon-caprolactone) micelles containing chemically conjugated and physically entrapped docetaxel: synthesis, characterization, and the influence of the drug on micelle morphology. Biomacromolecules 11: 1273–1280.
36. Zahedi P, Stewart J, De Souza R, Piquette-Miller M, Allen C (2012) An injectable depot system for sustained intraperitoneal chemotherapy of ovarian cancer results in favorable drug distribution at the whole body, peritoneal and intratumoral levels. J Control Release 158: 379–385.
37. Marusić M, Bajzer Z, Freyer JP, Vuk-Pavlović S (1994) Analysis of growth of multicellular tumour spheroids by mathematical models. Cell Prolif 27: 73–94.
38. Friedrich J, Eder W, Castaneda J, Doss M, Huber E, et al. (2007) A reliable tool to determine cell viability in complex 3-d culture: the acid phosphatase assay. J Biomol Screen 12: 925–937.
39. Joyce J a, Pollard JW (2009) Microenvironmental regulation of metastasis. Nat Rev Cancer 9: 239–252.
40. Hanahan D, Weinberg R a (2011) Hallmarks of cancer: the next generation. Cell 144: 646–674.
41. Okubo H, Matsushita M, Kamachi H, Kawai T, Takahashi M, et al. (2002) A novel method for faster formation of rat liver cell spheroids. Artif Organs 26: 497–505.
42. Del Duca D, Werbowetski T, Del Maestro RF (2004) Spheroid preparation from hanging drops: characterization of a model of brain tumor invasion. J Neurooncol 67: 295–303.
43. Friedrich J, Seidel C, Ebner R, Kunz-Schughart L a (2009) Spheroid-based drug screen: considerations and practical approach. Nat Protoc 4: 309–324.

44. Hardelauf H, Frimat J-P, Stewart JD, Schormann W, Chiang Y-Y, et al. (2011) Microarrays for the scalable production of metabolically relevant tumour spheroids: a tool for modulating chemosensitivity traits. Lab Chip 11: 419–428.

45. Ong S-M, Zhao Z, Arooz T, Zhao D, Zhang S, et al. (2010) Engineering a scaffold-free 3D tumor model for in vitro drug penetration studies. Biomaterials 31: 1180–1190.

46. Kyle AH, Huxham L a, Yeoman DM, Minchinton AI (2007) Limited tissue penetration of taxanes: a mechanism for resistance in solid tumors. Clin Cancer Res 13: 2804–2810.

47. Han M, Oba M, Nishiyama N, Kano MR, Kizaka-Kondoh S, et al. (2009) Enhanced percolation and gene expression in tumor hypoxia by PEGylated polyplex micelles. Mol Ther 17: 1404–1410.

48. Fonseca C, Simões S, Gaspar R (2002) Paclitaxel-loaded PLGA nanoparticles: preparation, physicochemical characterization and in vitro anti-tumoral activity. J Control Release 83: 273–286.

49. Nemati F, Dubernet C, Poupon M, Couvreur P (1996) Reversion of multidrug resistance using nanoparticles in vitro: influence of the nature of the polymer. Int J Pharm: 237–246.

50. Cavallaro G, Fresta M, Giammona G, Puglisi G, Villari A (1994) Entrapment of B-lactams antibiotics in polyethylcyanoacrylate nanoparticles: Studies on the possible in vivo application of this colloidal delivery system. Int J Pharm 111: 31–41.

51. Panyam J, Labhasetwar V (2003) Dynamics of endocytosis and exocytosis of poly(D,L-lactide-co-glycolide) nanoparticles in vascular smooth muscle cells. Pharm Res 20: 212–220.

52. Yoo HS, Park TG (2004) Folate-receptor-targeted delivery of doxorubicin nano-aggregates stabilized by doxorubicin-PEG-folate conjugate. J Control Release 100: 247–256.

53. Yoo HS, Park TG (2004) Folate receptor targeted biodegradable polymeric doxorubicin micelles. J Control Release 96: 273–283.

Analysis of the Murine Immune Response to Pulmonary Delivery of Precisely Fabricated Nano- and Microscale Particles

Reid A. Roberts[1], Tammy Shen[2], Irving C. Allen[3], Warefta Hasan[4¤], Joseph M. DeSimone[2,4,5,6,7,8,9,10,11,12*], Jenny P. Y. Ting[1,5*]

1 Department of Microbiology and Immunology, University of North Carolina, Chapel Hill, North Carolina, United States of America, 2 Eshelman School of Pharmacy, University of North Carolina, Chapel Hill, North Carolina, United States of America, 3 Department of Biomedical Sciences and Pathobiology, Virginia-Maryland Regional College of Veterinary Medicine, Virginia Polytechnic Institute and State University, Blacksburg, Virginia, United States of America, 4 Department of Chemistry, University of North Carolina, Chapel Hill, North Carolina, United States of America, 5 Lineberger Comprehensive Cancer Center, University of North Carolina, Chapel Hill, North Carolina, United States of America, 6 Department of Biochemistry and Biophysics, University of North Carolina, Chapel Hill, North Carolina, United States of America, 7 Carolina Center of Cancer Nanotechnology Excellence, University of North Carolina, Chapel Hill, North Carolina, United States of America, 8 Department of Chemical and Biomolecular Engineering, North Carolina State University, Raleigh, North Carolina, United States of America, 9 Department of Pharmacology, University of North Carolina, Chapel Hill, North Carolina, United States of America, 10 Institute for Advanced Materials, University of North Carolina, Chapel Hill, North Carolina, United States of America, 11 Institute for Nanomedicine, University of North Carolina, Chapel Hill, North Carolina, United States of America, 12 Sloan-Kettering Institute for Cancer Research, Memorial Sloan-Kettering Cancer Center, New York, New York, United States of America

Abstract

Nanomedicine has the potential to transform clinical care in the 21st century. However, a precise understanding of how nanomaterial design parameters such as size, shape and composition affect the mammalian immune system is a prerequisite for the realization of nanomedicine's translational promise. Herein, we make use of the recently developed Particle Replication in Non-wetting Template (PRINT) fabrication process to precisely fabricate particles across and the nano- and micro-scale with defined shapes and compositions to address the role of particle design parameters on the murine innate immune response in both *in vitro* and *in vivo* settings. We find that particles composed of either the biodegradable polymer poly(lactic-co-glycolic acid) (PLGA) or the biocompatible polymer polyethylene glycol (PEG) do not cause release of pro-inflammatory cytokines nor inflammasome activation in bone marrow-derived macrophages. When instilled into the lungs of mice, particle composition and size can augment the number and type of innate immune cells recruited to the lungs without triggering inflammatory responses as assayed by cytokine release and histopathology. Smaller particles (80×320 nm) are more readily taken up *in vivo* by monocytes and macrophages than larger particles (6 μm diameter), yet particles of all tested sizes remained in the lungs for up to 7 days without clearance or triggering of host immunity. These results suggest rational design of nanoparticle physical parameters can be used for sustained and localized delivery of therapeutics to the lungs.

Editor: Salik Hussain, National Institute of Health (NIH), United States of America

Funding: This work was supported by National Institutes of Health grants 8-VP1-CA174425-04 (awarded to Joseph M. DeSimone), U19-AI077437 and U54 AI057157 (Jenny P.Y. Ting) and NC Tracs 100K1202 (Joseph M. DeSimone and Jenny P.Y. Ting). The funders had no role in study design, data collection and analysis, decision to publish, or preparation of the manuscript.

Competing Interests: Joseph M. DeSimone is a founder, member of the board of directors, and maintains a financial interest in Liquidia Technologies. PRINT and Fluorocur are registered trademarks of Liquidia Technologies, Inc. One author, Warefta Hasan, is currently employed by a commercial company but she was an employee of the University of North Carolina at the time of the studies detailed in this manuscript. She does not have any competing interests.

* E-mail: jpyting@gmail.com (JPYT); desimone@email.unc.edu (JMDS)

¤ Current address: Aurasense Therapeutics, Skokie, Illinois, United States of America

Introduction

The application of nanoparticles in medicine for disease treatment is a potentially transformative area of research. The possibility of potent instruction and modulation of host physiology through nanomaterials has been abundantly demonstrated. These efforts include modulation of cell-specific gene expression through delivery of antisense oligonucleotides, dose-sparing and targeted delivery of pharmacologics, as well as enhanced multi-functional imaging diagnostics through nanoformulations [1–3]. Another area where nanotechnology may revolutionize clinical care is the ability to direct immune responses in defined manners. The most obvious benefit is in the design of next-generation vaccines against microbial pathogens, whereby antigen specific immune responses can be elicited at levels far more potent than existing vaccines, commensurate with the immune response engendered by live organisms [4–6]. These latter advances owe much to our recent understanding of the critical role of innate immunity in contextualizing an appropriate adaptive immune response. This context is triggered through endogenous host receptor signaling pathways, most well characterized by the TLR family of pattern recognition receptors (PRR), but hallmarked by a panoply of such PRRs including C-type lectin receptors, RIG-like helicases and the burgeoning understanding of Nod-like receptors (NLRs) as sentinels of the intracellular environment [7].

Researchers in the emerging field of immune-engineering are capitalizing on these exciting advances to open up the possibility to direct and instruct immunological outcomes to a variety of pathological conditions [8]. These include improving current pathogen vaccines, to the more nascent fields of cancer immunotherapy, tolerance induction in the setting of autoimmunity and organ transplantation, as well as general immunological rebalancing in diseased settings, such as the chronic inflammation associated with type 2 diabetes. The implications of such technology are profound and potentially represent a paradigm shift in clinical practice across a broad swath of medicine. However, efforts to use nanotechnology and material sciences engineering to modulate human biology *in situ* require a comprehensive understanding of the immune response, or lack thereof, engendered to introduced nanocarriers [9,10].

In order for the gamut of potential downstream therapeutic applications of nanomedicine to be realized, we must first understand how the physical properties of nanomaterials augment host immune responses. These principles will then enable the appropriate design of nano- and micro-scale interventions for specific purposes. For example, the use of nanoparticles to deliver potent biological molecules, such as oligonucleotides or small molecules augmenting intracellular signaling pathways, may squander the regulatory opportunity to reach the clinic if the nanocarriers for such entities initiate off-target events that activate host immune responses. Conversely, vaccines can be made more potent if nanocarriers are designed to activate the appropriate innate immune response to tailor adaptive immune responses to delivered antigens. As an example, the current state of the art is to use Alum as a non-specific immunomodulatory adjuvant in vaccine formulations. This could explain, in part, why some of the most pressing pathogens do not yet have useful vaccines because Alum is known to engender a mixed Th1/Th2-biased humoral immune response that does not reflect the immune activation which occurs during an actual microbial infection [11]. It is conceivable that nano-carriers designed to elicit the appropriate adaptive immune response to an antigen of interest-i.e., Th17 against fungal pathogens-may enable the generation of vaccine-induced immune responses that more closely mimic the natural immune response elicited by infection with a given pathogen as opposed to the non-tailored immunity induced by Alum.

A great issue in advancing nanotechnology from a laboratory pursuit into a component of clinical care is a robust understanding of how physical particle properties augment biological outcomes. There is a wealth of literature promoting the use of nanotechnology in modern medicine, but much of this literature relies on particle fabrication methods, such as oil-in-water emulsion, that generate heterogeneous populations of particles that can vary widely between batches and across labs [12,13]. In addition, most studies related to biomedical applications of nanotechnology have not addressed the role of the immune system in the host response to particulate delivery, a critical issue that threatens to diminish the utility of such particles if these are rapidly cleared by innate immune cells and/or induce localized or systemic immune responses that pose unintended complications for clinical development. In cases where the immunological parameters of nanotechnology are being addressed, great variance is seen based on size, composition and even surface modification of particles,

highlighting the tremendous complexity and exquisite sensitivity of the immune system to nanoscale events [9,10,14].

Most currently employed fabrication methods do not allow precise control over particle physical parameters, thus it is difficult to draw conclusions as to how size, shape and composition affect the innate immune response to particles in the nano and micron range. To address this lack of knowledge, we employed our recently developed top-down nanofabrication technique termed Particle Replication in Non-Wetting Templates (PRINT) [15–18]. Using soft lithography techniques adopted from the semi-conductor industry, PRINT enables the production of monodisperse nano- and microparticles with well-defined control over particle size, shape, composition, modulus and surface chemistry. Therefore, the role of these physical parameters in augmenting biological responses can be reproducibly probed using PRINT technology.

To lend both clinical and field relevance to our findings, particles were designed with either the F.D.A approved polymer Poly-lactic co-gloycolic acid (PLGA) or derivatives of the commonly published hydrogel polymer polyethylene glycol (PEG). While PLGA is an attractive polymer given its long history of clinical use, our study was aimed in part to clarify discrepancy in the literature as to whether particles made of PLGA trigger inflammation. As our main purpose was to define the 'baseline' status of whether particles of defined size, shape and composition triggered inflammation, we did not augment particles in this study to include additional biologically active molecules, such as oligonucleotides, small molecules or adjuvants as previously published by our group and others [19–22]. To this end, we used *in vitro* assays with murine derived macrophages and *in vivo* delivery of particles to the lungs of mice to test the inflammatory potential of these particles. The lung is a highly desired site for therapeutic delivery of nanomedicine and we chose it for both clinical relevance and its sensitivity as an immunological organ [23]. Our findings imply that the delivery of PRINT nano- and microparticles do not engender systemic or localized inflammatory responses and may not be impeded by host immune responses to the polymers used in this study. Future design strategies for the panoply of therapeutic opportunities made available by nanoengineering are likely available, as particles across broad size ranges can thus be rationally designed from an inert state.

Materials and Methods

Particle Materials

Poly(ethylene glycol) diacrylate (M_n 700) (PEG$_{700}$DA), 2-aminoethyl methacrylate hydrochloride (AEM), Diphenyl (2,4,6-trimethylbenzoyl)-phoshine oxide (TPO), and poly lactic co-glycolic acid (PLGA; 85:15 lactic acid/glycolic acid, MW = 55 000 g/mol) were purchased from Sigma-Aldrich. Tetraethylene glycol monoacrylate (HP4A) was synthesized in-house as previously described [24]. Thermo Scientific Dylight 650 maleimide, PTFE syringe filters (13 mm membrane, 0.220 μm pore size), dimethylformamide (DMF), triethanolamine (TEA), pyridine, sterile water, borate buffer (pH 8.6), Dulbecco's phosphate buffered saline (DPBS) (pH 7.4), 1X phosphate buffered saline (PBS) (pH 7.4), acetic anhydride and methanol were obtained from Fisher Scientific. Conventional filters (2 μm) were purchased from Agilent and polyvinyl alcohol (Mw 2000) (PVOH) was purchased from Acros Organics. All PRINT molds used in these studies (80 nm×320 nm, 1 μm cylinder, 1.5 μm and 6 μm donuts) were kindly provided by Liquidia Technologies.

PRINT PLGA Particle Fabrication

The PRINT process for fabricating particles has been described previously [21,22]. Briefly, to fabricate PLGA particles, a preparticle solution containing PLGA was prepared in a DMSO/DMF/water solvent mixture (4:16:1) and cast on a poly-(ethylene teraphthalate) (PET) sheet (delivery sheet) using a #5 Mayer Rod (R.D. Specialties). The delivery sheet was placed in contact with a PRINT mold with desired features patterned (e.g., 80×320 nm). The delivery sheet and mold were passed through a heated laminator (150°C, 5.5×105 Pa) and separated at the nip. This heating process enables the PLGA polymer solution to fill the molds, thereby forming nanoparticles of desired size and shape. Nanoparticles were then harvested from the PRINT mold by placing it in contact with a PET sheet coated with a layer (400 nm cast from water) of poly(vinyl alcohol) (PVA, MW = 2000 g/mol). This mold/PET-PVA ensemble was then passed through the laminator (150°C, 5.5×105 Pa) to transfer the nanoparticles to the PVA sheet. Both laminator steps, the filling of the mold and transfer of particles onto the PVA-coated PET sheet, were performed at low humidity (~20–30%). Particles were released from the PET/PVA sheet by delivering ~1 ml of sterile water via a bead harvester to dissolve the PVA layer and remove the particles from the PET sheet. A typical yield of 80×320 nm PLGA particles was ~0.4 mg particles/ft of PRINT mold, though this depended on the particle feature size of the mold. To remove excess PVA and concentrate the particles, tangential flow filtration (TFF; Spectrum Labs) was used to concentrate particles in sterile water (1–2 mg/ml). For later use in particle characterization assays and experiments, particles were lyophilized by adding 10× mannitol and 8× sucrose (10× and 8× to mass of particles) using a tree lyophilizer. Mannitol and sucrose were used as cryoprotectants.

PRINT Hydrogel Fabrication

The process of fabricating 80×320 nm hydrogel particles was conceptually similar to PLGA fabrication, but with important differences. The pre-particle solution (PPS) contained a composition of 67.5 wt% HP_4A, 20 wt% AEM (functional monomer), 10 wt% $PEG_{700}DA$ (crosslinker), 1 wt% TPO (photo initiator) and 1.5 wt% Dylight 650 maleimide. This composition was then dissolved at 3.5 wt% in methanol and drawn as a thin film using a # 3 Mayer rod (R.D. Specialties) onto a roll of corona treated PET using an in-house custom-made roll-to-roll lab line (Liquidia Technologies) running at 12 ft/min. The solvent was evaporated from this delivery sheet by exposing the film to heat guns. The delivery sheet was laminated (80 PSI, 12 ft/min) to the patterned side of the mold, followed by delamination at the nip. Particles were cured by passing the filled mold through a UV-LED (Phoseon, 395 nm, 3 SCFM N_2, 12 ft/min). A PVOH harvesting sheet was hot laminated to the filled mold (140°C, 80 PSI, 12 ft/min). Upon cooling to room temperature, particles were removed from the mold by splitting the PVOH harvesting sheet from the mold. Particles were then harvested by dissolving the PVOH in a bead of water (1 mL of water per 5 ft of harvesting sheet). The particle suspension was passed through a 2 µm filter (Agilent) to remove any large particulates. To remove the excess PVOH, particles were centrifuged (Eppendorf Centrifuge 5417R) at 14000 rpm for 15 min, the supernatant was removed and the particles were re-suspended in sterile water. This purification process was repeated 4 times prior to lyophilization as detailed above. The 80×320 nm particles were acetylated prior to experimental use to match negative charge of micron sized hydrogel particles.

The 1.5 and 6 µm donut shaped hydrogel particles were fabricated using a dropcast method. The pre-particle solution (PPS) was composed of 20% PEG_{700}-DA, 78% HP_4A, 1% TPO (photoinitiator), and 1% Dylight 650. The solution was spread onto a fluorocur mold and a poly(ethylene terephthalate) (PET) sheet was laminated on top of the mold and polymer mixture and run through a heated, pressurized laminator to fill the molds. The mold was then cured with a UV LED lamp for 30 seconds. Particles were transferred out of the mold onto a Luvitec harvesting layer by laminating the mold and Luvitec sheet together and running them through a heated laminator nip. The mold and harvesting sheet was separated, leaving free particles on the harvest layer. Particles were collected from the harvest sheet by bead harvesting with water and pelleted by centrifugation. The particles were re-suspended in tert-butanol and lyophilized overnight.

Particle Characterization

Thermogravimetric analysis (TGA) was used to determine stock particle concentrations (TA Instruments Q5000 TGA). Briefly, 20 µL of the stock nanoparticle solution was pipetted into a tared aluminum sample pan. The sample was heated at 30°C/min to 130°C and held at this temperature for 10 minutes. The sample was then cooled at 30°C/min down to 30°C and held for 2 minutes. A Hitachi S-4700 scanning electron microscope (SEM)-was used to visualize particles. Prior to imaging, the SEM samples were coated with 1.5 nm of gold-palladium alloy using a Cressington 108 auto sputter coater. Particle size and zeta potential were measured by dynamic light scattering (DLS) on a Zetasizer Nano ZS (Malvern Instruments, Ltd.).

Experimental animals

All studies were conducted in accordance with National Institutes of Health guidelines for the care and use of laboratory animals and approved by the Institutional Animal Care and Use Committee (IACUC) of the University of North Carolina at Chapel Hill. All animals were maintained in pathogen-free facilities at the University of North Carolina at Chapel Hill.

In vitro confocal analysis of hydrogel particle uptake

MH-S murine alveolar macrophages were plated in complete DMEM at 20,000 cells per well in 8-well chamber slides (LabTek) 48 hours prior to treatment with particles. Particles were resuspended in DMEM at 20 µg/ml and 300 µl of particle solution were added to each well. Particles were incubated with cells at 37°C for 4 hours. Cells were then washed twice with PBS and cells fixed with 4% Paraformaldehyde (PFA) solution and later stained with Alexa Fluor 488 Phalloidin (Invitrogen) and DAPI (Vectashield, Vector Labs). Fluorescent imaging of stained cells was performed on a Zeiss 710 laser scanning confocal imaging system (Zeiss).

In vitro inflammation assays

Bone marrow macrophages were isolated from the femurs of C57Bl/6 and BALB/c mice using standard procedures. Bone marrow-derived macrophages were cultured for six days in DMEM supplemented with 10% fetal bovine serum, L-Glutamine, pen/strep and 20% L929-conditioned medium prior to use in particle experiments. Adherent cells were isolated and plated in complete Dulbecco's Modifed Eagle Medium (Gibco) with 10% fetal calf serum, 1% penicillin/streptomycin and 1% L-glutamine at 200,000 cells per well in a 96-well dish for 24 hours prior to treatment with particles for up to 24 hours. Some cells were primed with LPS (50 ng/ml) for 24 prior to particle treatment to provide signal 1 for inflammasome activation. Monosodium urate (MSU; Invivogen) treated at 300 µg/ml or ATP (5 mM; Sigma-Aldrich) was used as a positive control for inflammasome activation. Particles were resuspended in PBS prior to dosing at various concentrations. After 24 hours of particle treatment in triplicate, supernatants were harvested and analyzed by murine IL-1β, TNF-α and IL-6 ELISA (BD Biosciences). Limit of detection for ELISAs was 31.3 pg/ml (IL-1β) and 15.6 pg/ml (TNF-α and IL-6). Lactate dehydrogenase release was used to measure particle-induced cytotoxicity (Roche Applied Sciences).

Assessment of airway Inflammation

To assess whether PRINT particles induce airway inflammation, 10–12 week old female C57Bl/6 mice were anesthetized with isofluorane inhalation and particles were instilled via intratracheal (i.t.) administration. 50 µg of particles were dosed in 50 µl of PBS. Intratracheal administration of PBS (50 µl) or LPS (20 µg in 50 µl PBS) served as negative and positive controls for airway inflammation, respectively, as previously described [25]. Mice were euthanized and airway inflammation was assessed 48 hours or 7 days post treatment.

Serum was collected from animals by cardiac puncture and centrifuged at 15,000 RPM for 10 minutes. The serum supernatant was collected and used for ELISA analysis of inflammatory markers. Bronchoalveolar lavage fluid (BALF) was also collected to evaluate local leukocyte and cytokine levels in the lungs. For this purpose, lungs were lavaged three times with 1 ml Hanks Balanced Salt Solution (HBSS; Gibco). After centrifugation at 1500 RPM for 5 minutes, cell-free supernatants were collected and used to assess cytokine levels of IL-1β, TNF-α and IL-6 via ELISA (BD Biosciences). RBC were lysed via brief hypotonic saline treatment and the cell pellet was resuspended in PBS. Total BALF cellularity was assessed with a hemacytometer. The cellular composition was determined by cytospin of BALF aliquots onto slides and staining with Diff-Quik (Dade Behring) for differential cell counts. Leukocytes were identified based on the morphology of ≥200 cells per sample. Following BALF harvest, the lungs were fixed by inflation (20-cm pressure) and immersed in 10% buffered formalin.

Histopathological examination

Inflammation was evaluated in 5 µm sections of the left lung lobe after hematoxylin and eosin (H&E) staining. Serial paraffin-embedded sections were set and cut to reveal maximum longitudinal visualization of the intrapulmonary main axial airway and inflammation was scored by one of the authors (I.C.A.) who was blinded to genotype and treatment. As previously described, histology images were evaluated on each of the following inflammatory parameters and scored between 0 (absent) to 3 (severe): mononuclear cell infiltration, polymorphonuclear cell infiltration, airway epithelial cell hyperplasia/injury, extravasation, perivascular cuffing, and estimated percentage of the lung involved with inflammation [26,27]. Scores for each parameter were averaged for a total histology score.

Particle uptake in BALF

BALF aliquots from PEG treated mice were fixed in 2% paraformaldehyde and stained with DAPI (nuclei) and Phalloidin 488 (actin) and then viewed via epifluorescence microscopy for particle uptake (Dylight 650). Five distinct fields of view (FOV) were captured for each slide. The percentage of cell uptake was determined by dividing the number of cells showing particle internalization by the total number of cells in each field of view.

Statistical Analysis

GraphPad Prism 5 software was used to identify statistical significance. Single data point comparisons were evaluated by Student's two-tailed t-test, whereas multiple comparisons were evaluated for statistical significance using Analysis of Variance (ANOVA) followed by Tukey-Kramer HSD post-test. All cytokine and cell count data are presented as mean +/− standard deviation (SD) or standard error of the mean (SEM), respectively, with a p-value less than 0.05 considered statistically significant.

Results

PRINT enables the fabrication of monodisperse and homogenous particles

We employed Particle Replication in Non-Wetting Templates (PRINT) in an effort to address whether particles of defined size, shape and composition trigger an inflammatory response in mice. This fabrication platform enables production of homogenous and monodisperse particles with user-defined physical parameters. As a large amount of literature shows crucial biological differences depending on size and shape, the PRINT technique enables reproducible probing of basic cell biology with nearly complete control of design parameters [10,28–30].

For the purposes of our studies, we fabricated particles across the nano and micron range to reflect biologically relevant sizes. These include 80×320 nm particles (commensurate with the sizes of small bacteria and large viruses), 1 µm and 1.5 µm particles (commensurate with bacteria and platelets), and 6 µm particles (akin to a red blood cell in size) [28,31]. To characterize the fabricated particles, we performed dynamic light scattering (DLS) and zeta potential measurements as shown in Figure 1A. Note that DLS measurements are quantified based on particles with a perfect sphere shape, so that the size ranges detected are in line with non-spherous shapes of the molds used. The poly-dispersity index (PDI), a measure of heterogeneity in a particle population, indicates we were able to fabricate monodisperse particles of the same size and shape. This is further evidenced by scanning electron microscopy images (Figure 1B). As the surface charge of particles has been shown to play a role in biological outcomes, such as protein adsorption and cell uptake, we measured the Zeta potential of particles to quantify net surface charge [30,32,33]. For all PLGA particles, surface charge was negative and decreased with increasing particle size (Figure 1A).

While our initial studies used particles fabricated from the F.D.A-approved biocompatible and biodegradable polymer *poly(-lactic-co-glycolic acid)* PLGA, we also incorporated studies using hydrogel particles fabricated with derivatives of biocompatible poly (ethylene glycol) (PEG). Chemical modification of PEG is more feasible than with PLGA and thus it is often used to add increased functionality to nanocarriers, such as decoration of cell-targeting ligands, imaging agents, and pharmacologic cargo incorporation. Characteristics of fabricated PEG particles were similar to that of the PLGA particles, with low PDI and negative surface charge (Figure 1C) and monodisperisty as evidenced by SEM (Figure 1D). These PEG particles were fabricated with a dye (DyLight 650) to enable fluorescent imaging of particles in downstream assays and may also be referred to as hydrogels hereafter. To validate hydrogel particle uptake in a pertinent pulmonary immune cell population, we performed confocal analysis using the MH-S murine alveolar macrophage cell line. As shown in Figure 1E, all particle sizes are taken by four hours after treatment.

A

PLGA Particle Type	Size (nm)	PDI	Charge (mV)
80x320nm	226	0.03	-4.21
1µm cylinder	1465	NA	-18

B

1µm PLGA cylinder   80x320 nm PLGA rod

C

Hydrogel Particle Type	Composition (solids, wt%)					Size (nm)	PDI	Charge (mV)
	PEG$_{700}$-DA	HP$_4$A	Dye	TPO (PI)	AEM			
6 µm Donut	20	78	1	1	-	-	-	-6.87
1.5 µm Donut	20	78	1	1	-	-	-	-8.96
80x320 nm (acetylated)	10	68	1	1	20	236.1	0.06	-41.5

D

6 µm Donut   1.5 µm Donut   80x320 nm rod

E

6 µm Donut   1.5 µm Donut   80x320 nm rod

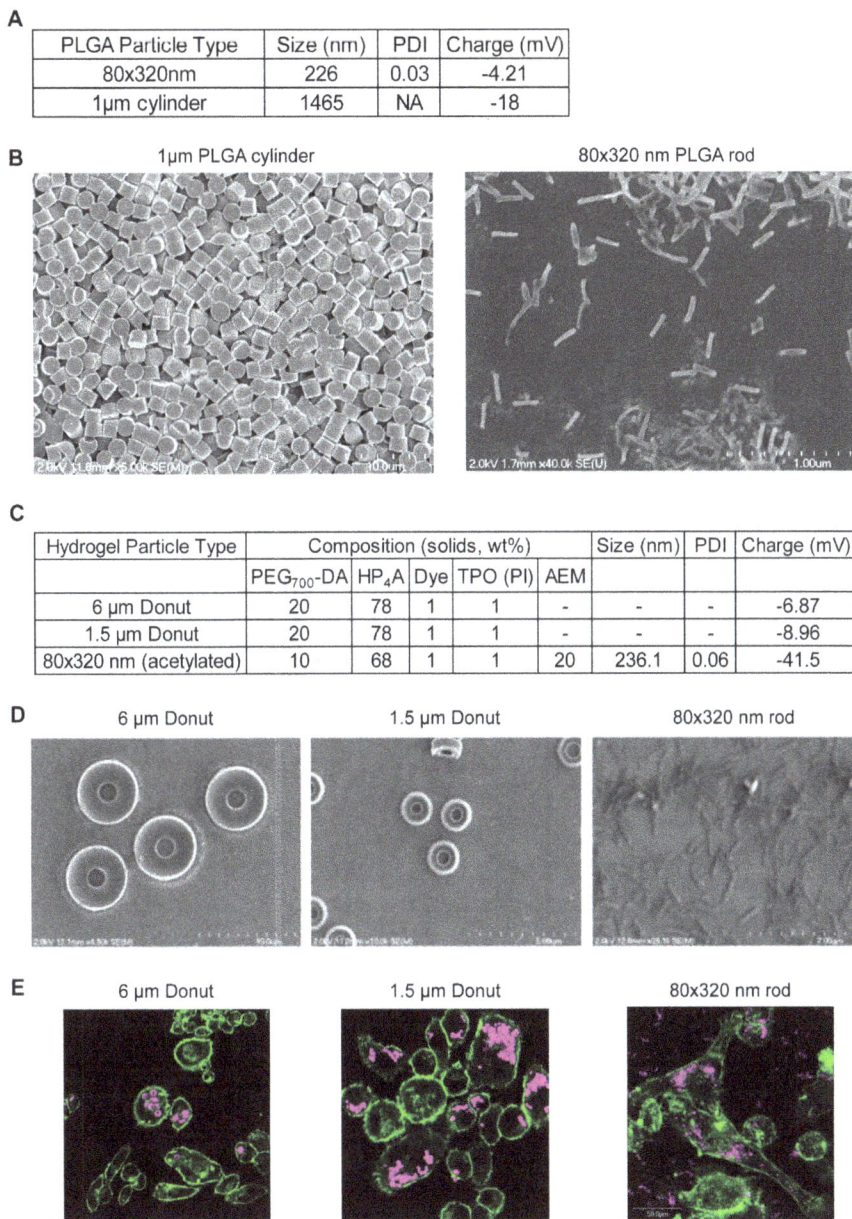

Figure 1. PRINT Particle Characterization. A) Dynamic light scattering (DLS) and zeta potential measurements of PLGA particles used in studies. Particle charge decreases with increasing size. **B)** Scanning electron microscope (SEM) images of PLGA particles. **C)** PEG particle composition and characterization. **D)** SEM of PEG particles. **E)** Confocal images of hydrogel particle uptake in MH-S alveolar macrophage cells after 4 hours of treatment. Scale bar is 50 µm.

PLGA particles do not induce inflammation by bone marrow-derived macrophages

Much work has been done *in vitro* to assess the potential use of PLGA nanoparticles in a variety of therapeutic modalities, from delivery of chemotherapeutics and siRNA to imaging agents for improved diagnostics. However, less work has been done to characterize the innate immune response to such particles and whether physical parameters of particles can augment the immune response. As a primary sentinel of host homeostasis, the innate immune system is tasked with identifying foreign matter in the body and initiating an appropriate response. Subsequent activation of the innate immune response is hallmarked by release of soluble protein messengers like cytokines that serve to recruit other immune cells to the area to participate in defense and repair of the host [7]. This inflammatory response is initiated by release of pro-inflammatory cytokines, including TNF-α, IL-6 and IL-1β, from innate immune cells, such as macrophages.

The field of environmental toxicology has long studied the role of nanoparticulates in inducing inflammation, in particular in the lung [34]. Attempting to synthesize work by other groups using a range of particle compositions and sizes suggest that there is no clear correlation between the physical parameters of a particle and the ensuing inflammatory response to it. Generally speaking, the composition of a particle has greater bearing on the inflammatory response than its size or shape. As an example, titanium dioxide and silica dioxide nanoparticles trigger inflammation, whereas zinc oxide nanoparticles do not, even though all particles were of

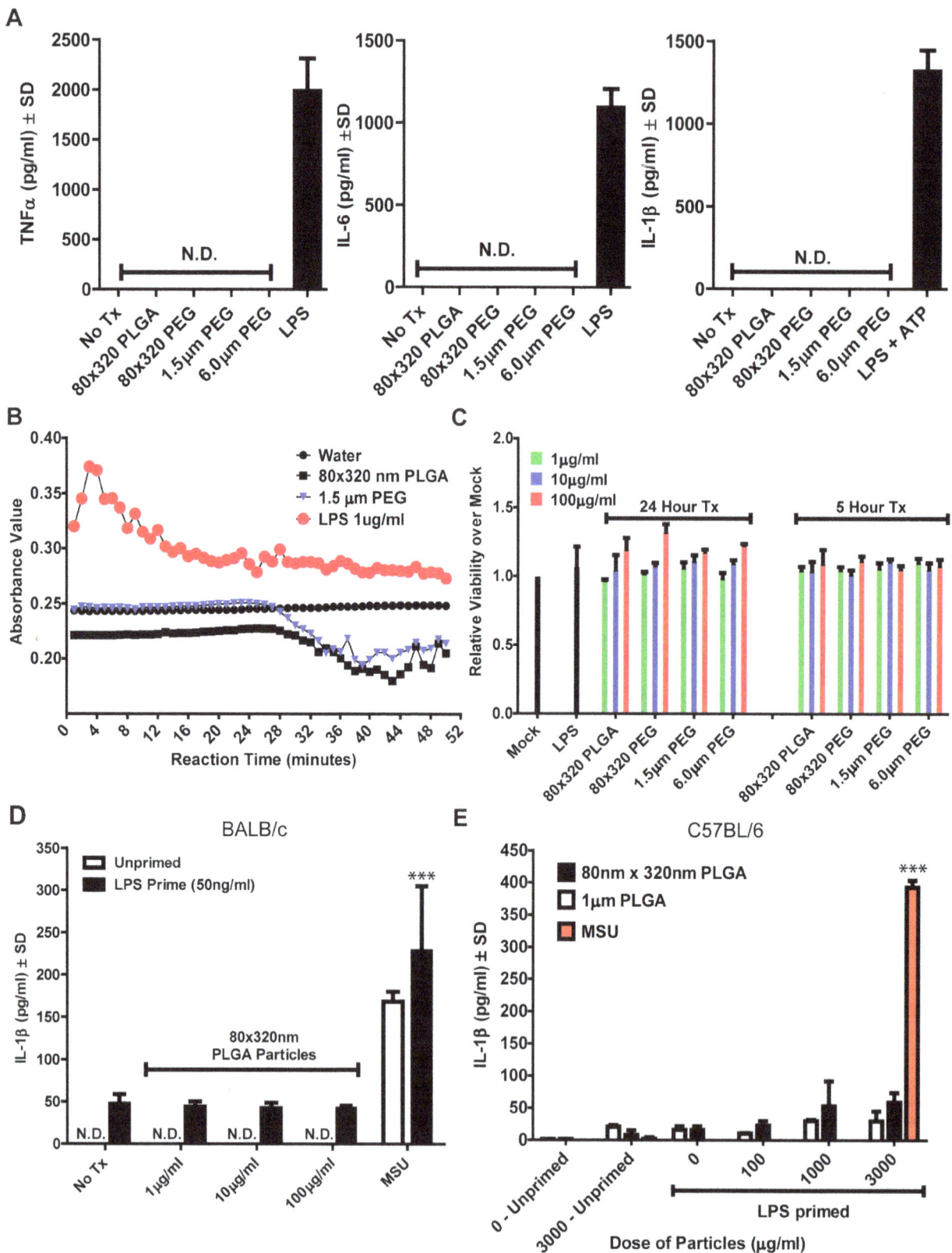

Figure 2. PRINT particles do not cause inflammation in bone marrow-derived macrophages from BALB/c or C57BL/6 mice. A) Overnight stimulation with a panel of PRINT PLGA and hydrogel particles (PEG) at 100 µg/ml does not cause TNF-α, IL-6, or IL-1β release from bone marrow-derived macrophages from C57BL/6 mice as measured by ELISA. **B)** Both PLGA and hydrogel PRINT particles (PEG) tested negative for endotoxin contamination using a Limulus amebocyte lysate assay. **C)** PRINT particles are not cytotoxic in bone-marrow derived macrophages as determined by lactate dehydrogenase (LDH) release. **D)** 80×320 nm PLGA particles do not synergize with LPS to induce inflammasome activation as measured by IL-1β ELISA in BALB/c bone-marrow derived macrophages. **E)** Neither 80×320 nm nor 1 µm PLGA particles synergize with LPS to induce

inflammasome activation as measured by IL-1β ELISA in C57BL/6 bone-marrow derived macrophages. MSU was dosed at 300 µg/ml. *** = p<0.001. Experiments were performed in triplicate. Data shown are representative of at least three independent experiments.

similar size (15–20 nm) [35]. Others have identified size-dependent inflammation and cell death that could be inhibited simply with surface modification of silica particles with common chemical groups such as aldehydes [36–38]. These findings highlight the sensitivity of the innate immune system as each particle may engender unique responses depending on its size, shape and composition.

We initially tested the inflammatory potential of PRINT particles in an *in vitro* cell culture system with bone marrow-derived macrophages from C57BL/6 mice. We used a panel of PRINT particles that differed in composition and size (Figure 1). After either a 5 hour or 24 hour incubation with a panel of PRINT particles comprised of either PLGA or PEG derivatives, we saw no detectable levels of any tested pro-inflammatory cytokine (TNF-α, IL-6, IL-1β) across a range of doses (1–100 µg/ml) (Figure 2A and data not shown). Lipopolysaccharide (LPS), a cell wall component of gram-negative bacteria, was used as a positive control for inflammation induction. The lack of cytokine induction was in line with data from endotoxin assays indicating our fabrication process was endotoxin-free (Figure 2B). In addition, across all doses of particles tested, we did not observe any particle-induce cytotoxicity as measured by LDH assay (Figure 2C).

Given the recent discovery of the inflammasome as a mediator of the innate immune response to particulate challenge, we also sought to address whether PRINT-fabricated PLGA particles could cause inflammasome activation [35,39,40]. The inflammasome is a multi-protein complex that is formed in response to variety of environmental stimuli, including asbestos, silica and monosodium urate crystals (MSU), that results in the activation of caspase-1 and subsequent maturation and secretion of the pro-inflammatory cytokines IL-1β and IL-18 [41]. As our initial results did not indicate any particle induction of IL-1β (Figure 2A), we next assessed whether priming macrophages with LPS would cause particles to induce inflammasome activation. LPS priming is thought to provide signal 1 to inflammasome formation by upregulating the protein levels of pro-IL-1β and NLRP3, a main component of the inflammasome complex [42,43]. As assessed by IL-1β release, we did not see PLGA particle-induced activation of the inflammasome in the presence or absence of LPS-priming when tested in either BALB/c (Figure 2D) or C57BL/6 macrophages (Figure 2E). Importantly, we tested particles across a range of doses (100 ng–3000 µg/ml) and sizes (80×320 nm and 1 µm cylinders). These results suggest that PLGA particles across the nano and micron range do not synergize with TLR ligands (i.e., LPS) to induce inflammasome activation *in vitro* and lend further credence to the use of PLGA particles for *in vivo* applications.

PLGA particles do not induce lung inflammation

Bolstered by our *in vitro* findings, we next were interested in whether PLGA particles could be delivered to the lungs of mice without causing overt signs of immune activation as hallmarked by inflammation. The lung was chosen as a highly sensitive mucosal organ with clearly defined markers of inflammation that is the sight of numerous therapeutically relevant diseases, from allergies and asthma to chronic obstructive pulmonary disorder (COPD) and respiratory infections by microbial pathogens such as tuberculosis and influenza [23,44]. As such, therapeutic modulation of lung biology is a highly desired clinical goal with relevance

to the vast majority of the human population. We used intra-tracheal (i.t.) delivery to determine whether 80×320 nm PLGA particles (50 µg) caused inflammation in the lungs, with PBS (50 µl) and LPS (20 µg) used as negative and positive controls, respectively. Forty eighthours after i.t. installation, mice (n = 5 per group) were harvested and lung inflammation was assessed via field standards used in respiratory infection models [26,27]. Broncheoalveolar lavage fluid (BALF) cellularity indicated no recruitment of immune cells to the lungs after particle treatment, as cell numbers were no different than the PBS control (Figure 3A). LPS-treated mice revealed a robust accumulation of leukocytes as is expected during inflammatory responses. Assessing the composition of leukocyte populations in the BALF revealed no significant recruitment of immune cells to the lungs of particle-treated mice. Conversely, LPS-treated mice had high levels of both monocytes and neutrophils, key mediators of the innate immune system's inflammatory response (Figure 3B).

While BALF cellularity is widely used as a marker of lung inflammation, lung histopathology enables a deeper understanding of inflammatory effects on the lung parenchyma. We examined representative sections of histopathology slides of the main bronchi of the left lobe to further delineate leukocyte infiltration around lung vasculature, parenchyma and the large and small airways (Figure 3C). Whereas LPS treatment caused a clear accumulation of leukocytes throughout the lung, treatment with 80×320 nm PLGA particles showed no difference as compared to PBS controls (Figure 3D). To further verify the non-inflammatory nature of these particles, pro-inflammatory cytokine levels were assessed in the BALF and serum of treated mice. No significant release of IL-1β (Figure 3E) or IL-6 (Figure 3F) was seen in the BALF. Serum measurements for these same cytokines and TNF-α were undetectable (data not shown). In total, these results are in agreement with our *in vitro* findings and suggest that 80×320 nm PLGA particles can be delivered to the lungs without causing innate immune activation and inflammation.

PEG particles stably remain within the lungs for 7 days without causing lung inflammation

To broaden the implication of our *in vivo* findings, we fabricated a series of particles using PEG polymers and their derivatives (hydrogels) that incorporated fluorescent dyes which enabled us to track them *in vivo* over time after lung instillation. The hydrogel particles ranged in size from 80×320 nm to 1.5 µm and 6 µm as characterized in Figure 1. *In vitro* experiments indicated they did not elicit inflammatory cytokines or cell death from bone marrow-derived macrophages (Figure 2). Using the same experimental approach as outlined above, we instilled 50 µg of particles i.t. into C57BL/6 mice and assessed lung inflammation at two time points, 48 hours and 7 days post-particle instillation. As shown in Figure 4A, total BALF cellularity does not increase in the presence of hydrogel particles as compared to PBS at 48 hours, which is in marked contrast to LPS-induced cell recruitment to the lungs. Breaking down the BALF cell types revealed a similar number of monocytes in the lungs PBS and particle-treated mice, whereas LPS-treatment induced a marked influx of both monocytes and neutrophils. At 7 days post-particle treatment, there was no significant increase in the total BALF cellularity or composition in mice treated with any hydrogel particles (Figure 4C and 4D). Histopathology analyses indicated neither lung architecture disruption nor leukocyte infiltration into the lungs or airways of

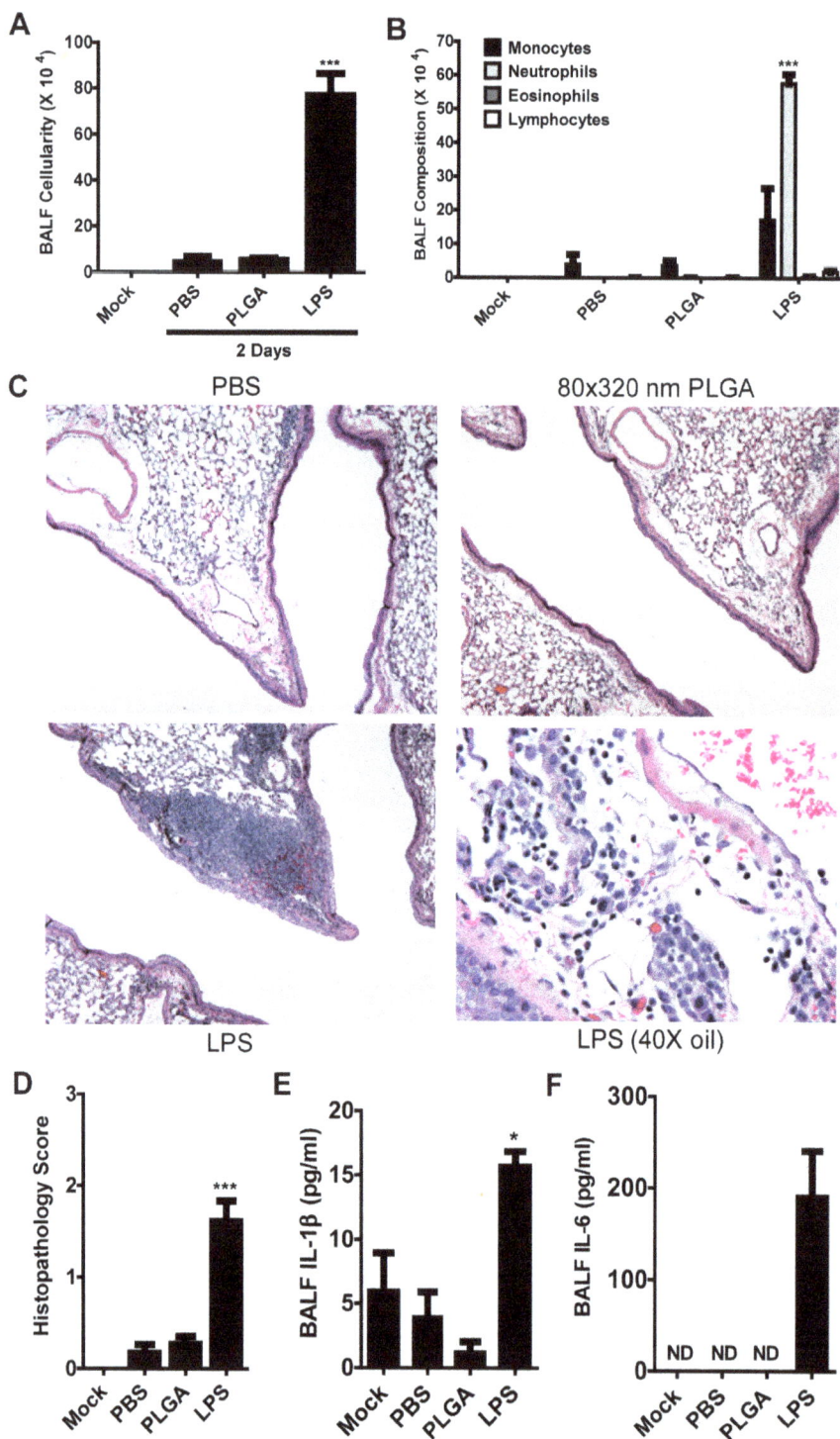

Figure 3. 80×320 nm PLGA particles do not cause lung inflammation in mice. Mice were challenged with either 50 μg of 80×320 nm PLGA particles or 20 μg LPS i.t. and airway inflammation was assessed 48 hours post-challenge. **A)** Total cellularity of bronchoalveolar lavage fluid (BALF) in treated C57BL/6 mice is no different after 48 hours than PBS-treated mice and is significantly less than the inflammatory cell recruitment seen in LPS-treated mice. **B)** PLGA particle treatment does not induce any appreciable immune cell recruitment to the lungs of mice, as opposed to the heightened levels of monocytes and neutrophils seen in the lungs of LPS-treated mice. **C)** Histopathology revealed no significant differences in lung architecture between PBS- and 80×320 nm PLGA particle-treated mice. This is in stark contrast to the airway occlusion and significant innate immune cell recruitment seen in LPS-treated mice. **D)** Histopathology scoring confirmed that no significant differences were seen between the lungs of PBS and PLGA particle treated mice. **E–F)** The increased lung levels of pro-inflammatory IL-1β and IL-6 seen in LPS-treated mice is not found in PLGA-treated mice. PBS, n = 3; 80×320 nm PLGA particle-treated, n = 5; LPS-treated, n = 3. ND = Not Detected. * = p<0.05, *** = p<0.001. Experiments were performed using 3–5 mice per group. Data shown are representative of at least two independent experiments.

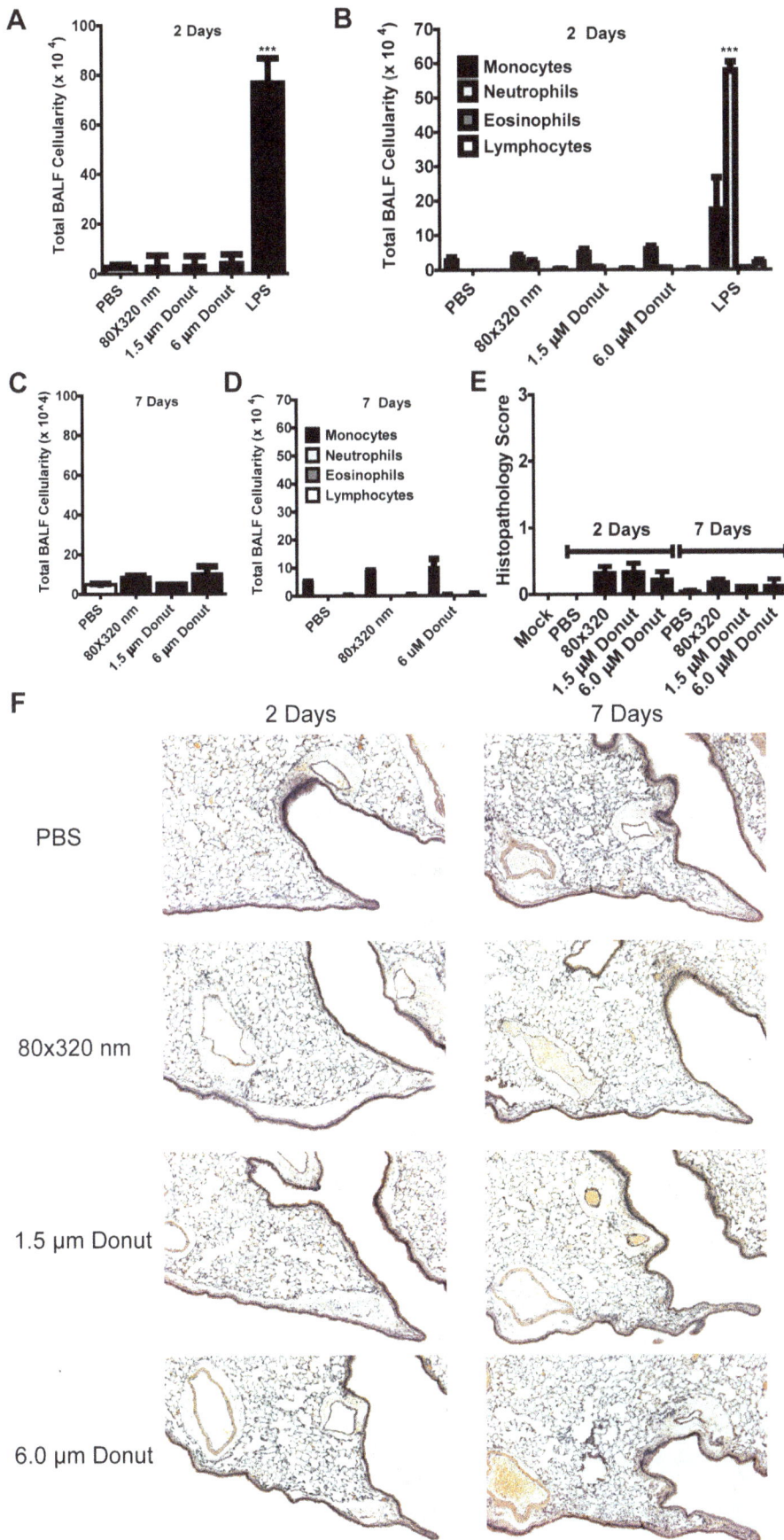

Figure 4. Hydrogel particles do not cause lung inflammation in mice. Mice were challenged with 50 µg of hydrogel particles (80×320 nm, 1.5 µm, or 6.0 µm donuts) i.t. and airway inflammation was assessed 48 hours and 7 days post-challenge. **A)** BALF analysis indicated no increased cellularity 48 hours after hydrogel particle treatment, whereas a significant cellular influx was seen in LPS-treated controls. **B)** At 48 hours, BALF cellular composition does not show any significant trend for immune cell recruitment in hydrogel particle-treated mice. **C–D)** BALF cellularity and composition was not significantly augmented seven days after hydrogel particle treatment. **E)** Histopathology analysis revealed no significant differences in lung architecture between PBS- and hydrogel particle-treated mice at either 2 or 7 days post-treatment. **F)** Histopathology scoring confirmed that no significant differences were seen between the lungs of PBS and hydrogel particle treated mice at any time points. *** = p<0.001. Experiments were performed using 2–5 mice per group. Data shown are representative of at least two independent experiments.

particle-treated mice as compared to PBS controls at either the 48 hour or 7 day time point (Figure 4E and 4F).

Remarkably, and despite the absence of overt signs of inflammation, 6 µm particles with their hallmark donut appearance could be viewed within the lung spaces of multiple mice by H&E staining 2 and 7 days post-challenge (Figure 5A–B). The lack of immune cell recruitment or disruption of tissue architecture around these particles may suggest an immunologically inert deposition of particles within the alveolar spaces. Such a depot may provide sustained localized delivery of therapeutically attractive molecules. Because the 80×320 nm and 1.5 µm particles were too small to see in lung histology samples, we also performed immunofluorescence imaging on BALF samples to determine whether lung-localized cells took up particles. As shown in Figure 5C, BALF cells contained hydrogel particles of all sizes at 48 hrs (magnified view in Figure 5D). We also noted that the percentage of cells with particles decreases as particles size increases (Figure 5E). Whether this is due to the quantitatively higher number of 80×320 nm particles at the same dose weight of larger particles, the relatively easier ability for a cell to take up smaller particles as compared to larger ones, or an as yet unidentified size-dependent biological effect remains unanswered. We were also interested to find that BALF cells 7 days after particle instillation show particle uptake, albeit to a lesser extent than the 48 hour time point (Figure 5F and magnified view in Figure 5G). Finally, we quantified the levels of pro-inflammatory cytokines released into the BALF and serum of PEG particle-treated mice. At both the 48 hour and 7 day time points we were unable to detect IL-1β, IL-6, or TNF-α for any particle treatment (data not shown). In total, these data highlight the ability of PRINT particles to remain localized to the lung for long periods of time in an immunologically inert manner.

Discussion

Given the diverse therapeutic potential of nano- and microscale particles, this study sought to define whether particles composed of either PLGA or PEG-derivatives induced inflammatory responses in an *in vitro* and *in vivo* setting. By making use of the highly controlled PRINT fabrication method, we were also able to determine whether particle size affected any ensuing innate immune responses. Our findings reveal that PRINT particles do not cause any obvious activation of the innate immune response in murine macrophages or the murine lung and maintain long term (7 day) immunologic stability in the lungs of mice.

The wide array of polymers and particle fabrication techniques used in nanomedicine studies makes it difficult to reach definitive conclusions regarding particle effects on innate immune functions. We initially used particles fabricated from PLGA as this is a commonly used polymer with attractive clinical potential given its F.D.A approval. There is some discrepancy in the literature as to whether PLGA particles are inflammatory *in situ*. Some groups suggest PLGA particles are inflammatory *in vitro* and *in vivo*, whereas others have not found this to be the case [39,45–48]. Our study reveals that PLGA particles of nano and micron range

fabricated by PRINT technology do not synergize with a TLR ligand to cause inflammasome activation nor inflammation in general, and that *in vivo* delivery does not trigger an inflammatory reaction, contrary to a previous report [39]. The discrepancy between these findings may be due to differences in particle fabrication or experimental settings [49]. However, given the long clinical history of PLGA and the broad literature reporting PLGA particle uses for biomedical applications, it seems unlikely that particles derived from PLGA would trigger potent inflammatory responses, yet this confusion is precisely why more research must be carried out to ensure such unwanted side-effects are avoided as fabrication methods or material sourcing may impact immune responses significantly [50–52].

In addition, our studies using PEG particles enabled us to broaden our understanding of innate immune activation by particles comprised of a polymer composition that enables wide-ranging chemical modifications for enhanced functionality, such as cell targeting, pH-specific cargo release and siRNA incorporation as previously reported by our group and others [19,20,53,54]. Interestingly, although these PEG particles are not considered biodegradable, they did not induce lung inflammation as seen with other non-degradable particles such as those comprised of polystyrene [46]. This suggests our PEG polymer composition may also be an attractive alternative from an environmental toxicology perspective in applications currently employing polystyrene particles.

The issue of innate immune activation by particles is of central relevance to the translational application of nanotechnology. While particulate vaccines against some pathogens and cancers will likely be designed to trigger localized inflammation as part of the general innate immune activation required for robust adaptive immune responses, most other biomedical applications for nano- and microparticles will benefit by avoiding such responses. Additionally, a strong immune response might lead to the undesirable outcome of rapid particle clearance as well as hypersensitivity responses. Drug delivery, diagnostic imaging and physiological bio-mimicry are examples of nanoengineering applications that may be impeded by innate immune activation. Importantly, many advances in immune modulation made available through rationally designed nano- and microscale particles such as tolerance induction in the setting of autoimmunity or organ transplantation, direct targeting of immune cell subsets and immune-skewing of pathological microenvironments such as tumors or sites of chronic inflammation, require that particles be designed initially from an inert immunological state [55–61]. To wit, if particles alone trigger inflammatory responses that skew towards any type of adaptive response (e.g., Th1, Th2) then many of these therapeutic goals will not be achieved. For these reasons, we feel it is of utmost importance that baseline innate immune responses to particles be assessed as part of field standards [9].

Our *in vivo* studies reveal that particle size augments uptake into innate immune cells of the lungs, with larger particles taken up less than smaller particles. This finding suggests a duality in design considerations depending on therapeutic application. For exam-

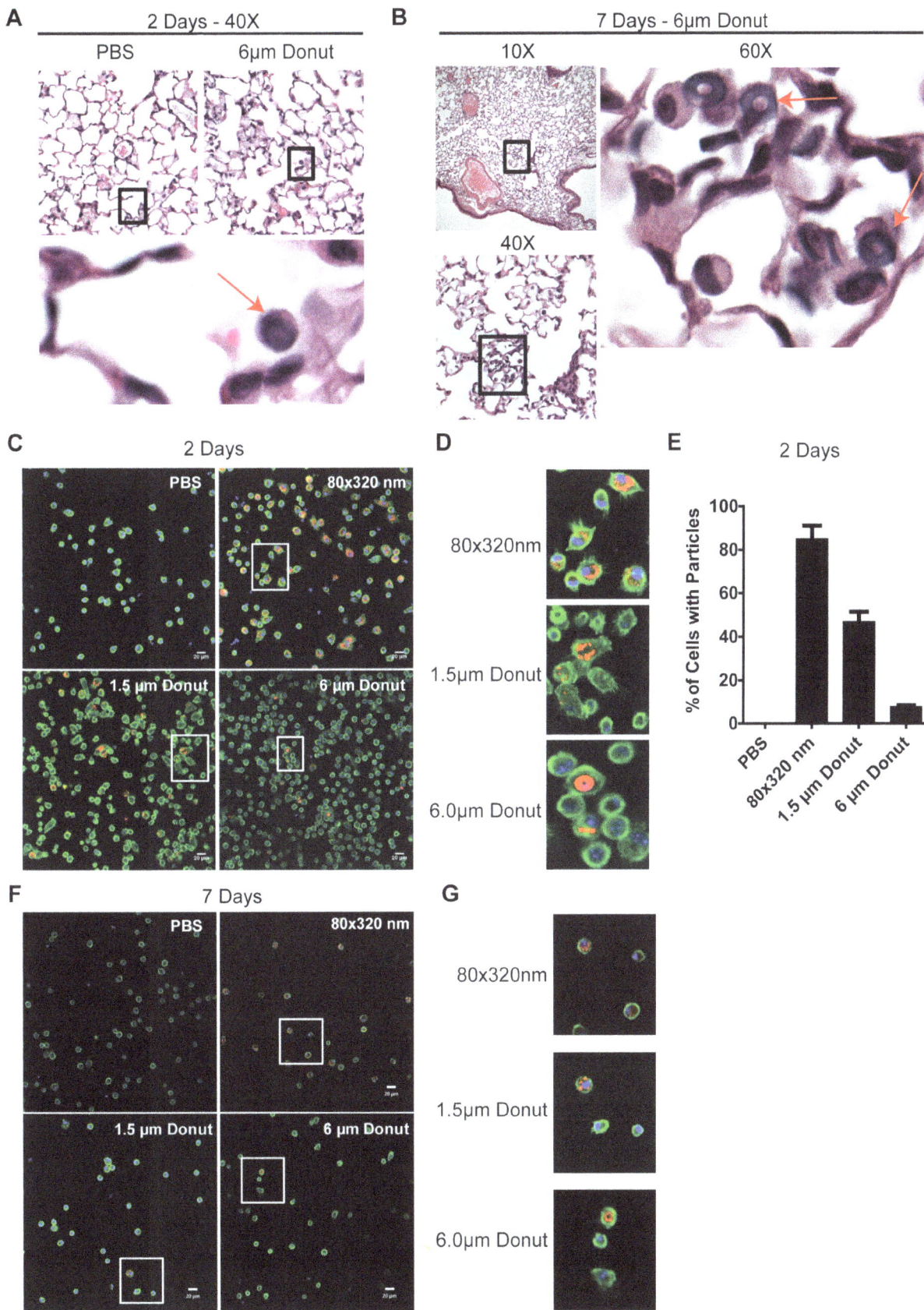

Figure 5. Hydrogel particles remain in the lungs for multiple days without overt signs of inflammation. A) 6 μm hydrogel particles (denoted by red arrows) are visible in the alveolar spaces 2 days after intratracheal installation. Lower insets are a magnified view of black bounding box. PBS treated mice are shown as control. **B)** Multiple 6 μm hydrogel particles (denoted by black bounding box and red arrows) are visible in the

alveolar spaces 7 days after intratracheal installation. **C)** Two days after treatment with hydrogel particles, BALF cells were stained and visualized for particle uptake via epifluorescence microscopy. Particles (Dylight 650, red); nuclei (DAPI, blue); F-actin (Phalloidin 488, green). **D)** Magnified views of BALF cells taking up hydrogel particles as denoted by white bounding boxes in Figure C. **E)** Quantification of particle uptake indicates smaller particles are more readily taken up in BALF cells than larger particles. **F)** All types of hydrogel particles can still be seen in BALF cells seven days after treatment, though there is a marked decrease in the number of particles present as compared to the 2 day time point. **G)** Magnified views of BALF cells taking up hydrogel particles 7 days after treatment as denoted by white bounding boxes in Figure C. Scale bar is 20 μm. Data shown are representative of at least two independent experiments.

ple, drug delivery to the lungs to ameliorate asthma would likely be best served by larger particles that can release their cargo to extracellular spaces. Conversely, if trying to deliver a respiratory vaccine, smaller particles that are more readily taken up by antigen presenting cells and traffic to lymph nodes would be more appropriate. Our finding that particles of all tested sizes remain in the lungs up to 7 days post instillation also suggest the ability to provide sustained localized delivery of therapeutically attractive molecules via particulate formulations. This is far different than the rapid clearance seen for smaller particles (<50 nm diameter) and reflects the importance of particle design parameters when considering therapeutic interventions [62].

The lung serves as an attractive route for therapeutic delivery due to its ease of access and its large absorptive surface area. There are several important particle characteristics that need to be considered for effective pulmonary delivery such as size, shape, surface charge, toxicity, and potential inflammatory effects. Inhaled particles with mass median aerodynamic diameters (MMAD) larger than 5 μm tend to be deposited in the upper conducting airway while particles with MMAD between 1–5 μm deposit in the lower respiratory airways [63]. Using PRINT, we have the ability to design particles with aerodynamically relevant deposition characteristics while having distinct non-spherical geometries enabling different deposition profiles in the lung [64,65]. The investigation of the safety profile and inflammatory response of these inhaled polymeric particles is important to support their use as drug delivery vehicles as highlighted in this study.

Having identified particles across the nano- and microscale that do not trigger inflammatory responses in mice while remaining in

the lungs, we plan to next use these particles as delivery devices for a range of biologically relevant molecules, including siRNAs, anti-inflammatory agents, and immune-skewing compounds. It should also be noted that mucoadhesive components may be incorporated within the PRINT particle system to enhance adsorption in the mucosal region which may enable differential deposition and enhanced temporal localization to the lungs. These studies will test the hypothesis that targeted modulation of lung immunology via nanoengineering may enable a new class of therapeutics for lung disorders that avoid systemic side-effects while also reducing administration doses. While we have tested a 50 μg dose of inert particles in this manuscript, PRINT enables a high weight percent loading of bioactive molecules and thus local and sustained pulmonary delivery may show therapeutic efficacy at low particle doses [19–22,66–68].

Importantly, we will move studies into human cells to provide much needed data regarding the immunological response to nanomaterials in our own species. Using the design control inherent to PRINT technology, we will also be able to systematically address the role of particle size and shape during delivery of bioactive molecules. Such results may be crucial to advancing next-generation respiratory vaccines and treatments for asthma, allergies and chronic disorders of lung function.

Author Contributions

Supervised the overall research project: JPYT JMDS. Conceived and designed the experiments: RAR. Performed the experiments: RAR ICA TS. Analyzed the data: RAR ICA TS. Contributed reagents/materials/analysis tools: TS WH. Wrote the paper: RAR.

References

1. Leuschner F, Dutta P, Gorbatov R, Novobrantseva TI, Donahoe JS, et al. (2011) Therapeutic siRNA silencing in inflammatory monocytes in mice. Nat Biotechnol 29: 1005–1010.
2. Dhar S, Kolishetti N, Lippard SJ, Farokhzad OC (2011) Targeted delivery of a cisplatin prodrug for safer and more effective prostate cancer therapy in vivo. Proc Natl Acad Sci U S A 108: 1850–1855.
3. Kircher MF, de la Zerda A, Jokerst JV, Zavaleta CL, Kempen PJ, et al. (2012) A brain tumor molecular imaging strategy using a new triple-modality MRI-photoacoustic-Raman nanoparticle. Nat Med 18: 829–834.
4. Kasturi SP, Skountzou I, Albrecht RA, Koutsonanos D, Hua T, et al. (2011) Programming the magnitude and persistence of antibody responses with innate immunity. Nature 470: 543–547.
5. Moon JJ, Suh H, Li AV, Ockenhouse CF, Yadava A, et al. (2012) Enhancing humoral responses to a malaria antigen with nanoparticle vaccines that expand Tfh cells and promote germinal center induction. Proc Natl Acad Sci U S A 109: 1080–1085.
6. Bershteyn A, Hanson MC, Crespo MP, Moon JJ, Li AV, et al. (2012) Robust IgG responses to nanograms of antigen using a biomimetic lipid-coated particle vaccine. J Control Release 157: 354–365.
7. Pulendran B, Ahmed R (2011) Immunological mechanisms of vaccination. Nat Immunol 12: 509–517.
8. Hubbell JA, Thomas SN, Swartz MA (2009) Materials engineering for immunomodulation. Nature 462: 449–460.
9. Dobrovolskaia MA, McNeil SE (2007) Immunological properties of engineered nanomaterials. Nat Nanotechnol 2: 469–478.
10. Zolnik BS, Gonzalez-Fernandez A, Sadrieh N, Dobrovolskaia MA (2010) Nanoparticles and the immune system. Endocrinology 151: 458–465.
11. Schijns VE, Lavelle EC (2011) Trends in vaccine adjuvants. Expert Rev Vaccines 10: 539–550.
12. Jain RA (2000) The manufacturing techniques of various drug loaded biodegradable poly(lactide-co-glycolide) (PLGA) devices. Biomaterials 21: 2475–2490.
13. Chan JM, Valencia PM, Zhang L, Langer R, Farokhzad OC (2010) Polymeric nanoparticles for drug delivery. Methods Mol Biol 624: 163–175.
14. Dobrovolskaia MA, Germolec DR, Weaver JL (2009) Evaluation of nanoparticle immunotoxicity. Nat Nanotechnol 4: 411–414.
15. Perry JL, Herlihy KP, Napier ME, Desimone JM (2011) PRINT: a novel platform toward shape and size specific nanoparticle theranostics. Acc Chem Res 44: 990–998.
16. Jeong W, Napier ME, DeSimone JM (2010) Challenging nature's monopoly on the creation of well-defined nanoparticles. Nanomedicine (Lond) 5: 633–639.
17. Canelas DA, Herlihy KP, DeSimone JM (2009) Top-down particle fabrication: control of size and shape for diagnostic imaging and drug delivery. Wiley Interdiscip Rev Nanomed Nanobiotechnol 1: 391–404.
18. Rolland JP, Maynor BW, Euliss LE, Exner AE, Denison GM, et al. (2005) Direct fabrication and harvesting of monodisperse, shape-specific nanobiomaterials. J Am Chem Soc 127: 10096–10100.
19. Parrott MC, Finniss M, Luft JC, Pandya A, Gullapalli A, et al. (2012) Incorporation and controlled release of silyl ether prodrugs from PRINT nanoparticles. J Am Chem Soc 134: 7978–7982.
20. Dunn SS, Tian S, Blake S, Wang J, Galloway AL, et al. (2012) Reductively responsive siRNA-conjugated hydrogel nanoparticles for gene silencing. J Am Chem Soc 134: 7423–7430.
21. Hasan W, Chu K, Gullapalli A, Dunn SS, Enlow EM, et al. (2012) Delivery of multiple siRNAs using lipid-coated PLGA nanoparticles for treatment of prostate cancer. Nano Lett 12: 287–292.
22. Enlow EM, Luft JC, Napier ME, DeSimone JM (2011) Potent engineered PLGA nanoparticles by virtue of exceptionally high chemotherapeutic loadings. Nano Lett 11: 808–813.

23. Roy I, Vij N (2010) Nanodelivery in airway diseases: Challenges and therapeutic applications. Nanomedicine: Nanotechnology, Biology and Medicine 6: 237–244.

24. Guzmán J, Iglesias MT, Riande E, Compañ V, Andrio A (1997) Synthesis and polymerization of acrylic monomers with hydrophilic long side groups. Oxygen transport through water swollen membranes prepared from these polymers. Polymer 38: 5227–5232.

25. Allen IC, Pace AJ, Jania LA, Ledford JG, Latour AM, et al. (2006) Expression and function of NPSR1/GPRA in the lung before and after induction of asthma-like disease. Am J Physiol Lung Cell Mol Physiol 291: L1005–1017.

26. Allen IC, Scull MA, Moore CB, Holl EK, McElvania-TeKippe E, et al. (2009) The NLRP3 inflammasome mediates in vivo innate immunity to influenza A virus through recognition of viral RNA. Immunity 30: 556–565.

27. Willingham SB, Allen IC, Bergstralh DT, Brickey WJ, Huang MT, et al. (2009) NLRP3 (NALP3, Cryopyrin) facilitates in vivo caspase-1 activation, necrosis, and HMGB1 release via inflammasome-dependent and -independent pathways. J Immunol 183: 2008–2015.

28. Bachmann MF, Jennings GT (2010) Vaccine delivery: a matter of size, geometry, kinetics and molecular patterns. Nat Rev Immunol 10: 787–796.

29. Rice-Ficht AC, Arenas-Gamboa AM, Kahl-McDonagh MM, Ficht TA (2010) Polymeric particles in vaccine delivery. Curr Opin Microbiol 13: 106–112.

30. Gratton SE, Ropp PA, Pohlhaus PD, Luft JC, Madden VJ, et al. (2008) The effect of particle design on cellular internalization pathways. Proc Natl Acad Sci U S A 105: 11613–11618.

31. Merkel TJ, Jones SW, Herlihy KP, Kersey FR, Shields AR, et al. (2011) Using mechanobiological mimicry of red blood cells to extend circulation times of hydrogel microparticles. Proc Natl Acad Sci U S A 108: 586–591.

32. Capriotti AL, Caracciolo G, Cavaliere C, Foglia P, Pozzi D, et al. (2012) Do plasma proteins distinguish between liposomes of varying charge density? J Proteomics 75: 1924–1932.

33. Karmali PP, Simberg D (2011) Interactions of nanoparticles with plasma proteins: implication on clearance and toxicity of drug delivery systems. Expert Opin Drug Deliv 8: 343–357.

34. Sayes CM, Reed KL, Warheit DB (2011) Nanoparticle toxicology: measurements of pulmonary hazard effects following exposures to nanoparticles. Methods Mol Biol 726: 313–324.

35. Yazdi AS, Guarda G, Riteau N, Drexler SK, Tardivel A, et al. (2010) Nanoparticles activate the NLR pyrin domain containing 3 (Nlrp3) inflammasome and cause pulmonary inflammation through release of IL-1alpha and IL-1beta. Proc Natl Acad Sci U S A 107: 19449–19454.

36. Morishige T, Yoshioka Y, Inakura H, Tanabe A, Narimatsu S, et al. (2012) Suppression of nanosilica particle-induced inflammation by surface modification of the particles. Arch Toxicol.

37. Morishige T, Yoshioka Y, Inakura H, Tanabe A, Yao X, et al. (2010) Cytotoxicity of amorphous silica particles against macrophage-like THP-1 cells depends on particle-size and surface properties. Pharmazie 65: 596–599.

38. Morishige T, Yoshioka Y, Inakura H, Tanabe A, Yao X, et al. (2010) The effect of surface modification of amorphous silica particles on NLRP3 inflammasome mediated IL-1beta production, ROS production and endosomal rupture. Biomaterials 31: 6833–6842.

39. Sharp FA, Ruane D, Claass B, Creagh E, Harris J, et al. (2009) Uptake of particulate vaccine adjuvants by dendritic cells activates the NALP3 inflammasome. Proc Natl Acad Sci U S A 106: 870–875.

40. Dostert C, Petrilli V, Van Bruggen R, Steele C, Mossman BT, et al. (2008) Innate immune activation through Nalp3 inflammasome sensing of asbestos and silica. Science 320: 674–677.

41. Martinon F, Mayor A, Tschopp J (2009) The inflammasomes: guardians of the body. Annu Rev Immunol 27: 229–265.

42. Bauernfeind FG, Horvath G, Stutz A, Alnemri ES, MacDonald K, et al. (2009) Cutting edge: NF-kappaB activating pattern recognition and cytokine receptors license NLRP3 inflammasome activation by regulating NLRP3 expression. J Immunol 183: 787–791.

43. Qiao Y, Wang P, Qi J, Zhang L, Gao C (2012) TLR-induced NF-kappaB activation regulates NLRP3 expression in murine macrophages. FEBS Lett 586: 1022–1026.

44. Barnes PJ (2008) Immunology of asthma and chronic obstructive pulmonary disease. Nat Rev Immunol 8: 183–192.

45. Demento SL, Eisenbarth SC, Foellmer HG, Platt C, Caplan MJ, et al. (2009) Inflammasome-activating nanoparticles as modular systems for optimizing vaccine efficacy. Vaccine 27: 3013–3021.

46. Dailey LA, Jekel N, Fink L, Gessler T, Schmehl T, et al. (2006) Investigation of the proinflammatory potential of biodegradable nanoparticle drug delivery systems in the lung. Toxicol Appl Pharmacol 215: 100–108.

47. Demento SL, Bonafe N, Cui W, Kaech SM, Caplan MJ, et al. (2010) TLR9-targeted biodegradable nanoparticles as immunization vectors protect against West Nile encephalitis. J Immunol 185: 2989–2997.

48. Nicolete R, dos Santos DF, Faccioli LH (2011) The uptake of PLGA micro or nanoparticles by macrophages provokes distinct in vitro inflammatory response. Int Immunopharmacol 11: 1557–1563.

49. Vaine CA, Patel MK, Zhu J, Lee E, Finberg RW, et al. (2013) Tuning Innate Immune Activation by Surface Texturing of Polymer Microparticles: The Role of Shape in Inflammasome Activation. J Immunol.

50. Danhier F, Ansorena E, Silva JM, Coco R, Le Breton A, et al. (2012) PLGA-based nanoparticles: An overview of biomedical applications. J Control Release 161: 505–522.

51. Holgado MA, Alvarez-Fuentes J, Fernandez-Arevalo M, Arias JL (2011) Possibilities of poly(D,L-lactide-co-glycolide) in the formulation of nanomedicines against cancer. Curr Drug Targets 12: 1096–1111.

52. Lu JM, Wang X, Marin-Muller C, Wang H, Lin PH, et al. (2009) Current advances in research and clinical applications of PLGA-based nanotechnology. Expert Rev Mol Diagn 9: 325–341.

53. Parrott MC, Luft JC, Byrne JD, Fain JH, Napier ME, et al. (2010) Tunable bifunctional silyl ether cross-linkers for the design of acid-sensitive biomaterials. J Am Chem Soc 132: 17928–17932.

54. Wang J, Tian S, Petros RA, Napier ME, Desimone JM (2010) The complex role of multivalency in nanoparticles targeting the transferrin receptor for cancer therapies. J Am Chem Soc 132: 11306–11313.

55. Fahmy TM, Fong PM, Park J, Constable T, Saltzman WM (2007) Nanosystems for simultaneous imaging and drug delivery to T cells. Aaps J 9: E171–180.

56. Stephan MT, Stephan SB, Bak P, Chen J, Irvine DJ (2012) Synapse-directed delivery of immunomodulators using T-cell-conjugated nanoparticles. Biomaterials 33: 5776–5787.

57. Tsai S, Shameli A, Yamanouchi J, Clemente-Casares X, Wang J, et al. (2010) Reversal of autoimmunity by boosting memory-like autoregulatory T cells. Immunity 32: 568–580.

58. Moon JJ, Suh H, Bershteyn A, Stephan MT, Liu H, et al. (2011) Interbilayer-crosslinked multilamellar vesicles as synthetic vaccines for potent humoral and cellular immune responses. Nat Mater 10: 243–251.

59. Park J, Gao W, Whiston R, Strom TB, Metcalfe S, et al. (2011) Modulation of CD4+ T lymphocyte lineage outcomes with targeted, nanoparticle-mediated cytokine delivery. Mol Pharm 8: 143–152.

60. Park J, Wrzesinski SH, Stern E, Look M, Criscione J, et al. (2012) Combination delivery of TGF-beta inhibitor and IL-2 by nanoscale liposomal polymeric gels enhances tumour immunotherapy. Nat Mater.

61. Getts DR, Martin AJ, McCarthy DP, Terry RL, Hunter ZN, et al. (2012) Microparticles bearing encephalitogenic peptides induce T-cell tolerance and ameliorate experimental autoimmune encephalomyelitis. Nat Biotechnol 30: 1217–1224.

62. Choi HS, Ashitate Y, Lee JH, Kim SH, Matsui A, et al. (2010) Rapid translocation of nanoparticles from the lung airspaces to the body. Nat Biotechnol 28: 1300–1303.

63. Yang W, Peters JI, Williams R 3rd (2008) Inhaled nanoparticles–a current review. Int J Pharm 356: 239–247.

64. Gebril A, Alsaadi M, Acevedo R, Mullen AB, Ferro VA (2012) Optimizing efficacy of mucosal vaccines. Expert Rev Vaccines 11: 1139–1155.

65. Garcia A, Mack P, Williams S, Fromen C, Shen T, et al. (2012) Microfabricated engineered particle systems for respiratory drug delivery and other pharmaceutical applications. J Drug Deliv 2012: 941243.

66. Petros RA, Ropp PA, DeSimone JM (2008) Reductively labile PRINT particles for the delivery of doxorubicin to HeLa cells. J Am Chem Soc 130: 5008–5009.

67. Chen K, Merkel TJ, Pandya A, Napier ME, Luft JC, et al. (2012) Low modulus biomimetic microgel particles with high loading of hemoglobin. Biomacromolecules 13: 2748–2759.

68. Chu KS, Hasan W, Rawal S, Walsh MD, Enlow EM, et al. (2012) Plasma, tumor and tissue pharmacokinetics of Docetaxel delivered via nanoparticles of different sizes and shapes in mice bearing SKOV-3 human ovarian carcinoma xenograft. Nanomedicine.

14

Detergent/Nanodisc Screening for High-Resolution NMR Studies of an Integral Membrane Protein Containing a Cytoplasmic Domain

Christos Tzitzilonis[1,2]☉, **Cédric Eichmann**[1]☉, **Innokentiy Maslennikov**[2], **Senyon Choe**[2], **Roland Riek**[1,2]*

1 Laboratory of Physical Chemistry, Swiss Federal Institute of Technology, ETH-Hönggerberg, Zürich, Switzerland, **2** Structural Biology Laboratory, The Salk Institute, La Jolla, California, United States of America

Abstract

Because membrane proteins need to be extracted from their natural environment and reconstituted in artificial milieus for the 3D structure determination by X-ray crystallography or NMR, the search for membrane mimetic that conserve the native structure and functional activities remains challenging. We demonstrate here a detergent/nanodisc screening study by NMR of the bacterial α-helical membrane protein YgaP containing a cytoplasmic rhodanese domain. The analysis of 2D $[^{15}N,^{1}H]$-TROSY spectra shows that only a careful usage of low amounts of mixed detergents did not perturb the cytoplasmic domain while solubilizing in parallel the transmembrane segments with good spectral quality. In contrast, the incorporation of YgaP into nanodiscs appeared to be straightforward and yielded a surprisingly high quality $[^{15}N,^{1}H]$-TROSY spectrum opening an avenue for the structural studies of a helical membrane protein in a bilayer system by solution state NMR.

Editor: Paul C. Driscoll, MRC National Institute for Medical Research, United Kingdom

Funding: The authors have no funding or support to report.

Competing Interests: The authors have declared that no competing interests exist.

* E-mail: roland.riek@phys.chem.ethz.ch

☉ These authors contributed equally to this work.

Introduction

It is well known that detergent-solubilized membrane proteins for 3D structure determination by X-ray crystallography or NMR spectroscopy often loose in part their function, such as enzymatic activity or ligand binding, along with structural changes [1–3]. The nature of this phenomenon is attributed to distortion of an extramembrane domain by detergents or the not well membrane imitation of detergent micelles. For example, a significant surface curvature of the water/micelle interface and less ordered nature of the detergent micelles may result in a deterioration of the 3D structure of both transmembrane and extramembrane domains and increased magnitudes of conformational exchange dynamics and motions [4]. This can lead to the study of a non-relevant protein conformation and non-relevant dynamics [5]. Recently, self-assembling lipid bilayer nanodiscs [6,7] [5] have been introduced as an alternative approach for reconstitution of integral membrane proteins (IMPs) in a bilayer-mimetic. The nanodisc recreates a native-like lipid bilayer without using detergents. To highlight one of the many membrane mimicking properties of nanodiscs they possess a phase transition from gel to liquid-crystal state similar to that of a pure phospholipid bilayer, though broadened and shifted by a few degrees to higher temperatures attributed to the presence of the membrane scaffold protein (MSP) [6]. Furthermore, the 10–12 nm in diameter large nanodiscs have a thickness of ~4 nm, which corresponds to the thickness of the biological membrane [4,6,8,9]. In addition, lipid bilayer nanodiscs apparently provide enhanced sample stability when compared with conventional membrane mimetic, which may be a critical property for the long duration of measurements required for the structural and dynamical studies of membrane proteins by solution state NMR [10–12]. Furthermore, nanodiscs lack detergents that may destabilize or unfold extra-membranous segments of membrane proteins such as long loops, periplasmic, extra-cellular or cytoplasmic domains, [1] [7] (see also this work below). Finally, reconstitution of membrane proteins in nanodiscs appears to yield a functional entity [10] [13] [4] [14]. The major shortcoming of IMPs reconstituted in nanodiscs for NMR studies is the relatively broad line width of the signals, which are expected to be twice as large as for IMPs embedded in conventional membrane mimetic because of the size of the nanodiscs (~10 nm). Hence, nanodiscs have been previously used so far mainly as a reference medium to confirm native-like protein fold during detergent screening of membrane proteins for solution NMR studies [14].

In this communication, we describe a NMR-based detergent/nanodisc screening study for high-resolution solution state NMR of the **α**-helical dimeric IMP YgaP from *E. coli*. YgaP has a molecular weight of 18.6 kDa and is composed of 174 residues, of which 119–174 are predicted by the TMHMM [15] to form two transmembrane helices, while residues 1–118 are predicted to form a cytoplasmic rhodanese domain with sulfurtransferase activity [16–17]. The function of YgaP is however so far unknown. As we shall see, the analysis of 2D $[^{15}N,^{1}H]$-TROSY spectra indicates that detergent solubilization of YgaP perturbed the structural integrity of the cytoplasmic rhodanese domain unless a careful usage of low amounts of mixed detergents was used. In contrast, the incorporation of YgaP into nanodiscs appeared to be straightforward and yielded a surprisingly well $[^{15}N,^{1}H]$-TROSY spectrum of full-length protein with properly folded cytoplasmic domain.

Materials and Methods

Expression of Full Length YgaP and N-terminal Rhodanese Domain

The pET3a-LIC based plasmid containing the full length YgaP (SwissProt ID P55734) gene was provided by Robert Stroud lab at UCSF [18]. The construction of the N-terminal rhodanese domain plasmid was achieved by inserting, with site directed mutagenesis (Agilent), a stop codon between residues Q109 and P110 of the full-length protein. YgaP C159S mutant (YgaP⁻) was obtained by site-directed mutagenesis (Agilent). YgaP⁻ was expressed, purified and reconstituted in nanodiscs using the same methods, as the ones described below for the wild-type YgaP. YgaP was expressed in $E.\ coli$ strain BL21(DE3) pLysS Star (Invitrogen, Carlsbad, USA). ^2H, ^{15}N labeled YgaP was produced using standard M9 minimal medium [19] based on ^2H water (Isotec) and 1g/L of $^{15}NH_4Cl$ (Isotec). Cells were grown at 37°C until $OD_{600} = 0.8$, and transferred to 18°C for 45 min while shaking. YgaP was expressed by induction with 0.5 mM IPTG (Invitrogen) at 18°C for 12 hours. Expression of the N-terminal rhodanese domain was performed using the same procedure as for the full length YgaP replacing D_2O by H_2O.

Purification of Full Length YgaP

After expression cells were centrifuged at 5,000 g for 10 minutes and resuspended in the lysis buffer (50 mM Tris-HCl pH 8, 300 mM NaCl, 10 mM DTT, 0.5 mg/ml lysozyme (Sigma), protease inhibitor cocktail tablets (Roche)) and incubated at 4°C with gentle stirring for 30 minutes. Cells were lysed by two Microfluidizer (Microfluidics) cycles at 80,000 psi. The lysed cells were centrifuged at 8,000 g for 15 minutes. The supernatant was collected and centrifuged at 100,000 g for 2 hours. The pelleted membrane fraction was resuspended in extraction buffer, 50 mM Tris-HCl pH 8, 300 mM NaCl, 10 mM Imidazole, supplemented with 15 mM DHPC-7 (1,2-diheptanoyl-sn-glycero-3-phosphocholine) plus 1 mM LMPG (1-Myristoyl-2-Hydroxy-sn-Glycero-3-Phospho-(1'-rac-Glycerol), 5 mM TCEP (Tris(2-carboxyethyl)phosphine), 10% w/v Glycerol and protease inhibitor cocktail tablets (Roche). YgaP was extracted over night by gentle stirring at 4°C. The solubilized membrane protein fraction was separated from large aggregates by centrifugation at 100,000 g for 45 minutes. The supernatant was loaded on 5 ml Ni Sepharose 6 Fast Flow resin (GE Healthcare), pre-equilibrated with 50 mM Tris-HCl pH 8, 300 mM NaCl, 10 mM Imidazole, 3 mM DHPC-7, 1 mM LMPG, and 1 mM TCEP. The resin was washed with the latter buffer, followed by elution of YgaP with 50 mM Tris-HCl pH 8, 300 mM NaCl, 500 mM Imidazole, 6 mM DHPC-7, 1 mM LMPG, and 5 mM TCEP. Fractions containing the protein were pooled together and the buffer was exchanged using a PD10 desalting column to 20 mM bis-Tris-HCl pH 7, 6 mM DHPC-7, 1 mM LMPG, and 5 mM TCEP. For NMR measurements YgaP was concentrated not more than 10 times [20] using a 30 kDa molecular weight cut-off Centricon (Amicon) concentrator to a final sample concentration of 0.3–0.5 mM. 3% of D_2O was added to the final sample for the NMR measurements. Size-exclusion chromatography was carried out at a flow rate of 0.5 ml/min on a Superdex 200 10/300GL gel filtration column (GE Healthcare) equilibrated with 20 mM bis-Tris-HCl pH 7, 150 mM NaCl, 3 mM DHPC-7, 1 mM LMPG, and 5 mM TCEP.

The YgaP purification protocol described above yielded the best NMR spectra (see below). For detergent screening the only alteration of the purification protocols were the use of different detergent types and concentrations. In the case of the sample preparation of YgaP in FC12 (Fos-Choline 12) only, the extraction buffer contained 15 mM FC12, while the wash, elution and desalting buffer contained 3 mM FC12. For the sample preparation of YgaP in DHPC-7 only, the extraction buffer contained 15 mM DHPC-7, while the wash, elution and desalting buffer contained 3 mM DHPC-7. For the mixed micelle samples the following condition were used: (A) For the sample preparation of YgaP in DHPC-7/FC12 mixed micelle, the extraction buffer contained 15 mM DHPC-7 and 3 mM FC12, while the wash buffer contained 3 mM DHPC-7 and 3 mM FC12, and the elution and desalting buffer 15 mM DHPC-7 and 3 mM FC12, respectively. (B) For the sample preparation of YgaP in DHPC-7/LMPG, the extraction buffer contained 15 mM DHPC-7 and 1 mM LMPG, the wash buffer 3 mM DHPC-7 and 1 mM LMPG, while the elution and desalting buffer contained 15 mM DHPC-7 and 1 mM LMPG, respectively.

Purification of the N-terminal Rhodanese Domain of YgaP

The rhodanese domain comprising residues 1–109 of YgaP was purified using the same procedure as described above for YgaP with modifications that include the absence of detergents in the buffers used. In short, after cell lysis and centrifugation the supernatant was loaded on 5 ml Ni⁺ Sepharose 6 Fast Flow resin (GE Healthcare), pre-equilibrated with 50 mM Tris-HCl pH 8, 300 mM NaCl, 10 mM Imidazole, and 1 mM TCEP. The resin was washed with the above buffer and the rhodanese domain was eluted with 50 mM Tris-HCl pH 8, 300 mM NaCl, 500 mM Imidazole, and 5 mM TCEP. Fractions containing the protein were pooled together and the buffer was exchanged using a PD10 desalting column to 20 mM bis-Tris-HCl pH 7, and 5 mM TCEP. For NMR measurements, the rhodanese domain was concentrated using a 10 kDa molecular weight cut-off Centricon (Amicon) concentrator to a concentration of 1–1.5 mM and 3% of D_2O was added.

Expression and Purification of MSP1

The plasmid (pET-28a) with the coding sequence of MSP1 fused to a TEV (Tobacco Etch Virus) protease cleavable N-terminal hexahistidine tag was a generous gift of the Arseniev lab. Expression and purification of MSP1 was carried out following the protocol of Shenkarev et al., (2009) with modifications. In short, MSP1 was expressed in $E.\ coli$ strain BL21(DE3) Star (Invitrogen, Carlsbad) in Terrific Broth. Cells were grown at 37°C until $OD_{600} = 0.8$ and protein production was initiated with the addition of 0.3 mM IPTG (Invitrogen, Carslbad). The cultures were incubated initially for 1 hour at 37°C and then the temperature was lowered to 28°C for further 2 hours.

Cells were harvested by centrifugation at 5000 g for 10 min. The cell pellet was resuspended in 100 ml of buffer A (20 mM Tris-HCl, pH 8.0, 0.5 M NaCl) containing 5 mg of DNAse deoxyribonuclease I (Sigma, DN-25), protease inhibitor cocktail tablets and incubated at 4°C with gentle stirring for 20 minutes. 1% Triton X-100 was added to the cell suspension and the solution was incubated with gentle stirring for further 20 minutes at room temperature. The lysate was clarified by centrifugation at 30,000 g for 30 min and applied to a Ni²⁺ resin (25 ml) (Qiagen), which was equilibrated with buffer A containing 1% Triton X-100. The column was sequentially washed with six column volumes of buffer A containing 1% Triton X-100, six volumes of buffer A containing 50 mM of sodium cholate, four volumes of buffer A, and finally eight column volumes of buffer A containing 50 mM imidazole. MSP1 was eluted with buffer A containing 0.5

M imidazole. Purified MSP1 was dialyzed against 10 mM Tris-HCl, pH 7.4, containing 100 mM NaCl and 1 mM EDTA (Ethylenediaminetetraacetic acid). Protein purity was checked by SDS-PAGE.

Proteolytic cleavage of the N-terminal His-tag from MSP1 was performed with the addition of TEV protease for 16 hours at room temperature. The buffer of the reaction mixture was exchanged to 20 mM Tris-HCl, 100 mM NaCl, 50 mM sodium cholate, 10 mM imidazole, pH 8.0 (buffer B) using a PD-10 column (GE Healthcare). The sample was applied to a Ni^{2+} column (volume 15 ml) (Qiagen), which was equilibrated with the same buffer. The flow through containing MSP1$^-$ without His tag (MSP1$^-$) was collected and dialyzed against 10 mM Tris-HCl, 100 mM NaCl, 1 mM EDTA, pH 7.4. MSP1$^-$ was stored for several days at 4°C.

Reconstitution of YgaP into DMPC Nanodiscs

DMPC (1,2-dimyristoyl-*sn*-glycero-3-phosphocholine) was solubilized in sodium cholate (cholate/DMPC 2:1 molar ratio) in a glass vial with a teflon-lined screw cap. MSP1$^-$ in 10 mM Tris-HCl, 100 mM NaCl, 1 mM EDTA, pH 7.4 buffer was mixed with DMPC cholate solution and incubated for 30 minutes at 27°C while shaking at 150 rpm. YgaP solubilized in detergent micelles was added to the MSP1$^-$/DMPC/Cholate mixture and incubated overnight at 27°C while shaking at 150 rpm. Incorporation of YgaP into nanodiscs was initiated with addition of 80% w/v Biobeads SM2 (Biorad) for 2 hours at 27°C while shaking at 150 rpm. The optimal reconstitution of YgaP into nanodiscs was achieved with molar ratio 1:8:320:640, YgaP: MSP1$^-$:DMPC: Cholate. Size-exclusion chromatography of YgaP incorporated in DMPC nanodiscs was carried out at a flow rate of 0.5 ml/min on a Superdex 200 10/300GL gel filtration column (GE Healthcare) equilibrated with 20 mM bis-Tris-HCl pH 7, 150 mM NaCl and 5 mM TCEP. For NMR measurements that required high concentration of protein sample, the solution of YgaP incorporated into nanodics was transferred to a 3 kDa molecular weight cut-off dialysis cassette (Slide-A-Lyzer®, Thermo Scientific, USA) and dialyzed against a solution of 10% Polyethylene glycol (PEG 10K) (Sigma) until the YgaP/nanodisc assembly sample volume was reduced to ~ 300 μl. YgaP-free nanodiscs were separated from nanodiscs containing N-terminal His-tagged YgaP by Ni^{2+} affinity chromatography.

NMR Spectroscopy

2D [^{15}N,^{1}H]-TROSY spectra with a time domain data size of 256*2048 complex points and with t_{1max}(^{15}N) = 44.8 ms, and t_{2max}(^{1}H) = 204.9 ms were recorded at 30°C on a Bruker Avance 700 MHz spectrometer equipped with a triple-resonance cryoprobe. The spectra were processed with the program PROSA [21] and analyzed with the program XEASY [22]. The concentration of ^{2}H,^{15}N- labeled YgaP in NMR samples was ~80 μM, and ~300 μM in DMPC and d54-DMPC nanodiscs, respectfully), while the concentration of YgaP in NMR samples with detergents was between 200 and 500 μM.

Results and Discussion

Detergent Screening for YgaP

YgaP is an integral *E. coli* membrane protein that has been predicted to have two trans-membrane α -helices using the TMHMM algorithm [15] and an N-terminal cytoplasmic domain. According to Protein Knowledgebase (UniprotKB/Swiss-Prot) the N-terminal domain of YgaP belongs to the rhodanese family of sulfultransferases [23]. YgaP has been previously over-expressed in *E. coli* as well as in a cell free system [18], and has been successfully solubilised by two different detergents n-dodecyl-β-D-maltopyranoside (DDM) and n-Octyl-β-D-glucopyranoside (OG). Based on these findings we developed a protocol for large-scale production of YgaP for solution NMR studies (Figure 1). Expression of YgaP in *E. coli* BL21 cells yielded 15–20 mg of purified protein per liter of M9 isotope enriched minimal media as estimated by absorbance at 280 nm and SDS electrophoresis (Figure 2C). The overexpressed protein was predominantly found in the inner membrane of *E. coli* as tested by cell fractionation of the bacterial membrane, detergent extraction and purification (Figure 2C), suggesting correct targeting and folding. Several detergents (i.e. FC12, DHPC-7 and mixed micelle systems, see Methods) were tested for extraction from the membrane fraction. In all the detergents tested YgaP was extracted from the membrane fraction of *E. coli* with efficiency close to 90%. After a successful purification with Ni affinity chromatography the protein purity was estimated to be higher than 95% (Figure 2C). Since it is our aim to study the structure of YgaP by NMR, we decided to test the homogeneity and integrity of the membrane protein detergent complex by recording 2D [^{15}N,^{1}H]-TROSY spectra [24] under various detergent conditions and concentrations directly after the affinity purification and buffer exchange without further purification as exemplified in Figure 1. This approach [25] allows us to screen several protein/detergent combinations and correlate the quality of the NMR spectra with the effect that different detergents may have to the conformational state of YgaP. The method is based on the fact that chemical shifts are very sensitive to changes in the chemical environment and therefore can be considered as excellent probes for structural heterogeneity and slow conformational exchange dynamics. Furthermore, using this approach the stability of the protein-detergent complex in relatively high concentrations (~0.3 mM) and high temperatures (30°C and above), both of which are critical for NMR structure determination, can be evaluated.

Two different zwitterionic detergents with different alkyl chain, FC12 and DHPC-7 were used for recording 2D [^{15}N,^{1}H]-TROSY spectra of ^{15}N, ^{2}H- labeled YgaP (Figure 1A–B). Both spectra had significant fewer cross-peaks than expected (i.e. approximately 70 peaks instead of the expected 165) and the peak intensities were not homogeneous. Furthermore, because YgaP has a soluble cytoplasmic rhodanese domain with expected β-sheets and α-helices [17], a broader dispersion of chemical shifts was expected. On top of that YgaP in DHPC-7 aggregated after a few hours at 30°C.

In order to explore the origin of the low sample quality and in particular the lack of the expected chemical shift dispersion, we decided to overexpress and purify the N-terminal cytoplasmic rhodanese domain of YgaP and collect 2D [^{15}N,^{1}H]-TROSY spectra. It was further rationalized that this spectrum may be used as a reference for the detergent screening towards a wellbehaved YgaP – detergent complex. The 2D [^{15}N,^{1}H]-TROSY spectrum of ^{15}N-labeled N-terminal rhodanese domain (Figure 1 E) shows a chemical shift dispersion expected for a folded protein containing α-helixes and β-sheets. Subsequent sequential assignment (indicated in Figure 1E) using standard triple resonance experiments [26] on a [^{15}N,^{13}C]-labeled sample of the N-terminal cytoplasmic domain of YgaP further indicated that it is indeed composed of a rhodanese fold [5] (to be published elsewhere).

Comparing the spectra of the rhodanese domain with those of YgaP in FC12 and DHPC-7 it was assumed that the presence of detergents may perturb the cytoplasmic rhodanese domain of YgaP and that FC12 is even worse doing so than DHPC-7. In contrast YgaP in DHPC-7 was not stable as mentioned above, precipitating within hours. Based on these findings, we decided to

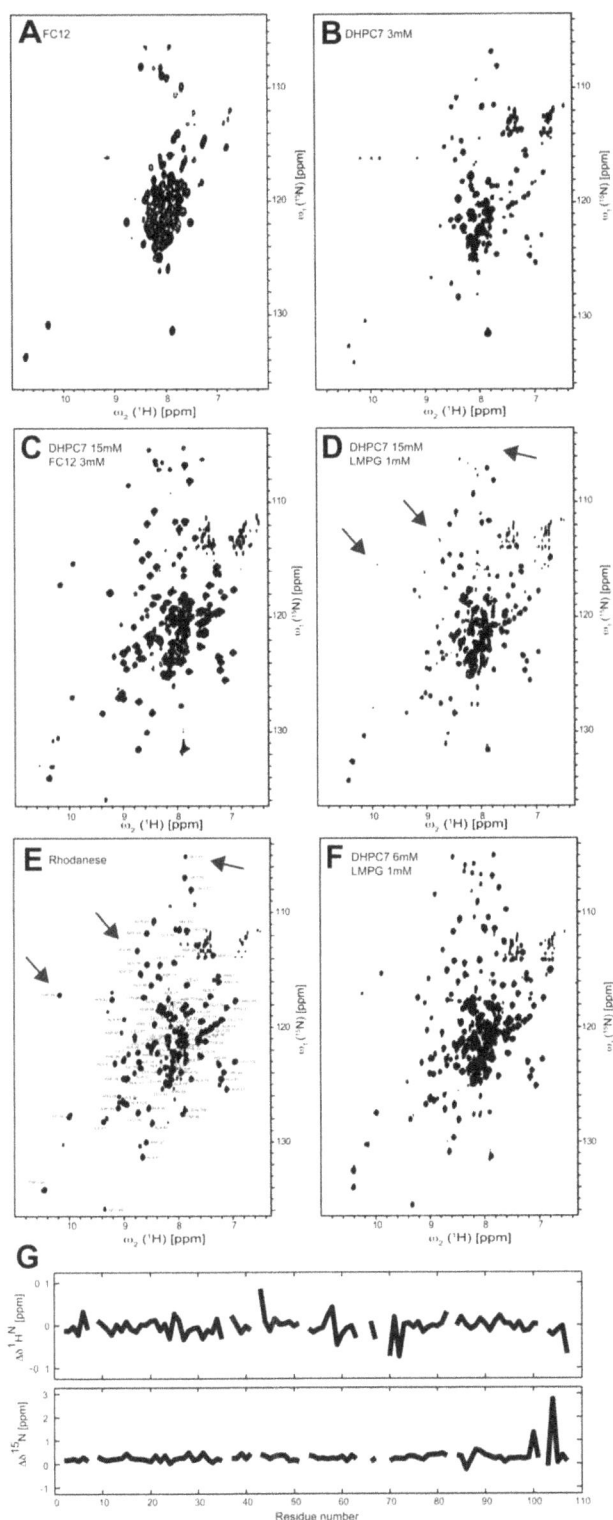

Figure 1. NMR spectra of YgaP in various micellar systems as indicated. 2D [^{15}N,^{1}H]-TROSY spectra of ^{2}H,^{15}N-labeled YgaP^{-} in (**A**) FC12, (**B**) DHPC-7, (**C**) DHPC-7 and FC12, (**D, F**) DHPC-7 and LMPG. (**E**) 2D [^{15}N,^{1}H]-TROSY of the N-terminal rhodanese domain of YgaP. The individual cross peaks are labeled according to the sequential assignment. (**G**) ^{1}H and ^{15}N chemical shift differences (labeled $\Delta\delta^{1}$H^{N} and $\Delta\delta^{15}$N) between the N-terminal rhodanese domain in solution and the N-terminal rhodanese domain of full length YgaP^{-} in the optimized mixed micellar conditions (i.e. 6 mM DHPC, 1 mM LMPG). The lack of profound up- or down-filed $\Delta\delta$ ^{1}H^{N} and $\Delta\delta^{15}$N chemical shift

differences indicates the same tertiary structure of the rhodanese domain in solution and in presence of mixed micelles.

screen for a mixture of detergents based on DHPC-7 using little amounts of detergents. The concept of mixed micelles has been used before for solving X-ray structures of membrane proteins [27] and also for solution NMR studies of membrane proteins [25]. Hence, ^{15}N,^{2}H-labeled YgaP was purified and NMR samples prepared in DHPC-7/FC12 and DHPC-7/LMPG mixed micelles. As shown in Figure 1C and 1D the 2D [^{15}N,^{1}H]-TROSY spectra have a good chemical shift dispersion of the ^{1}H^{N} backbone signals (from 6.8 to 10.4 ppm). Furthermore, most of the cross peaks present in the [^{15}N,^{1}H]-TROSY spectrum of the rhodanese domain (Figure 1E) have their counterpart in the spectra of the full-length YgaP unlike the spectra acquired of YgaP in FC12 only (Figure 1A). Actually, no significant chemical shift deviations of the sequentially assigned ^{15}N-^{1}H moieties of the rhodanese domain are observed between the spectra acquired with the rhodanese domain only and full-length YgaP solubilized in mixed detergent micelles (Figure 1G). These findings indicate the structural conservation of the rhodanese domain of YgaP in presence of mixed micelles. In addition, cysteine 159 was replaced by serine in the YgaP construct in order to interfere with potential aggregation through disulfide formation (note, residue 159 is located in the predicted second trans-membrane helix). This mutated construct entitled YgaP^{-} was used for all the subsequent studies.

Although the use of mixed micelles improved significantly the spectra quality of YgaP^{-}, a few resonances of the rhodanese domain where still absent (a few examples are indicated by an arrow in Figure 1). Hence, the optimization of the spectra quality was continued by changing the ratio of detergents in mixed micelles and by reducing further the overall amount of detergent in the NMR sample. In parallel, a known amount of both DHPC-7/LMPG detergents (i.e. 9 mM DHPC-7 and 2 mM LMPG) was added to the ^{15}N-labeled rhodanese domain accompanied with NMR measurements in order to study directly the effect of detergents on the rhodanese domain (Figure 2A). Indeed, at detergent concentration of 9 mM of DHPC-7 and 2 mM of LMPG, the detergent induced peak losses and/or chemical shift changes in the rhodanese sample (Figure 2A). A similar effect was evident in the [^{15}N,^{1}H]-TROSY spectrum of full-length YgaP^{-} collected under the same conditions (Figure 2B) although some chemical shift changes and peak losses differ between the two samples. Based on these findings the detergent concentration was lowered further while maintaining the mixed micelle approach yielding an excellent [^{15}N,^{1}H]-TROSY spectrum at a detergent concentration of 6 mM DHPC-7 and 1 mM LMPG (Figure 1F). In this spectrum cross peaks of the ^{15}N-^{1}H moieties of the rhodanese domain have their counterpart in the full-length YgaP^{-} mixed micelle complex and no significant chemical shift deviations of the sequentially assigned ^{15}N-^{1}H moieties of the rhodanese domain are observed between the spectra acquired with the rhodanese domain only and full-length YgaP^{-} solubilized in mixed detergent micelles (Figure 1G). This indicates that the rhodanese domain of the full-length YgaP^{-} in presence of mixed micelles is folded and has the same conformation as that of the free rhodanese domain in solution. In summary, many samples were prepared and many experiments were undertaken to improve the NMR spectra of YgaP, focusing on the inhibition of the unfolding of the cytoplasmic domain of YgaP by the detergents added. This approach yielded finally a high quality [^{15}N,^{1}H]-TROSY spectrum of YgaP being an excellent starting point for the NMR structure determination of YgaP.

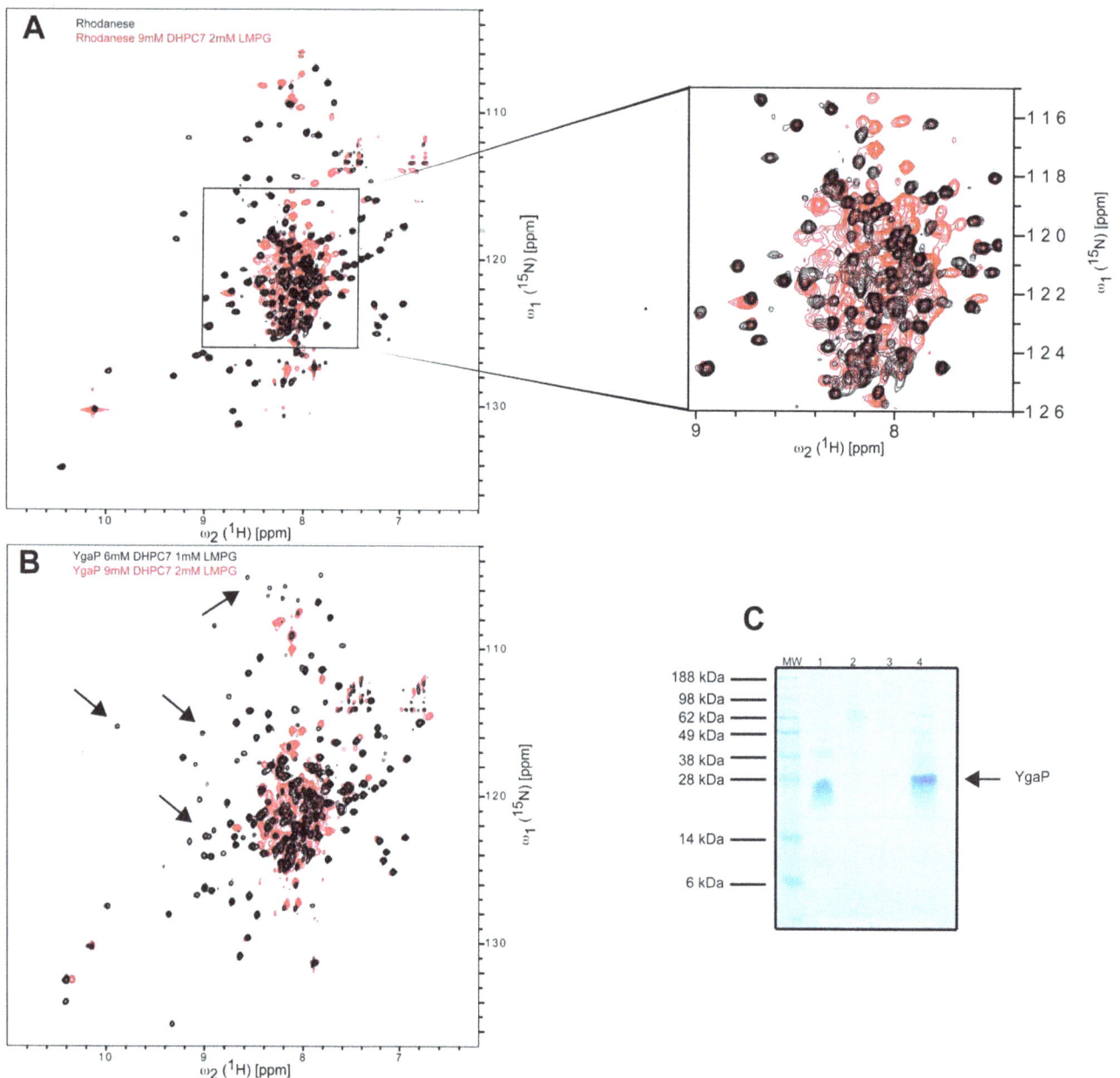

Figure 2. Effects of DHPC-7/LMPG mixed micelles on the N-terminal rhodanese domain and full length YgaP⁻. (**A**) 2D [^{15}N,^{1}H]-TROSY spectra of the N-terminal rhodanese domain in absence (Black) and in presence of 9 mM DHPC-7, 2 mM LMPG (Red). For better clarity a portion of the spectrum is magnified as indicated. (**B**) 2D [^{15}N,^{1}H]-TROSY spectra of ^{2}H,^{15}N-labeled of YgaP with optimum detergent concentration (i.e. 6 mM DHPC, 1 mM LMPG) (Black) and in presence of 9 mM DHPC-7, and 2 mM LMPG, respectively (Red). Black arrows indicate regions of the red spectrum where resonances are missing, indicating the effect of detergent excess in the quality of the spectrum. (**C**) SDS-PAGE of the nickel affinity purification of YgaP⁻ in DHPC-7/LMPG. 4–12% NuPAGE Bis-Tris gel (Invitrogen, Carslbad). Lanes: (MW) SeeBlue plus2 prestained (Invitrogen, Carslbad), (1) YgaP⁻ after membrane extraction in DHPC-7/LMPG micelles, (2) Loading flow-through fraction of Nickel resin, (3) Washing of Nickel resin, (4) Elution of YgaP⁻ with buffer containing 500 mM imidazole (details of the buffer used are given in the Material and Methods section).

YgaP⁻ Incorporation into Nanodiscs

Because membrane proteins incorporated in nanodiscs are surrounded by a lipid bilayer, they have been suggested to be a much more attractive membrane mimetic than detergent micelles [1]. This includes structural and dynamical studies by solution state NMR even though nanodiscs are rather large having a diameter of 10–12 nm yielding a long rotational correlation time, which may result in relatively broad peaks in the NMR spectra [13,14]. In addition to the membrane-alike character of nanodiscs,

they lack detergent that may unfold cytoplasmic domains as exemplified above for YgaP. Hence, we decided to test the applicability of nanodiscs as an alternative media for NMR studies of YgaP⁻.

The protein was purified as described above using the optimized conditions (i.e. 6 mM DHPC-7 and 1 mM LMPG), and incorporated into nanodiscs, following the published protocol of nanodisc assembly, which is based on the gradual removal of detergents with Biobeads from the sample containing detergent

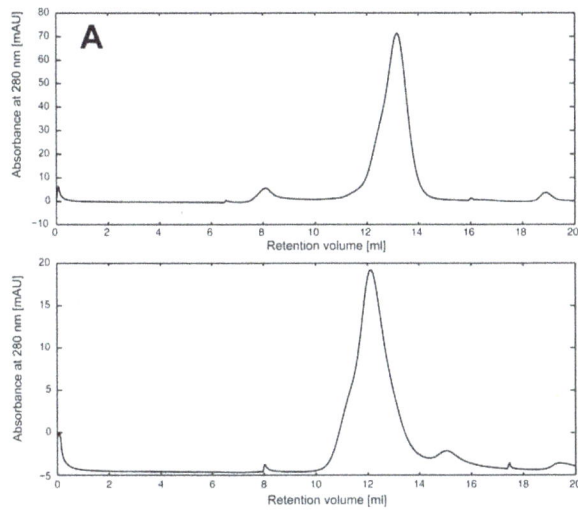

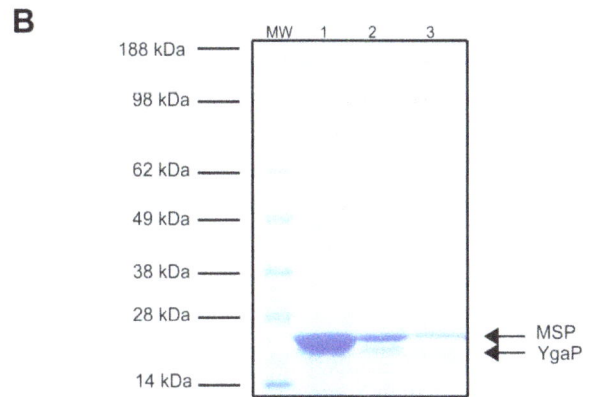

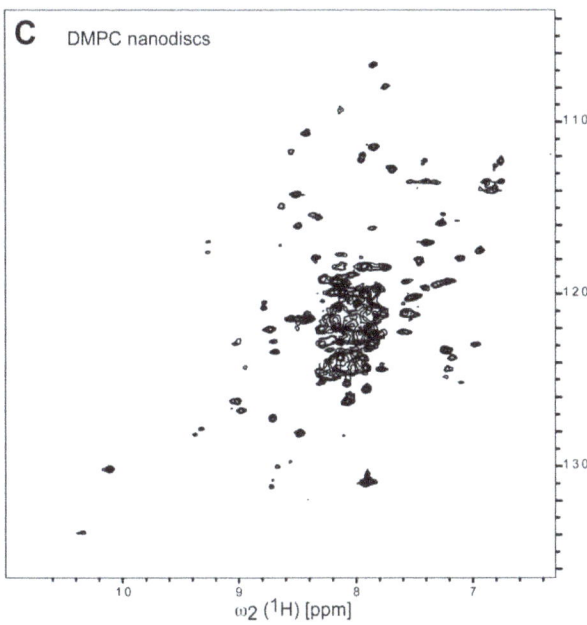

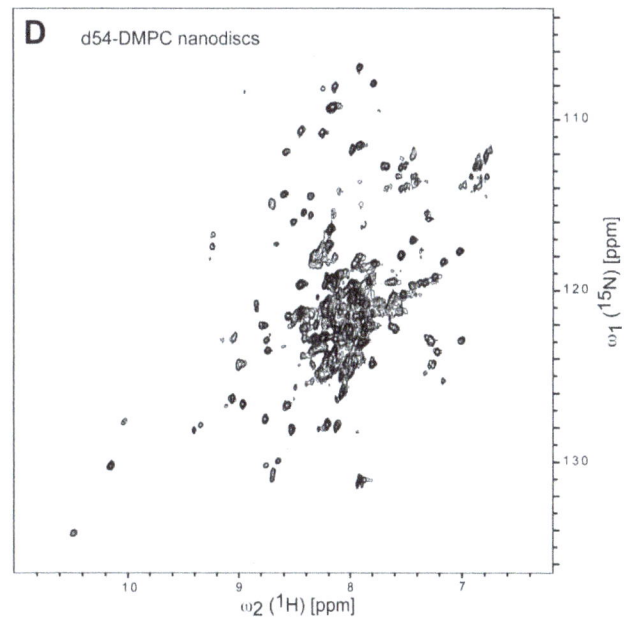

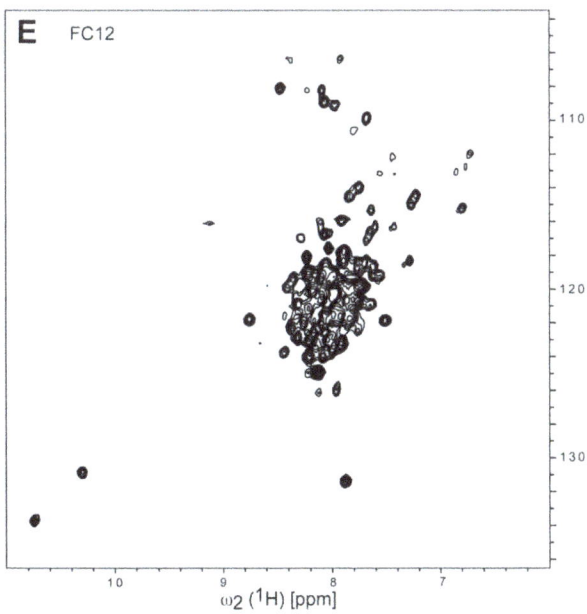

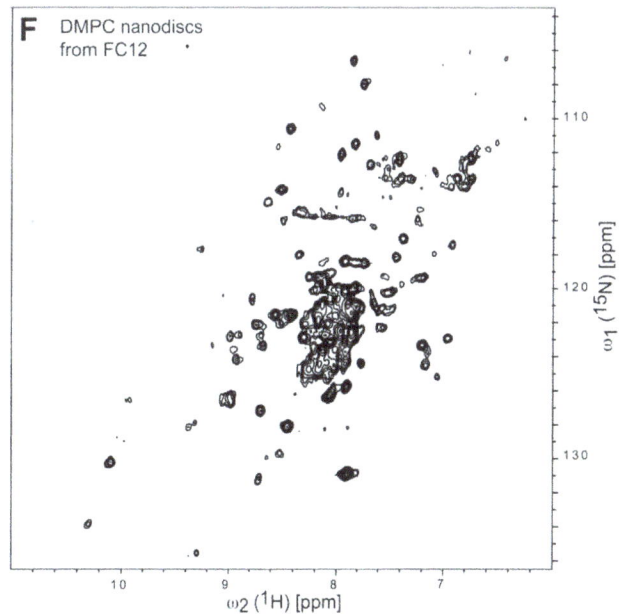

Figure 3. YgaP⁻ incorporation into DMPC nanodiscs. (**A**) Size exclusion chromatography (Superdex 200 10/300GL) of YgaP⁻ in 20 mM bis-Tris-HCl pH 7, 150 mM NaCl, 3 mM DHPC-7, 1 mM LMPG, and 5 mM TCEP. (upper panel) and YgaP⁻ in MSP1/DMPC nanodiscs. (**B**) SDS-PAGE of YgaP⁻ in DMPC nanodiscs. 12% NuPAGE Bis-Tris gel (Invitrogen, Carslbad). Lanes: (MW) SeeBlue plus2 prestained (Invitrogen, Carslbad), (1)-(3) Different dilutions of the YgaP/DMPC nanodisc reaction mixture in the SDS sample buffer after the removal of detergents by Biobeads to resolve the partial overlap due to apparent over-staining in lane 1 for the individual identification of YgaP and MSP1 as indicated. (**C**)–(**D**) 2D [^{15}N,^{1}H]-TROSY spectra of ^{2}H,^{15}N-labeled YgaP purified in 6 mM DHPC-7 and 1 mM LMPG and (**C**) incorporated in DMPC nanodiscs or (**D**) in nanodiscs with deuterated d-54 DMPC. (**E**) 2D [^{15}N,^{1}H]-TROSY spectrum of ^{2}H,^{15}N-labeled YgaP purified in 3 mM FC12. The sample of (E) was used for a DMPC nanodisc preparation as shown in (**F**): 2D [^{15}N,^{1}H]-TROSY spectrum of ^{2}H,^{15}N-labeled YgaP.

solubilized lipids (i.e DMPC), the membrane scaffold protein MSP and YgaP⁻ in 6 mM DHPC-7 and 1 mM LMPG (see Methods). Several ratios of MSP, DMPC and YgaP⁻ were tested because incorporation of an integral membrane protein in nanodiscs appears to depend mainly on the correct stoichiometry [6]. The incorporation of YgaP⁻ into DMPC nanodiscs was monitored by size-exclusion chromatography (Figure 3A–B), and SDS gel electrophoresis [13], while NMR was used to measure the depletion of the detergents used (i.e 6 mM DHPC-7 and 1 mM LMPG). A ratio of 1:8:320 YgaP⁻: MSP: DMPC was found to yield the best NMR spectra (Figure 3C). The cross peaks of the 2D [^{15}N,^{1}H]-TROSY spectrum (measured for 4 hours) of ~80 µM YgaP⁻ reconstituted in DMPC nanodiscs are well-dispersed (Fig. 3C). A total of 70 ^{15}N-^{1}H cross peaks of the rhodanese domain and the C-terminus of YgaP⁻ were unambiguously identified and were found at almost identical positions as in the 2D [^{15}N,^{1}H]-TROSY spectra of the rhodanese domain only (Figure 1E) with a negligible average change in the chemical shifts. These findings indicate the structural integrity of the cytoplasmic rhodanese domain in YgaP⁻ reconstituted in DMPC nanodiscs.

The [^{15}N,^{1}H]-TROSY spectrum of YgaP⁻ reconstituted in DMPC nanodiscs shown in Figure 3 was recorded at a protein concentrations of ~80 ∫ since attempts to make a more concentrated sample using Amicon concentrators failed. In order to increase the protein concentration to ~0.3 mM concentration, which is a required concentration if structural and dynamical studies by solution state NMR are attempted, an alternative method of concentration was adopted. The entire assembly reaction after removal of detergents with Biobeads was dialysed against 10% w/v PEG 10K using a dialysis membrane with a 3 kDa molecular weight cut-off. This approach resulted in a several fold reduction of the volume of the YgaP⁻/nanodisc sample and thus an increase in protein concentration yielding an NMR sample of ~0.3 mM YgaP⁻ assembled in nanodiscs containing deuterated d54-DMPC lipids (Figure 3D). As expected the 2D [^{15}N,^{1}H]-TROSY of the concentrated sample was similar to the previous diluted sample of YgaP⁻ in DMPC nanodiscs, but has an improved signal to noise. A more detailed analysis reveals the presence of ~80% of the cross peaks of the rhodanese domain. Furthermore, there are additional ~40 cross peaks in the [^{15}N,^{1}H]-TROSY spectrum. If these cross peaks are tentatively assigned to the transmembrane segment of YgaP, also around 80% of the backbone ^{15}N,^{1}H-moieties of the latter segment show a cross peak in the [^{15}N,^{1}H]-TROSY spectrum.

In an attempt to further improve the quality of the NMR spectra His-tag YgaP⁻ in DMPC nanodiscs was separated from YgaP⁻ free nanodiscs using metal affinity chromatography (Figure S1). Comparison of the size exclusion chromatography profile (Figure S1B) of the purified YgaP⁻/DMPC nanodisc complex with the one obtained from the unpurified YgaP⁻/DMPC complex (Figure 3B) shows that the complex in both cases have similar molecular weights, and that removal of YgaP⁻ free nanodiscs resulted in a more symmetrical peak shape in the size exclusion profile, indicating a more homogeneous sample. However, the

[^{15}N,^{1}H]-TROSY spectrum does not improve qualitatively upon removal of free nanodiscs (compare Figure S1C with Figure 3D).

While the preparation of YgaP⁻ in nanodiscs yielded straightforwardly a sample with a folded rhodanese domain contrasting the tedious sample preparation optimization in the case of detergents it must be noted that we used already optimized detergent conditions for the extraction of YgaP⁻ from the *E. coli* membrane and purification. In order to test whether the optimized detergent conditions are necessary for the nanodisc preparation of YgaP⁻ we decided to make a sample preparation using non-optimized conditions. YgaP⁻ was purified using 3 mM FC12 yielding a bad quality [^{15}N,^{1}H]-TROSY spectrum as shown in Figure 3E. The lack of well-resolved cross peaks from the rhodanese domain indicates that FC12 does perturb the structural integrity of the rhodanese domain (Figure 3E). Upon incorporation of this particular sample into DMPC nanodiscs a [^{15}N,^{1}H]-TROSY spectrum (Figure 3F) is obtained with a significant number of well resolved cross-peaks of the rhodanese domain (note, the difference in the quality of this spectrum with the corresponding one shown in Figure 1E is attributed to the low concentration of ~50 µM used here). This finding indicates that the rhodanese domain is folded upon YgaP⁻ incorporation into nanodiscs irrespective whether (partial) unfolding happens during the purification procedure caused by the unfolding properties of detergents (Figure 2A).

Conclusions

The present analysis of the sample preparation of the IMP YgaP indicated that while detergent-micelle preparations require a careful screening of conditions in order to prevent the unfolding of the cytoplasmic rhodanese domain by the detergent used, the incorporation of YgaP into nanodiscs appeared to be straightforward. The obtained high resolution [^{15}N,^{1}H]-TROSY of YgaP in the optimized mixed micelles on the one hand is considered a promising starting point towards a NMR structure determination. On the other hand the corresponding [^{15}N,^{1}H]-TROSY spectrum of YgaP incorporated in nanodiscs opens an avenue for solution state NMR towards structural and dynamical studies of a helical integral membrane protein embedded in a lipid bilayer. Both paths are taken in order to exploit the structure and dynamics of YgaP.

Supporting Information

Figure S1   Purification of YgaP⁻ incorporated in DMPC nanodiscs. (**A**) SDS-PAGE of Ni^{2+} affinity purification of YgaP⁻ in DMPC nanodiscs. 12% NuPAGE Bis-Tris gel (Invitrogen, Carslbad). Lanes: (MW) SeeBlue plus2 prestained (Invitrogen, Carslbad), (1) YgaP⁻/DMPC nanodisc reaction mixture before Ni⁺ purification. (2)–(4) Fractions containing MSP1⁻ collected during the loading and washing of the Ni^{2+} affinity column. (5)-(6) Fractions of YgaP⁻/DMPC nanodisc complex eluted from the Ni^{2+} affinity column. (**B**) Size exclusion chromatography (Superdex 200 10/300GL) of Ni^{2+} affinity purified YgaP/DMPC nanodisc complex. (C) 2D [^{15}N,^{1}H]-TROSY spectra of ^{2}H,^{15}N-labeled YgaP in DMPC nanodiscs after Ni^{2+} affinity purification of the complex.

References

1. Raschle T, Hiller S, Etzkorn M, Wagner G (2010) Nonmicellar systems for solution NMR spectroscopy of membrane proteins. Curr Opin Struct Biol 20: 471–479.
2. Kawai T, Caaveiro JM, Abe R, Katagiri T, Tsumoto K (2011) Catalytic activity of MsbA reconstituted in nanodisc particles is modulated by remote interactions with the bilayer. FEBS Lett 585: 3533–3537.
3. Tate CG (2010) Practical considerations of membrane protein instability during purification and crystallisation. Methods Mol Biol 601: 187–203.
4. Shenkarev ZO, Lyukmanova EN, Solozhenkin OI, Gagnidze IE, Nekrasova OV, et al. (2009) Lipid-protein nanodiscs: possible application in high-resolution NMR investigations of membrane proteins and membrane-active peptides. Biochemistry (Mosc) 74: 756–765.
5. Brewer KD, Li W, Horne BE, Rizo J (2011) Reluctance to membrane binding enables accessibility of the synaptobrevin SNARE motif for SNARE complex formation. Proc Natl Acad Sci U S A 108: 12723–12728.
6. Bayburt TH, Sligar SG (2010) Membrane protein assembly into Nanodiscs. FEBS Lett 584: 1721–1727.
7. Ritchie TK, Grinkova YV, Bayburt TH, Denisov IG, Zolnerciks JK, et al. (2009) Chapter 11 - Reconstitution of membrane proteins in phospholipid bilayer nanodiscs. Methods Enzymol 464: 211–231.
8. Shih AY, Denisov IG, Phillips JC, Sligar SG, Schulten K (2005) Molecular dynamics simulations of discoidal bilayers assembled from truncated human lipoproteins. Biophys J 88: 548–556.
9. Wu Z, Wagner MA, Zheng L, Parks JS, Shy JM, 3rd, et al. (2007) The refined structure of nascent HDL reveals a key functional domain for particle maturation and dysfunction. Nat Struct Mol Biol 14: 861–868.
10. Lyukmanova EN, Shenkarev ZO, Paramonov AS, Sobol AG, Ovchinnikova TV, et al. (2008) Lipid-protein nanoscale bilayers: a versatile medium for NMR investigations of membrane proteins and membrane-active peptides. J Am Chem Soc 130: 2140–2141.
11. Gluck JM, Wittlich M, Feuerstein S, Hoffmann S, Willbold D, et al. (2009) Integral membrane proteins in nanodiscs can be studied by solution NMR spectroscopy. J Am Chem Soc 131: 12060–12061.
12. Yu TY, Raschle T, Hiller S, Wagner G (2012) Solution NMR spectroscopic characterization of human VDAC-2 in detergent micelles and lipid bilayer nanodiscs. Biochim Biophys Acta 1818: 1562–1569.
13. Raschle T, Hiller S, Yu TY, Rice AJ, Walz T, et al. (2009) Structural and functional characterization of the integral membrane protein VDAC-1 in lipid bilayer nanodiscs. J Am Chem Soc 131: 17777–17779.
14. Shenkarev ZO, Lyukmanova EN, Paramonov AS, Shingarova LN, Chupin VV, et al. (2010) Lipid-protein nanodiscs as reference medium in detergent screening for high-resolution NMR studies of integral membrane proteins. J Am Chem Soc 132: 5628–5629.
15. Krogh A, Larsson B, von Heijne G, Sonnhammer EL (2001) Predicting transmembrane protein topology with a hidden Markov model: application to complete genomes. J Mol Biol 305: 567–580.
16. Gueune H, Durand MJ, Thouand G, DuBow MS (2008) The ygaVP genes of Escherichia coli form a tributyltin-inducible operon. Appl Environ Microbiol 74: 1954–1958.
17. Li H, Yang F, Kang X, Xia B, Jin C (2008) Solution structures and backbone dynamics of Escherichia coli rhodanese PspE in its sulfur-free and persulfide-intermediate forms: implications for the catalytic mechanism of rhodanese. Biochemistry 47: 4377–4385.
18. Savage DF, Anderson CL, Robles-Colmenares Y, Newby ZE, Stroud RM (2007) Cell-free complements in vivo expression of the E. coli membrane proteome. Protein Sci 16: 966–976.
19. Sambrook J, Fritsch EF, Maniatis T (1989) Molecular Cloning: A Laboratory Manual: Cold Spring Harbor Laboratory, Cold Spring Harbor, NY.
20. Maslennikov I, Kefala G, Johnson C, Riek R, Choe S, et al. (2007) NMR spectroscopic and analytical ultracentrifuge analysis of membrane protein detergent complexes. BMC Struct Biol 7: 74.
21. Guntert P, Dotsch V, Wider G, Wuthrich K (1992) Processing of Multidimensional Nmr Data with the New Software Prosa. Journal of Biomolecular Nmr 2: 619–629.
22. Bartels C, Xia TH, Billeter M, Guntert P, Wuthrich K (1995) The Program Xeasy for Computer-Supported Nmr Spectral-Analysis of Biological Macromolecules. Journal of Biomolecular Nmr 6: 1–10.
23. Cipollone R, Ascenzi P, Tomao P, Imperi F, Visca P (2008) Enzymatic detoxification of cyanide: clues from Pseudomonas aeruginosa Rhodanese. J Mol Microbiol Biotechnol 15: 199–211.
24. Pervushin K, Riek R, Wider G, Wuthrich K (1997) Attenuated T2 relaxation by mutual cancellation of dipole-dipole coupling and chemical shift anisotropy indicates an avenue to NMR structures of very large biological macromolecules in solution. Proc Natl Acad Sci U S A 94: 12366–12371.
25. Columbus L, Lipfert J, Jambunathan K, Fox DA, Sim AY, et al. (2009) Mixing and matching detergents for membrane protein NMR structure determination. J Am Chem Soc 131: 7320–7326.
26. Grzesiek S, Bax A (1992) An Efficient Experiment for Sequential Backbone Assignment of Medium-Sized Isotopically Enriched Proteins. Journal of Magnetic Resonance 99: 201–207.
27. Koronakis V, Sharff A, Koronakis E, Luisi B, Hughes C (2000) Crystal structure of the bacterial membrane protein TolC central to multidrug efflux and protein export. Nature 405: 914–919.

Author Contributions

Conceived and designed the experiments: CT CE RR. Performed the experiments: CT CE IM. Analyzed the data: CT CE IM RR. Contributed reagents/materials/analysis tools: CT CE IM SC RR. Wrote the paper: CT CE IM RR SC.

15

Regulation of Hematopoietic Stem Cell Behavior by the Nanostructured Presentation of Extracellular Matrix Components

Christine Anna Muth[1,2], **Carolin Steinl**[3], **Gerd Klein**[3], **Cornelia Lee-Thedieck**[1,2,4]*

1 Department of New Materials and Biosystems, Max Planck Institute for Intelligent Systems, Stuttgart, Germany, 2 Department of Biophysical Chemistry, University of Heidelberg, Heidelberg, Germany, 3 Section for Transplantation Immunology and Immunohematology, Center for Medical Research, University of Tübingen, Tübingen, Germany, 4 Institute of Functional Interfaces, Karlsruhe Institute of Technology (KIT), Eggenstein-Leopoldshafen, Germany

Abstract

Hematopoietic stem cells (HSCs) are maintained in stem cell niches, which regulate stem cell fate. Extracellular matrix (ECM) molecules, which are an essential part of these niches, can actively modulate cell functions. However, only little is known on the impact of ECM ligands on HSCs in a biomimetic environment defined on the nanometer-scale level. Here, we show that human hematopoietic stem and progenitor cell (HSPC) adhesion depends on the type of ligand, i.e., the type of ECM molecule, and the lateral, nanometer-scaled distance between the ligands (while the ligand type influenced the dependency on the latter). For small fibronectin (FN)–derived peptide ligands such as RGD and LDV the critical adhesive interligand distance for HSPCs was below 45 nm. FN-derived (FN type III 7–10) and osteopontin-derived protein domains also supported cell adhesion at greater distances. We found that the expression of the ECM protein thrombospondin-2 (THBS2) in HSPCs depends on the presence of the ligand type and its nanostructured presentation. Functionally, THBS2 proved to mediate adhesion of HSPCs. In conclusion, the present study shows that HSPCs are sensitive to the nanostructure of their microenvironment and that they are able to actively modulate their environment by secreting ECM factors.

Editor: Marc Tjwa, University of Frankfurt - University Hospital Frankfurt, Germany

Funding: The project was supported by contract research 'Biomaterialien/Biokompatibilität' of the Baden-Württemberg Stiftung (P-LS-Biomat/22). CL-T gratefully acknowledges support by the Brigitte Schlieben-Lange-Programme and the 'Käthe und Josef Klinz-Stiftung'. The funders had no role in study design, data collection and analysis, decision to publish, or preparation of the manuscript.

Competing Interests: The authors have declared that no competing interests exist.

* E-mail: cornelia.lee-thedieck@kit.edu

Introduction

Hematopoietic stem cells (HSCs) are located in specific environments in the bone marrow, i.e. the stem cell niches. Specialized niche cells, extracellular matrix (ECM) and soluble factors play essential roles in regulating HSC function and maintenance. However, to which extent those factors contribute to the functionality of the bone marrow stem cell niches and how they are regulated remains uncertain [1].

Evidence has been found that in addition to the niche regulating stem cell behavior, stem cells and their progeny also actively modulate their niche [2,3]. Several HSC niches have been described in mice: (i) the endosteal niche containing osteoblastic cells as the major HSC-supporting cell type [4,5,6], (ii) the perivascular niche, in which HSCs are influenced by vascular and perivascular cells [7,8,9], and (iii) a niche formed by nestin+ mesenchymal stem cells [10]. In all niches mesenchymal stem cells play an important role [11]. To date, the relevance of each of these niches for individual HSC functions is under debate.

The ECM is of particular interest, because it can induce diverse cell responses [12]. The bone marrow ECM is a complex composition of collagens, proteoglycans, glycosaminoglycans, and glycoproteins such as fibronectin (FN), osteopontin (OPN), laminins and thrombospondins (THBS) [1,13]. Different ECM molecules can influence adhesion, proliferation, survival, migration and differentiation of

stem cells [14]. In elegant studies the Werner group has shown in a biomimetic setup that HSC fate decisions depend on the delicate balance of adhesive interactions with ECM components and stimulation by soluble factors [15,16,17]. Among the studied ECM-derived ligands FN proved to mediate the strongest adhesion of HSCs to material surfaces [15].

Many ECM proteins mediate cell adhesion via integrin receptors [18]. Integrins are heterodimers containing two distinct subunits, referred to as the α and β chain. So far, 18 α and 8 β chains have been described. Twenty-four unique receptors types, generated through different subunit combinations, are known [18,19]. Integrins can be categorized into different subsets according to which α or β chain they contain and the class of ECM proteins to which they bind [20]. It has been shown that the integrin $\beta 1$ chain is essential for hematopoietic stem and progenitor cell (HSPC) homing and migration [21,22], and that blocking the integrin α_5 chain inhibits HSPC adhesion to FN [15]. Integrin $\alpha_4\beta_1$- and $\alpha_5\beta_1$-mediated binding to FN has also been described to impact HSC growth [23,24].

FN is a well-characterized ECM protein, which is present in the bone marrow. Because FN has been described to promote as well as to inhibit HSC proliferation [17,25,26,27], its significance for HSC function remains controversial. One possible explanation for these contradictory findings may be the existence of different conformations of the FN molecules, depending on the type of surface they are immobilized on [28]. The FN molecule is composed of three types of modules (type I, II, III). Short peptide motifs within these modules have been identified as key elements of the integrin receptor recognition sites of FN. The 10th type III module contains an RGD sequence (Fig. S1A). This motif is the best-studied minimal cell adhesive sequence [29]. It is situated in a flexible loop structure between two strands, in close proximity to the synergy sequence PHSRN of the 9th type III module [30]. This synergy sequence enhances the interaction stability of $\alpha_5\beta_1$ integrin with FN [31]. In addition, the alternatively spliced V region in the C-terminal cell binding domain of FN contains an LDV motif that is recognized by hematopoietic cells via their $\alpha_4\beta_1$ integrin receptor [32]. Figure S1 shows a schematic cartoon of FN and FN-derived ligands that were applied in this study, including the location of the cell-binding domains.

Another important ECM molecule of the hematopoietic microenvironment is OPN, which has been described as a negative regulator of the murine HSC pool size [33,34]. The N-terminal thrombin fragment of OPN (amino acids 17–168) contains an RGD sequence and thrombin-cleaved OPN can regulate HSPC functions (e.g., migration and homing) through interactions with $\alpha_9\beta_1$ and $\alpha_4\beta_1$ integrins [35].

The geometric arrangement of ligands on the nanometer scale and the matrix elasticity are as important for HSC function as the composition of the ECM [36,37,38,39]. It is well known that the nanopatterned spatial presentation of ECM ligands influences adhesion, migration and focal adhesion assembly of fully differentiated tissue cells such as fibroblasts and osteoblasts [40,41]. However, only little is known about the impact of ECM ligands on HSCs in a biomimetic environment defined at the nanometer-scale level. It has been shown that synthetic nanostructured environments influence HSC adhesion, lipid raft clustering and expansion [36,42]. The aim of the current study was to identify ECM signals that guide HSC function in the context of a nanostructured environment. In order to mimic the natural microenvironment of cells, which is structured from the micro- to the nanometer-scale, biocompatible materials allowing the control over ligand choice, ligand orientation and receptor clustering in the nanometer range are essential [43,44]. These requirements were fulfilled in the current study by using hydrogel-supported gold nanopatterns equipped with bioactive molecules. Quasi-hexagonally ordered gold nanoparticle (NP) arrays were produced using block copolymer micelle nanolithography (BCML). The distances between the NPs were adjusted to values between 20 and 110 nm by varying the production parameters [45]. NP diameters ranged from 6 to 8 nm. The gold NP arrays were embedded in polyethylene glycol (PEG) hydrogels [46]. PEG is a nontoxic, biologically inert and protein-repellent material, making it highly useful for biological applications [47,48]. Since a single gold NP is smaller than a singular integrin receptor ($\sim$ 10 nm in diameter [49]), theoretically only one receptor should be able to bind per biofunctionalized NP due to steric hindrance. Thus, using this nanopatterned PEG hydrogel system, an one-to-one interaction between a single receptor and a single biofunctionalized gold NP is possible and lateral receptor clustering can be controlled [50].

In order to unravel the significance of matrix nanostructure for the HSC niche function, we investigated the influence of different ECM ligands and nanopatterns on HSPC adhesion, proliferation, differentiation and gene expression.

Results

Characterization of Biofunctionalized, Nanopatterned Hydrogels

Gold nanopatterned PEG hydrogels were produced and biofunctionalized with ligands that are present in the bone marrow ECM to mimic properties of the ECM environment. The basis for ligand presentation was set by gold NPs, which were evenly spread in a predefined pattern with a lateral particle distance between 36 ± 7 and 110 ± 18 nm (Fig. 1A–C). N-terminally His-tagged protein domains were bound to the gold NPs via a thiolated NTA-linker in a site-specific and oriented manner [51]. The applied protein domains were derived from FN or OPN. The OPN domain, called OPNs, corresponds to the thrombin-cleaved OPN fragment (amino acids 17–168) containing a C-terminal RGD sequence. The FN domain, called FNRGD, consists of the type III modules 7–10 and contains the characteristic integrin-binding motif RGD. This protein domain was used in a functionally intact version (FNRGD) equipped with the RGD sequence and in a mutated version (FNΔRGD) lacking this crucial integrin-binding motif. Fluorescent labeling of nanostructured hydrogels functionalized with FNRGD showed the immobilization of the domains to be selective for the nanostructured part of the hydrogel, not for the unstructured internal control (Fig. 1D). This suggests a specific functionalization of the gold NPs and not of the PEG hydrogel itself. As expected, when using EDTA instead of Ni^{2+} during coupling, binding did not occur and no fluorescence above background noise could be detected (Fig. 1E). This indicates His-tag-mediated oriented and site-specific binding. Controls where the primary antibody was omitted showed no unspecific antibody binding (Fig. 1F).

Contacts of Hematopoietic Cells to Biofuntionalized Gold NPs

Nanostructured PEG hydrogels were functionalized with a cyclic RGD peptide (cRGD) featuring the RGD sequence that is usually located in a loop region of FN (Fig. S1). KG-1a cells were allowed to adhere to these substrates for 1 h. Scanning electron microscopy (SEM) revealed that KG-1a cells kept their round morphology upon adhesion and were in close contact with the cRGD-functionalized nanostructured substrates through their filopodia (Fig. S2). Higher magnification imaging of cells on hydrogel substrates with sufficient resolution to monitor cellular structures and the substrate nanostructure at the same time was impossible due to electrical charging of the samples during SEM measurements. To overcome this limitation BCML-nanostructured glass substrates were employed. The glass background of these substrates was passivated with PEG in order to prevent non-specific cell adhesion and protein absorption and the gold NPs were functionalized with cRGD similar to the hydrogels. The morphology of cells imaged on glass and hydrogels appeared to be identical in SEM (Fig. 2A, B and S2). Higher magnification of the filopodia showed that the cells were in contact with the biofunctionalized gold NPs and not with the PEG-passivated area in between the NPs (Fig. 2C, D). These results indicate that hematopoietic cells are able to sense nanostructures on cRGD-functionalized surfaces.

Impact of ECM Ligand Type and Spacing on KG-1a and HSPC Adhesion

The combined influence of nanostructural and biochemical parameters on cell adhesion *in vitro* was studied on nanopatterned, biofunctionalized PEG hydrogels. The nanostructured hydrogels

NP spacing: 36 ± 7 nm 　　　　60 ± 11 nm 　　　　110 ± 18 nm

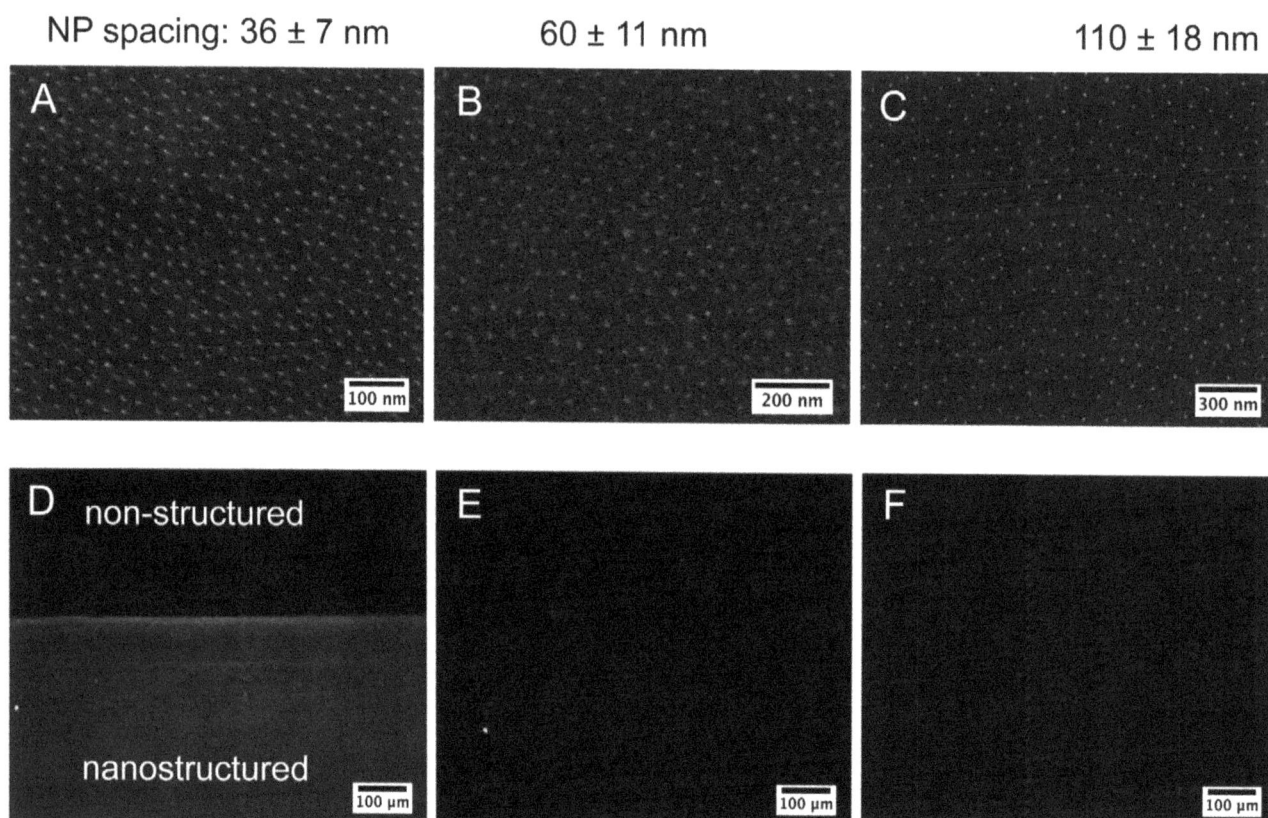

Figure 1. Nanopatterned and biofunctionalized PEG hydrogels. (A–C) Cryo-SEM images of the quasi-hexagonally ordered gold NP patterns on PEG hydrogels with interparticle distances of (A) 36±7 nm, (B) 60±11 nm and (C) 110±18 nm. (D–F) Micrographs of fluorescently labeled, FNRGD-functionalized nanostructured hydrogels. Images of the border between the nanostructured area in the lower part of the micrographs and the unstructured area visible in the upper part are shown. (D) The FNRGD domain on biofunctionalized hydrogels was detected with the help of specific primary antibodies. Controls were produced by (E) substituting nickel with EDTA during functionalization and by (F) omitting the primary antibody during the staining procedure. One representative experiment out of 3 is shown.

were biofunctionalized with cRGD, cLDV, FNRGD, OPNs or the non-functional control ligands (cRGE, FNΔRGD). Micrographs of KG-1a cells on representative cRGD-, cRGE-, FNRGD- or FNΔRGD-functionalized nanostructured hydrogels are shown in Fig. 3A. In general, KG-1a cells exclusively adhered to the nanostructured, rather than the unstructured, area of the hydrogels. Furthermore, adhesion was only observed to the functional ligands cRGD and FNRGD, while adhesion to the control ligands (cRGE, FNΔRGD) was negligible to non-existent (Fig. 3A upper panel). KG-1a cells adhered to surfaces with cRGD and FNRGD at 36±7 nm interparticle spacing. On surfaces with 60±11 nm interparticle spacing KG-1a cells were unable to adhere to cRGD, while adhesion to the FNRGD domain was similar to that on the surfaces with an interparticle spacing of 36±7 nm (Fig. 3A lower panel).

To compare hematopoietic cell adhesion to nanostructured surfaces varying in interparticle distance and ligand type an assay based on the accurate determination of DNA content was applied. The number of adhering cells was normalized to the number of cells adhering to full-length FN (defined as 100% adhesion). On hydrogels functionalized with cRGD, FNRGD or OPNs at 36±7 nm NP spacing, the number of adhering HSPC and KG-1a cells was comparable to the number adhering to a surface functionalized with the full-length FN protein (100%). On hydrogels functionalized with cLDV cell adhesion was generally lower compared to that of the other adhesive ligands (Fig. 3B, C).

Only 1–22% adhesion could be observed on surfaces with peptide ligands spaced apart 60±11 nm in comparison to full-length FN. This value is comparable to that observed for the non-adhesive control ligands. In contrast, cells were able to successfully adhere to hydrogels biofunctionalized with protein domains at 60±11 nm spacing.

In summary, hematopoietic cells were able to bind to all investigated adhesive ligands when these were located close enough to each other (36±7 nm). At 60±11 nm interparticle spacing only the protein domains elicited substantial cell adhesion. The critical interparticle distance (i.e., the maximal distance), which was still able to support KG1a cell adhesion, was determined to be between 37±7 nm and 45±9 nm for cRGD (Fig. S3), between 85±14 nm and 110±18 nm for the FNRGD domain (Fig. S4A) and between 75±13 nm and 85±14 nm for the OPNs domain (Fig. S4B). Based on these results, for the following experiments we focused on the ligands mediating the highest cell adhesion, which were cRGD, the FN domains and OPNs.

Integrin Mediated HSPC Adhesion to FN

Eleven different integrins with the ability to bind to FN have been described [52]. Eight of these bind to the 9th and 10th type III modules, which were contained in the FN domains we used. To investigate which integrins are responsible for HSPC binding to surfaces biofunctionalized with FN we used RGD to block all RGD-binding integrins, specific antibodies to block particular

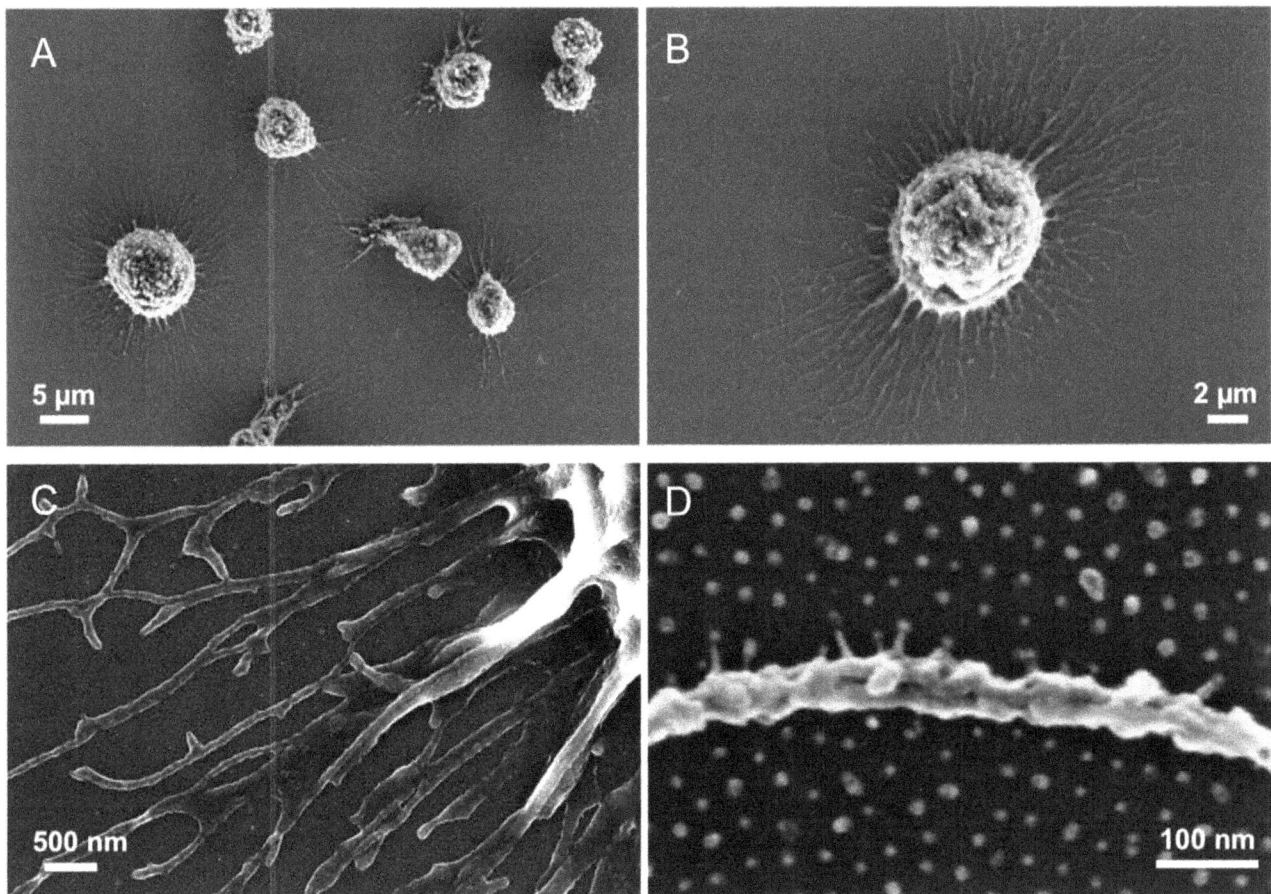

Figure 2. Cell morphology on nanostructured glass substrates. SEM images of critical point dried KG-1a cells on nanostructured, PEG-passivated, cRGD-functionalized substrates with interparticle distances of 36±7 nm. Magnification increases from A to D.

integrin chains (anti-β_1, anti-α_4 and anti-α_5) and an isotype control without specificity for the target cells.

The addition of soluble RGD prevented HSPC adhesion, confirming that HSPCs bind to the RGD motif in FN. Furthermore the addition of function-blocking anti-β_1 integrin antibodies also prevented HSPC adhesion to FN (Fig. 4A, C). Three of the RGD-binding integrins ($\alpha_5\beta_1$, $\alpha_V\beta_1$ and $\alpha_8\beta_1$) contain the β_1 chain [18,52], but only integrin $\alpha_5\beta_1$ is known to be expressed by HSPCs. This suggests that HSPCs bind to the RGD sequence in FN via the integrin $\alpha_5\beta_1$ receptor. Interestingly, the effect of α_5 integrin chain inhibition was weaker than β_1 integrin chain inhibition (Fig. 4B, C), indicating that additional β_1 integrins besides $\alpha_5\beta_1$ are involved in HSPC adhesion to FN. HSPCs also express $\alpha_4\beta_1$ integrin, which binds to the carboxy-terminal cell binding domain (containing the LDV motif) of FN [32]. The effect of the combined inhibition of α_4 and α_5 integrin chains was similar to the result of inhibiting only the β_1 chain, leading to similarly weak cell adhesion to FN. The contribution of each of the two α integrins to HSPC adhesion was donor-dependent (Fig. 4B). As observed for full-length FN, adhesion to the FNRGD domain could be blocked by the addition of β_1 integrin antibodies (Fig. S5).

Effect of FN-derived and OPN-derived Ligands on HSPC Proliferation and Differentiation

To investigate the influence of ligands and nanostructure on HSPC fate, colony forming assays were performed on adhesive cRGD- and non-adhesive cRGE-functionalized nanostructured PEG hydrogels. No effect of the ligand on the number or type of the formed colonies could be observed (Fig. S6). In order to identify ligands, which are potent in influencing HSPC proliferation and differentiation, we performed CFSE-proliferation experiments and differentiation analyses with continuous layers of ligands. In this setting the ligand density is much higher than on nanopatterned substrates and the resulting surfaces are better comparable to the ones used in other studies investigating the impact of ECM ligands on HSPCs [15,25,26,27,34].

Continuous gold surfaces were biofunctionalized with cRGD, cRGE, FNRGD, FNΔRGD or OPNs. Cell proliferation on these surfaces was compared to cells growing on unfunctionalized gold surfaces. Freshly isolated HSPCs were labeled with CFSE and incubated on the biofunctionalized gold surfaces. After 4 and 7 days of pre-culture on the different surfaces the amount of retained CFSE and CD34 protein expression were determined by flow cytometry. All 5 biofunctionalized surfaces gave similar results and were comparable to the gold control surface with regard to cell expansion (Fig. S7A, C) and CD34 expression (Fig. S7B, D).

Colony forming assays were performed after 7 to 10 days of pre-cultivation on the different surfaces to test the influence of the ligands on HSPC differentiation. No significant differences in the number or type of colonies were found (Fig. 5), indicating that the immobilized ECM-derived ligands had no impact on HSPC differentiation.

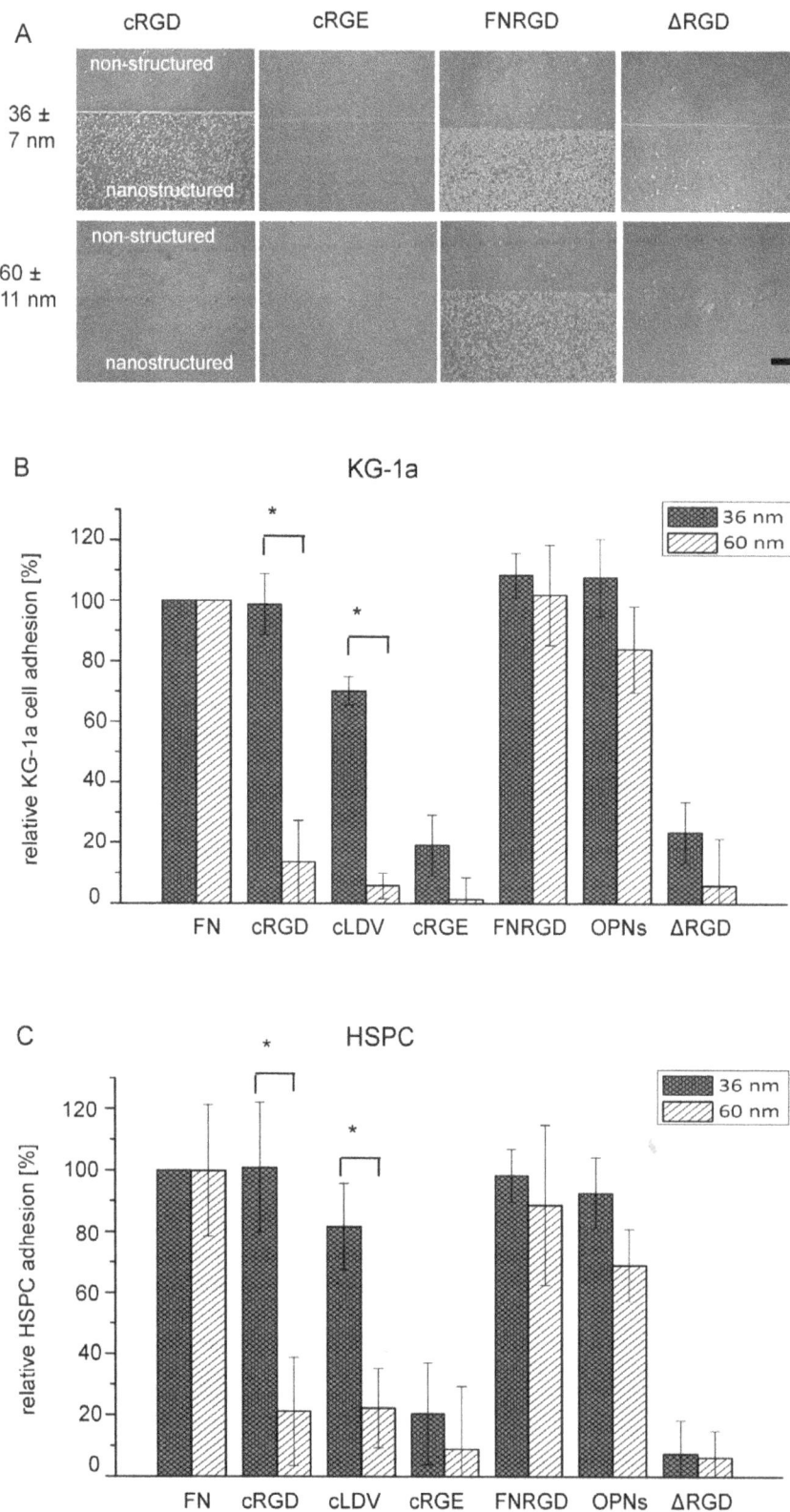

Figure 3. Adhesion of KG-1a cells and HSPC to nanopatterned, biofunctionalized PEG hydrogels. Cells were plated on hydrogels in adhesion media and evaluated by phase contrast microscopy or Cy Quant cell quantification after 1 h of incubation. (A) Microscopic images of KG-1a cells on biofunctionalized PEG hydrogels. Cells appear as bright spots on a gray background. The area in the lower part of each image is nanostructured and the upper area is unstructured (internal control). The name of the applied ligand is given above each image. Cell adhesion to NP arrays with a distance of 36 ± 7 nm or 60 ± 11 nm are shown in the upper and lower row of images, respectively. FNΔRGD is abbreviated with "ΔRGD". One representative experiment out of 4 is shown. Scale bar = 200 µm. (B) Relative quantification of KG-1a cell adhesion and (C) HSPC adhesion to different ligands on NP arrays with 36 ± 7 nm (filled columns) or 60 ± 11 nm (diagonally striped columns) spacing. On the y-axis the number of

adherent cells normalized to the value for adhesion to full-length FN is plotted. The different ligands are indicated on the x-axis. $N_{independent\ experiments} = 4$, each experiment carried out in technical duplicates; error bars = standard deviation of the mean; * = significant p value <0.05 in Wilcoxon rank sum test.

Effects of Ligand Type and Nanostructure on the Expression of THBS2

The investigated ligands and their nano-scale spacing influenced integrin-mediated cell adhesion. Quantitative RT-PCR screening for the expression of integrin ligands revealed thrombospondin-2 (THBS2) as a gene that is regulated in its expression by the FN-derived ligands and the lateral interligand distance.

HSPCs expressed THBS2 mRNA on blank, ligandless PEG hydrogels, which were unable to elicit integrin signaling (relative quantification value (RQ) set to 1, Fig. 6A). THBS2 expression was similar (RQ = 0.97) on FNΔRGD functionalized hydrogels with interparticle spacings of 35 ± 7 nm, but significantly lower (RQ = 0.16) on 35 ± 7 nm hydrogels functionalized with FNRGD. On 60 ± 11 nm FNRGD-functionalized hydrogels THBS2 expression was at an RQ of 0.58. This value lies between the relative expression on 35 ± 7 nm hydrogels functionalized with FNRGD and those functionalized with FNΔRGD, but is not significantly different to either value. This indicates that the lateral spacing of FNRGD and the presence or absence of integrin activation regulates THBS2 mRNA expression.

In a next step, these findings were verified on the protein level by immunofluorescence staining of THBS2 in HSPCs that were incubated on nanostructured hydrogels. THBS2 protein expression on nanostructured hydrogels was regulated similar to the mRNA expression (Fig. 6B, Fig. S8). The highest protein expression was found on blank or FNΔRGD functionalized hydrogels. On 35 ± 7 nm nanostructured FNRGD-functionalized hydrogels none or only minor THBS2 expression could be observed. Significantly more THBS2 protein expression was detected on FNRGD-functionalized hydrogels with interparticle distances of 65 ± 11 nm, and even more at 85 ± 14 nm. On these matrices the lateral distance between the FNRGD-functionalized particles was increased, resulting in a higher distance between the targeted cellular integrins. This shows that THBS2 protein expression was enhanced in the absence of the RGD sequence in FN and by increasing the lateral distance between integrins.

To reveal a possible function of THBS2 in the context of the HSC microenvironment, adhesion assays comparing cell surfaces bearing THBS2, FN or BSA were performed. HSPC adhesion was significantly higher on THBS2 (68% of all applied cells bound) and FN (88%) functionalized surfaces than on control surfaces

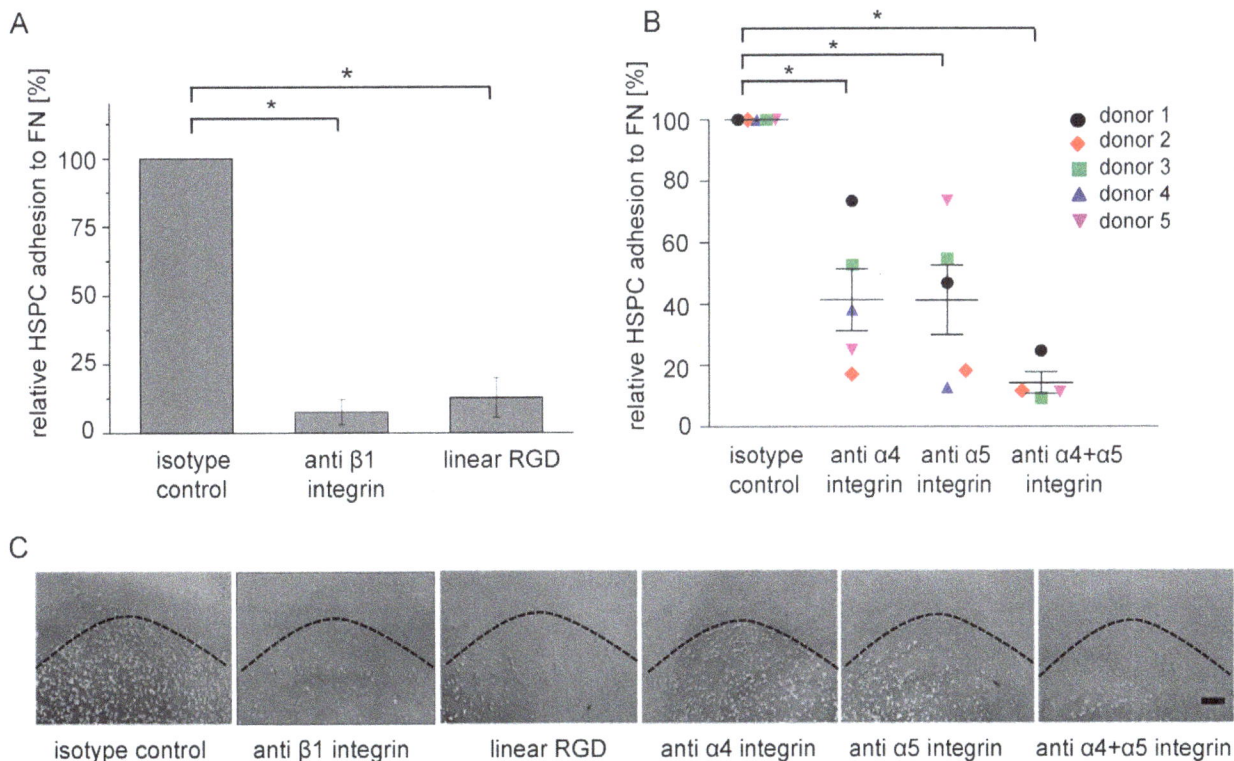

Figure 4. Integrin-mediated HSPC adhesion to FN. HSPCs were preincubated with integrin-specific antibodies or a linear RGD peptide for 1 h and plated onto an adsorbed FN spot. After nonadherent cells were removed, the spots were imaged and the adherent cells counted. (A) Relative HSPC adhesion to FN was either inhibited by preincubation with antibodies blocking the β1 integrin chain or with a linear RGD peptide. The isotype control (set to 100%) shows cell adhesion to FN after preincubation with isotype control antibodies. $N_{independent\ experiments} = 4$. (B) HSPC adhesion to FN was significantly reduced when inhibiting either the α_4 integrin chain, the α_5 integrin chain or both. $N_{independent\ experiments} = 5$. (A, B) Error bars = standard error of the mean; * = significant p value <0.05 in Wilcoxon rank sum test. (C) Representative microscopic images of HSPC adhesion to FN spots. Cells are visible as white spots in the lower part of each image, the dashed black line indicates the border of the FN spot. Scale bar = 200 μm.

A

B

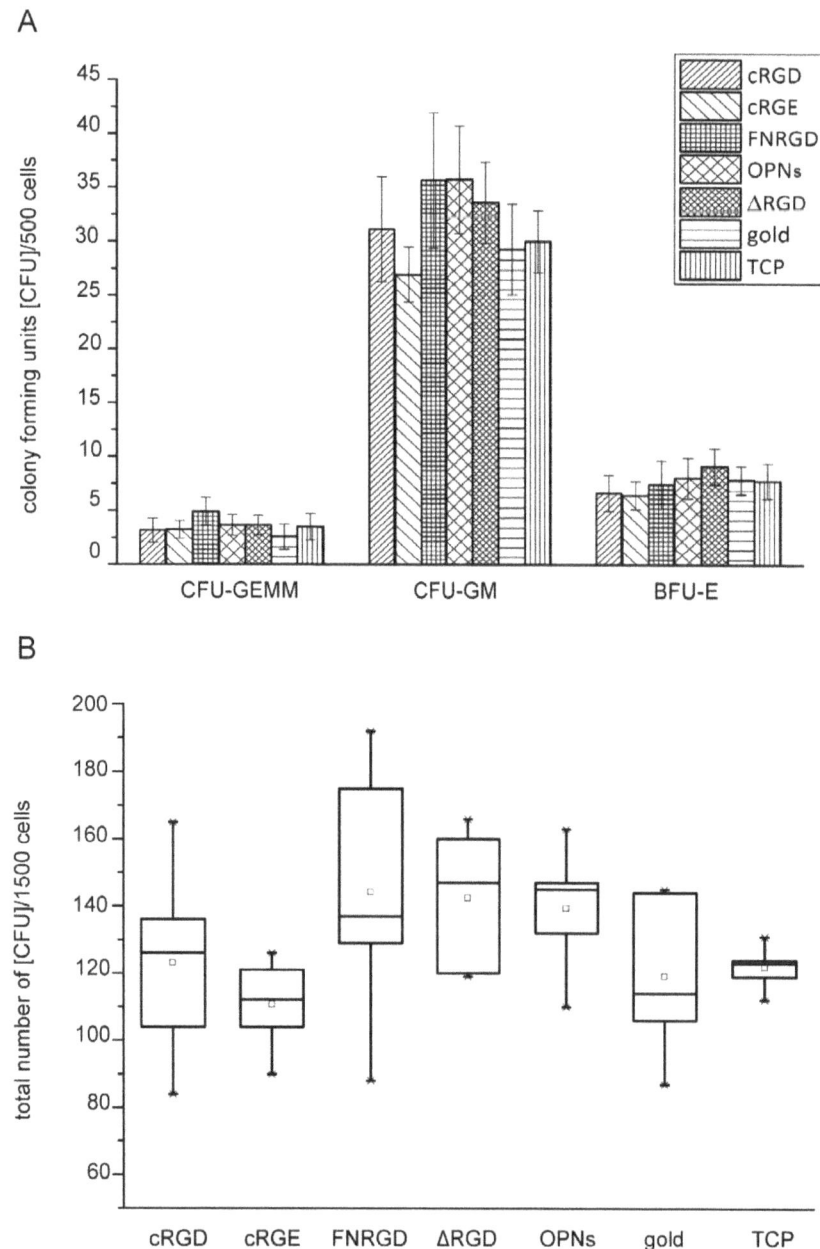

Figure 5. Determination of HSPC differentiation in colony forming assays. (A) Colony forming units of precultured HSPCs on glass slides biofunctionalized with different ligands. After 7–10 days preculture, colony forming assays were performed in triplicates. Colonies were distinguished into CFU-GEMM (*colony forming unit granulocyte, erythroid, macrophage, megakaryocyte*), CFU-GM (*colony forming unit granulocyte, macrophage*) and BFU-E (*burst forming unit erythroid*) after 2 weeks. $N_{independent\ experiments} = 5$ in technical triplicates; error bars = standard error of the mean. (B) Box Plot of the total number of colony forming units formed by 1500 precultured HSPCs on glass slides biofunctionalized with different ligands. $N_{independent\ experiments} = 5$; TCP = tissue culture plastic, gold = continuous gold film on glass, FNΔRGD is abbreviated with "ΔRGD".

with BSA (22%) (Fig. 6C). This indicates that THBS2 functions as an adhesion-mediating molecule for HSPCs.

Discussion

To mimic signals of the bone marrow ECM, three important features of the ECM have to be taken into consideration: (i) (bio)chemistry, (ii) mechanical properties and (iii) (nano)structure. In principle, all three properties can be controlled in the nanostructured hydrogel system we applied, making it highly suitable for the investigation of biomimetic cell-matrix interac-

tions. In the current study we focused on ligand identity and nanostructure. We determined that nanostructured hydrogels biofunctionalized with different ECM-derived ligands, all of which are present in the bone marrow HSC niche, can influence the adhesive behavior of HSPCs and KG-1a cells depending on the ligand type and nanostructured presentation. Furthermore, THBS2 expression by HSPCs was regulated by the presence of a functional FN-derived ligand as well as the nanometer-scaled interligand distance. We identified THBS2 as an adhesive protein for human HSPCs.

A

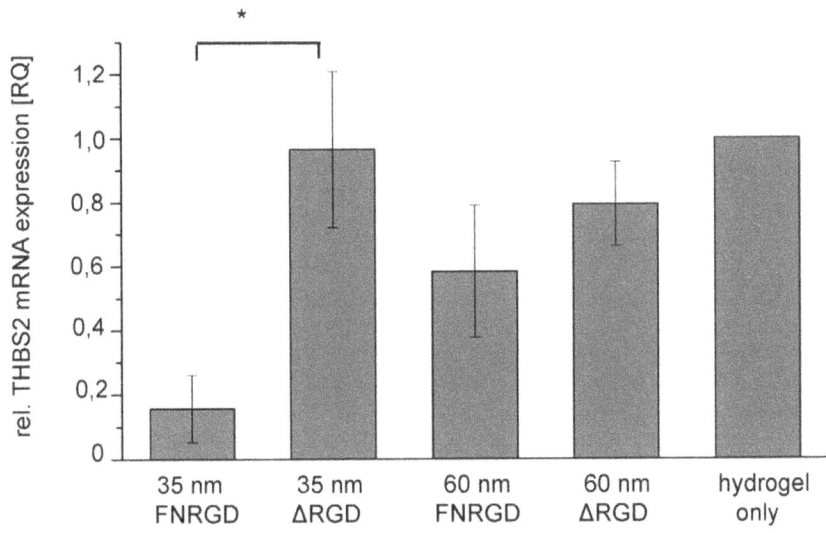

B

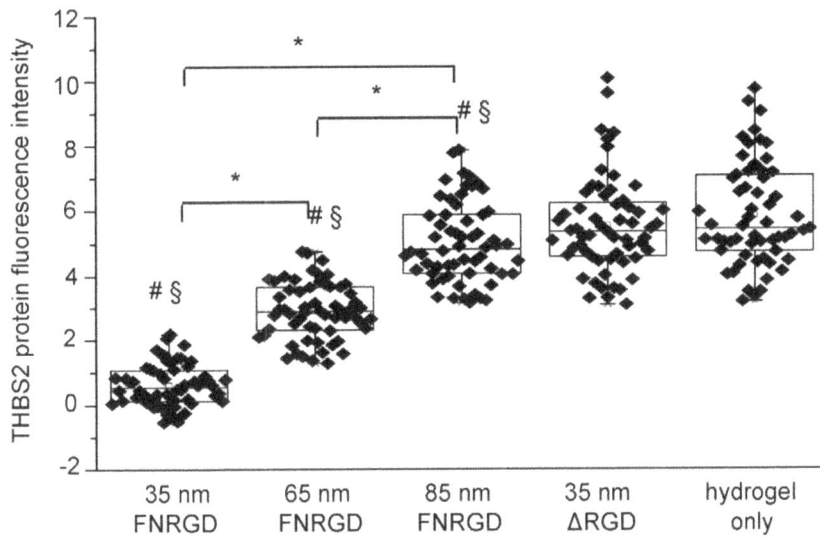

C

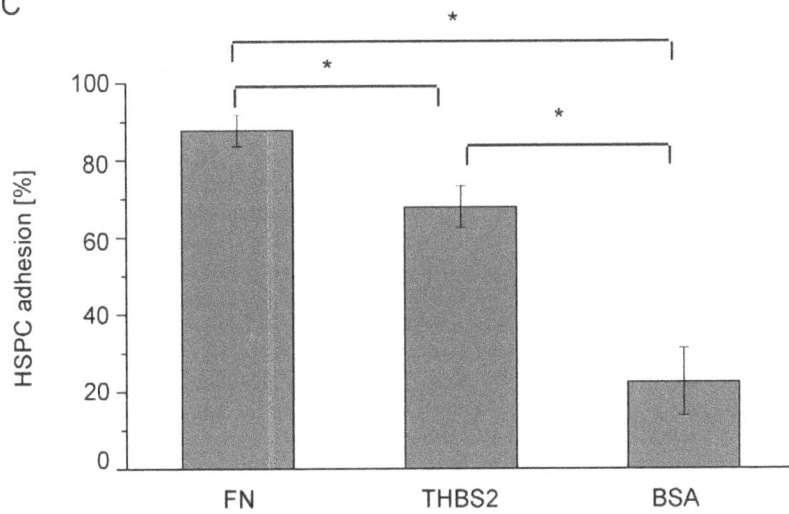

Figure 6. Influence of nanostructured matrices on THBS2 expression by HSPC. (A) Relative gene expression of THBS2 in HSPC incubated for 140 min on nanostructured PEG hydrogels biofunctionalized with FNRGD or FNΔRGD (abbreviated with "ΔRGD"). RQ values were normalized to the unstructured hydrogel controls and are plotted on the y-axis. The different functionalized and nanostructured substrates are indicated on the x-axis. $N_{independent\ experiments} = 4$; error bars = standard error of the mean; * = significant p value <0.05 in Wilcoxon rank sum test. (B) Immunofluorescence staining of THBS2 protein in HSPCs incubated for 13 h on nanostructured, biofunctionalized PEG hydrogels. Fluorescence intensity was measured per cell by applying Image J software and is plotted on the y-axis. $N_{cells} = 60$ from 3 donors (3×20); data are presented as box plots overlaid with individual data points (black squares/diamonds); * = significant p value <0.001; # = significant p value <0.001 to cells on hydrogel only; § = significant p value <0.001 to cells on 35±7 nm FNΔRGD (abbreviated with "ΔRGD") biofunctionalized hydrogels. Two-tailed, unpaired Student's t-test. (C) HSPC adhesion to full-length FN, BSA or recombinant THBS2 protein in percent of the total applied cell number per well. $N_{independent\ experiments} = 5$, each experiment carried out in technical triplicates; error bars = standard deviation of the mean; * = significant p value <0.01 in Wilcoxon rank sum test.

To date the *in vivo* niche is the only environment known which allows HSC proliferation under full maintenance of their stem cell potential. Therefore, biomimetic approaches for culturing HSPCs are promising and are pursued with different approaches [53]. Besides investigating the impact of medium composition and ligand coatings on HSPCs, the influence of environmental parameters such as matrix stiffness [37], or scaffold microstructure on HSPC proliferation and differentiation have been investigated. For example, spatial restriction of HSPCs in ECM-coated microcavities (15 to 80 μm in diameter) has been shown to support the maintenance of immature HSPCs in quiescence [17]. With our study we addressed structural elements one scale length smaller – the nanometer scale. Our finding that HSPCs are sensitive to nanostructural surface features is in accordance with previous studies showing that the expansion and adhesion of HSPCs depend on the nanoscale topography of nanofiber substrates [42] and that adhesion and lipid raft clustering of HSPCs are influenced by the nanostructured presentation of cRGD-ligands on PEG-passivated glass surfaces [36].

We provide evidence that the nanometer-scaled lateral distance of several different ECM-derived ligands is an important determinant of HSPC adhesion. The critical maximum interligand distance at which HSPC adhesion was supported depended on the ligand type (cRGD and cLDV <45 nm, OPN ~75 nm, FNRGD ~110 nm). The observed differences in tolerated distances between the peptides and the protein domains might be due to ligand size or the presence of synergy sequences. The ligands (immobilized on the NPs) are flexible, rather than rigid, structures. The FNRGD domain, with an estimated length of 12–14 nm, might be able to "shorten" the distance between two NPs by bending towards each other, and thereby enabling integrins to cluster. In contrast, the peptide length of the cRGD peptide (estimated at ~ 3.5 nm) limits their ability to converge and enable integrin binding. Additionally, the protein domains provide synergy sequences such as PHSRN that are able to enhance the adhesive capacity of the ligand [31].

The lateral cRGD distance sufficient for KG1a and HSPC adhesion was different to that found in previous studies using REF52-fibroblasts, 3T3-fibroblasts, MC3T3-osteoblasts and B16-melanocytes [50,54,55]. One possible explanation may be differences in the adhesive behavior of these cell types, related to whether they are anchorage-dependent (such as the cells used in former studies) or non-anchorage-dependent cells (such as the cells of the current study). Cell size, morphology and signaling vary greatly between the previously investigated anchorage-dependent and the non-anchorage-dependent hematopoietic cells and may contribute to differences in cell adhesion to nanopatterned surfaces. The molecular mechanisms underlying the cellular sensitivity to lateral ligand spacing at the nanometer scale (leading to differences in cell adhesion) are thought to depend on the composition and action of the protein network involved in the formation of the adhesion plaque and its interaction with the actin

cytoskeleton [56]. While the integrin-mediated focal adhesion sites of anchorage-dependent cells such as fibroblasts and osteoblasts are well studied, relatively little is known about these adhesions in HSPCs. Nevertheless, it is clear that the integrin-linked multiprotein complexes of HSPCs and anchorage-dependent cells differ from each other, e.g., in the principal adhesion kinase, which is Pyk2 in HSPCs and FAK in anchorage-dependent cells [57,58]. Furthermore, the actin cytoskeleton displays fundamental differences with its fibrous appearance in adherent and spreading anchorage-dependent cells and a ring-like structure in hematopoietic cells that keep more or less their round shape also upon adhesion to a solid support. Such differences in the composition of the integrin-linked multiprotein complex, the central signaling molecules such as kinases and the cytoskeleton, might cause the cell type dependence of the maximum tolerated ligand spacing. However, elucidating the mechanisms underlying the observed differences remains an ongoing challenge.

In our previous study using KG-1a cells and HSPCs, the maximum tolerated distance for adhesion to cRGD functionalized, nanostructured, PEG-passivated glass surfaces was determined at 32±6 nm [36]. In the present study, the maximum tolerated cRGD spacing for KG-1a cells on cRGD-functionalized PEG hydrogels was observed between 37±7 nm and 45±8 nm. Such variances may be a result of differences in the physical and chemical properties of the applied substrates, such as organization of the PEG. For the two different nanostructured surface types (passivated glass and hydrogel) passivation against unspecific protein adsorption and cell adhesion was achieved by using PEG as a background. However, different PEG molecules that vary for example in the chain length were used for the two surface types ($M_r = 700$ for hydrogels and $M_r = 2000$ for the PEG passivation layer on glass). These differences in chain length could lead to different molecular conformations of the PEG [59]. While the PEG molecules within the PEG monolayer on the glass are highly ordered, presenting a terminal methyl group at the surface [60], the PEG molecules in the hydrogels are randomly orientated, which means that on average all structural elements of the PEG molecules are exposed at the hydrogel surface to an equal extent. Such alterations in PEG molecules, orientation and cross-linking result in different chemical structures that are exposed to the cells. These chemical dissimilarities plus additional factors with the potential to influence cell adhesion such as the surface charge [61], hydrophobicity and surface roughness [62] might contribute to the slightly different HSPC adhesive behavior on nanostructured hydrogels and passivated glass surfaces. It was described previously that material stiffness influences cell-adhesion and stem cell proliferation, differentiation or self-renewal [37,63,64,65]. We could recently show that HSPCs are mechanosensitive to hydrogels with Young's moduli below 100 kPa [38]. This is in line with other reports on the biologically relevant stiffness range [63]. Since the Young's moduli of the applied PEG hydrogels (6 MPa [46]) and of glass (64 GPa [66]) are much higher, and lie in a range in which cells are not mechanosensitive, we conclude that both substrate types appear very stiff to cells and that the mechanical properties do not contribute to the observed small differences. In summary, our adhesion studies demonstrate that the cellular sensitivity to nanometer scale lateral spacing depends on (i) the cell type, (ii) the targeted receptor and (iii) the provided ligand.

Hematopoietic cells were in contact with the nanostructured matrices via filopodia, but at the same time kept their round morphology. Because low conductivity and the swelling behavior of PEG hydrogels represent limitations for imaging cells on hydrogels by electron microscopy, higher resolution images were obtained on nanopatterned, PEG-passivated glass slides using scanning electron microscopy. The results show that cells were in contact with the immobilized cRGD ligands on the gold NP arrays, but did not bind to the PEG layer between the NPs. This is in line with previous findings for other cell types, e.g., fibroblasts [41].

HSPC adhesion to FN was mediated by a donor-dependent combination of $\alpha_4\beta_1$ and $\alpha_5\beta_1$ integrins. The integrin receptors $\alpha_4\beta_1$ and $\alpha_5\beta_1$ are important for mediating adhesion as well as for migration and homing of HSCs in the niche and for cell survival in co-culture with osteoblasts [67,68]. It has been shown before that HSPC binding to FN is mediated through integrin $\alpha_4\beta_1$ (which binds to the LDV motif) and through integrin $\alpha_5\beta_1$ (which binds to the RGD motif) [69,70]. In these studies integrins were additionally activated through antibodies or cytokines, which can both have an impact on cell adhesion. In the current study adhesion to FN was investigated only in the presence of ions that are necessary for integrin activation. However, our result differs from a previous study, in which HSPC adhesion to FN could be prevented by inhibiting only the integrin $\alpha5$ chain in similar experiments with integrin blocking antibodies [15]. Possible explanations for this discrepancy might be differences in the study design including the immobilization of FN, the source and isolation of HSPCs (CD133$^+$ cells from peripheral blood of G-CSF-treated healthy volunteers [15] and CD34$^+$ cells from umbilical cord blood in the present study) and the individual variability from donor to donor.

Blocking the RGD recognition site by pre-incubation with RGD peptides was sufficient to completely block cell adhesion to FN. This indicates that the presence of the RGD molecule within FN is essential for HSPC adhesion. RGD also binds to $\alpha_4\beta_1$ integrins, in addition to $\alpha_5\beta_1$ integrins, at higher concentrations [71]. This may explain the complete loss of cell adhesion, including adhesion to the LDV motif.

The donor-dependent differences that we found when inhibiting cell adhesion to FN by blocking α_4 or α_5 integrin may be the result of different integrin expression levels in cells derived from different donors. In HSPCs isolated from human cord blood $\alpha_v\beta_3$ integrin is not expressed or only expressed in very low amounts [38]. In accordance with this finding, we did not find evidence for $\alpha_v\beta_3$ integrin involvement in HSPC adhesion to FN.

Immobilized FN-derived and OPN-derived ligands did not affect HSPC proliferation or differentiation. This is in contrast to previous publications that showed a negative influence of OPN on murine HSPC proliferation [33,34]. One possible explanation for this discrepancy may be that the OPN constructs used in those studies were different to the thrombin-cleaved OPN fragment that we applied. Another possible explanation might be the intrinsic differences between the murine and the human system. Both positive and negative influences on HSPC proliferation have been described for FN [17,25,26,27]. In these studies the full-length proteins were immobilized on substrates in random orientation,

which is in contrast with our study where the FN domain type III 7–10 was offered in an oriented manner. Possible explanations for divergent findings concerning the influence of FN on cell proliferation are: (i) different surfaces (plastic, glass or gold) might lead to different conformations of FN, which can have an impact on cell proliferation and differentiation [28], (ii) differences in the orientation of the immobilized ligands and (iii) the availability of additional cell interaction sites on the full-length protein in comparison to the much shorter domain. Since, as we could show, $\alpha_5\beta_1$ integrin mediates cell binding to the FN domain type III 7–10 and this domain does not affect HSPC differentiation, we conclude that in our setup $\alpha_5\beta_1$ integrin had no impact on HSPC differentiation. This is in consensus with other publications showing that the loss of β_1 integrin in HSCs does not influence differentiation into blood cells [22]. Yokota et al. did not find any influence of the FN domains type 8–10 on HSPC differentiation, which is also in good agreement with our findings [24].

Interestingly, we found that both THBS2 mRNA and protein are expressed by human HSPCs in vitro. We determined the presence of (or lack of) the RGD sequence within the FN domains type III 7–10 to influence THBS2 expression in adhering HSPCs, suggesting that THBS2 expression is regulated by $\alpha_5\beta_1$ integrin-mediated signaling pathways. Integrin signaling induced by FN domain type III 7–10 led to a reduction in THBS2 expression. Furthermore, we could show that THBS2 expression depends on lateral ligand spacing and, in consequence, on lateral integrin clustering. THBS2 expression increased with increasing distance between FNRGD-functionalized NPs and the respective integrin receptors. Future investigations will reveal whether the defined lateral distances or a FN dose-dependent mechanism trigger THBS2 expression.

THBS2 is an ECM protein secreted by fibroblasts, osteoblasts and mesenchymal stem cells, and has, to our knowledge, not been described in HSCs as of yet [72,73]. It acts as a matricellular protein that interacts with cell surface receptors such as integrins and with other extracellular components including matrix proteins, growth factors and proteases [74,75]. THBS2 knockout mice showed connective tissue abnormalities, such as disordered collagen fibrillogenesis, abnormal bone formation with increased total density, cortical thickness of long bones, and a reduced marrow cavity [76]. THBS2 has been shown to inhibit marrow stromal cell proliferation, which indicates a function in the bone marrow microenvironment [72].

In the present study, we show that HSPCs can express THBS2 and are able to adhere to it. This indicates that THBS2 is involved in HSPC regulation and that HSPCs can influence their ECM composition. Furthermore, HSPC adhesion to THBS2 suggests that, in addition to having paracrine effects on the surrounding cells and matrix, THBS2 also has autocrine effects on HSPCs themselves. One possible explanation for the FN-dependent THBS2 expression pattern might be that HSPCs produce the matrix protein in the absence of or at low FN concentrations (such as on a non-adhesive PEG-background), in order to generate a surrounding matrix. Since THBS2 can interact with other ECM proteins to organize the ECM and can also influence other niche cells such as mesenchymal stem cells [72] and osteoblasts [77], it is likely that HSPCs can actively modulate their environment. This hypothesis is supported by previous findings, which describe a regulation of the niche by HSPCs [2,3]. Here we report that HSPCs can directly influence the composition and organization of the ECM, and thereby have the ability to modulate the niche by a mechanism that functions independently of cell-cell contacts.

In conclusion, we show that HSPCs are sensitive to the nanostructured presentation of several ECM derived ligands.

Although no influence of FN-derived or OPN-derived ligands on HSPC proliferation and differentiation was observed, THBS2 was expressed by HSPCs in an integrin-mediated, ligand-dependent and nanostructure-dependent manner. In addition, THBS2 mediates HSPC adhesion. Thus, we conclude that under unfavorable conditions, where ligands needed for adhesion are absent or spread too far apart from each other, HSPCs are able to modulate their environment and actively participate in the formation and regulation of the HSC niche.

Materials and Methods

Nanopatterning

Glass slide nanopatterning by block-copolymer micelle nano-lithography (BCML) was performed as previously described [45,78]. With this technique quasi-hexagonally ordered nanoparticle arrays with lateral distances between gold NPs ranging from 20 ± 6 to 110 ± 18 nm were produced. The parameters of the nanopatterning processes for the different surfaces are given in Table 1. The gold nanopatterns were transferred to PEG diacrylate (M_r 700; PEG 700 DA) hydrogels according to a protocol published previously [46,79]. The applied PEG 700 DA hydrogels exhibit a Young's modulus of 6 MPa [46]. For high resolution microscopy, glass substrates were passivated with a protein repellent layer of PEG triethoxysilane that prevents unspecific protein adsorption to the glass [60], as described recently [36].

Protein Expression and Purification of Recombinant Proteins

The pET15b plasmids encoding for the fibronectin domain type III 7–10 with the RGD sequence (FNRGD) and the deletion mutant (FNΔRGD) were kindly provided by Prof. Dr. R. Fässler (Max Planck Institute for Biochemistry, Martinsried, Germany). A pET15b plasmid encoding amino acids 17–168 of osteopontin (the N-terminal thrombin fragment, hitherto referred to as shortened OPN or OPNs) was obtained from GenScript (Piscataway, NJ, USA). The *E.coli* strain BL21 (Invitrogen) was used for protein expression, and expression was induced according to a protocol described by Studier [80]. Bacterial cell pellets were resuspended in PBS and lysed by repeated freezing and thawing cycles and ultrasonic sound treatment. The N-terminal His-tagged fibronec-

tin type III 7–10 (containing the RGD sequence) and the deletion mutant FNΔRGD as well as OPNs were isolated and purified using a HisTrap™FF chromatography column containing pre-charged Ni-sepharose (GE Healthcare, Uppsala, Sweden) according to the instructor's manual. Eluates were desalted using PD-10 desalting columns (GE Healthcare) to remove all imidazole compounds and finally eluted with PBS. The purity of all recombinant proteins was controlled by silver stain analysis.

Biofunctionalization

Nanostructurally predefined hydrogels were functionalized with short peptide motifs (cLDV, cRGD or cRGE – a non-adhesive control peptide), larger protein domains (the N-terminal thrombin cleaved fragment of osteopontin or a cell-binding fibronectin domain containing the type III modules 7–10– the latter domain with or without the crucial RGD sequence) and full-length fibronectin. Details are given in Table 2. Prior to functionalization, nanostructured hydrogels were sterilized in 70% ethanol and washed. Biofunctionalization with FN-derived short peptide motifs (25 µM in aqueous solution) was achieved by incubating at room temperature for 2 h. Functionalization with His-tagged recombinant protein domains (12.5 µM solution in PBS) was accomplished using an NTA-thiol linker system, as described elsewhere [51,81]. Hydrogels were washed rigorously with PBS (3 times for 20 min) to remove unbound ligands. Before cell experiments were performed, biofunctionalized hydrogels were equilibrated for 30 min in cell-specific media under standard cell culture conditions.

Full-length FN protein was isolated from human plasma as described elsewhere [82] and adsorbed to nanostructured hydrogels for relative comparison purposes.

Immunofluorescence Staining of Hydrogels

Hydrogels functionalized with FN domains (FNRGD or FNΔRGD) were incubated with 10 µg/ml mouse-anti-human FN (Clone FN12-8, QED Bioscience Inc., San Diego, USA) overnight at 4°C. After washing with PBS the hydrogels were incubated with a secondary goat-anti-mouse IgG1 Alexa Flour 488 antibody (Invitrogen, Darmstadt, Germany) for 45 min. Because EDTA forms a complex with Ni^{2+} and other cations necessary for successful formation of NTA-His-tag complexes, one control was incubated with 25 µM EDTA instead of nickel chloride, hindering protein coupling via the His-tag. A second control was incubated without the primary antibody to test for unspecific antibody binding. The hydrogels were imaged using an Axiovert 200 M microscope (Zeiss, Jena, Germany).

Cell Culture

The human acute myeloid leukemia cell line KG-1a (DSMZ, Braunschweig, Germany), cultured in RPMI, 20% FBS (Invitrogen or Sigma, Taufkrichen, Germany) and 1% (v/v) penicillin/streptomycin (P/S, Gibco, Darmstadt, Germany) under standard cell culture conditions, was used as a model cell line for immature hematopoietic cells. Primary human CD34+ HSPCs were isolated from umbilical cord blood with a CD34 magnetic bead system (Miltenyi Biotec, Bergisch-Gladbach, Germany) according to the manufacturer's instructions and maintained in SFEM Media supplemented with 1% (v/v) cytokine mix cc100 (both Stemcell Technologies, Grenoble, France) and 1% (v/v) P/S at 37°C and 5% CO_2. Purity of the isolated cells was controlled by flow cytometry using CD34-PC5 or PC7 antibodies (Clone 581, Beckman Coulter, Krefeld, Germany) and the respective isotype controls (Beckman Coulter) on a Cytomics FC500 flow cytometer

Table 1. Parameters for producing nanostructured surfaces by BCML.

PS(x)-b-P2VP(y)	C [mg/ml]	V [mm/min]	d [nm]
PS(154)-b-P2VP(33)	5	24	20±6
PS(240)-b-P2VP(143)	5	24	30±6
	5	18	35±7, 36±7 or 37±7
	4	10	45±8
PS(1056)-b-P2VP(495)	5	30	60±11
	5	24	65±11
	4	24	75±13
	4	12	85±14
	3	16	110±18

PS: polystyrene units, P2VP: poly(2-vinylpyridine) units, C: concentration of the polymer, V: substrate retraction velocity, d: distance between gold NPs ± standard deviation.

Table 2. Short peptide binding motifs and protein domains used for biofunctionalization of gold NP arrays.

Name	Sequence/Description	Molecular weight
cRGD	Cys-PEG$_6$-Nε(Lys-Arg-Gly-Asp-D-Phe)$_{cyclo}$	1042 Da
cRGE	Cys-PEG$_6$-Nε(Lys-Arg-Gly-Glu-D-Phe)$_{cyclo}$	1056 Da
cLDV	Cys-PEG$_6$-Nε(Lys-Leu-Asp-Val-D-Phe)$_{cyclo}$	1046 Da
FNRGD	FN domain type III modules 7–10 N-terminal His-tag	40077 Da
FNΔRGD	FN domain type III modules 7–10 ΔRGD N-terminal His-tag	39731 Da
OPNs	OPN domain (amino acid 17–168) N-terminal His-tag	19426 Da

Peptides were purchased from Biosyntan, Berlin, Germany or synthesized by Dr. Hubert Kalbacher (University of Tübingen, Germany).

(Beckman Coulter). They were only used if more than 95% of the cells were CD34 positive.

Ethics Statement

Umbilical cord blood was obtained from the DKMS umbilical cord blood bank in Dresden and the University Hospital of Tübingen after written and informed consent of the parents and approval by the local ethics committee (Ethik-Kommission der Medizinischen Fakultät und am Universitätsklinikum Tübingen, project numbers 120/2012BO2 and 005/2012B02).

Scanning Electron Microscopy (SEM) of Cells on Nanostructured Surfaces

Adherent cells on PEG hydrogels or PEG-passivated glass slides were fixed with 4% paraformaldehyde (PFA; Alfa Aesar, Karlsruhe, Germany) in PBS for 20 min at room temperature. The aqueous liquid was exchanged with ethanol in a series of washing steps with increasing ethanol concentrations. The samples were critical point dried (Critical Point Drying Device Leica EM CPD030 Leica Microsystems, Wetzlar, Germany), coated with carbon (Modular High Vacuum Coating System MED 020, Leica Microsystems) and sample images were taken with an Ultra 55 field emission electron microscope (Zeiss). Hydrogel cryo SEM images were obtained under low temperature conditions (T ≈ −130°C). To cool the samples in liquid nitrogen and transfer them into the SEM chamber a BAL-TEC VLC 100 (BAL-TEC AG, Balzers, Lichtenstein) shuttle and a BAL-TEC MED 020 (BAL-TEC AG) loading device were used.

Cell Adhesion to Nanostructured Surfaces

2×10^6 KG1-a cells were incubated on biofunctionalized hydrogels in adhesion medium [RPMI supplemented with an ion mix (final concentration 1 mM CaCl$_2$, 1 mM MgCl$_2$, 25 μM MnCl$_2$)] at 37°C and 5% CO$_2$ for 1 h. The ions are essential for integrin activation [83]. After washing the hydrogels twice with PBS, surface images were taken at the interface between the nanostructured and the unstructured area using an Axiovert 40 CFL microscope (Zeiss).

Quantification of cell adhesion to different ligands on differently nanostructured hydrogels required measurements in a multiwell format. For this purpose, the nanostructured hydrogels were produced identical in size to a standard 24 mm x 60 mm microscope slide. The surface of these gels was divided into different growth areas (each one identical in size to a well of a 96-well plate) by using the flexiPERM system (Greiner Bio-One, Frickenhausen, Germany). After biofunctionalization and a washing step, 2×10^4 KG-1a cells or 1×10^4 freshly isolated HSPCs suspended in adhesion medium were added per well and

incubated for 1 h at 37°C and 5% CO$_2$. Nonadherent cells were carefully removed with the medium and the remaining adherent cells were stained using the Cy Quant NF Cell Proliferation Assay Kit (Invitrogen). Fluorescence intensity, which is proportional to the number of stained cells, was detected at 480 nm. The relative fluorescent intensity of each sample was normalized to the value obtained for cell adhesion to the adsorbed full-length FN protein for comparison. The fluorescence intensity of wells with cells on unstructured, not functionalized hydrogels was set as background fluorescence and subtracted from all other values.

Integrin Inhibition Adhesion Assay

One drop (1 μl) of 1 mg/ml FN was allowed to air dry on a tissue culture plastic dish. Unspecific cell binding to the dish was prevented by blocking with 1% BSA (albumin bovine fraction V, Serva Electrophoresis, Heidelberg, Germany) in PBS for 1 h and washing with PBS. For integrin inhibition assays, HSPCs were pre-cultured for 20 h, washed with PBS and incubated with the respective antibodies for 1 h at 37°C (details on the applied antibodies can be found in Table 3). Thereafter, the cells were distributed on to the prepared surfaces and incubated in adhesion medium for 1 h at 37°C and 5% CO$_2$. Finally, the dishes were carefully rinsed 3 times with PBS to remove unbound cells. The remaining adherent cells were imaged (Axiovert 40 CFL microscope). The cell number on each FN spot was determined by manual counting.

Thrombospondin Adhesion Assay

To quantify cell adhesion to THBS2 a 96-well plate was coated with 10 μg/ml THBS2 (recombinant human THBS2, R&D Systems, Wiesbaden, Germany), 10 μg/ml FN or BSA and incubated for 12 h at 4°C. To enhance its adhesive activity, THBS2 was reduced with 20 mM dithiothreitol [84] in PBS for 20 min, followed by 3 washing steps. 2.5×10^4 HSPCs (pre-cultured for 20 h) were applied per well and incubated for 1 h at 37°C and 5% CO$_2$ in adhesion media. The wells were carefully washed with PBS and the remaining adherent cells were stained using the Cy Quant NF Cell Proliferation Assay Kit, as described above. The reference value (100% value) was determined by staining 2.5×10^4 HSPCs.

Proliferation

A 5 nm adhesive titanium layer was sputtered on a glass coverslip using the Modular High Vacuum Coating System MED 020 (Leica Microsystems). An additional 50 nm gold layer was sputtered on top. Biofunctionalization of these continuous gold surfaces (same procedure as described for nanostructured substrates) resulted in the densest achievable packing of ligands.

Table 3. Antibodies for integrin inhibition assays.

Antibody	Clone	Concentration	Supplier
Mouse-anti-human CD29-RD1 (β_1 integrin)	4B4LDC9LDH8	1:20	Beckman Coulter
Mouse-anti-human CD49d (α_4 integrin)	2B4	5 µg/ml	R&D Systems
Goat-anti-human CD49e (α_5 integrin)	polyclonal	20 µg/ml	R&D Systems
IgG1-PE (isotype control)	679.1Mc7	1:5	Beckman Coulter

Isolated HSPCs were stained with CFSE (3.5 µM) for 10 min at 37°C in PBS with 0.1% FBS. Staining was stopped by incubating in PBS with 10% FBS for 5 min on ice. After washing with PBS, cells were cultured on the prepared sterilized surfaces in SFEM medium with 1% (v/v) cytokine mix cc100 and 1% (v/v) P/S. After 4 and 7 days cells were counted and stained with CD34-PC7 or IgG1-PC7 antibodies (Beckman Coulter, Krefeld, Germany) for 30 min on ice and analyzed with a FC 500 flow cytometer (Beckman Coulter).

Determination of HSPC Differentiation Using the Colony-forming Cell Assay

After the proliferation assay, an aliquot of the *in vitro* proliferated HSPC was subjected to differentiation analysis by colony-forming cell assay. This assay allows the enumeration and classification of formed colonies according to their morphology. 1500 cells were washed, resuspended in 300 µl IMDM media and vortexed with 3 ml MethoCult H4434 Classic methylcellulose-based Media (both Stemcell Technologies) supplemented with 1% (v/v) P/S. 1.1 ml of this mixture were plated in triplicates in 35 mm petri dishes and incubated for 13 days at 37°C and 5% CO_2. The different types of colonies were determined according to the "Atlas of human hematopoietic colonies from cord blood" (Stemcell Technologies) using an Axiovert 40 CFL microscope.

In order to study the influence of nanostructurally presented ligands on HSPC differentiation 500 freshly isolated HSPCs in 100 µl IMDM media were applied onto nanostructured, functionalized PEG hydrogels in 35 mm petri dishes and incubated for 1 h at 37°C and 5% CO_2 in a wet-chamber. Then 1 ml MethoCult H4434 classic methylcellulose-based media supplemented with 1% (v/v) P/S was added and the plates were incubated for 13 days at 37°C and 5% CO_2.

Real-time Reverse Transcription Polymerase Chain Reaction (qRT-PCR)

2×10^5 freshly isolated HSPCs were incubated on nanostructured, biofunctionalized hydrogels for 140 min in SFEM supplemented with 1% (v/v) P/S and the ion mix (1 mM $CaCl_2$, 1 mM $MgCl_2$, 25 µM $MnCl_2$) at 37°C and 5% CO_2. After removing the cells from the hydrogels, mRNA was isolated using the RNeasy MicroKit (Qiagen, Hilden, Germany) according to the manufacturer's protocol. 40 ng mRNA were reverse transcribed into cDNA applying the High Capacity cDNA Synthese Kit (Applied Biosystems, Darmstadt, Germany). After 10 cycles of cDNA pre-amplification using the TaqMan PreAmp Master Mix Kit (Applied Biosystems), quantitative real time PCR was performed with the TaqMan system and comparative $\Delta\Delta CT$ method (Applied Biosystems 7500 System). The relative quantification (RQ) values were calculated using the 7500 v2.0.1 software (Applied Biosystems; $RQ = 2^{-\Delta\Delta Ct}$) and give the change in expression of the test sample relative to the calibrator sample

(fold change). Primers and probes for human THBS2 (Hs011568063_m1) and the endogenous control human beta-2-microglobulin (B2M) were purchased from Applied Biosystems. The relative THBS2 gene expression was normalized to the value obtained for cells incubated on unstructured, unfunctionalized hydrogels. Undetectable gene expression was set to 0. To exclude unspecific gene amplifications, the size of the PCR product was controlled by agarose (Biozym, Hessisch Oldendorf, Germany) gel electrophoresis.

Immunofluorescence Staining of HSPCs

After incubation of HSPCs on nanostructured, biofunctionalized hydrogels for 13 h in SFEM supplemented with 1% (v/v) P/S and the ion mix at 37°C and 5% CO_2, the cells were removed from the hydrogels, centrifuged and resuspended in PBS. The cells were spinned onto a microscope slide using a cytospin (Cellspin II, Thermac, Waldsolms, Germany) and fixed with 4% PFA. After washing with PBS and permeabilisation with 0.1% Triton X100 (Carl Roth, Karlsruhe, Germany), the cells were stained with 2.5 µg/ml mouse-anti-human THBS2 antibody (clone 230927, R&D Systems) for 2 h, washed with PBS and incubated with 2.5 µg/ml secondary antibody goat-anti-mouse IgG1 Alexa Fluor 488 for 1 h. After washing with PBS, slides were sealed with mounting media (Prolong Gold antifade reagent with DAPI, Life Technologies, Darmstadt, Germany) and imaged with the Axiovert 200 M microscope. The fluorescence intensity of THBS2 (Alexa Flour 488) was analyzed with Image J Software (http://rsb.info.nih.gov/ij/National Institutes of Health, USA). The fluorescence of the negative controls (stained with the secondary antibody only) was set as background fluorescence and subtracted from the other values.

Statistical Analyses

Each experiment was independently repeated 3 to 5 times. For statistical analyses the Wilcoxon rank sum test or the two-tailed unpaired Student's t-test were performed using the MATLAB (The MathWorks, Inc, Natick, MA, USA) software and Microsoft Office Excel (Redmond, WA, USA), respectively. P values were regarded as statistically significant when smaller than 0.05. The type of test, exact numbers of independent experiments (N) and technical replicates in each experiment are given in the respective figure legends.

Supporting Information

Figure S1 Schematic representation of FN and FN-derived ligands applied in the present study. (A) Modular organization of the FN monomer: FN contains type I (green), type II (red) and type III (magenta) modules. The variable region containing the LDV sequence is shown in blue. The FN type III domains 7–10 containing the RGD and the PHSRN (a cell-binding domain that activates integrins) sequences are enlarged.

The FN type III 7–10 domain carries an N-terminal His-tag for biofunctionalization purposes. (B) The cRGD peptide with a PEG linker and a terminal cysteine. The thiol group of the cysteine side chain binds to a gold NP during biofunctionalization.

Figure S2 Cell morphology on nanostructured PEG hydrogels. SEM images of critical point dried KG-1a cells on nanostructured, cRGD-functionalized hydrogels with interparticle distances of 30 ± 6 nm. Magnification increases from A to C.

Figure S3 KG-1a cell adhesion to cRGD functionalized hydrogels with different nanoparticle distances. Microscopic images of the border between the structured (bottom) and the unstructured (top) part of the nanostructured, cRGD functionalized hydrogels are shown. The distances between the gold NP on the different substrates are depicted above the pictures. Cells can be observed as bright spots on a grey background. Scale bar = 200 μm.

Figure S4 Microscopic images of KG-1a cell adhesion to nanostructured hydrogels. The hydrogels were biofunctionalized with (A) FNRGD and (B) OPNs protein domains. NP distances are indicated above the panels. The images were taken at the border between the structured and the unstructured part of the substrates. One of 5 (A) or 3 (B) representative experiments is shown. Scale bar = 200 μm.

Figure S5 Microscopic images of HSPC adhesion to FNRGD spots. Adhesion to the FNRGD domain (left) was inhibited by addition of a function-blocking $\beta 1$ integrin antibody (right). Cells appear as bright spots on a dark background.

Figure S6 HSPC differentiation on nanostructured hydrogels. Differentiation of HSPCs on nanostructured hydrogels (37 nm) functionalized with two different peptide ligands. $N_{independent\ experiments} = 3$, error bars = standard deviation of the mean.

Figure S7 HSPC proliferation assays. (A) Cell proliferation was measured on day 4 and day 7 using a CFSE assay and is expressed as percentage in relation to the proliferation on unfunctionalized gold control surfaces. (B) The percentage of CD34 positive cells was determined after HSPC incubation for 4 or 7 days on glass slides biofunctionalized with different ligands. (C) Representative histograms of flow cytometry analyses of CFSE labeled cells after 4 days incubation on biofunctionalized glass surfaces. The respective ligands are named in the top left corner of each histogram and the number of cell divisions is indicated by vertical, dashed lines. (D) CD34 expression of HSPCs after 4 (red curve) and 7 (blue curve) days of incubation on biofunctionalized glass surfaces; The CD34 isotype control is shown in gray. $N_{independent\ experiments} = 4$; error bars = standard deviation of the mean; gold = homogeneous gold film on glass; FNΔRGD is abbreviated with "ΔRGD".

Figure S8 Immunofluorescence THBS2 staining of HSPCs. Representative microscopic images of HSPCs incubated for 13 h on nanostructured, biofunctionalized hydrogels. The top row of images shows bright field images, in the middle row THBS2 is made visible by Alexa Fluor 488 fluorescence staining (green), and in the bottom row cell nuclei are made visible by Dapi staining (blue). The negative control was incubated without the primary antibody. One representative experiment (based on one donor) of 3 is shown. 20 cells per donor were analyzed on each substrate and one cell per substrate is shown. Scale bar = 10 μm.

Acknowledgments

The authors thank Dr. Markus Axmann (MPI for Intelligent Systems, Stuttgart) for advice with the microscope and Dr. Melanie Hart (Department of Urology, University Medical Clinic, Tübingen) and Nina Grunze (MPI for Intelligent Systems, Stuttgart) for critically reviewing the manuscript. We are grateful to Prof. Dr. Reinhard Fässler (MPI for Biochemistry, Martinsried, Germany) for providing the vectors for the FN domains.

Author Contributions

Conceived and designed the experiments: CL-T GK. Performed the experiments: CM CS. Analyzed the data: CM CL-T. Contributed reagents/materials/analysis tools: CS GK. Wrote the paper: CL-T CM.

References

1. Hines M, Nielsen L, Cooper-White J (2008) The hematopoietic stem cell niche: what are we trying to replicate? J Chem Technol Biotechnol 83: 421–443.

2. Liao J, Hammerick KE, Challen GA, Goodell MA, Kasper FK, et al. (2011) Investigating the role of hematopoietic stem and progenitor cells in regulating the osteogenic differentiation of mesenchymal stem cells in vitro. J Orthop Res 29: 1544–1553.

3. Gillette JM, Lippincott-Schwartz J (2009) Hematopoietic progenitor cells regulate their niche microenvironment through a novel mechanism of cell-cell communication. Commun Integr Biol 2: 305–307.

4. Calvi LM, Adams GB, Weibrecht KW, Weber JM, Olson DP, et al. (2003) Osteoblastic cells regulate the haematopoietic stem cell niche. Nature 425: 841–846.

5. Zhang J, Niu C, Ye L, Huang H, He X, et al. (2003) Identification of the haematopoietic stem cell niche and control of the niche size. Nature 425: 836–841.

6. Arai F, Hirao A, Ohmura M, Sato H, Matsuoka S, et al. (2004) Tie2/angiopoietin-1 signaling regulates hematopoietic stem cell quiescence in the bone marrow niche. Cell 118: 149–161.

7. Sacchetti B, Funari A, Michienzi S, Di Cesare S, Piersanti S, et al. (2007) Self-renewing osteoprogenitors in bone marrow sinusoids can organize a hematopoietic microenvironment. Cell 131: 324–336.

8. Ding L, Saunders TL, Enikolopov G, Morrison SJ (2012) Endothelial and perivascular cells maintain haematopoietic stem cells. Nature 481: 457–462.

9. Kiel MJ, Yilmaz OH, Iwashita T, Yilmaz OH, Terhorst C, et al. (2005) SLAM family receptors distinguish hematopoietic stem and progenitor cells and reveal endothelial niches for stem cells. Cell 121: 1109–1121.

10. Mendez-Ferrer S, Michurina TV, Ferraro F, Mazloom AR, Macarthur BD, et al. (2010) Mesenchymal and haematopoietic stem cells form a unique bone marrow niche. Nature 466: 829–834.

11. Ehninger A, Trumpp A (2011) The bone marrow stem cell niche grows up: mesenchymal stem cells and macrophages move in. J Exp Med 208: 421–428.

12. Daley WP, Peters SB, Larsen M (2008) Extracellular matrix dynamics in development and regenerative medicine. J Cell Sci 121: 255–264.

13. Klein G (1995) The extracellular matrix of the hematopoietic microenvironment. Experientia 51: 914–926.

14. Ellis SJ, Tanentzapf G (2010) Integrin-mediated adhesion and stem-cell-niche interactions. Cell Tissue Res 339: 121–130.

15. Franke K, Pompe T, Bornhäuser M, Werner C (2007) Engineered matrix coatings to modulate the adhesion of CD133+ human hematopoietic progenitor cells. Biomaterials 28: 836–843.

16. Kurth I, Franke K, Pompe T, Bornhäuser M, Werner C (2011) Extracellular matrix functionalized microcavities to control hematopoietic stem and progenitor cell fate. Macromol Biosci 11: 739–747.

17. Kurth I, Franke K, Pompe T, Bornhäuser M, Werner C (2009) Hematopoietic stem and progenitor cells in adhesive microcavities. Integr Biol (Camb) 1: 427–434.

18. Humphries JD, Byron A, Humphries MJ (2006) Integrin ligands at a glance. J Cell Sci 119: 3901–3903.

19. Barczyk M, Carracedo S, Gullberg D (2010) Integrins. Cell Tissue Res 339: 269–280.

20. Hynes RO (2002) Integrins: bidirectional, allosteric signaling machines. Cell 110: 673–687.

21. Potocnik AJ, Brakebusch C, Fässler R (2000) Fetal and adult hematopoietic stem cells require beta1 integrin function for colonizing fetal liver, spleen, and bone marrow. Immunity 12: 653–663.

22. Hirsch E, Iglesias A, Potocnik AJ, Hartmann U, Fässler R (1996) Impaired migration but not differentiation of haematopoietic stem cells in the absence of β1 integrins. Nature 380: 171–175.

23. Schofield KP, Humphries MJ, de Wynter E, Testa N, Gallagher JT (1998) The effect of alpha4 beta1-integrin binding sequences of fibronectin on growth of cells from human hematopoietic progenitors. Blood 91: 3230–3238.

24. Yokota T, Oritani K, Mitsui H, Aoyama K, Ishikawa J, et al. (1998) Growth-supporting activities of fibronectin on hematopoietic stem/progenitor cells in vitro and in vivo: structural requirement for fibronectin activities of CS1 and cell-binding domains. Blood 91: 3263–3272.

25. Feng Q, Chai C, Jiang XS, Leong KW, Mao HQ (2006) Expansion of engrafting human hematopoietic stem/progenitor cells in three-dimensional scaffolds with surface-immobilized fibronectin. J Biomed Mater Res A 78: 781–791.

26. Kramer A, Horner S, Willer A, Fruehauf S, Hochhaus A, et al. (1999) Adhesion to fibronectin stimulates proliferation of wild-type and bcr/abl-transfected murine hematopoietic cells. Proc Natl Acad Sci U S A 96: 2087–2092.

27. Hurley RW, McCarthy JB, Verfaillie CM (1995) Direct adhesion to bone marrow stroma via fibronectin receptors inhibits hematopoietic progenitor proliferation. J Clin Invest 96: 511–519.

28. Garcia AJ, Vega MD, Boettiger D (1999) Modulation of cell proliferation and differentiation through substrate-dependent changes in fibronectin conformation. Mol Biol Cell 10: 785–798.

29. Pierschbacher MD, Ruoslahti E (1984) Variants of the cell recognition site of fibronectin that retain attachment-promoting activity. Proc Natl Acad Sci U S A 81: 5985–5988.

30. Leahy DJ, Aukhil I, Erickson HP (1996) 2.0 A crystal structure of a four-domain segment of human fibronectin encompassing the RGD loop and synergy region. Cell 84: 155–164.

31. Aota S, Nomizu M, Yamada KM (1994) The short amino acid sequence Pro-His-Ser-Arg-Asn in human fibronectin enhances cell-adhesive function. J Biol Chem 269: 24756–24761.

32. Wayner EA, Kovach NL (1992) Activation-dependent recognition by hemato-poietic cells of the LDV sequence in the V region of fibronectin. J Cell Biol 116: 489–497.

33. Stier S, Ko Y, Forkert R, Lutz C, Neuhaus T, et al. (2005) Osteopontin is a hematopoietic stem cell niche component that negatively regulates stem cell pool size. J Exp Med 201: 1781–1791.

34. Nilsson SK, Johnston HM, Whitty GA, Williams B, Webb RJ, et al. (2005) Osteopontin, a key component of the hematopoietic stem cell niche and regulator of primitive hematopoietic progenitor cells. Blood 106: 1232–1239.

35. Grassinger J, Haylock DN, Storan MJ, Haines GO, Williams B, et al. (2009) Thrombin-cleaved osteopontin regulates hemopoietic stem and progenitor cell functions through interactions with alpha9beta1 and alpha4beta1 integrins. Blood 114: 49–59.

36. Altrock E, Muth CA, Klein G, Spatz JP, Lee-Thedieck C (2012) The significance of integrin ligand nanopatterning on lipid raft clustering in hematopoietic stem cells. Biomaterials 33: 3107–3118.

37. Holst J, Watson S, Lord MS, Eamegdool SS, Bax DV, et al. (2010) Substrate elasticity provides mechanical signals for the expansion of hemopoietic stem and progenitor cells. Nat Biotechnol 28: 1123–1128.

38. Lee-Thedieck C, Rauch N, Fiammengo R, Klein G, Spatz JP (2012) Impact of substrate elasticity on human hematopoietic stem and progenitor cell adhesion and motility. J Cell Sci 125: 3765–3775.

39. Choi JS, Harley BA (2012) The combined influence of substrate elasticity and ligand density on the viability and biophysical properties of hematopoietic stem and progenitor cells. Biomaterials 33: 4460–4468.

40. Hirschfeld-Warneken VC, Arnold M, Cavalcanti-Adam A, Lopez-Garcia M, Kessler H, et al. (2008) Cell adhesion and polarisation on molecularly defined spacing gradient surfaces of cyclic RGDfK peptide patches. Eur J Cell Biol 87: 743–750.

41. Cavalcanti-Adam EA, Aydin D, Hirschfeld-Warneken VC, Spatz JP (2008) Cell adhesion and response to nanopatterned environments by steering receptor clustering and location. HFSP Journal 2: 276–285.

42. Chua KN, Chai C, Lee PC, Tang YN, Ramakrishna S, et al. (2006) Surface-aminated electrospun nanofibers enhance adhesion and expansion of human umbilical cord blood hematopoietic stem/progenitor cells. Biomaterials 27: 6043–6051.

43. Shekaran A, Garcia AJ (2011) Nanoscale engineering of extracellular matrix-mimetic bioadhesive surfaces and implants for tissue engineering. Biochim Biophys Acta 1810: 350–360.

44. Lepzelter D, Bates O, Zaman M (2012) Integrin clustering in two and three dimensions. Langmuir 28: 5379–5386.

45. Spatz JP, Mössmer S, Hartmann C, Möller M (2000) Ordered deposition of inorganic clusters from micellar block copolymer films. Langmuir 16: 407–415.

46. Aydin D, Louban I, Perschmann N, Blummel J, Lohmüller T, et al. (2010) Polymeric substrates with tunable elasticity and nanoscopically controlled biomolecule presentation. Langmuir 26: 15472–15480.

47. Lohmüller T, Aydin D, Schwieder M, Morhard C, Louban I, et al. (2011) Nanopatterning by block copolymer micelle nanolithography and bioinspired applications. Biointerphases 6: MR1–12.

48. Alcantar NA, Aydil ES, Israelachvili JN (2000) Polyethylene glycol-coated biocompatible surfaces. J Biomed Mater Res 51: 343–351.

49. Nermut MV, Green NM, Eason P, Yamada SS, Yamada KM (1988) Electron microscopy and structural model of human fibronectin receptor. EMBO J 7: 4093–4099.

50. Cavalcanti-Adam EA, Micoulet A, Blummel J, Auernheimer J, Kessler H, et al. (2006) Lateral spacing of integrin ligands influences cell spreading and focal adhesion assembly. Eur J Cell Biol 85: 219–224.

51. Wolfram T, Belz F, Schoen T, Spatz JP (2007) Site-specific presentation of single recombinant proteins in defined nanoarrays. Biointerphases 2: 44–48.

52. Leiss M, Beckmann K, Giros A, Costell M, Fässler R (2008) The role of integrin binding sites in fibronectin matrix assembly in vivo. Curr Opin Cell Biol 20: 502–507.

53. Lee-Thedieck C, Spatz JP (2012) Artificial niches: biomimetic materials for hematopoietic stem cell culture. Macromol Rapid Commun 33: 1432–1438.

54. Arnold M, Cavalcanti-Adam EA, Glass R, Blummel J, Eck W, et al. (2004) Activation of integrin function by nanopatterned adhesive interfaces. Chem Phys Chem 5: 383–388.

55. Cavalcanti-Adam EA, Volberg T, Micoulet A, Kessler H, Geiger B, et al. (2007) Cell spreading and focal adhesion dynamics are regulated by spacing of integrin ligands. Biophys J 92: 2964–2974.

56. Geiger B, Spatz JP, Bershadsky AD (2009) Environmental sensing through focal adhesions. Nat Rev Mol Cell Biol 10: 21–33.

57. Avraham H, Park SY, Schinkmann K, Avraham S (2000) RAFTK/Pyk2-mediated cellular signalling. Cell Signal 12: 123–133.

58. Melikova S, Dylla SJ, Verfaillie CM (2004) Phosphatidylinositol-3-kinase activation mediates proline-rich tyrosine kinase 2 phosphorylation and recruitment to beta1-integrins in human CD34+ cells. Exp Hematol 32: 1051–1056.

59. Zhu B, Eurell T, Gunawan R, Leckband D (2001) Chain-length dependence of the protein and cell resistance of oligo(ethylene glycol)-terminated self-assembled monolayers on gold. J Biomed Mater Res 56: 406–416.

60. Blümmel J, Perschmann N, Aydin D, Drinjakovic J, Surrey T, et al. (2007) Protein repellent properties of covalently attached PEG coatings on nanos-tructured SiO(2)-based interfaces. Biomaterials 28: 4739–4747.

61. van Wachem PB, Hogt AH, Beugeling T, Feijen J, Bantjes A, et al. (1987) Adhesion of cultured human endothelial cells onto methacrylate polymers with varying surface wettability and charge. Biomaterials 8: 323–328.

62. Lampin M, Warocquier-Clerout R, Legris C, Degrange M, Sigot-Luizard MF (1997) Correlation between substratum roughness and wettability, cell adhesion, and cell migration. J Biomed Mater Res 36: 99–108.

63. Engler AJ, Sen S, Sweeney HL, Discher DE (2006) Matrix elasticity directs stem cell lineage specification. Cell 126: 677–689.

64. Gilbert PM, Havenstrite KL, Magnusson KE, Sacco A, Leonardi NA, et al. (2010) Substrate elasticity regulates skeletal muscle stem cell self-renewal in culture. Science 329: 1078–1081.

65. Wells RG (2008) The role of matrix stiffness in regulating cell behavior. Hepatology 47: 1394–1400.

66. DURAN Group www.duran-group.com/de/ueber-duran-eigenschaften.html.

67. Jung Y, Wang J, Havens A, Sun Y, Wang J, et al. (2005) Cell-to-cell contact is critical for the survival of hematopoietic progenitor cells on osteoblasts. Cytokine 32: 155–162.

68. Papayannopoulou T, Priestley GV, Nakamoto B, Zafiropoulos V, Scott LM (2001) Molecular pathways in bone marrow homing: dominant role of alpha(4)beta(1) over beta(2)-integrins and selectins. Blood 98: 2403–2411.

69. Kerst JM, Sanders JB, Slaper-Cortenbach IC, Doorakkers MC, Hooibrink B, et al. (1993) Alpha 4 beta 1 and alpha 5 beta 1 are differentially expressed during myelopoiesis and mediate the adherence of human CD34+ cells to fibronectin in an activation-dependent way. Blood 81: 344–351.

70. Levesque JP, Leavesley DI, Niutta S, Vadas M, Simmons PJ (1995) Cytokines increase human hemopoietic cell adhesiveness by activation of very late antigen (VLA)-4 and VLA-5 integrins. J Exp Med 181: 1805–1815.

71. Mould AP, Komoriya A, Yamada KM, Humphries MJ (1991) The CS5 peptide is a second site in the IIICS region of fibronectin recognized by the integrin alpha 4 beta 1. Inhibition of alpha 4 beta 1 function by RGD peptide homologues. J Biol Chem 266: 3579–3585.

72. Hankenson KD, Bornstein P (2002) The secreted protein thrombospondin 2 is an autocrine inhibitor of marrow stromal cell proliferation. J Bone Miner Res 17: 415–425.

73. Bornstein P (2001) Thrombospondins as matricellular modulators of cell function. J Clin Invest 107: 929–934.

74. Bornstein P, Armstrong LC, Hankenson KD, Kyriakides TR, Yang Z (2000) Thrombospondin 2, a matricellular protein with diverse functions. Matrix Biol 19: 557–568.

75. Calzada MJ, Sipes JM, Krutzsch HC, Yurchenco PD, Annis DS, et al. (2003) Recognition of the N-terminal modules of thrombospondin-1 and thrombos-pondin-2 by alpha6beta1 integrin. J Biol Chem 278: 40679–40687.

76. Kyriakides TR, Zhu YH, Smith LT, Bain SD, Yang Z, et al. (1998) Mice that lack thrombospondin 2 display connective tissue abnormalities that are associated with disordered collagen fibrillogenesis, an increased vascular density, and a bleeding diathesis. J Cell Biol 140: 419–430.

77. Alford AI, Terkhorn SP, Reddy AB, Hankenson KD (2010) Thrombospondin-2 regulates matrix mineralization in MC3T3-E1 pre-osteoblasts. Bone 46: 464–471.

78. Glass R, Möller M, Spatz JP (2003) Block copolymer micelle nanolithography. Nanotechnology 14: 1153–1160.

79. Graeter SV, Huang J, Perschmann N, Lopez-Garcia M, Kessler H, et al. (2007) Mimicking cellular environments by nanostructured soft interfaces. Nano Lett 7: 1413–1418.

80. Studier FW (2005) Protein production by auto-induction in high density shaking cultures. Protein Expr Purif 41: 207–234.

81. Sigal GB, Bamdad C, Barberis A, Strominger J, Whitesides GM (1996) A self-assembled monolayer for the binding and study of histidine-tagged proteins by surface plasmon resonance. Anal Chem 68: 490–497.

82. Little WC, Smith ML, Ebneter U, Vogel V (2008) Assay to mechanically tune and optically probe fibrillar fibronectin conformations from fully relaxed to breakage. Matrix Biol 27: 451–461.

83. Johansson S, Svineng G, Wennerberg K, Armulik A, Lohikangas L (1997) Fibronectin-integrin interactions. Front Biosci 2: d126–146.

84. Sun X, Skorstengaard K, Mosher DF (1992) Disulfides modulate RGD-inhibitable cell adhesive activity of thrombospondin. J Cell Biol 118: 693–701.

16

Biomimetic Synthesis of Selenium Nanospheres by Bacterial Strain JS-11 and Its Role as a Biosensor for Nanotoxicity Assessment: A Novel Se-Bioassay

Sourabh Dwivedi[1], Abdulaziz A. AlKhedhairy[1], Maqusood Ahamed[2], Javed Musarrat[3]*

1 Department of Zoology, College of Science, King Saud University, Riyadh, Saudi Arabia, **2** King Abdullah Institute for Nanotechnology, King Saud University, Riyadh, Saudi Arabia, **3** Department of Agricultural Microbiology, Faculty of Agricultural Sciences, Aligarh Muslim University, Aligarh, India

Abstract

Selenium nanoparticles (Se-NPs) were synthesized by green technology using the bacterial isolate *Pseudomonas aeruginosa* strain JS-11. The bacteria exhibited significant tolerance to selenite (SeO_3^{2-}) up to 100 mM concentration with an EC_{50} value of 140 mM. The spent medium (culture supernatant) contains the potential of reducing soluble and colorless SeO_3^{2-} to insoluble red elemental selenium (Se^0) at 37°C. Characterization of red Se° product by use of UV-Vis spectroscopy, X-ray diffraction (XRD), atomic force microscopy (AFM) and transmission electron microscopy (TEM) with energy dispersive X-ray spectrum (EDX) analysis revealed the presence of stable, predominantly monodispersed and spherical selenium nanoparticles (Se-NPs) of an average size of 21 nm. Most likely, the metabolite phenazine-1-carboxylic acid (PCA) released by strain JS-11 in culture supernatant along with the known redox agents like NADH and NADH dependent reductases are responsible for biomimetic reduction of SeO_3^{2-} to Se° nanospheres. Based on the bioreduction of a colorless solution of SeO_3^{2-} to elemental red Se^0, a high throughput colorimetric bioassay (Se-Assay) was developed for parallel detection and quantification of nanoparticles (NPs) cytotoxicity in a 96 well format. Thus, it has been concluded that the reducing power of the culture supernatant of strain JS-11 could be effectively exploited for developing a simple and environmental friendly method of Se-NPs synthesis. The results elucidated that the red colored Se° nanospheres may serve as a biosensor for nanotoxicity assessment, contemplating the inhibition of SeO_3^{2-} bioreduction process in NPs treated bacterial cell culture supernatant, as a toxicity end point.

Editor: Asad U. Khan, Aligarh Muslim University, India

Funding: Financial support through the National Plan for Sciences and Technology (NPST Project No.10-NAN1115-02), King Saud University, Riyadh, for this study, is greatly acknowledged. (http://npst.ksu.edu.sa). The funders had no role in study design, data collection and analysis, decision to publish, or preparation of the manuscript.

Competing Interests: The authors have declared that no competing interests exist.

* E-mail: musarratj1@yahoo.com

Introduction

Selenium (Se^0) is a trace element commonly found in materials of the earth's crust, and belongs to group 16 (chalcogens) of the periodic table. Se° is well known for its photoelectric, semiconductor, free-radical scavenging, anti-oxidative and anti-cancer properties [1]. It occurs in different forms as red amorphous selenium (Se^0), highly water soluble selenate (SeO_4^{2-}) and selenite (SeO_3^{2-}), and as gaseous selenide (Se^{2-}). Amongst its various forms, the SeO_3^{2-} is highly toxic, which adversely affect the cellular respiration and antioxidant system causes protein inactivation and DNA repair inhibition [2,3,4]. Therefore, detoxification of SeO_3^{2-} has attracted a great deal of attention, particularly the reduction of this oxyanion by the microorganisms. The SeO_3^{2-} reducing bacteria are ubiquitous in diverse terrestrial and aquatic environments [5]. The ability to reduce the toxic SeO_4^{2-} and SeO_3^{2-} species into non-toxic elemental form Se° has been demonstrated under aerobic and anaerobic conditions [5,6,7,8]. However, the reduction of SeO_3^{2-} to Se^0, which is a common feature of many diverse microorganisms, is still not well understood. Earlier studies have suggested that SeO_3^{2-} reduction may involves the periplasmic nitrite reductase [8,9] in *Thauera*

selenatis [10] and *Rhizobium selenitireducens* strain B1 [11,12,13], nitrate reductase in *E. coli* [14], hydrogenase I in *Clostridium pasteurianum* [15] and arsenate reductase in *Bacillus selenitireducens* [16] or some of the non-enzymatic reactions [17]. Lortie et al. [6] have reported aerobic reduction of SeO_4^{2-} and SeO_3^{2-} to Se° by a *Pseudomonas stutzeri* isolate. Studies based on X-ray absorption spectroscopy also revealed that the soil bacterium *Ralstonia metallidurans* CH34, resistant to SeO_3^{2-} is capable of its detoxification, and localize the red Se° granules mainly in the cytoplasm [18]. Sarret et al. [19] investigated the kinetics of selenite and selenate accumulation and Se speciation to identify the chemical intermediates putatively appearing during reduction using X-ray absorption near-edge structure (XANES) spectroscopy. Furthermore, the NADPH/NADH dependent selenate reductase enzymes have been reported to catalyze the reduction of selenium oxyions [20], [21]. Most of the studies on the biogenesis of selenium nanoparticles (Se-NPs) are based on anaerobic systems. However, there are also few reports in literature on the aerobic formation of Se-NPs by microorganisms such as *Pseudomonas aeruginosa*, *Bacillus sp.* and *Enterobacter cloacae* [22,23,24].

With the overwhelming growth in the field of nanotechnology and a rapid stride in the synthesis and commercialization of nanomaterials, the occupational and inadvertent exposure to human population is imminent, which may pose serious hazards to human health and ecosystem. Therefore, improved characterization and reliable toxicity screening tools are required for exposure risk assessments. The commonly used cytotoxicity screening assays are mostly based on fluorescence or absorbance measurements following toxicant exposure and incubation with a colorimetric indicator dyes but has many limitations with nanotoxicity assessment [25,26,27,28]. Thus, Wang et al. [29] developed a novel bioluminescence inhibition assay exploiting *Photobacterium phosphoreum* to evaluate the toxicity of quantum dots. Also, a black and white method (Te-assay) for pre-screening of environmental samples based on reduction of tellurite (TeO_3^{2-}) to elemental tellurium has been reported [30]. Along the similar principle, we have attempted to exploit the selenite tolerant *Pseudomonas aeruginosa* strain JS-11 isolated from wheat rhizosphere for biosynthesis of Se-NPs and utilized its capacity of reducing of SeO_3^{2-} to Se^0, as a metabolic marker for visual assessment of the relative toxicity of several NPs in a single experiment in a 96-well format. Thus, the objectives of the study were to investigate the (i) metabolic potential of a SeO_3^{2-} tolerant *P. aeruginosa* strain JS-11 for green synthesis of elemental Se° nanospheres, (ii) characterization of Se-NPs by use of UV–Vis spectrophotometry, X-ray diffraction (XRD), dynamic light scattering, transmission electron microscopy (TEM), energy dispersive X-ray (EDX) analysis, Fourier transform infra red spectroscopy (FTIR) and atomic force microscopy, and (iii) development of a simple, colorimetric assay for toxicity assessment of NPs and other environmental pollutants.

Materials and Methods

Bacterial Growth and Resistance to SeO_3^{2-} Stress

The soil bacteria *P. aeruginosa* strain JS-11, initially isolated in our laboratory from wheat rhizosphere of herbicide contaminated soil by the enrichment culture technique [31], and maintained as glycerol cultures at −80°C, were used in this study. The strain JS-11 has already been well characterized based on its metabolic profile using BIOLOG GN plates (Biolog Inc., Hayward, CA, USA) and phylogenetic analysis based on 16SrDNA sequence homology [31]. In order to assess the tolerance of strain JS-11 for SeO_3^{2-} and its reduction to Se^0, the frozen culture was thawed and grown in Luria-Bertani (LB) broth. Cells from exponentially grown culture were streaked on to Luria agar (LA) plates supplemented with 12.5 mM sodium selenite ($Na_2SeO_3^{2-}$). The red color colonies developed on the plates after 18 hours (h) of incubation at 37°C were transferred to fresh LB medium containing 25 mM $Na_2SeO_3^{2-}$, and further sub-cultured for determining the SeO_3^{2-} tolerance limit. The effect of SeO_3^{2-} on bacterial growth was determined by culturing the cells ($\sim 2 \times 10^4$ CFUml^{-1}) in 250 ml Erlenmeyer flasks containing 100 ml of LB supplemented with increasing concentrations (12.5, 25, 50, 100 and 200 mM) of $Na_2SeO_3^{2-}$. The flasks were incubated at 37°C under constant shaking at 200 rpm for 24 h. The growth in each flask was determined by measuring the optical density (O.D.) at 600 nm. For growth kinetics, the bacterial growth was determined in the LB and M9 mineral salt medium ($Na_2HPO_4.7H_2O$, 42 mM; KH_2PO_4, 24 mM; NaCl, 9 mM; NH_4Cl, 19 mM; $MgSO_4$, 1 mM; $CaCl_2$, 0.1 mM and glucose 2%) supplemented with 12.5 mM $Na_2SeO_3^{2-}$, as a function of time of incubation and the O.D. was measured periodically at 600 nm.

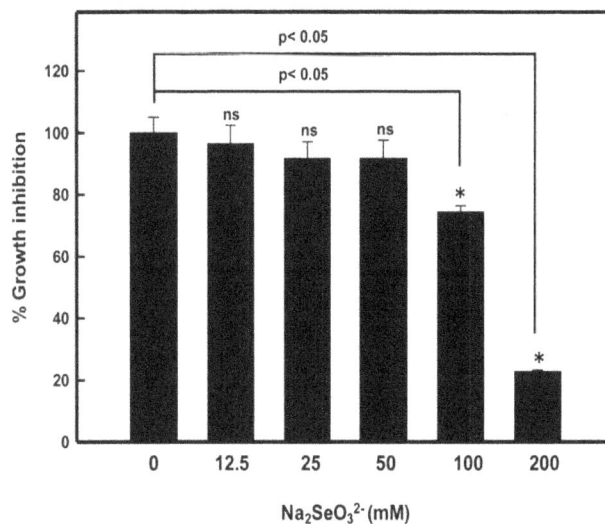

Figure 1. Selenite tolerance by *Pseudomonas aeruginosa* strain JS-11.

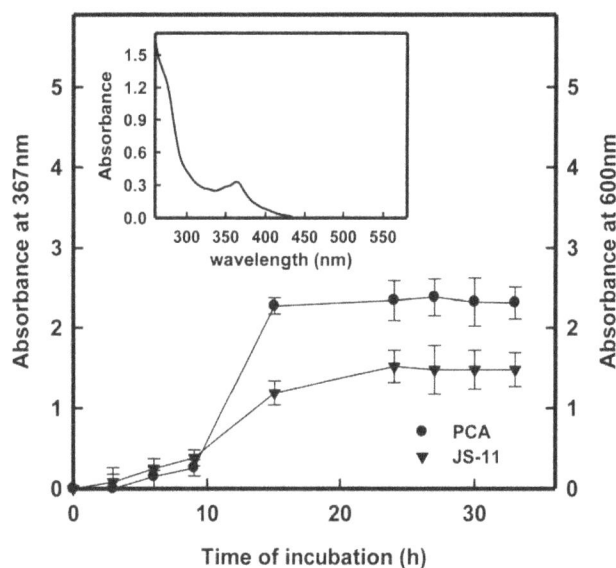

Figure 2. Growth curve and PCA production by strain JS-11. Bacterial growth in LB and the PCA present in culture supernatant were measured at 600 and 367 nm, respectively, as a function of time. Inset shows the absorption spectra of PCA with λmax at 367 nm, as determined by UV-visible spectrophotometer. The data represent the mean ± S.D of two independent experiments done in triplicate.

Determination of Phenazine Production by Strain JS-11 in Culture Supernatant

The phenazine-1-carboxylic acid (PCA) production in bacterial cell culture was determined following the method of Mavrodi et al. [32]. In brief, the cell culture of strain JS-11 was grown at 37°C in a 250 ml conical flask containing modified LB medium (LB +1 mM tryptophan) that favors the PCA production. Tryptophan facilitates the synthesis of PCA via anthranilate synthase II

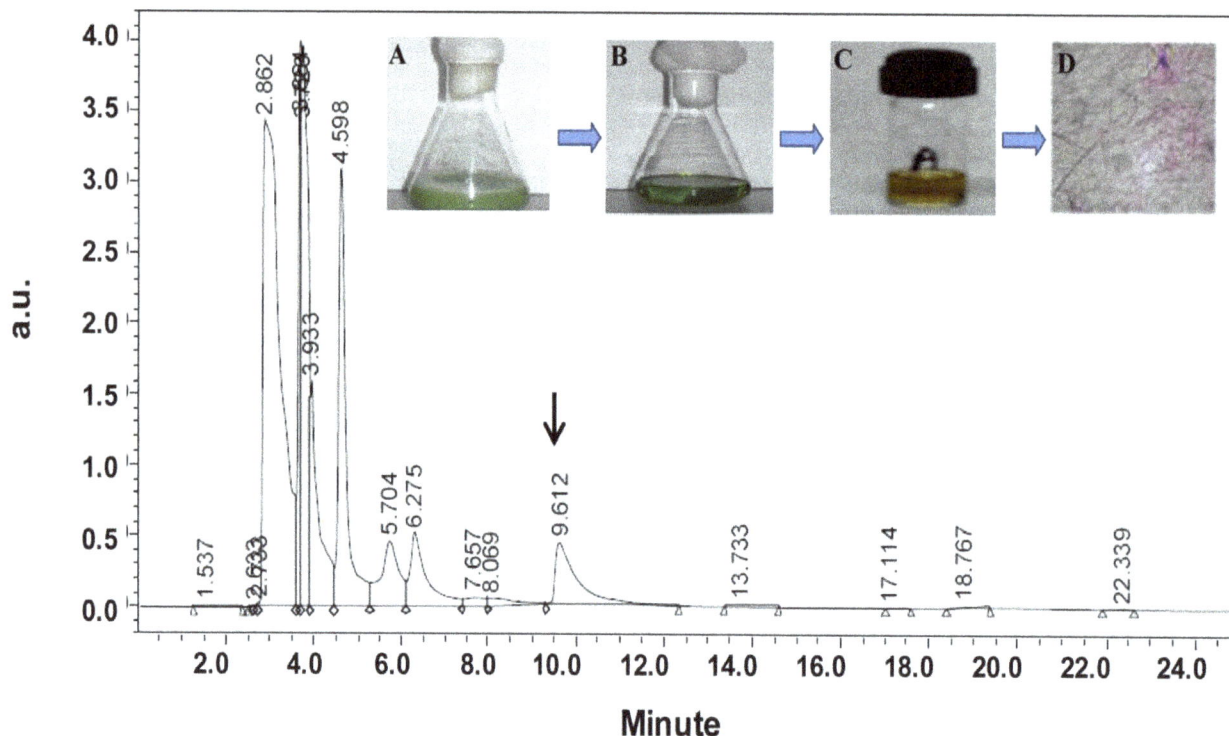

Figure 3. HPLC analysis of PCA produced by the strain JS-11. HPLC profile indicating the PCA peak at retention time of 9.6 min. Inset shows the (A): bacterial culture supernatant, (B): Benzene extract of PCA, (C): PCA after extraction and (D) PCA crystals.

pathway using anthranilate as a substrate [33,34]. The culture was grown for 72 h and the cells were centrifuged at 5000 rpm for 10 minutes (min.). The cell-free culture supernatant (100 ml) was transferred to another flask and acidified with concentrated hydrochloric acid to achieve a pH of 2.0. The acidified supernatant was extracted with equal volume of benzene. The organic phase was pooled and dried by evaporation. The dried pale yellow residue was dissolved in 1 ml of 0.1 M NaOH, and the absorbance was read at 367 nm against the benzene extract of acidified LB alone, used as a blank. Further, the HPLC analysis of PCA extract was performed by use of Waters HPLC System coupled with 2487 dual λ UV/visible detector using C-18 Novapak (4 μm) column (Waters Corp., Milford, MA, USA) with mobile phase of acetonitrile: water (70:30) at 254 nm. Fourier transform infra-red (FTIR) spectroscopic analysis was performed for examining the functional groups of PCA. The PCA was mixed with spectroscopic grade potassium bromide (KBr) in the ratio of 1:100 and spectrum recorded in the range of 400–4000 wavenumber (cm^{-1}) on FTIR spectrometer, Spectrum 100 (Perkin Elmer, USA) in the diffuse reflectance mode at a resolution of 4 cm^{-1} in KBr pellets.

Determination of Reducing Activity in Culture Supernatant by KMnO₄ Titrimetric Assay

Freshly grown cell of bacterial strain JS-11 were sub-cultured in LB broth containing the suspensions of silver nanoparticles (Ag-NPs, 27 nm), cadmium sulfide nanoparticles (CdS-NPs, 4 nm), titanium dioxide nanoparticles (TiO_2-NPs, 30.6 nm) and zinc ferrite nanoparticles ($ZnFe_2O_4$-NPs, 19 nm) at two different concentrations of 50 and 100 μgml^{-1}. The cultures of untreated and treated bacterial cells were grown for 24 h at 37°C. Cultures were centrifuged at 5000 rpm for 10 min and the supernatants of

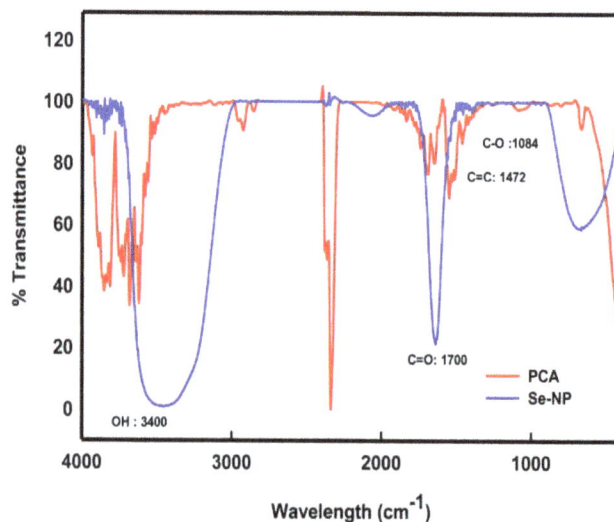

Figure 4. FTIR analysis of PCA. The spectra depict the changes in the peaks of PCA alone (red) and after treatment with 2 mM $Na_2SeO_3^{2-}$ solution (blue).

untreated and NPs treated bacteria were assessed for the innate reducing activity based on potassium permanganate ($KMnO_4$) back-titration, following the method of Fesharaki et al. [35]. Briefly, the supernatant (6 ml) was diluted by ultrapure water (20 ml) and acidified with 2 ml of 1.5 N phosphoric acid. The acidified supernatant was then oxidized with an excess of $KMnO_4$ (0.1 N) for 30 min at 60°C. The unreacted permanganate was titrated with a 0.04 N oxalic acid solution. The end-point was determined with the disappearance of violet color of the reactant solution, and the reducing activity of the supernatants was calculated considering the volume and normality of permanganate solution. A parallel set of experiment was performed with the un-inoculated culture medium for data normalization, and results reported as mg of $KMnO_4$ per ml of the supernatant.

Fluorescence Measurements

Fluorescence spectra of the supernatants obtained from the untreated bacterial culture and those treated with 100 μgml^{-1} of Ag-NPs, CdS-NPs, TiO_2-NPs and $ZnFe_2O_4$-NPs were measured in a 1 cm path length cell by use of Shimadzu spectrofluorophotometer, model RF5301PC (Shimadzu Scientific Instruments, Japan) equipped with RF 530XPC instrument control software, at ambient temperature. The excitation and emission slits were set at 5 nm each. The emission spectra were recorded in wavelength range of 290–380 nm and the excitation wavelength was set at 280 nm. The $Na_2SeO_3{}^{2-}$ solution was non-fluorescent at this wavelength range. The NADH fluorescence in the culture supernatant was also measured at 340 nm and 440 nm, as excitation and emission wavelengths, respectively.

Biosynthesis of Elemental Se-NPs

The cells of bacterial strain JS-11 were grown in LB broth in a 500 ml flask at 37°C with agitation at 200 rpm. After 24 h of incubation, the cell culture was centrifuged at 5000 rpm for 10 min. The supernatant was transferred to 250 ml flask to which $Na_2SeO_3{}^{2-}$ was added at a final concentration of 2 mM, and again incubated at 37°C, under constant agitation at 200 rpm for 72 h. The red colored (Se^0) product obtained in the supernatant was recovered and analyzed for the presence of Se-NPs.

Characterization of Se-NPs

UV–Visible spectral analysis. Color changes in the culture supernatant of strain JS-11 were monitored both by visual inspection and absorbance measurements using double beam UV–Vis spectrophotometer, (Labomed, U.S.A) as described earlier [36]. The spectra of the surface plasmon resonance of Se-NPs in the supernatants were recorded periodically at 2, 24, 48 and 72 h in wavelength range of 200 and 800 nm.

X ray diffraction analysis. The red colored bacterial culture supernatant containing Se° in the form of Se-NPs was freeze-dried on Heto Lyophilizer (Heto-Holten, Denmark) and stored in lyophilized powdered form until used for further characterization. The finely powdered sample was analyzed by X'pert PRO Panalytical diffractometer using CuK_α radiation ($\lambda = 1.54056$ Å) in the range of $20° \leq 2\theta \leq 80°$ at 40 keV. In order to calculate the particle size (D) of the sample, the Scherrer's relationship ($D = 0.9 \lambda/\beta cos\theta$) has been used [37], where λ is the wavelength of X-ray, β is the broadening of the diffraction line measured half of its maximum intensity in radians and θ is the Bragg's diffraction angle. The particle size of the sample was estimated from the line width of the (101) XRD peak.

Transmission Electron Microscopic (TEM) and Energy Dispersive X-ray (EDX) Analysis

Samples for TEM analysis were prepared by drop-coating Se-NPs solution onto carbon-coated copper TEM grids. The films on the TEM grids were allowed to stand for 2 min. The extra solution was removed using a blotting paper and the grid dried prior to measurement. Transmission electron micrographs were obtained on JEM-2100F (JEOL Inc., Japan) instrument with an accelerating voltage of 80 kV. To ascertain the reduction of $SeO_3{}^{2-}$ to elemental selenium (Se^0), the samples were processed by a method similar to that used for TEM studies. The selected areas within

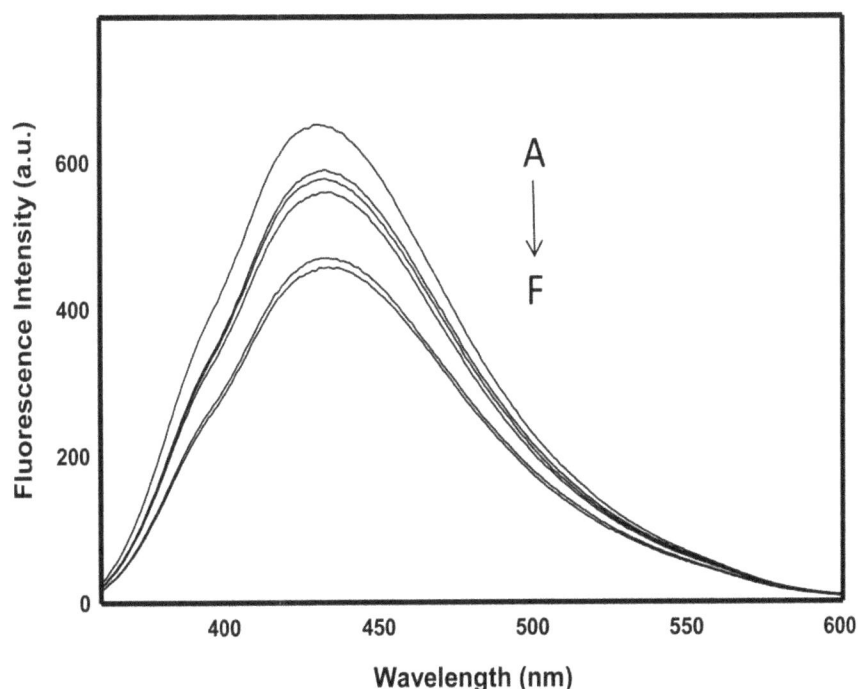

Figure 5. NADH fluorescence of bacterial supernatant alone and after treatment with NPs. The arrow represents the fluorescence quench of spectra A–F, where A is untreated control supernatant, and spectra B to F represent the supernatant treated with TiO_2-NPs, $ZnFe_2O_4$-NPs, CdS-NPs, Ag-NPs (100 μgml^{-1}) and EMS (2 mM) in a total volume of 3 ml, respectively.

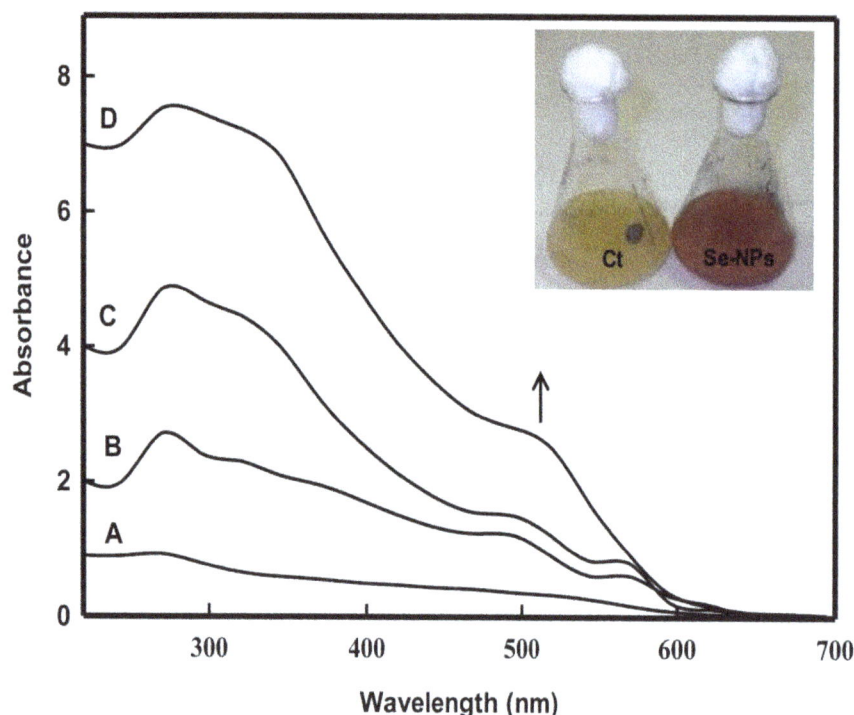

Figure 6. UV-Visible absorption spectra of extracellularly synthesized Se-NPs. The typical surface plasmon resonance (SPR) band is shown at 520 nm. The labels A–D represent 2, 24, 48 and 72 h of incubation, respectively. Inset depicts the change in color of culture supernatant from pale yellow to red after 24 h of incubation with 2 mM $Na_2SeO_3{}^{2-}$ solution.

TEM sections were subjected to elemental composition analysis using an EDX (JEOL Inc., Japan).

Dynamic Light Scattering and Zeta (ζ) Potential

Se-NPs powder was suspended in deionized ultrapure water to obtain a concentration of 50 μgml^{-1}, and sonicated at 40 W for 15 min. Hydrodynamic particle size and Zeta (ζ) potential of Se-NPs in an aqueous suspension were determined by measur-

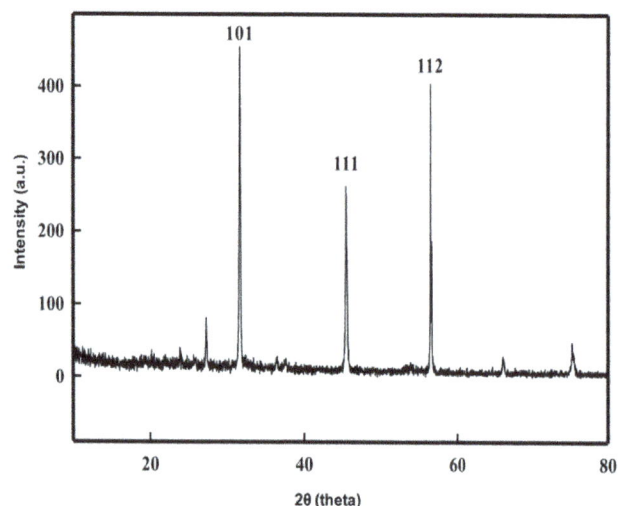

Figure 7. XRD pattern of the bacterial Se-NPs. The characteristic strong diffraction peak located at 31.64° is ascribed to the (101) facets of the face-centred cubic elemental Se° structure.

ing the dynamic light scattering by use of a ZetaSizer-HT (Malvern, UK).

Atomic Force Microscopic (AFM) Analysis

Bacterial Se-NPs were examined using Innova AFM (Veeco Instruments, Plainview, NY, USA) in a non-contact tapping mode, following the method described by Musarrat et al. [36]. The topographical images were obtained in tapping mode at a resonance frequency of 218 kHz. Tapping mode imaging was implemented in ambient air by oscillating the cantilever assembly at or near the cantilever's resonant frequency using a piezoelectric crystal. Characterization was done by observing the patterns on the surface topography and data analysis through WSXM software.

Selenium Reduction Bioassay (Se-Assay) for Toxicity Assessment

The bacterial strain JS-11 was grown for 24 h at 37°C. Freshly grown culture (50 ml) was centrifuged at 4500 rpm for 10 min. The culture supernatant (1 ml) was then aliquoted into 1.5 ml eppendorf tubes for treatment with toxicants. To each tube, increasing concentrations (6.25, 12.5, 25, 50, 75, 100 μgml^{-1}) of various analyte NPs viz. Ag-NPs, CdS-NPs, TiO_2-NPs and $ZnFe_2O_4$-NPs were added. A well known genotoxicant ethyl methane sulphonate (EMS) in concentration range of 0.125 - 2.0 mM, was used as a positive control. The tubes were incubated at 37°C for 24 h. Treated supernatants were again centrifuged for 10000 rpm for 10 min. Supernatants were collected separately in fresh tubes and 100 μl of each supernatant was then carefully transferred to all the wells (columns 1–12, rows A–F) in a 96-well microtitre plate. Subsequently, 100 μl of $Na_2SeO_3{}^{2-}$ solution (final concentration

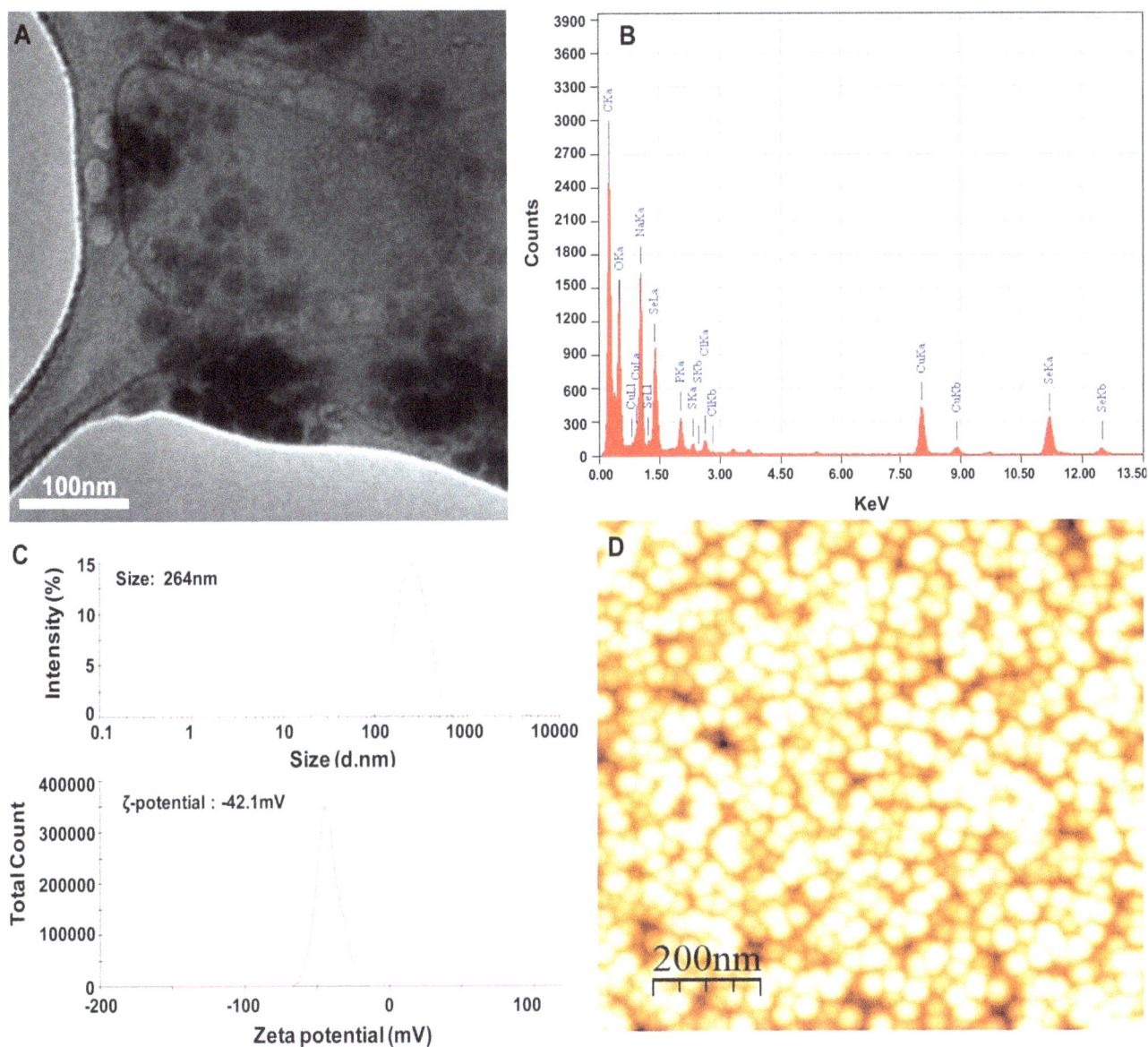

Figure 8. Microscopic analysis of Se-NPs produced by strain JS-11. Panel (A) shows the representative transmission electron micrograph recorded from a drop-coated film of the aqueous solution of Se-NPs; Panel (B) represents the energy dispersive X-ray spectrum of Se-NPs; (C) represents the average hydrodynamic size and zeta potential of Se-NPs; and (D) represents the 3D topography of Se-NPs in top view (scan size is 5×5 μm) by atomic force microscopic analysis.

20 mM) was added to each well, except the Lane 1 (untreated control). The mixture was then incubated at 37°C for 48 h. The plate was read at 520 nm on multi-well microplate reader (Thermo Scientific, USA). For quantitative assessment, the intensity of red color in wells of all lanes (columns 5–12) was compared with the intensity of red color in the control lanes 2 and 3 (wells F2 and F3) containing cell-free culture supernatant+SeO_3^{2-}, by considering their mean O.D. value as 100%. The decrease in redness by 50% upon treatment with the increasing concentrations of analytes (NPs/EMS), provided the IC_{50} of the microbial activity, which refers to 50% loss of metabolic activity compared to the control/no-analyte cells.

Results and Discussion

Bacterial Tolerance to SeO_3^{2-} in Culture Medium

The results in Fig. 1 show the SeO_3^{2-} tolerance of the *P. aeruginosa* strain JS-11 at increasing concentrations of $Na_2SeO_3^{2-}$. The bacteria exhibited substantial growth in LB medium supplemented with $Na_2SeO_3^{2-}$ in concentration range of 12.5 to 100 mM. Significant intensity of red color developed in culture medium after 24 h of growth due to reduction of SeO_3^{2-} to elemental Se^0, which suggested adequate metabolic activity in selenium oxyanions treated cells, as an indication of cell viability. The presence of SeO_3^{2-} up to 50 mM resulted in 8.5% growth inhibition, whereas 25.7% (p<0.05) and 77.3% (p<0.05) growth inhibition was observed at 100 and 200 mM SeO_3^{2-} concentrations, respectively, as compared to untreated control. Based on the extent of growth inhibition, the effective concentration (EC_{50}) of

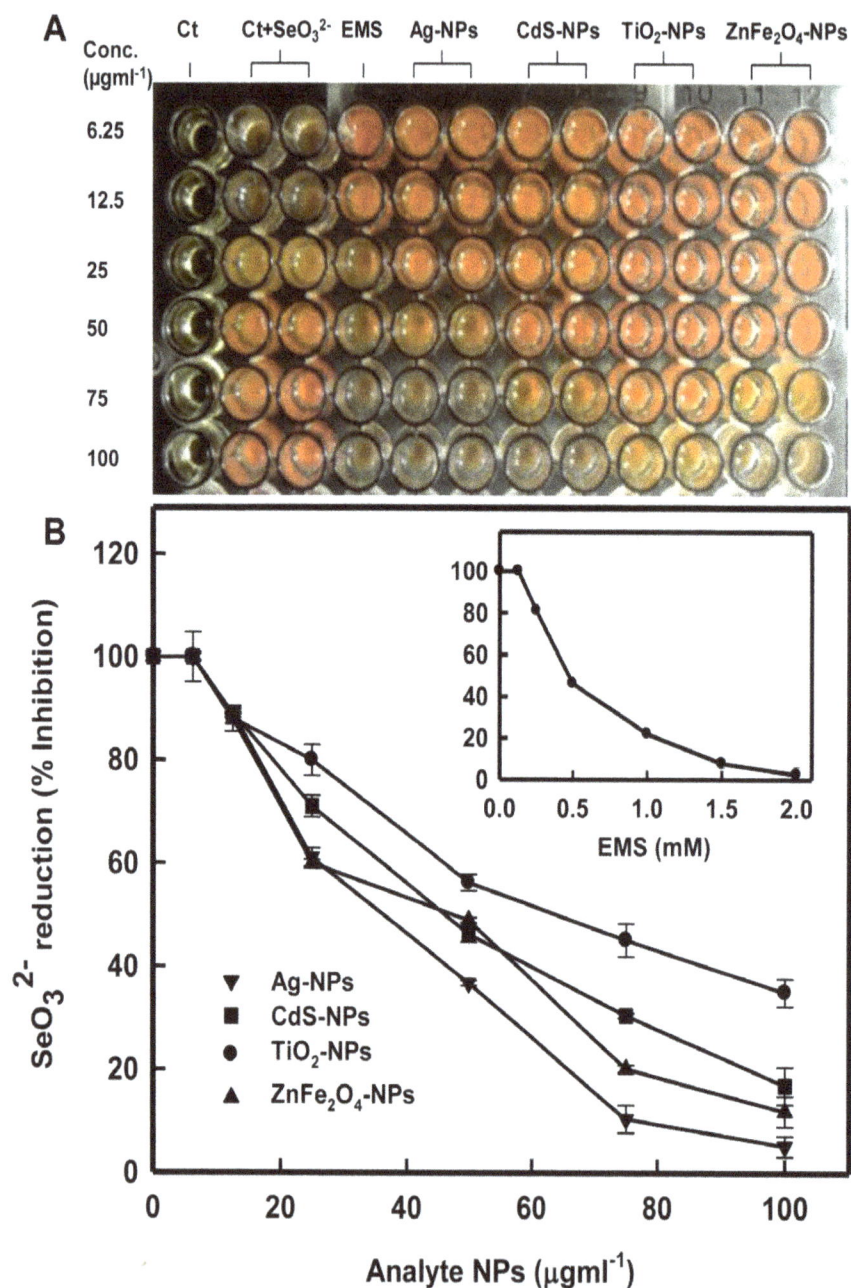

Figure 9. Se-Assay for toxicity assessment. Panel A: Colorimetric determination of toxicity based on inhibition of the reduction of SeO_3^{2-} (colorless solution) to Se° (red) in absence and presence of toxic analyte NPs. Lane 1: Control (untreated supernatant); Lane 2 and 3: Control (supernatant +20 mM SeO_3^{2-}); Lane 4: positive control (EMS 0.125 to 2 mM); Lanes 5–12: Ag-NPs, CdS-NPs, TiO_2-NPs and $ZnFe_2O_4$-NPs, in duplicate wells in 96 well microtitre plate. The density of the red color was read at 520 nm on multi-well microplate reader. Panel B: Percent inhibition of reduction in SeO_3^{2-} to Se° in presence of analyte NPs. Inset shows the percent inhibition of SeO_3^{2-} reduction by EMS, a known genotoxicant.

SeO_3^{2-} was determined to be 140 mM. Hunter and Manter [21] have also reported 57% and 66% reduced growth of *Pseudomonas* sp. strain CA5 at SeO_3^{2-} concentrations of 100 and 150 mM, respectively. Therefore, our results explicitly suggested the *P. aeruginosa* strain JS-11, as SeO_3^{2-} tolerant bacteria. Several earlier studies have suggested biogeochemical cycling of Se° through SeO_4^{2-}/SeO_3^{2-} reduction with the mixed cultures or sediment samples [38,39]. Furthermore, a decreased SeO_3^{2-} reduction by *Pseudomonas strutzeri* isolate have been reported at concentrations above 19.0 mM, due to greater toxicity of these oxyanions at

higher concentrations [6]. Thus, the non-toxic doses <12.5 mM were chosen for understanding the mechanistic aspects of extracellular SeO_3^{2-} reduction and biomimetic synthesis of elemental Se° nanospheres in this study using the spent medium (culture supernatant) of *P. aeruginosa* strain JS-11.

Phenazine-1-carboxylic Acid Production and Reducing Activity in Culture Supernatant

Fig. 2 shows the growth curve of *P. aeruginosa* strain JS-11 and release of a redox active metabolite phenazine-1-carboxylic acid

(PCA) in LB medium at 600 and 367 nm, respectively. Significant PCA synthesis occurred in the exponential phase, and reached to plateau at the stationary phase of bacterial growth (Fig. 2). PCA produced by strain JS-11 was crystallized in 0.1 N NaOH and the absorbance spectra of the yellow colored PCA solution in ultrapure water was obtained in the range of 250–550 nm (Fig. 2, inset). The characteristic absorption peak of PCA was obtained at 367 nm, which is identical to the PCA absorption wave length reported by Maddula et al. [40]. The PCA released in bacterial culture was further confirmed by HPLC analysis of PCA crystals dissolved in ultrapure water (Fig. 3). A typical peak appeared at a retention time of 9.6 min represents PCA, which conforms to earlier report [32], and reaffirmed the presence of this reducing agent in the culture supernatant. Incubation of 100 µl PCA solution with 2 mM $Na_2SeO_3^{2-}$ resulted in appearance of red color $Se°$ (Se-NPs) within 2 h. The data shown in Fig. 4 provide the information on the FTIR analysis of PCA alone and in presence of 2 mM $Na_2SeO_3^{2-}$. The FTIR spectra exhibited two clear peaks at 1,700 and 3400 cm^{-1}, whereas the absorbance peaks of PCA appeared at 1,693.8 and 1,605.7 cm^{-1} (Fig. 4). The shift in O–H broad absorbance peak between 3,000 and 3,500 cm^{-1} in the FTIR spectra of PCA indicated significant modification of PCA molecules upon interaction with $Na_2SeO_3^{2-}$ and its consequent reduction into Se-NPs. Moreover, the disappearance of PCA peaks at 1084 cm^{-1} suggests the occurrence of coordination between oxygen atom in the –C–O–C group of PCA and the Se atom.

Also, the reducing potential of the cell-free culture supernatant, based on $KMnO_4$ oxidation through the back titration assay was determined to be 1 $mgml^{-1}$ in LB medium (data not shown), which corresponds with the reduction ability of *K. pneumoniae* (0.96 $mgml^{-1}$) in LB medium [33]. Indeed, the reduction ability of cell-free culture supernatant varies with the types of culture media used for growing the organisms [33]. The reducing potential of the supernatant was further validated by measuring the native fluorescence of extracellular nicotinamide adenine dinucleotide (NADH) released by strain JS-11 in culture medium (Fig. 5). The fluorescence of NADH was observed to be quenched by 9.0, 10.8, 13.25, 27.0 and 29.1% with the addition of TiO_2-NPs, $ZnFe_2O_4$-NPs, CdS-NPs, Ag-NPs and EMS, respectively. These results signify the interaction of metal NPs with NADH, and suggested that the extracellular NADH could also be one of the important factors in reducing SeO_3^{2-} to elemental Se^0. NADH is known to play a fundamental role in the conversion of chemical energy to useful metabolic energy [39,41]. It is a well known reduced co-enzyme in redox reaction, and can be used as reducing agents by several enzymes [42,43]. Thus it is concluded that the reducing activity of strain JS-11 culture supernatant is mainly attributed to soluble redox active agents like PCA and NADH, besides the already known NADH dependent reductases.

Extracellular Biosynthesis of Se-NPs

The culture supernatant of bacterial strain JS-11 when challenged with 2 mM $Na_2SeO_3^{2-}$ solution, exhibited a change in color of the solution from light yellow to red (Fig. 6). The appearance of the red color indicated the occurrence of the reaction resulting in the formation of elemental $Se°$ in solution. The characteristic red color of the reaction solution was due to excitation of the surface plasmon vibrations of the Se-NPs and provided a convenient spectroscopic signature of their formation [44]. The synthesis of Se-NPs in solution was monitored by measuring the time dependent changes in the onset absorbance at an interval of 2, 24, 48, and 72 h. The onset absorption of NPs was centered at 520 nm (Fig. 6) with a red shift with increasing time

period. A time dependent increase was noticed in the absorption peaks of Se-NPs. No absorption peak corresponding to the control supernatant (without $Na_2SeO_3^{2-}$) or Se ion solution in the range of measurement was observed. Also, the symmetric plasmon band implies that the solution does not contain much of aggregated particles.

The application of the biological systems for synthesis of Se-NPs has been reported earlier [21,45]. However, the exact reaction mechanism leading to the formation of Se-NPs by the organisms has not yet been elucidated. Earlier studies suggested that NADH and NADH dependent nitrate reductase enzyme are important factors in the biosynthesis of metal nanoparticles. Ahmad et al. [46] reported that certain NADH dependent reductases are involved in reduction of metal ions in case of *F. oxysporum*. Also, Wang et al. [29] suggested the reduction of metal ions by the nitrate-dependent reductase and a shuttle quinine extracellular process. The reductases may also function as a capping agent, and ascertain the formation of thermodynamically stable nanostructures [47]. Hunter and Manter [21] have demonstrated the presence of selenite reductase (~ Mol. Wt. 115 kD) and a 700 kD protein capable of reducing both selenate and nitrate in cell free extracts of *Pseudomonas* sp. strain CA5. Etezad et al. [20] have also reported the role of NADPH and NADPH-dependent selenate reductases in SeO_4^{2-}/SeO_3^{2-} reduction. Therefore, it is likely that a multi-component redox system including PCA, NADH and most likely NADH-dependent reductases in the culture supernatant of strain JS-11 might act independently and/or in conjunction in catalyzing the biomimetic synthesis of Se-NPs in aqueous medium.

Physical Attributes of Se-NPs

The XRD pattern obtained for the extracellular Se-NPs with three intense peaks in the whole spectrum of 2θ values ranging from 20 to 80 is shown in Fig. 7. The diffractions at 31.64°, 45.35° and 56.41° can be indexed to the (101), (111) and (112) planes of the face-centered cubic (fcc) Se-NPs, respectively. The lattice parameters calculated by the Powder × software revealed that the maximum deviation that occurred between the observed and calculated values of interplanar spacing (d) remains below 0.002 Å. The full-width-at-half-maximum (FWHM) values measured for 101 planes of reflection were used to calculate the size of the NPs. The calculated average particle size of the extracellularly produced Se-NPs was determined to be 21 nm. A representative TEM image recorded from Se-NPs film deposited on a carbon-coated copper grid is shown in Fig. 8(A). The image shows the individual $Se°$ particles as well as some aggregates. The morphology of the Se-NPs was predominantly spherical. The EDS spectra derived from a nanosphere indicated that it was composed entirely of selenium (Fig. 8B). The Cu peaks were associated with the TEM grid, the Na and Cl peaks reflected the high salt content of the medium, and the C and O peaks most likely were associated with cellular exudate. The lack of any other metal peaks in the spectrum suggested that the selenium occurred in the elemental state $Se°$ rather than as a metal selenide (Se^{2-}). The results of hydrodynamic size of the Se-NPs obtained with dynamic light scattering are shown in Fig. 8 (C). The distribution curves show the Se-NPs aggregates of 264 nm in deionized ultrapure water. DLS is widely used to determine the size of brownian NPs in colloidal suspensions in the nano and submicron ranges. The higher size of nanoparticles in aqueous suspension as compared to TEM size might be due to the tendency of particles to agglomerate in aqueous state. The results also corroborate with the finding of Tran and Webster [48]. Since, the primary and secondary sizes of the NPs are regarded as important parameters, therefore, the behavior of Se-NPs in MQ water was evaluated through dynamic light scattering (DLS), to understand the extent of aggregation and secondary size of these NPs. However, the zeta potential of Se-NPs in aqueous solution was estimated to be -42 mV (Fig. 8C). The large negative or positive zeta potential indicates the repulsion between particles with little tendency to come together. On the contrary, if the particles have low zeta potential values then there is propensity of particles to come together and form aggregates. Thus, the high negative charge on Se° nanospheres is probably responsible for their higher stability without forming very large aggregates over a prolonged period of time. The morphology and size of the nanoparticles were further validated by AFM analysis. Fig. 8(D) shows the AFM image of Se-NPs obtained on scanning probe microscope in tapping mode, under ambient conditions. The average size of the nanoparticles and roughness (Ra) of surface were determined to be 23 nm and 18 nm, respectively using the WSXM and SPIP softwares.

Se-Assay for Nanotoxicity Assessment

A simple bacterial cell-free bioassay (Se-assay) was developed based on the reduction of SeO_3^{2-} to a red colored elemental Se° in culture supernatant of metabolically competent bacterial strain JS-11 cells. The Se-assay could be effectively used for visual assessment of the relative toxicity of a variety test chemicals including nanomaterials. The assay is based on the ability of a toxicant to inhibit the innate reducing power of the bacterial supernatant, and thereby impedes the SeO_3^{2-} reduction to Se^0. It is a phenomenon that does not occur in culture supernatant of dead or compromised cells, which maintains the original transparent and colorless state of SeO_3^{2-} solution. The development of red colored elemental Se° upon SeO_3^{2-} reduction is attributed either due to the PCA, NADH or NADH reductases, released in the culture supernatant, and regarded as metabolic markers. The data in Fig. 9 revealed the percent inhibition of SeO_3^{2-} reduction at the highest concentration (100 μgml^{-1}) of the Ag-NPs, $ZnFe_2O_4$-NPs, CdS-NPs, and TiO_2-NPs as 94.8%, 88%, 83% and 65%, respectively, which has suggested the differential toxicity of analyte NPs in the order as Ag>$ZnFe_2O_4$>CdS>TiO_2. The IC_{50} values based on the concentration of NPs that inhibits 50% of the SeO_3^{2-} reduction were determined to be 35, 47, 47 and 63 μgml^{-1} for Ag-NPs, $ZnFe_2O_4$-NPs, CdS-NPs and TiO_2-NPs, respectively. The percent inhibition with a known environmental toxicant EMS was observed to be 97.5% with an IC_{50} value of 24 μgml^{-1}. The IC_{50} or EC_{50} values vary with the nature of NPs and the assay types used for the assessment. Aruoja et al. [49] reported the EC_{50} values of ZnO-NPs, TiO_2-NPs and CuO-NPs to *Pseudokirchneriella subcapitata* in algal growth inhibition test as 0.04 mgl^{-1}, 5.83 mgl^{-1} and 0.71 mg l^{-1}, respectively. Whereas, the EC_{50} value of Ag-NPs has been reported to be 45–47 mgl^{-1} in *Vibrio fischeri*, based on inhibition of bioluminescence [50]. Thus, any comparisons of IC_{50} or EC_{50} values will be difficult due to differences in sensitivities of the assay systems. Nevertheless, it is suggested that in Se-assay, the SeO_3^{2-} reduction

is closely linked to the metabolic status of a bacterial cell and regarded as a measure of cell viability and integrity, which could be qualitatively or quantitatively determined based on the intensity of the red colored Se° product. Since, more active cells are able to generate and release more reducing factors in culture supernatant, therefore, a more intense red color develops due to greater reduction of SeO_3^{2-} to Se° [21]. Whereas, the treated cells or their culture supernatant under the influence of toxic effects of NPs or any other toxicant may lose their SeO_3^{2-} reduction ability and remains colorless. Thus, the Se-Assay entails a sharp red or colorless output, which could be easily applicable for prescreening of a variety of environmental toxicants including nanoparticles prior to intensive toxicity investigations.

Conclusions

In this study, the Se-NPs were synthesized employing the green technology, which involves the biological reduction process by the SeO_3^{2-} tolerant bacteria *Pseudomonas aeruginosa* strain JS-11. The strain JS-11 maintains the characteristics encompassing the (i) rapid and easy growth with greater tolerance to higher concentrations of SeO_3^{2-}, (ii) capability of viable cells to perform the intracellular and extracellular reduction of SeO_3^{2-} to elemental Se^0, (iii) susceptibility to target analytes, such as metal NPs and EMS, and (iv) ability of non-viable or dead cells to lose the ability of SeO_3^{2-} reduction. The Se-NPs were characterized by UV-Vis, XRD, TEM, EDX, FTIR and AFM analyses. It is envisaged that the metabolically active culture supernatant of SeO_3^{2-} tolerant strain JS-11 can be exploited as a redox active system for economically viable and environmental friendly production of Se-NPs. Furthermore, it is elucidated that the formation of red Se° from SeO_3^{2-} could serve as a molecular marker, whereas the inhibition of critical bioreduction step was considered as a toxicity end point for the qualitative and quantitative toxicity assessment. Nevertheless, further studies are warranted to better understand the stoichiometry of SeO_3^{2-} reduction as function of temperature and pH, to optimize the efficiency of Se-NPs production. Also, the Se-assay needs further validation with known xenobiotics comprising a wider spectrum of toxicants, to firmly establish this method as a broad spectrum and low cost eco-toxicity assay for pre-screening of an array of environmental toxicants.

Author Contributions

Conceived and designed the experiments: SD JM. Performed the experiments: SD MA. Analyzed the data: JM SD AAA. Contributed reagents/materials/analysis tools: AAA. Wrote the paper: JM SD.

References

1. Zhang J, Zhang SY, Xu JJ, Chen HY (2004) A new method for the synthesis of selenium nanoparticles and the application to construction of H_2O_2 biosensor. Chinese Chem Lett 15: 1345–1348.
2. Dong Y, Zhang H, Hawthorn L, Ganther HE (2003) Delineation of the molecular basis for selenium-induced growth arrest in human prostate cancer cells by oligonucleotide array. Cancer Res 63: 52–59.
3. Eustice DC, Kull FJ, Shrift A (1981) Selenium toxicity: aminoacylation and peptide bond formation with selenomethionine. Plant Physiol 67: 1954–1958.
4. Turner RJ, Weiner JH, Taylor DE (1998) Selenium metabolism in *Escherichia coli*. Biometals 11: 223–227.
5. Narasingarao P, Haggblom MM (2007) Identification of anaerobic selenate respiring bacteria from aquatic sediments. Appl Environ Microbiol 73: 3519–3527.
6. Lortie L, Gould WD, Rajan S, Meeready RGL, Cheng KJ (1992) Reduction of elemental selenium by a *Pseudomonas stutzeri* isolate. Appl Environ Microbiol 58: 4042–4044.
7. Oremland RS, Blum JS, Culbertson CW, Visscher PT, Miller LG, et al. (1994) Isolation, growth and metabolism of an obligately anaerobic, selenate-respiring bacterium, strain SES-3. Appl Environ Microbiol 60: 3011–3019.

8. Sabaty M, Avazeri C, Pignol D, Vermeglio A (2001) Characterization of the reduction of selenate and tellurite by nitrate reductases. Appl Environ Microbiol 67: 5122–5126.

9. DeMoll-Decker H, Macy JM (1993) The periplasmic nitrite reductase of *Thauera selenatis* may catalyze the reduction of selenite to elemental selenium. Arch Microbiol 160: 241–247.

10. Bledsoe TL, Cantafio AW, Macy JM (1999) Fermented whey- an inexpensive feed source for a laboratory-scale selenium-bioremediation reactor system inoculated with *Thauera selenatis*. Appl Microbiol Biotechnol 51: 682–685.

11. Euzeby J (2008) Validation List no. 121. List of new names and new combinations previously effectively, but not validly, published. Int J Syst Evol Microbiol 58: 1057.

12. Hunter WJ, Kuykendall LD (2007) Reduction of selenite to elemental red selenium by *Rhizobium* sp. strain B1. Curr Microbiol 55: 344–349.

13. Hunter WJ, Kuykendall LD, Manter DK (2007) *Rhizobium selenireducens* sp. nov.: a selenite reducing α-Proteobacteria isolated from a bioreactor. Curr Microbiol 55: 455–460.

14. Avazeri C, Turner RJ, Pommier J, Weiner JH, Giordano G, et al. (1997) Tellurite and selenate reductase activity of nitrate reductases from *Escherichia coli*: correlation with tellurite resistance. Microbiol 143: 1181–1189.

15. Yanke LJ, Bryant RD, Laishley EJ (1995) Hydrogenase (I) of *Clostridium pasteurianum* functions a novel selenite reductase. Anaerobe 1: 61–67.

16. Afkar E, Lisak J, Saltikov C, Basu P, Oremland RS, et al. (2003) The respiratory arsenate reductase from *Bacillus selenitireducens* strain MLS10. FEMS Microbiol Lett 226: 107–112.

17. Tomei FA, Barton LL, Lemanski CL, Zocco TG (1992) Reduction of selenate and selenite to elemental selenium by *Wolinella succinogenes*. Can J Microbiol 38: 1328–1333.

18. Roux M, Sarret G, Pignot-Paintrand I, Fontecave, Covès J (2001) Mobilization of selenite by *Ralstonia metallidurans* CH34. Appl Environ Microbiol 67: 769–773.

19. Sarret G, Avoscan L, Carrière M, Collins R, Geoffroy N, et al. (2005) Chemical Forms of Selenium in the Metal-Resistant Bacterium *Ralstonia metallidurans* CH34 Exposed to Selenite and Selenate. Appl Env Microbiol 71: 2331–2337.

20. Etezad SM, Khajeh K, Soudi M, Ghazvini PTM, Dabirmanesh B (2009) Evidence on the presence of two distinct enzymes responsible for the reduction of selenate and tellurite in *Bacillus* sp. STG-83. Enzyme Microb Tech 45: 1–6.

21. Hunter WJ, Manter DK (2009) Reduction of selenite to elemental red selenium by *Pseudomonas* sp. strain CA5. Curr Microbiol 58: 493–498.

22. Tejo Prakash N, Sharma N, Prakash R, Raina KK, Fellowes J, et al. (2009) Aerobic microbial manufacture of nanoscale selenium: exploiting nature's bio-nanomineralization potential. Biotechnol Lett 31: 1857–1862.

23. Yadav V, Sharma N, Prakash R, Raina KK, Bharadwaj LM, et al. (2008) Generation of selenium containing nano-structures by soil bacterium, *Pseudomonas aeruginosa*. Biotechnol 7: 299–304.

24. Losi M, Frankenberger WT (1997) Reduction of selenium by *Enterobacter cloacae* SLD1a-1: isolation and growth of bacteria and its expulsion of selenium particles. Appl Environ Microbiol 63: 3079–3084.

25. Casey A, Davoren M, Herzog E, Lyng FM, Byrne HJ, et al. (2007) Probing the interaction of single walled carbon nanotubes within cell culture medium as a precursor to toxicity testing. Carbon 45: 34–40.

26. Hurt RH, Monthioux M, Kane A (2006) Toxicology of carbon nanomaterials: status, trends, and perspectives on the special issue. Carbon 44: 1028–1033.

27. Monteiro-Riviere NA, Inman AO (2006) Challenges for assessing carbon nanomaterial toxicity to the skin. Carbon 44: 1070–1078.

28. Wörle-Knirsch JM, Pulskamp K, Krug HF (2006) Oops they did it again! Carbon nanotubes hoax scientists in viability assays. Nano Lett 6: 1261–68.

29. Wang B, Yu G, Hu H, Wang L (2007) Quantitative structure-activity relationships and mixture toxicity of substituted benzaldehydes to *Photobacterium phosphoreum*. Bull Environ Contam Toxicol 78(6): 503–9.

30. Lloyd-Jones G, Williamson WM, Slootweg T (2006) The Te-Assay: A black and white method for environmental sample pre-screening exploiting tellurite reduction. J Microbiol Meth 67: 549–556.

31. Dwivedi S, Singh BR, Al-Khedhairy AA, Musarrat J (2011) Biodegradation of isoproturon using a novel *Pseudomonas aeruginosa* strain JS-11 as a multi-functional bioinoculant of environmental significance. J Hazard Mat 185: 938–944.

32. Mavrodi DV, Bonsall RF, Delaney SM, Soule MJ, Phillips G, et al. (2001) Functional analysis of genes for biosynthesis of pyocyanin and phenazine-1-carboxamide from *Pseudomonas aeruginosa* PAO1. J Bacteriol 183: 6454–65.

33. Anjaiah V, Koedam N, Thompson BN, Loper JE, Höfte M, et al. (1998) Involvement of Phenazines and Anthranilate in the Antagonism of *Pseudomonas aeruginosa* PNA1 and Tn5 Derivatives Toward Fusarium spp. And Pythium spp. MPMI 11: 847–854.

34. Tjeerd van Rij E, Wesselink M, Chin-A-Woeng TFC, Bloemberg GV, Lugtenberg BJJ (2004) Influence of Environmental Conditions on the Production of Phenazine-1-Carboxamide by *Pseudomonas chlororaphis* PCL1391. MPMI 17: 557–566.

35. Fesharaki PJ, Nazari P, Shakibaie M, Rezaie S, Banoee M, et al. (2010) Biosynthesis of selenium nanoparticles using *Klebsiella pneumoniae* and their recovery by a simple sterilization process. Brazilian J Microbiol 41: 461–466.

36. Musarrat J, Dwivedi S, Singh BR, Al-Khedhairy AA, Azam A, et al. (2010) Production of antimicrobial silver nanoparticles in water extracts of the fungus *Amylomyces rouxii* strain KSU-09. Biores Technol 101: 8772–8776.

37. Patterson AL (1939) The Scherrer formula for X-ray particle size determination. Phys Rev 699 (56): 978–82.

38. Oremland RS, Hollibaugh JT, Maest AS, Presser TS, Miller L, et al. (1989) Selenate reduction to elemental selenium by anaerobic bacteria in sediments and culture: biogeochemical significance of a novel, sulfate-independent respiration. Appl Environ Microbiol 55: 2333–2343.

39. Oremland RS, Steinberg NA, Maest AS, Miller LG, Hollibaugh JT (1990) Measurement of in situ rates of selenate removal by dissimilatory bacterial reduction in sediments. Environ Sci Technol 24: 1157–1164.

40. Maddula VSRK, Pierson EA, Pierson LS (2008) Altering the ratio of phenazines in *Pseudomonas chlororaphis* (*aureofaciens*) Strain 30–84: Effects on biofilm formation and pathogen inhibition. J Bacteriol 190: 2759–2766.

41. Hull RV, Conger PS, Hoobler RJ (2001) Conformation of NADH studied by fluorescence excitation transfer spectroscopy. Biophys Chem 90: 9–16.

42. Dudev T, Lim C (2010) Factors controlling the mechanism of NAD$^+$ non-redox reactions. J Am Chem Soc 132: 16533–16543.

43. Lin H (2007) Nicotinamide adenine dinucleotide: Beyond a redox coenzyme. Org Biomol Chem 5: 2541–2554.

44. Lin ZH, Wang CRC (2005) Evidence on the size-dependent absorption spectral evolution of selenium nanoparticles. Mater Chem Phys 92: 591–594.

45. Dhanjal S, Cameotra SS (2010) Aerobic biogenesis of selenium nanospheres by *Bacillus cereus* isolated from coalmine soil. Microb Cell Fact 9: 52.

46. Ahmad A, Mukherjee P, Senapati S, Mandal D, Khan MI, et al. (2003) Extracellular biosynthesis of silver nanoparticles using the fungus *Fusarium oxysporum*. Colloids Surf B 28: 313–318.

47. He S, Guo Z, Zhang Y, Zhang S, Wang J, et al. (2007) Biosynthesis of gold nanoparticles using the bacteria *Rhodopseudomonas capsulate*. Mater Lett 61: 3984.

48. Tran PA, Webster TJ (2011) Selenium nanoparticles inhibit *Staphylococcus aureus* growth. Int J Nanomed 6: 1553–1558.

49. Aruoja V, Dubourguier HC, Kasemets K, Kahru A (2009) Toxicity of nanoparticles of CuO, ZnO and TiO$_2$ to microalgae *Pseudokirchneriella subcapitata*. Sci Total Environ 407: 1461–1468.

50. Binaeian E, Rashidi AM, Attar H (2012) Toxicity study of two different synthesized silver nanoparticles on bacteria *Vibrio Fischeri*. World Acad Sci Eng Technol 67: 1219–1225.

17

Nanopore Analysis of Wild-Type and Mutant Prion Protein (PrP^C): Single Molecule Discrimination and PrP^C Kinetics

Nahid N. Jetha[1]*, **Valentyna Semenchenko**[2], **David S. Wishart**[2,3], **Neil R. Cashman**[4], **Andre Marziali**[1]

1 Department of Physics and Astronomy, University of British Columbia, Vancouver, British Columbia, Canada, **2** National Institute for Nanotechnology, Edmonton, Alberta, Canada, **3** Departments of Computing Science and Biological Sciences, University of Alberta, Edmonton, Alberta, Canada, **4** Brain Research Centre, University of British Columbia, Vancouver, British Columbia, Canada

Abstract

Prion diseases are fatal neurodegenerative diseases associated with the conversion of cellular prion protein (PrP^C) in the central nervous system into the infectious isoform (PrP^{Sc}). The mechanics of conversion are almost entirely unknown, with understanding stymied by the lack of an atomic-level structure for PrP^{Sc}. A number of pathogenic PrP^C mutants exist that are characterized by an increased propensity for conversion into PrP^{Sc} and that differ from wild-type by only a single amino-acid point mutation in their primary structure. These mutations are known to perturb the stability and conformational dynamics of the protein. Understanding of how this occurs may provide insight into the mechanism of PrP^C conversion. In this work we sought to explore wild-type and pathogenic mutant prion protein structure and dynamics by analysis of the current fluctuations through an organic α-hemolysin nanometer-scale pore (nanopore) in which a single prion protein has been captured electrophoretically. In doing this, we find that wild-type and D178N mutant PrP^C, (a PrP^C mutant associated with both Fatal Familial Insomnia and Creutzfeldt-Jakob disease), exhibit easily distinguishable current signatures and kinetics inside the pore and we further demonstrate, with the use of Hidden Markov Model signal processing, accurate discrimination between these two proteins at the single molecule level based on the kinetics of a single PrP^C capture event. Moreover, we present a four-state model to describe wild-type PrP^C kinetics in the pore as a first step in our investigation on characterizing the differences in kinetics and conformational dynamics between wild-type and D178N mutant PrP^C. These results demonstrate the potential of nanopore analysis for highly sensitive, real-time protein and small molecule detection based on single molecule kinetics inside a nanopore, and show the utility of this technique as an assay to probe differences in stability between wild-type and mutant prion proteins at the single molecule level.

Editor: Maria Gasset, Consejo Superior de Investigaciones Cientificas, Spain

Funding: This work was funded by PrioNet Canada (http://www.prionetcanada.ca/). NNJ received scholarship funding from the Natural Sciences and Engineering Research Council of Canada (http://www.nserc-crsng.gc.ca/index_eng.asp). The funders had no role in study design, data collection and analysis, decision to publish, or preparation of the manuscript.

Competing Interests: The authors have declared that no competing interests exist.

* E-mail: nahid@phas.ubc.ca

Introduction

Prion diseases are a class of fatal neurodegenerative diseases affecting both humans and animals that are associated with the accumulation of PrP^{Sc} in the central nervous system [1,2,3,4], a pathological isoform of normal cellular prion protein (PrP^C). The widely accepted protein-only hypothesis of prion disease pathogenesis implicates PrP^{Sc} as the principal and possibly sole infectious agent, capable of self-replication by post-translational interaction with PrP^C stimulating its conversion into PrP^{Sc} (a process known as template-directed conversion) [1,2,3]. The mechanics of template-directed conversion, however, are almost entirely unknown, stymied by the lack of an atomic-level structure for PrP^{Sc}, primarily due to its insolubility and tendency to aggregate [5,6]. A number of pathogenic PrP^C mutants exist that are characterized by an increased propensity for conversion into PrP^{Sc} and that differ from wild-type by only a single amino-acid point mutation in their primary structure. These mutations are known to perturb the stability properties and conformational

dynamics of the protein [7,8,9]. Understanding of how this occurs may provide insight into the mechanism of PrP^C conversion in disease. In this work we sought to explore prion protein structure and dynamics, for both wild-type and pathogenic mutant PrP^C by analysis of the current fluctuations through a nanometer-scale pore in which a single prion protein has been captured. This method of biomolecule analysis, known as single-molecule nanopore analysis, has emerged as a powerful tool for investigating and characterizing the structure and dynamics of individual biomolecules [10,11,12,13]. The technique involves electrophoretically driving an individual biomolecule (immersed in an electrolyte) into a nanometer-scale pore (nanopore) formed in an insulating membrane (e.g. an organic nanopore formed in a lipid bilayer). Direct monitoring of the ionic current through the pore enables detection of individual biomolecule capture events which are characterized by a substantial reduction in the nanopore current relative to an open-pore. Once captured, the ionic current through the pore serves as an extremely sensitive metric and rich source of information on the conformational dynamics and structural properties of the captured molecule [11,12,13]. To date, single-molecule nanopore analysis has been predominantly applied towards characterizing the structure and properties of individual DNA molecules. This has been primarily due to the enormous potential of nanopore analysis to form the basis of a low-cost, high-throughput DNA sequencing technology [14,15]. Analysis of proteins and polypeptides by comparison has only recently begun and has shown much promise as a means by which to probe the unfolding kinetics of proteins [16,17], characterize protein-pore interactions [18,19,20] and to study the transport properties of proteins through pores [18,21,22]. A major challenge, however, with nanopore protein analysis (in comparison to nucleic acid analysis for example) is that in contrast to heavily charged biopolymers such as DNA, the charge distribution of proteins and polypeptides can be highly irregular, positive, negative or neutral, significantly affecting the ability to capture proteins in the pore. Here we present results demonstrating capture of individual wild-type and D178N mutant PrP^C molecules into an organic α-hemolysin nanopore (mutant D178N is a pathogenic PrP^C mutant associated with both Fatal Familial Insomnia and Creutzfeldt-Jakob disease [23,24]). We show that these two proteins, which differ from each other by only a single amino-acid point mutation in their primary structure, exhibit easily distinguishable current signatures and kinetics inside the pore and we further demonstrate, with the use of Hidden Markov Model signal processing, accurate detection and discrimination between these two proteins at the single molecule level based on the kinetics of a single PrP^C capture event. In addition, we present a four-state model to describe wild-type PrP^C kinetics in the pore which represents a first step in our investigation into characterizing the differences in kinetics and conformational dynamics between wild-type and D178N mutant PrP^C.

Materials and Methods

PrP^C Constructs

PrP^C (both wild-type and mutant) was expressed and purified by the PrioNet Prion Protein & Plasmid Production Platform Facility (refer to File S1 for details on expression and purification protocol).Truncated Syrian Hamster PrP^C (residues 120–232–designated ShPrP(120–232)) was used in order to investigate the structure and dynamics of the PrP^C structural core. To facilitate capture of PrP^C inside the nanopore, the N-terminus of ShPrP(120–232) was adapted with four positively charged amino-acid residues (KKRR) (designated KKRR-ShPrP(120–232)). We expect these additional residues to have a minimal effect on the overall structure and stability of PrP^C based on previous studies whereby truncated PrP^C (of various lengths) was adapted with a 22 residue N-terminal fusion tag, which was found to have no influence on PrP^C structural stability [25,26,27]. Three PrP^C constructs were investigated in this study, namely: ShPrP(120–232), KKRR-ShPrP(120–232) and KKRR-ShPrP(120–232)-D178N (i.e. mutant PrP^C).

Nanopore Experiments

α-hemolysin (α-HL) nanopores were formed using a method adapted from that of Akeson et al. [28]. Briefly, a black lipid membrane of 1,2-diphytanoyl-sn-glycero-3-phosphocholine (Avanti Polar Lipids Inc., Alabaster, AL) and hexadecane (Sigma-Aldrich, St. Louis, MO) is formed across a 25 μm PTFE aperture connecting two baths filled with electrolyte (Figure 1. Details on lipid bilayer formation are described elsewhere [29]). Owing to the low charge density of KKRR-ShPrP(120–232) (relative to heavily charged biopolymers such as DNA and RNA),

and thus decreased ability to capture PrP^C in the pore, experiments were conducted under asymmetric salt conditions which, through a combination of electric field enhancement around the entrance of the pore and osmotic flow [30,31], substantially increases the nanopore-capture rate of small molecules in solution (relative to symmetric salt conditions) [30,31]. PrP^C was preloaded on the *trans* side of the pore in 0.3 M KCl, 45 mM NaPO$_4$, 10 mM HEPES, pH 8.0 solution (final PrP^C concentration was ~13.4 μM). We do not expect oligomerization, aggregation or precipitation of PrP^C under these solution conditions based on mass spec results of KKRR-ShPrP(120–232) in solution and on previous studies with various full-length and truncated forms of PrP^C in high salt [32,33]. Our mass spec data confirms the presence of monomeric PrP^C in solution and the absence of dimers (data not shown), suggesting the absence of higher-order oligomers as well. Evidence in the literature indicates that the aggregation propensity of PrP^C in the presence of high salt is due to interactions of anions in solution with glycine groups in the glycine-rich unstructured PrP^C N-terminus (i.e. residues 23–119), and is therefore a property of full-length PrP^C (i.e. PrP^C(23–232)) [32,33]. Moreover, studies of truncated PrP^C (i.e. PrP^C(120–232)) in high salt buffer (0.5 M NaCl) find no change in secondary structure content compared to salt-free buffer and do not report of aggregation or precipitation of PrP^C in solution [32]. In contrast to the *trans* side of the pore the *cis* side contained 3 M KCl, 10 mM HEPES, pH 8.0 solution. All experiments were conducted (and maintained) at a temperature of 20°C ±0.1°C. An Axon Axopatch 200 B patch clamp amplifier is used to measure the ionic current. Data is low-pass filtered at 10 kHz by a 4-pole Bessel filter and sampled at 100 kHz. Experiments were conducted with a single α-HL pore incorporated in the lipid bilayer. Formation of an α-HL pore was done under symmetric salt conditions (i.e. 0.3 M KCl on both the *cis* and *trans* sides of the pore) by injection of free subunits into solution (on the *cis* side of the pore) which subsequently self-assemble into heptameric, membrane-spanning pores or by injection of preformed heptameric α-HL into solution which spontaneously forms into a transmembrane pore in a lipid bilayer

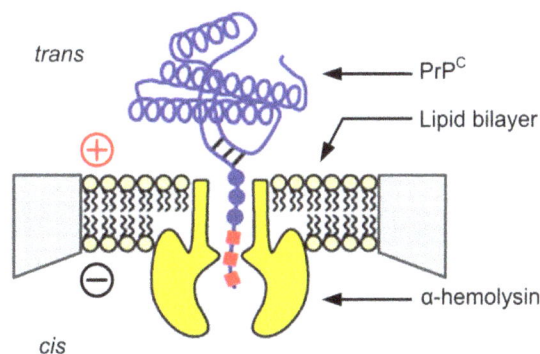

Figure 1. Cartoon illustrating the capture of an individual PrP^C molecule into an α-hemolysin nanopore. PrP^C is electrophoretically driven into the α-HL nanopore (voltage polarity given by the plus and minus signs) via its positively charged N-terminus. The *trans* chamber contains 0.3 M KCl, 45 mM NaPO$_4$, 10 mM HEPES, pH 8.0 solution at a PrP^C concentration of 13.4 μM. The *cis* chamber contains 3 M KCl, 10 mM HEPES, pH 8.0 solution. The salt-concentration gradient across the pore generates an osmotic flow from *trans*-to-*cis* and enhances the electric field around the entrance of the *trans*-side of the pore [30,31] thereby substantially increasing the nanopore-capture rate of PrP^C in solution relative to symmetric salt conditions [30,31]. Experiments were conducted (and maintained) at a temperature of 20°C ±0.1°C.

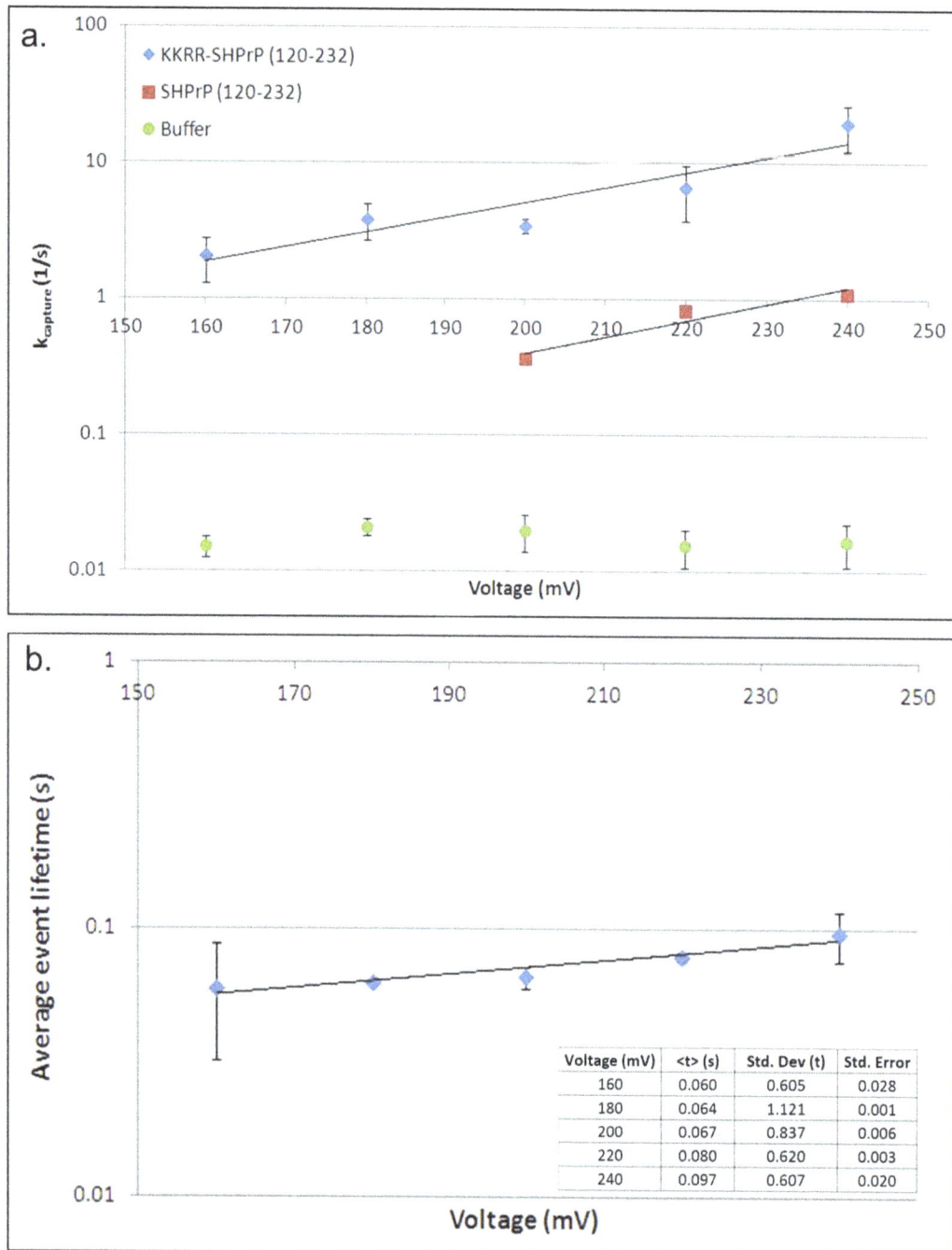

Figure 2. Nanopore capture rate and KKRR-ShPrP(120–232) average event lifetime as a function of voltage (3 M KCl *cis*, 0.3 M KCl *trans*). (**a**) Shown is the capture rate as a function of voltage for KKRR-ShPrP(120–232) (blue), ShPrP(120–232) (red) and the buffer-only control (green). In the case of the buffer-only control the capture rate represents the rate of pore gating as a function of voltage. Both KKRR-ShPrP(120–232) and ShPrP(120–232) exhibit capture kinetics that are exponentially dependent on voltage consistent with the applied voltage acting on the positively charged residues at the N-terminus (five in the case of KKRR-ShPrP(120–232) and one in the case of ShPrP(120–232)) to decrease the energy barrier height for entry into the pore, thereby exponentially increasing the capture rate. In addition, the capture rate for KKRR-ShPrP(120–232) is between one and one-and-a-half orders of magnitude higher than ShPrP(120–232) indicating that the large majority of captures of KKRR-ShPrP(120–232) involve threading of the N-terminus through the pore. Error bars represent the standard error on the mean, and were determined via bootstrapping in the case of ShPrP(120–232) and the buffer-only control, whereas in the case of KKRR-ShPrP(120–232) error bars were determined based on two separate datasets. (**b**) The average event lifetime for KKRR-ShPrP(120–232) increases exponentially with voltage consistent with PrP^C escape from the pore (as opposed to translocation) over an electrostatic energy barrier (governed by the applied voltage). The standard deviation of the event lifetime distribution indicates the presence of both short and very long time events (i.e. >1 s). Error bars represent standard error on the mean and were determined based on two separate datasets.

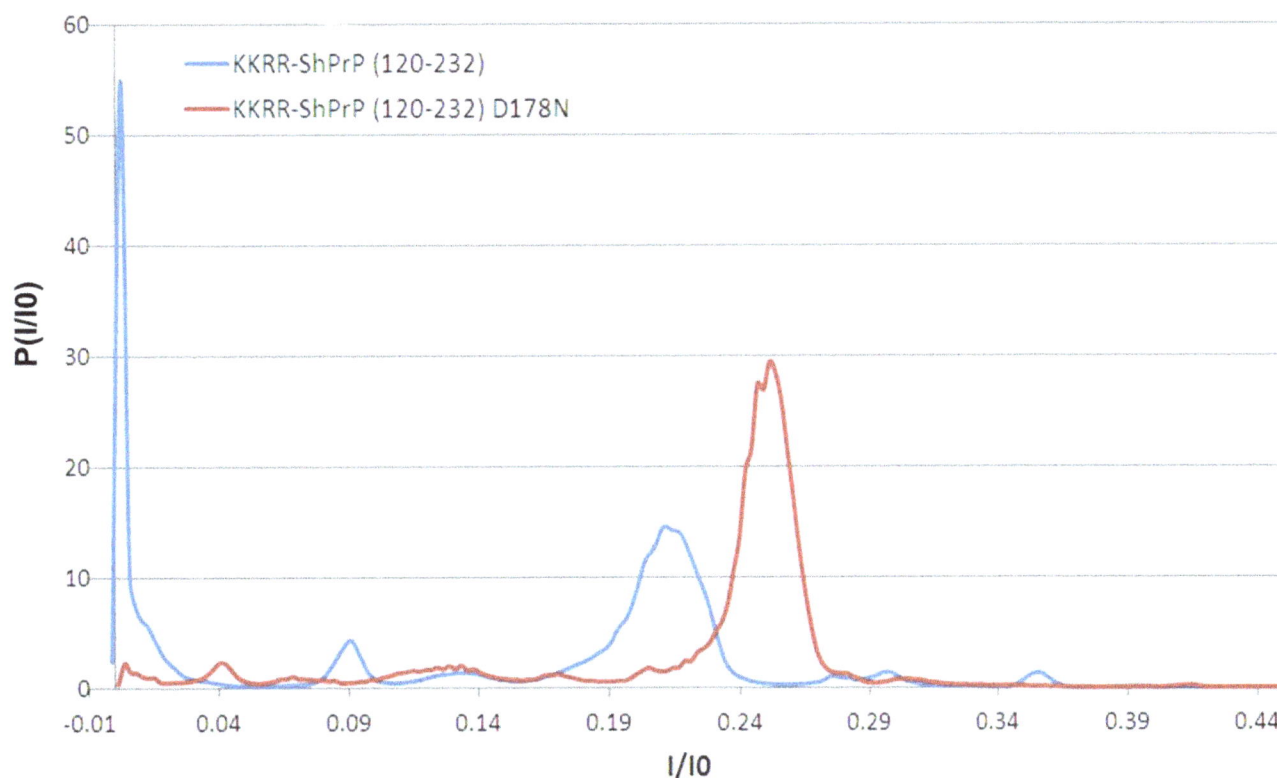

Figure 3. All-point current histogram for KKRR-ShPrP(120–232) and KKRR-ShPrP(120–232)-D178N. Ionic current histogram of all PrP^C capture events from all voltages, with ionic current normalized by the open-pore current (I0) at a given voltage, for KKRR-ShPrP(120–232) (blue) and KKRR-ShPrP(120–232)-D178N (red). Ionic current is median filtered to 2.99 ms per data-point. The histograms exhibit multiple peaks with varying degrees of overlap indicative of complex PrP^C kinetics in the pore. Moreover, the histograms exhibit clear distinguishable features (e.g. the near absence of the peak at I/I0 ~0 pA with respect to KKRR-ShPrP(120–232)-D178N).

[34]. Confirmation of a single pore incorporation is achieved by applying a +100 mV electric potential across the membrane (*trans* side positive) and observing a specific step-wise increase in the current (~+28 pA at +100 mV and ~−20 pA at −100 mV). Once this is observed, the RMS current noise on the pore (5 kHz bandwidth using a Butterworth filter – independent of the 10 kHz low-pass Bessel filter) is then probed at +100 mV and +200 mV to confirm that a heptameric pore has incorporated into the bilayer (as opposed to an anomalous structure, e.g. a hexameric pore). At +100 mV the upper limit for the 5 kHz noise (indicative of a heptameric pore), for our experimental setup, is ~0.80 pA RMS, and at +200 mV the upper limit is ~1.20 pA RMS. If both the current and noise are within specification, the *cis* side of the pore is then perfused with 3 M KCl, 10 mM HEPES, pH 8.0 solution several times to ensure that any free, unbound α-HL has been removed from solution and to ensure that the salt concentration on the *cis*-side of the pore has been increased to 3 M KCl. Custom-built data acquisition software (described elsewhere [35]) is used to apply large positive voltages (e.g. +160 mV or greater) across the pore (and record the corresponding current and voltage) thereby electrophoretically driving individual PrP^C molecules in solution into the pore. Several thousand individual PrP^C capture events were recorded at each voltage in the range of +160 mV to +240 mV for each PrP^C construct.

PrP^C Capture Rate & KKRR-ShPrP (120–232) Event Lifetime

To confirm that KKRR-ShPrP(120–232) enters the pore N-terminal first, we characterized and compared the nanopore capture rate as a function of voltage for KKRR-ShPrP(120–232), ShPrP(120–232), and a buffer only control (indicative of the pore gating rate under the given buffer conditions – i.e. asymmetric salt concentration, Figure 2A). The capture rate for KKRR-ShPrP(120–232) and ShPrP(120–232) is exponentially dependent on voltage, consistent with the capture process being dominated by an energy barrier whereby, according to classical Kramer's theory, the applied voltage acts on the N-terminal positive charges (five in the case of KKRR-ShPrP(120–232) and one in the case of ShPrP(120–232)), decreasing the energy barrier height for entry into the pore, thereby exponentially increasing the rate of PrP^C capture. Moreover, the nanopore-capture rate of KKRR-ShPrP(120–232) is between one and one-and-a-half orders of magnitude higher (depending on voltage) than the capture rate of ShPrP(120–232), reflecting the greater charge density at the N-terminus in the case of KKRR-ShPrP(120–232). These results indicate that the large majority of captures of KKRR-ShPrP(120–232) involve threading of the N-terminus through the pore.

The average event lifetime for KKRR-ShPrP(120–232) as a function of voltage is shown in Figure 2B. The event lifetime increases exponentially with voltage, consistent with PrP^C escape from the pore (as opposed to translocation) being the dominant mode of termination of an event, and requiring crossing an

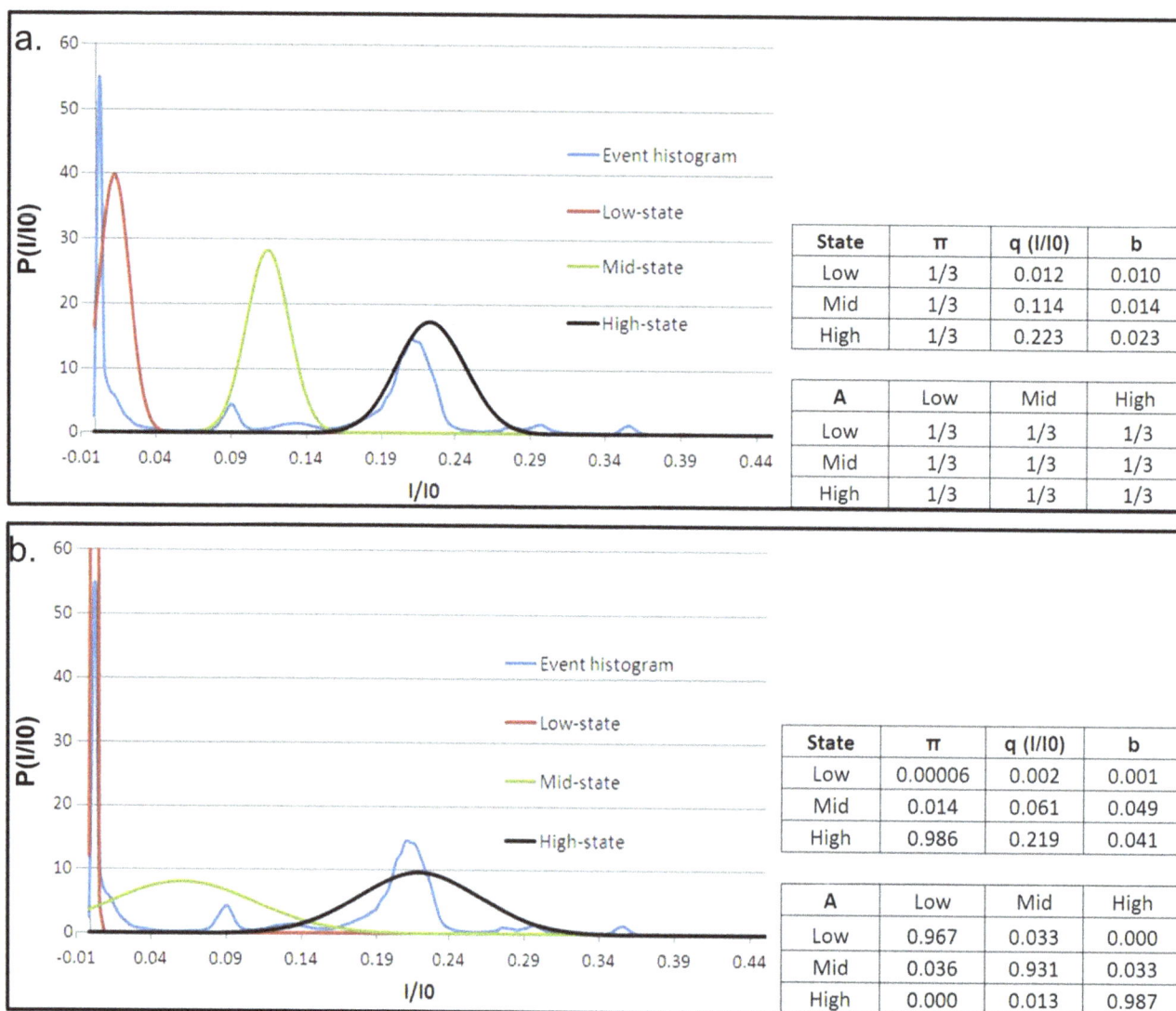

a.

State	π	q (I/I0)	b
Low	1/3	0.012	0.010
Mid	1/3	0.114	0.014
High	1/3	0.223	0.023

A	Low	Mid	High
Low	1/3	1/3	1/3
Mid	1/3	1/3	1/3
High	1/3	1/3	1/3

b.

State	π	q (I/I0)	b
Low	0.00006	0.002	0.001
Mid	0.014	0.061	0.049
High	0.986	0.219	0.041

A	Low	Mid	High
Low	0.967	0.033	0.000
Mid	0.036	0.931	0.033
High	0.000	0.013	0.987

Figure 4. KKRR-ShPrP(120–232) event histogram and initial and optimal HMM models. (a) (Left) The KKRR-ShPrP(120–232) event histogram (blue) is divided into three regimes/states a high-state (black), mid-state (green) and low-state (red). The location of the peak and width of the distribution for each state in our initial model represents our best guess of the location and size of a given regime. **(Right)** The model parameters: π (i.e. the initial condition or probability that an event begins in a given state), q (the location of the peak of the Gaussian distribution, in terms of I/I0, for a given state), b (the standard deviation on the Gaussian of each state, which defines a state's noise properties), and A (the state-to-state transition probability matrix). In our initial model we assume ignorance of the probabilities and therefore assume π to be uniformly distributed (i.e. an event is assumed equally likely to begin in any of the three states). Similarly with the transition probability matrix A, we assume all transitions to be equally likely (e.g. if in the low-state there is an equal probability for remaining in the low-state as there is for transitioning into the mid-state or the high-state). **(b) (Left)** After 40 iterations of the EM algorithm the optimal three-state model that best describes the data (i.e. the maximum likelihood model estimate) is converged upon. The low-state, far from encompassing all of the low current (as was presumed in our initial model) is very narrow and well defined, while the mid and high states both broaden out (the peak of the mid-state also shifts to a deeper conductance level relative to the initial model). **(Right)** The corresponding optimal model parameters.

electrostatic energy barrier (governed by the applied voltage). The standard deviation of the event lifetime distribution shows the lifetime spanning several orders of magnitude indicating the presence of both short and very long time events (i.e. >1 s) (refer to File S1 regarding details on how the capture rate and average event lifetime are determined).

Results and Discussion

Figure 3 shows the ionic current histograms for all PrP^C capture events from all voltages, with ionic current normalized by the open-pore current at a given voltage, for KKRR-ShPrP(120–232)

and KKRR-ShPrP(120–232)-D178N. Both histograms display multiple peaks, with varying degrees of overlap, indicative of complex PrP^C kinetics inside the pore. Moreover, the histograms exhibit clear distinguishable features (e.g. the near absence of the peak at I/I0 ∼0 pA with respect to mutant PrP^C), indicating differences in conformational dynamics between the two proteins in the pore.

In order to model PrP^C kinetics in the pore and characterize signal statistics we developed a signal processing algorithm based on Hidden Markov Models (HMMs). HMM signal processing is a powerful technique by which to extract and characterize low-level signals buried in background noise [36,37,38]. The approach has

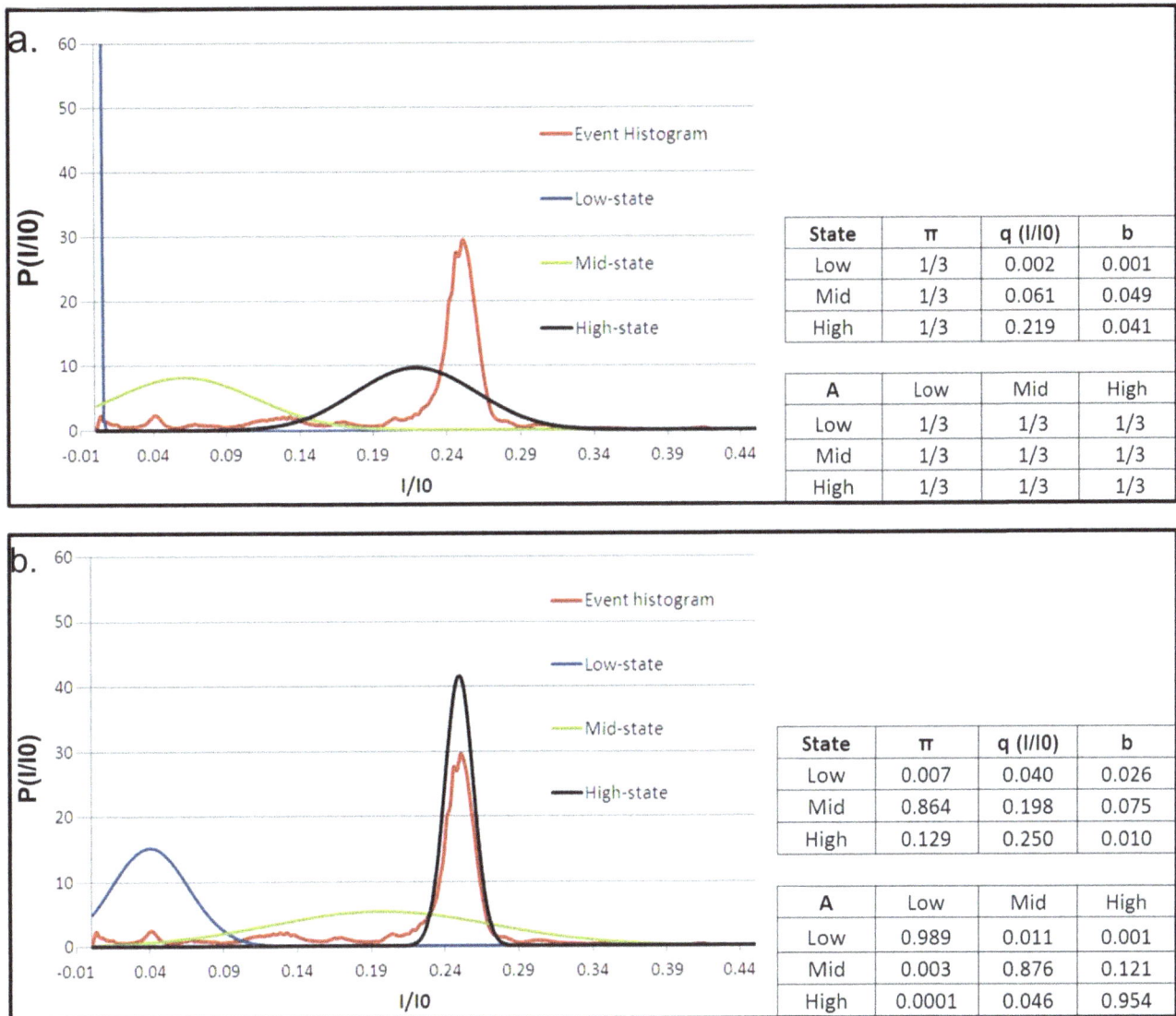

a.

State	π	q (I/I0)	b
Low	1/3	0.002	0.001
Mid	1/3	0.061	0.049
High	1/3	0.219	0.041

A	Low	Mid	High
Low	1/3	1/3	1/3
Mid	1/3	1/3	1/3
High	1/3	1/3	1/3

b.

State	π	q (I/I0)	b
Low	0.007	0.040	0.026
Mid	0.864	0.198	0.075
High	0.129	0.250	0.010

A	Low	Mid	High
Low	0.989	0.011	0.001
Mid	0.003	0.876	0.121
High	0.0001	0.046	0.954

Figure 5. KKRR-ShPrP(120–232)-D178N event histogram and initial and optimal HMM models. (a) (Left) The KKRR-ShPrP(120–232)-D178N event histogram (red), and the corresponding high-state (black), mid-state (green) and low-state (blue) that make up the initial model for HMM analysis. The location of the peak (q) and width of the distribution (b) for each state are the same as for the optimal KKRR-ShPrP(120–232) model (Figure 4B). This choice for q and b serves to highlight how the individual states evolve and differ from that of wild-type PrP^C. **(Right)** The corresponding initial model parameters. Similar to the initial model for KKRR-ShPrP(120–232) (Figure 4A) we assume ignorance of the probabilities and therefore assume π to be uniformly distributed. Likewise, with the transition probability matrix (A), we assume all transitions to be equally likely. **(b) (Left)** After 36 iterations of the EM algorithm the optimal three-state model that best describes the data (i.e. the maximum likelihood model estimate) is converged upon. The individual states (properties and kinetics) are significantly different from wild-type PrP^C (Figure 4B), highlighting the importance of amino-acid residue D178 to the dynamics and structural stability of PrP^C **(Right)** The corresponding optimal model parameters.

been previously applied in characterizing the complex kinetics of DNA hairpins trapped in the α-HL nanopore [39,40]. HMM analysis requires an initial model for the system defined by the following parameters: State levels in terms of I/I0 (q), the initial condition (π) (i.e. the probability that an event begins in a given state), the transition probabilities between states (A), and the noise properties (the standard deviation on a Gaussian) of each state (b). These parameters represent best guesses and can be estimated, with respect to q and b, from the event histogram. Once the initial model is developed and provided with the corresponding data, the HMM operates through an expectation maximization (EM) algorithm improving upon the model parameters with each iteration until convergence to an optimal set of parameters that

best describes the data, based on model likelihood, i.e. HMM analysis converges to the maximum likelihood model estimate (we refer the reader to [36,41] for details regarding HMM theory and its implementation). We model PrP^C kinetics in the pore as a three-state system. Our choice of three states to describe PrP^C kinetics is based on parsimony i.e. selecting a model with the fewest parameters that describes the data well. Moreover, our choice is based on wild-type PrP^C kinetics and in particular on a qualitative assessment of the form of the wild-type histogram. As mentioned previously, the histogram displays multiple peaks, with varying degrees of overlap. We find, however, that the peaks are concentrated into roughly three regimes. A simple description of the histogram, therefore, is one in which the current is split into a

Table 1. Wild-type and mutant predictive value.

	Case 1	Case 2
# of events in 50:50 mix	1000	176
Wild-type predictive value	0.71	0.85
Mutant predictive value	0.86	0.90

Case 1 refers to the situation whereby all events, regardless of event lifetime, are analyzed. The simulated 50:50 mix dataset forms the individual events to be protein-called by which wild-type and mutant predictive value is determined. These events are randomly selected from the total number of wild-type and mutant events. After selection (and removal from the total number of events), the remaining events form the training sets for both wild-type and mutant. Wild-type predictive value refers to the likelihood that an event is in fact a wild-type event given that the protein-calling algorithm has called it as wild-type. Similarly the mutant predictive value refers to the likelihood that an event is in fact a mutant event given that the protein-calling algorithm has called it as mutant. Case 2 refers to the case whereby only those events that have an event lifetime of ≥1 s are analyzed (i.e. only long time events makeup the training sets for both wild-type and mutant and the generated 50:50 mix dataset). The predictive value, not surprisingly, improves when only considering long-time events which is primarily due to the fact that long-time events can be better assessed in terms of their kinetics than short-time events (i.e. the amount of data available to characterize an event is proportional to the event lifetime) thereby improving protein-calling accuracy.

high, mid and low regime (i.e. three states - refer to Figure 4A, which shows our initial wild-type PrP^C HMM model and best guess at the location and size of each regime). For ease of comparison between wild-type and D178N mutant PrP^C we model mutant kinetics in the pore as a three-state system as well. The initial and optimal (post-HMM processed) models are shown in Figures 4 and 5 for KKRR-ShPrP(120–232) and KKRR-ShPrP(120–232)-D178N respectively.

The optimal three-state model for wild-type and mutant PrP^C (Figures 4B and 5B respectively) reveal significant differences in both state properties and kinetics between the two proteins, highlighting the importance of amino acid residue D178 to the dynamics and structural stability of PrP^C. It is known that residue D178 in wild-type PrP^C stabilizes the protein through salt-bridge interactions with R164 (the C-terminus of β-strand 2) and by hydrogen bonding to Y128 (the N-terminal Tyr of β-strand 1) [42,43,44,45]. In mutant D178N therefore, these stabilizing forces are no longer present, the loss of which appears to significantly affect the conformational dynamics of mutant PrP^C in the pore.

The clear distinction between these two proteins also highlights the sensitivity of nanopore analysis in detecting changes in biomolecule structure and demonstrates the potential of using this technique for detection and identification of small molecules and proteins in solution based on differences in kinetics in a nanopore. In order to explore this potential we characterized protein-calling accuracy between wild-type and D178N mutant PrP^C at the single event level, based on kinetic differences in the pore (i.e. given a single event we characterized the accuracy with which it can be determined which protein, either wild-type or mutant, produced the event based on kinetics). In this regard we investigated two cases in particular:

1) Where all events are analyzed, regardless of event lifetime and

2) Where only those events that have a lifetime of ≥1 s are analyzed

Case 2 allows us to characterize how protein-calling accuracy changes as we limit our study to long-time events (i.e. those events

with long observation times and therefore better statistics for discriminating between the two proteins). In order to make individual calls given an event we developed a protein-calling algorithm based on HMM model likelihood. The method by which we characterize protein-calling accuracy and the results obtained are described in the following.

Case 1

Of all wild-type and mutant PrP^C events we first simulate a 50:50 mix dataset. Therefore 500 wild-type events and 500 mutant events are randomly selected and combined to form a simulated 50:50 mix of 1000 events. The remaining events for both wild-type and mutant form the training sets for the corresponding protein by which to build optimal HMM models (refer to Figures S1 and S2 in File S1 for the determined optimal wild-type and mutant models, respectively). Protein-calling accuracy is assessed by calling the individual events from the 50:50 mix dataset (i.e. the blind, unanalyzed events). Individual events are called as either wild-type or mutant using our HMM protein-calling algorithm. The algorithm works as follows: Given an individual event and the optimal wild-type and mutant HMM models (as determined via HMM analysis of the respective training data) the algorithm calculates the likelihood function of the event given each model (i.e. $P(E|\lambda_{\text{wild-type}})$ and $P(E|\lambda_{\text{mutant}})$, where E is the event and λ represents a given model). The algorithm then calls an event as either wild-type or mutant based on maximum likelihood (i.e. the protein-call is based on whichever of the two determined likelihood functions is largest). The protein-calling results are given in Table 1. The results are given in terms of the wild-type and mutant predictive value. This is defined as the likelihood that an event which is called as either wild-type or mutant is called correctly (e.g. if the algorithm calls an event as mutant there is a 0.86 likelihood that the event is a mutant event).

Case 2

In characterizing protein-calling accuracy in case 2 the method is the same as described for case 1 with the difference being that in case 2 we are only interested in events that have an event lifetime of ≥1 s. In other words, all events making up the training sets (wild-type and mutant) and the generated 50:50 mix dataset have an event lifetime of ≥1 s (refer to Figures S3 and S4 in File S1 for the determined optimal wild-type and mutant models in this case, respectively). The total number of events for the 50:50 mix dataset is 176. The results, in terms of predictive value are given in Table 1.

These results show, not surprisingly, that predictive value (i.e. protein calling accuracy) improves when considering only long-time events. This is primarily due to the fact that long-time events can be better assessed in terms of their kinetics than short-time events (i.e. the amount of data available to characterize an event increases proportionately with the event lifetime) thereby improving protein-calling accuracy. In particular, the mutant at short times has a greater propensity for being called as wild-type than at long-times (i.e. mutant kinetics at short-times is less distinguishable from wild-type than at long-times). We note here that of all wild-type events, ~92% of them have an event lifetime of <0.1 s. Similarly of all D178N mutant events, ~66% have an event lifetime of <0.1 s. Therefore the majority of observed events (the large majority in the case of wild-type PrP^C) are short-lived. Even in the case of short-lived events which are between 1 and 33 datapoints long (i.e. events are filtered to 2.99 ms per datapoint), with a large proportion of events having an event lifetime of ≤0.01 s (i.e. between 1 and 3 datapoints long − ~30% of events in the case of wild-type PrP^C), the results show that wild-type and

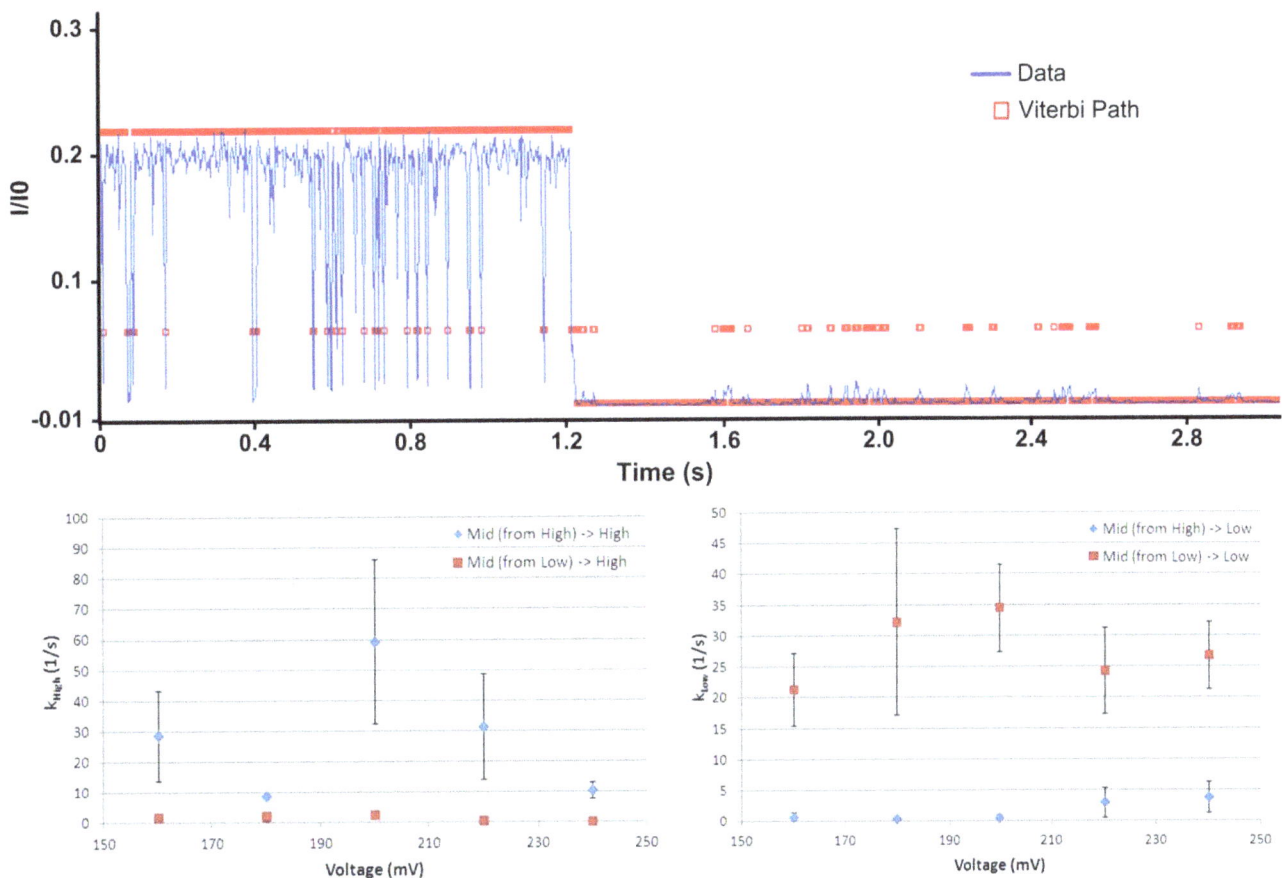

Figure 6. KKRR-ShPrP(120–232) mid-state statistics as a function of voltage. (Top) A sample KKRR-ShPrP(120–232) event (blue) with the most likely state sequence (i.e. Viterbi path) overlayed (red) as determined by a Viterbi analysis of the event. The sample event highlights the dependence of mid-state statistics (i.e. the transition rates out of the mid-state) on how the mid-state is entered. The event qualitatively shows that if the mid-state is entered from the high state then a transition back to the high-state is more likely than a transition into the low-state. Likewise, transitions into the mid-state from the low-state are more likely to return to the low-state as opposed to entering the high-state. **(Bottom left)** Mid-state transition rate into the high-state as a function of voltage, depending on how the mid-state is entered. If the mid-state is entered from the high-state (blue) the transition rate back into the high-state is between one and two orders of magnitude higher (depending on voltage) than the transition rate into the high-state when the mid-state is entered from the low-state (red). **(Bottom right)** Mid-state transition rate into the low-state as a function of voltage, depending on how the mid-state is entered. If the mid-state is entered from the low-state (red) the transition rate back into the low-state is between one and two orders of magnitude higher (depending on voltage) than the transition rate into the low-state when the mid-state is entered from the high-state (blue).

mutant PrP^C are easily distinguished based on their kinetics in the pore. The results substantially improve, particularly in the case of wild-type predictive value, when only long-time events are considered. These results demonstrate that nanopore analysis in combination with HMM signal processing can be used to detect

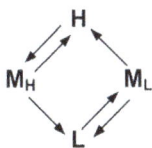

Figure 7. Four-state model characterizing KKRR-ShPrP(120–232) kinetics in the pore. H, M$_H$, L, and M$_L$ refer to the high, mid-high, low, and mid-low states respectively (N.B. PrP^C can escape from the pore from each state, which is not explicitly shown in the four-state model). HMM analysis of KKRR-ShPrP(120–232) in combination with the mid-state analysis (refer to text) yields the information on how the states are connected.

and discriminate between wild-type and mutant PrP^C at the single event level based on their kinetics in the pore. These results therefore show the potential of using this technique as an assay to probe differences in stability between wild-type and mutant prion proteins at the single molecule level, which opens up the possibility of studying small molecule-PrP^C interactions and the effects of these molecules on PrP^C stability as a possible screen for small molecules that improve the stability properties of the protein. Moreover, the ability to discriminate between two proteins that differ by only single-amino acid point mutation demonstrates the sensitivity of this approach in detecting subtle changes in biomolecule structure, and points to the possibility of developing this technique for highly sensitive, real-time detection and identification of small molecules and proteins in solution, with potential applications in disease biomarker and pathogen detection.

We return now to a more detailed analysis of PrP^C kinetics in the pore with the goal of characterizing the kinetic differences between wild-type and D178N mutant PrP^C. We limit our discussion here specifically to the kinetics of wild-type PrP^C (see below for a discussion D178N mutant kinetics). The optimal model shown in Figure 4B is a model of the kinetics of PrP^C in the pore over all voltages. To characterize the voltage-dependence of PrP^C kinetics the voltage-specific optimal model must be determined. This is done by HMM analysis of only those events at a given voltage whereby the optimal model (Figure 4B) serves as the initial model for the voltage-specific HMM analysis, with the caveat that the state levels (q) and noise properties of each state (b) remain constant during the analysis (i.e. only π and A are updated during the voltage-specific HMM analysis). The voltage-specific HMM analysis therefore, improves the estimate of the initial condition and the transition probabilities for a given voltage. Given the voltage-specific optimal model and an individual event the most likely state sequence for the event is then determined (i.e. the event Viterbi path) via the Viterbi algorithm (refer to [41] regarding the theory and implementation of the Viterbi algorithm). State properties, such as the lifetime distribution of each state, and state-to-state transition rates are then determined by analyzing the Viterbi path for all events (refer to File S1 for details on how the state-to-state transition rates are determined). Shown in Figure 6 are the statistics of the mid-state (i.e. the transition rate from the mid-state to the high and low states as a function of voltage), highlighting the dependence of mid-state statistics on how it is entered. For example, the transition rate from the mid-state to the low-state differs by one-to-two orders of magnitude depending on if the mid-state is entered from the high-state versus if it is entered from the low-state. In general for a single state, we would expect the transition rates in Figure 6 (bottom left) and Figure 6 (bottom right) to be within an error bar of each other (assuming the rates are Gaussian distributed). Given their degree of separation, between two and four error bars depending on voltage, these results indicate that the mid-state is more accurately modeled as two separate states a mid-high and a mid-low state, to distinguish between transitions into the mid-state from the high-state (mid-high state) versus mid-state transitions from the low-state (mid-low state). Given this, together with the results from the HMM analysis of the data (i.e. the state-to-state transition probabilities), we can model KKRR-ShPrP(120–232) kinetics in the pore as a four-state system (Figure 7).

Given the detailed kinetics of wild-type PrP^C in the pore, the voltage-dependence of all the state-to-state transition rates can be determined, which may yield information on the different conformations of PrP^C in the pore. For example, transitions that exhibit an exponential dependence on voltage (i.e. Arrhenius kinetics) indicate energy barrier crossing processes and therefore yield clues on the types of conformations and conformational transitions that can makeup said processes. This together with computational modeling of PrP^C trapped inside the pore should reveal detailed information on the specific conformations of PrP^C in the pore. With respect to D178N mutant kinetics and how it compares with wild-type we find that given the substantial difference in state properties between these two constructs (i.e. Figures 4B and 5B) no simple comparisons can be made. As mentioned previously, residue D178 plays an important role in maintaining the structural stability of PrP^C. This loss of stability, in the case of the mutant, likely enables it to adopt a variety of conformations inside the pore that are inaccessible to wild-type, which complicates the comparison between these two constructs, and hints at the need for additional states in a description of

mutant kinetics in the pore. In order to make meaningful comparisons with wild-type PrP^C, therefore, a more detailed analysis of mutant kinetics is required.

Conclusions

We probed wild-type and D178N mutant PrP^C structure and dynamics by analyzing the current fluctuations through an α-HL nanopore in which a single PrP^C molecule has been captured electrophoretically. We have shown that these two proteins (proteins that differ by only a single amino-acid point mutation) exhibit easily distinguishable current signatures and kinetics inside the pore and have demonstrated, with the use of HMM signal processing, accurate detection and discrimination between these two proteins at the single molecule level based on the kinetics of a single PrP^C capture event. This method of protein analysis may be useful as an assay to probe differences in stability between wild-type and mutant prion proteins at the single molecule level, opening up the possibility to study small molecule-PrP^C interactions and their effects on PrP^C stability as a possible screen for small molecules that improve the stability properties of the protein. Moreover, our results demonstrate the sensitivity of nanopore analysis in detecting subtle changes in biomolecule structure and show its potential for highly sensitive, real-time protein and small molecule detection and identification based on single molecule kinetics inside a nanopore with potential applications in disease biomarker and pathogen detection. In addition, we developed a four-state model to characterize wild-type PrP^C kinetics in the pore which represents a first step in our investigation on characterizing the differences in kinetics and conformational dynamics between wild-type and D178N mutant PrP^C, a comparison of which may ultimately yield clues into the molecular mechanism of PrP^C conversion in disease. These results demonstrate the ability of nanopore analysis to probe the detailed kinetics and conformational dynamics of a single biomolecule and point to the potential of using this technique in probing the molecular properties of other clinically relevant proteins (e.g. Aβ oligomers, α-synuclein, etc…).

Acknowledgments

We thank Kate R. Lieberman, Joseph M. Dahl and Gerald M. Cherf for supplying purified preformed heptameric α-hemolysin for experiments. We thank Oxford Nanopore Technologies (ONT) and in particular James Clarke of ONT for supplying heptameric α-hemolysin. We thank past members of the University of British Columbia Applied Biophysics Lab, in particular Matthew Wiggin, Christopher Feehan, Dhruti Trivedi, Jason R. Dwyer and Vincent Tabard-Cossa. We thank Will C. Guest and Michael T. Woodside for helpful discussions.

Author Contributions

Conceived and designed the experiments: NNJ AM NRC. Performed the experiments: NNJ. Analyzed the data: NNJ. Contributed reagents/materials/analysis tools: VS DSW. Wrote the paper: NNJ VS AM.

References

1. Collinge J (2001) Prion diseases of humans and animals: Their causes and molecular basis. Annu. Rev. Neurosci. 24: 519–550.

2. Prusiner SB (1982) Novel proteinaceous infectious particles cause scrapie. Science. 216: 136–144.
3. Prusiner SB (1998) Prions. Proc. Natl. Acad. Sci. USA. 95: 13363–13383.

4. Coulthart MB, Cashman NR (2001) Variant Creutzfeldt-Jakob disease: A summary of current scientific knowledge in relation to public health. Can. Med. Assoc. J. 165: 51–58.

5. Caughey B, Kocisko DA, Raymond GJ, Lansbury PT Jr (1995) Aggregates of scrapie-associated prion protein induce the cell-free conversion of protease-sensitive prion protein to the protease-resistant state. Chem. Biol. 2: 807–817.

6. Caughey B, Raymond GJ, Kocisko DA, Lansbury PT Jr (1997) Scrapie infectivity correlates with converting activity, protease resistance, and aggregation of scrapie-associated prion protein in guanidine denaturation studies. J. Virol. 71: 4107–4110.

7. van der Kamp MW, Daggett V (2010) Pathogenic Mutations in the Hydrophobic Core of the Human Prion Protein Can Promote Structural Instability and Misfolding. J. Mol. Biol. 404: 732–748.

8. Vanik DL, Surewicz WK (2002) Disease-associated F198S Mutation Increases the Propensity of the Recombinant Prion Protein for Conformational Conversion to Scrapie-like Form. J. Biol. Chem. 277: 49065–49070.

9. Meli M, Gasset M, Colombo G (2011) Dynamic diagnosis of familial prion diseases supports the β2-α2 loop as a universal interference target. PLoS One. 6: e19093.

10. Ma L, Cockroft SL (2010) Biological nanopores for single molecule biophysics. ChemBioChem. 11: 25–34.

11. DeGuzman VS, Lee CC, Deamer DW, Vercoutere WA (2006) Sequence-dependent gating of an ion channel by DNA hairpin molecules. Nucleic Acids Res. 34: 6425–6437.

12. Manrao EA, Derrington IM, Laszlo AH, Langford KW, Hopper MK, et al. (2012) Reading DNA at single-nucleotide resolution with a mutant MspA nanopore and phi29 DNA polymerase. Nat. Biotechnol. 30: 349–353.

13. Lin J, Kolomeisky A, Meller A (2010) Helix-coil kinetics of individual polyadenylic acid molecules in a protein channel. Phys. Rev. Lett. 104: 158101.

14. Venkatesan BM, Bashir R (2011) Nanopore sensors for nucleic acid analysis. Nat. Nanotechnol. 6: 615–624.

15. Branton D, Deamer DW, Marziali A, Bayley H, Benner SA, et al. (2008) The potential and challenges of nanopore sequencing. Nat. Biotechnol. 26: 1146–1153.

16. Payet L, Martinho M, Pastoriza-Gallego M, Betton JM, Auvray L, et al. (2012) Thermal unfolding of proteins probed at the single molecule level using nanopores. Anal. Chem. 84: 4071–4076.

17. Merstorf C, Cressiot B, Pastoriza-Gallego M, Oukhaled A, Betton JM, et al. (2012) Wild-type, mutant protein unfolding and phase transition detected by single nanopore recording. ACS Chem. Biol. 7: 652–658.

18. Movileanu L, Schmittschmitt JP, Scholtz JM, Bayley H (2005) Interactions of peptides with a protein pore. Biophys. J. 89: 1030–1045.

19. Mohammad MM, Movileanu L (2008) Excursion of a single polypeptide into a protein pore: Simple physics but complicated biology. Eur. Biophys. J. 37: 913–925.

20. Mohammad MM, Prakash S, Matouschek A, Movileanu L (2008) Controlling a single protein in a nanopore through electrostatic traps. J. Am. Chem. Soc. 130: 4081–4088.

21. Soskine M, Biesemans A, Moeyaert B, Cheley S, Bayley H, et al. (2012) An Engineered ClyA Nanopore Detects Folded Target Proteins by Selective External Association and Pore Entry. Nano Lett. 12: 4895–4900.

22. Pastoriza-Gallego M, Rabah L, Gibrat G, Thiebot B, Gisou van der Gout F, et al. (2011) Dynamics of unfolded protein transport through an aerolysin pore. J. Am. Chem. Soc. 133: 2923–2931.

23. Goldfarb LG, Petersen RB, Tabaton M, Brown P, LeBlanc AC, et al. (1992) Fatal familial insomnia and familial Creutzfeldt-Jakob disease: disease phenotype determined by a DNA polymorphism. Science 258: 806–808.

24. Monari L, Chen SG, Brown P, Parachi P, Petersen RB, et al. (1994) Fatal familial insomnia and familial Creutzfeldt-Jakob disease: different prion proteins determined by a DNA polymorphism. Proc. Natl. Acad. Sci. USA. 91: 2839–2842.

25. Julien O, Chatterjee S, Bjorndahl TC, Sweeting B, Acharya S, et al. (2011) Relative and regional stabilities of the hamster, mouse, rabbit, and bovine prion proteins toward urea unfolding assessed by nuclear magnetic Resonance and circular dichroism spectroscopies. Biochemistry. 50: 7536–7545.

26. Bjorndahl TC, Zhou G, Liu X, Perez-Pineiro R, Semenchenko V, et al. (2011) Detailed biophysical characterization of the acid-induced PrP^C to PrP^β conversion process. Biochemistry. 50: 1162–1173.

27. Perez-Pineiro R, Bjorndahl TC, Berjanskii MV, Hau D, Li L, et al. (2011) The prion protein binds thiamine. FEBS J. 278: 4002–4014.

28. Akeson M, Branton D, Kasianowicz JJ, Brandin E, Deamer DW (1999) Microsecond timescale discrimination among polycytidylic acid, polyadenylic acid, and polyuridylic acid as homopolymers or as segments within single RNA molecules. Biophys. J. 77: 3227–3233.

29. Jetha N, Wiggin M, Marziali A (2009) Forming an α-hemolysin nanopore for single molecule analysis. In: Lee JW, Foote RS, editors. Micro and Nano Technologies in Bioanalysis. Humana Press, Totowa NJ. 113–127.

30. Wanunu M, Morrison W, Rabin Y, Grosberg AY, Meller A (2010) Electrostatic focusing of unlabelled DNA into nanoscale pores using a salt gradient. Nat. Nanotechnol. 5: 160–165.

31. Hatlo MM, Panja D, Roij RV (2011) Translocation of DNA molecules through nanopores with salt gradients: The role of osmotic flow. Phys. Rev. Lett. 107: 068101.

32. Nandi PK, Leclerc E, Marc D (2002) Unusual property of prion protein unfolding in neutral salt solution. Biochemistry. 41: 11017–11024.

33. Apetri AC, Surewicz WK (2003) Atypical effect of salts on the thermodynamic stability of human prion protein. J. Biol. Chem. 278: 22187–22192.

34. Movileanu L, Cheley S, Howorka S, Braha O, Bayley H (2001) Location of a constriction in the lumen of a transmembrane pore by targeted covalent attachment of polymer molecules. J. Gen. Physiol. 117: 239–251.

35. Jetha N, Wiggin M, Marziali A (2009) Nanopore force spectroscopy on DNA duplexes. In: Lee JW, Foote RS, editors. Micro and Nano Technologies in Bioanalysis. Humana Press, Totowa NJ. 129–150.

36. Chung SH, Moore JB, Xia L, Premkumar LS, Gage PW (1990) Characterization of single channel currents using digital signal processing techniques based on hidden markov models. Phil. Trans. R. Soc. Lond. B. 329: 265–285.

37. Churbanov A, Winters-Hilt S (2008) Clustering ionic flow blockade toggles with a mixture of HMMs. BMC Bioinf. 9: S13.

38. Venkataramanan L, Sigworth FJ (2002) Applying hidden markov models to the analysis of single ion channel activity. Biophys. J. 82: 1930–1942.

39. Winters-Hilt S, Vercoutere W, DeGuzman VS, Deamer D, Akeson M, et al. (2003) Highly accurate classification of watson-crick basepairs on termini of single DNA molecules. Biophys. J. 84: 967–976.

40. Vercoutere WA, Winters-Hilt S, DeGuzman VS, Deamer D, Ridino SE, et al. (2003) Discrimination among individual Watson-Crick base pairs at the termini of single DNA hairpin molecules. Nucleic Acids Res. 31: 1311–1318.

41. Rabiner LR (1989) A tutorial on hidden markov models and selected applications in speech recognition. Proc. IEEE. 77: 257–286.

42. Alonso DO, DeArmond SJ, Cohen FE, Daggett V (2001) Mapping the early steps in the pH-induced conformational conversion of the prion protein. Proc. Natl. Acad. Sci. USA. 98: 2985–2989.

43. Riek R, Wider G, Billeter M, Hornemann S, Glockshuber R, et al. (1998) Prion protein NMR structure and familial human spongiform encephalopathies. Proc. Natl. Acad. Sci. USA. 95: 11667–11672.

44. Zuegg J, Gready JE (1999) Molecular dynamics simulations of human prion protein: Importance of correct treatment of electrostatic interactions. Biochemistry. 38: 13862–13876.

45. Barducci A, Chelli R, Procacci P, Schettino V (2005) Misfolding pathways of the prion protein probed by molecular dynamics simulations. Biophys. J. 88: 1334–1343.

18

Cyclooxgenase-2 Inhibiting Perfluoropoly (Ethylene Glycol) Ether Theranostic Nanoemulsions—*In Vitro* Study

Sravan Kumar Patel[1], Yang Zhang[1], John A. Pollock[2], Jelena M. Janjic[1]*

1 Graduate School of Pharmaceutical Sciences, Mylan School of Pharmacy, Duquesne University, Pittsburgh, Pennsylvania, United States of America, **2** Department of Biological Sciences, Bayer School of Natural and Environmental Sciences, Duquesne University, Pittsburgh, Pennsylvania, United States of America

Abstract

Cylcooxgenase-2 (COX-2) expressing macrophages, constituting a major portion of tumor mass, are involved in several pro-tumorigenic mechanisms. In addition, macrophages are actively recruited by the tumor and represent a viable target for anticancer therapy. COX-2 specific inhibitor, celecoxib, apart from its anticancer properties was shown to switch macrophage phenotype from tumor promoting to tumor suppressing. Celecoxib has low aqueous solubility, which may limit its tumor inhibiting effect. As opposed to oral administration, we propose that maximum anticancer effect may be achieved by nanoemulsion mediated intravenous delivery. Here we report multifunctional celecoxib nanoemulsions that can be imaged by both near-infrared fluorescence (NIRF) and ^{19}F magnetic resonance. Celecoxib loaded nanoemulsions showed a dose dependent uptake in mouse macrophages as measured by ^{19}F NMR and NIRF signal intensities of labeled cells. Dramatic inhibition of intracellular COX-2 enzyme was observed in activated macrophages upon nanoemulsion uptake. COX-2 enzyme inhibition was statistically equivalent between free drug and drug loaded nanoemulsion. However, nanoemulsion mediated drug delivery may be advantageous, helping to avoid systemic exposure to celecoxib and related side effects. Dual molecular imaging signatures of the presented nanoemulsions allow for future *in vivo* monitoring of the labeled macrophages and may help in examining the role of macrophage COX-2 inhibition in inflammation-cancer interactions. These features strongly support the future use of the presented nanoemulsions as anti-COX-2 theranostic nanomedicine with possible anticancer applications.

Editor: Jian-Xin Gao, Shanghai Jiao Tong University School of Medicine, China

Funding: JMJ, SKP, YZ and JAP research is supported by a grant from the Pittsburgh Tissue Engineering Initiative Interface Seed Grant Fund 2012. JMJ and SKP were also supported by Pennsylvania State Health Formula Research Grants Program (C.U.R.E. Award) and Faculty Development Funds from Duquesne University. JAP acknowledges support from NSF/Multi-User Instrumentation grant NSF# 0400776. The funders had no role in study design, data collection and analysis, decision to publish, or preparation of the manuscript.

Competing Interests: The authors have declared that no competing interests exist.

* E-mail: janjicj@duq.edu

Introduction

Inflammation processes are involved in all stages of cancer development [1]. The tumor environment contains a wide variety of inflammatory cells such as mast cells, dendritic cells, natural killer cells and macrophages [2]. Macrophages, constituting up to 50% of tumor mass, are actively recruited during cancer development and play an important role in tumor angiogenesis and metastasis [3]. Cyclooxygenase-2 (COX-2) is an inducible pro-inflammatory enzyme implicated in tumor development and progression [4]. Recruitment of COX-2 expressing macrophages can create an inflammatory environment that strongly promotes tumor growth and angiogenesis [5]. COX-2 is involved in the synthesis of prostaglandin E_2 (PGE_2) which is necessary for the development of immunosuppressive cells (tumor associated suppressive macrophages and myeloid-derived suppressor cells) [6]. Therefore, we hypothesize that inhibiting COX-2 in tumor recruited macrophages can be a viable anticancer strategy.

Celecoxib, a COX-2 selective inhibitor is reported to reduce cancer risk and suppress tumor growth in preclinical and clinical studies [4,7–9]. It acts as a multifunctional drug that simultaneously induces COX-2 independent apoptosis, inhibits PGE_2 mediated anti-apoptotic proteins and inhibits angiogenesis [10].

Recently, celecoxib has shown to alter the phenotype of macrophages from protumor (M2) to antitumor (M1) subtype via COX-2 inhibition [11]. However, celecoxib, classified as a BCS (Biopharmaceutics classification system) class II drug, has very poor aqueous solubility of 7 µg/mL [12] and 22–40% oral bioavailability in dogs [13] (to our knowledge absolute bioavailability in humans has not been reported). Celecoxib is also rapidly eliminated from the plasma further lowering drug levels at the tumor site [14,15]. In clinical cancer studies, celecoxib is administered orally at high doses (200–400 mg, twice daily) for several months leading to cardiovascular side effects, which may be severe [16]. To overcome these limitations, nanoparticle formulation of celecoxib was recently reported for colon cancer treatment in a human xenograft mouse model [15]. Based on these findings, we propose that the celecoxib loaded theranostic nanomedicine can suppress COX-2 activity in the circulating macrophages and allow us to track the macrophages tumor infiltration dynamics by molecular imaging (^{19}F magnetic resonance and near-infrared fluorescence).

Integration of diagnosis with therapy (theranostics) in a single nanocarrier could facilitate visualization of nanocarrier biodistribution and treatment response. This ultimately enables assessment of safety, toxicity and efficacy of the therapeutic intervention [17] leading to personalized medicine. Multiple imaging approaches are being investigated for this purpose such as: using optical probes, radioactive ligands, magnetic resonance imaging (MRI) and ultrasound contrast agents [18–21]. Near-infrared fluorescence (NIRF) imaging is a promising technique due to low near-infrared (NIR) absorbance by living tissues, high detection sensitivity and minimal autofluorescence [22–24]. However, *in vivo* NIRF imaging is semi-quantitative with limited tissue penetration [25]. ^{19}F MRI has unlimited tissue penetration and is a quantitative technique [26,27]. ^{19}F MRI is widely used to track the *in vivo* behavior of *ex vivo* perfluorocarbon (PFC) labeled cells [28,29]. ^{19}F magnetic resonance (MR) signal provides *in vivo* localization of exogenously introduced PFCs while conventional ^{1}H MRI provides the anatomical context [29–31]. However, for effective imaging with ^{19}F MRI, relatively large amounts of ^{19}F nuclei (minimum of 7.5×10^{16} atoms per voxel) at the target site is required in preclinical models [27]. By coupling NIRF and ^{19}F MR imaging modalities, sensitivity, specificity and high tissue penetration can be obtained [24].

Aspects of dual mode imaging of nanoemulsion have been previously reported [19,24]. ^{1}H MRI contrast agents in combination with NIRF imaging agents have been used as theranostic nanomedicine [19]. We recently reported a tyramide conjugated PFPE nanoemulsion with dual mode imaging capabilities [32]. In recent studies, macrophages were labeled *in vivo* by intravenously (i.v.) injected PFC nanoemulsions and their migration to the inflammation sites was monitored by ^{19}F MRI [33,34].

Here, we report for the first time theranostic nanomedicine integrating ^{19}F MRI and NIRF imaging agents for simultaneous drug delivery and macrophage tracking. The presented theranostic PFC nanoemulsion design is innovative in that: 1) It incorporates a selective COX-2 inhibitor; 2) It can serve as a multimodal biological probe for studying the role of COX-2 in macrophage-tumor interaction; and 3) Can be imaged by two complimentary molecular imaging techniques-NIRF and ^{19}F MR. Achieving the balance between imaging (^{19}F MRI and NIRF) and therapeutic functionalities (COX-2 inhibition) in a single nanocarrier is critical. Using ^{19}F NMR labeling, NIRF signal and COX-2 inhibition we achieved this balance successfully in *in vitro* cell culture studies. Targeting COX-2 in macrophages with a dual mode theranostic (^{19}F MRI/NIRF capabilities) is shown for the first time. We report detailed *in vitro* characterization and *ex vivo* biological testing of the PFPE theranostic nanoemulsion in mouse macrophages.

Materials and Methods

Materials

Celecoxib was purchased from LC Laboratories® (Woburn, MA, USA). Miglyol 810N was generously donated by Croda® International Plc. Pluronic® P105 was obtained from BASF Corporation. Cremophor® EL was purchased from Sigma-Aldrich. Perfluoropoly (ethylene glycol) ether (produced by Exfluor Research Corp., Roundrock, TX, USA) was generously provided by Celsense Inc., Pittsburgh, PA, USA and used without further purification. CellVue® NIR815 (786 nm/814 nm) and CellVue® Burgundy (683 nm/707 nm) Fluorescent Cell Linker Kit was purchased from Molecular Targeting Technologies, Inc. (MTTI), West Chester, PA, USA. 0.4% Trypan blue solution was obtained from Sigma-Aldrich. CellTiter-Glo® Luminescent Cell Viability Assay was obtained from Promega Corporation, WI, USA. Prostaglandin E_2 enzyme-linked immunosorbent assay (ELISA) kit was purchased from Cayman Chemical Company, MI, USA. Adherent mouse macrophage cell line (RAW 264.7) was obtained

from American Type Culture Collection (ATCC), Rockville, MD, USA and cultured according to the instructions. Dulbecco's modified eagle medium (DMEM; GIBCO-BRL, Rockville, MD, USA) for cell culture experiments was supplemented with 10% fetal bovine serum (FBS), Penicillin/Streptomycin (1%), L-Glutamine (1%), HEPES (2.5%) and 45% D(+) glucose (1%). Trypsin EDTA, 1× was obtained from Mediatech, Inc., VA, USA. All cells were maintained in 37°C incubator with 5% carbon dioxide. Purified mouse anti-mouse CD45.1 monoclonal antibody conjugated to FITC (fluorescein isothiocyanate) used for cell labeling was obtained from BD Pharmingen™, Material No. 553775. Antifade ProLong® Gold (Invitrogen) was used as the mounting medium. Lysotracker® Green DND-26 and Hoechst 33342 were obtained from Invitrogen.

Preparation of PFPE nanoemulsions

PFPE nanoemulsions were prepared using a mixture of nonionic surfactants, Pluronic® P105 (P105) and Cremophor® EL (CrEL). A premade aqueous solution of mixed surfactants was used.

Preparation of CrEL/P105 surfactant mixture. A solution containing mixed surfactants was prepared as follows: P105 (4 g) was dissolved in 100 mL water by stirring slowly at room temperature for the final concentration of 4% w/v (weight/volume). CrEL 6% w/v in water was prepared by magnetic stirring at room temperature. The two solutions were gently mixed at room temperature in 1:1 v/v (volume/volume) ratio in a 500 mL round bottomed flask. The flask was placed in a water bath preheated to 45°C and slowly rotated for 20 min. The solution was then chilled on ice for 15 min, and stored in the refrigerator until use. The final concentration of this mixed surfactant solution was 5% w/v, where 2% w/v was P105 and 3% w/v was CrEL.

General procedure for the preparation of nanoemulsions using microfluidization. PFPE formulations contained 1.38% w/v CrEL, 0.92% w/v P105, 7.24% w/v PFPE, 3.8% w/v Miglyol 810N, 0.02% w/v celecoxib, 0.24 μM NIRF dye (Cellvue® NIR815 or Burgundy) and deionized water (final volume to 25 mL). Celecoxib (5 mg) was first dissolved in 0.95 g of Miglyol 810N by overnight stirring while 6 μL of NIRF dye stock solution (1 mM in EtOH) was added before blending with PFPE. PFPE oil (1.81 g) was transferred to a 500 mL round bottomed flask containing celecoxib, NIRF dye and Miglyol 810N and stirred at 1200 rpm, room temperature for 15 min. To this 11.5 mL (0.575 g of mixed surfactant) of mixed surfactant solution was added and stirred at 1200 rpm for additional 15 min. To this mixture, 11.5 mL of deionized water was added and stirred under ice cold conditions for 5 min at 1200 rpm. The coarse emulsion was microfluidized on a Microfluidics M110S for 30 pulses under recirculation mode (inlet air pressure ~80 psi; operating liquid pressure ~17500 psi) and temperature was noted. The nanoemulsion was sterilized using sterile 0.22 μm cellulose filter (Millex® - GS, 33 mm). Filtered nanoemulsion samples (1.5 mL) were stored at 4°C and 25°C to assess the stability. The bulk of the nanoemulsion was stored at 4°C until use. Nanoemulsion without celecoxib and NIRF dye was prepared in the same way to serve as the control. Table 1 show components of all the nanoemulsions (**A**, **B** and **C**) formulated. PFPE used in the nanoemulsions is a clear liquid (d = 1.81 g/mL) represented by the formula $CF_3O(CF_2CF_2O)_nCF_3$, where n = 4–16, with the average molecular weight of 1380 g/mol.

Table 1. Composition of nanoemulsions.

Nanoemulsion Component	A mg/mL	B mg/mL	Cᵃ mg/mL
Celecoxib	0	0.2	0.2
PFPE	72	72	72
Miglyol 810N	38	38	38
Cremophor® EL	13.8	13.8	13.8
Pluronic® P105	9.2	9.2	9.2
NIRF Dye	µM	µM	µM
Cellvue® NIR815	0	0.24	0
Cellvue® Burgundy	0	0	0.24

ᵃNanoemulsion **C** is used for confocal microscopy of labeled macrophages.

Characterization

Nanoemulsions were characterized by dynamic light scattering (DLS) measurements (Zetasizer Nano, Malvern, UK), ^{19}F NMR (nuclear magnetic resonance) (Bruker, 470 MHz) and NIRF imaging (Odyssey® Infrared Imaging System, LI-COR Biosciences, NE, USA).

Droplet size and zeta potential measurements by DLS. The size distribution of the nanoemulsion droplets in aqueous medium was determined by DLS using Zetasizer Nano. Measurements were taken after diluting the nanoemulsion in water (1:39 v/v). Measurements were made at 25°C and 173° scattering angle with respect to the incident beam. The stability of nanoemulsions was assessed by measuring the hydrodynamic diameter (Z average) and half width of polydispersity index (PDIw/2) at different time points (days). The stability of nanoemulsions incubated (37°C, 5% CO_2) in cell culture medium (DMEM with 10% FBS) for 24 h was tested under same conditions. Nanoemulsions were monitored by DLS at two storage temperatures, 4°C and 25°C. Zeta potential was measured at same dilution using specialized zeta cells with electrodes following the manufacturer instructions.

^{19}F NMR measurements of nanoemulsions. ^{19}F NMR was recorded on nanoemulsions (and dilutions in water) with trifluoroacetic acid (TFA) as the internal standard in borosilicate NMR tubes (5 mm diameter). Briefly, nanoemulsion and 0.02% v/v TFA in water solution were mixed in 1:1 v/v ratio (200 µL each) and spectra recorded (Bruker, 470 MHz). ^{19}F NMR peak around −91.5 ppm corresponding to 40 fluorine nuclei was integrated with TFA (set at -76.0 ppm) as reference. Amount of PFPE per mL nanoemulsion was quantified based on the number of ^{19}F under PFPE peak at −91.5 ppm (see Equation S1 for calculation).

NIRF imaging of nanoemulsions. NIRF images of the above prepared NMR samples were recorded on Odyssey® Infrared Imaging System. Nanoemulsion **B** loaded with celecoxib and NIRF dye was imaged. The NMR tubes with nanoemulsions were aligned and carefully taped to a paper, placed in the sample compartment and imaged. Images at 785 nm excitation wavelength and emission above 810 nm were collected. Imaging parameters include an intensity setting of 2 and 2.5 mm focus offset. NIRF signal was quantified from the obtained images using the instrument software (Odyssey® Imager v.3). Nanoemulsion **A** was used to correct for the fluorescence background. The total area corresponding to the nanoemulsion (with aqueous TFA) in the NMR tube was carefully selected for quantification after setting the nanoemulsion **A** fluorescence as background in the instrument software.

Drug content in nanoemulsion. A validated high performance liquid chromatography (HPLC) method was used to assess celecoxib content in nanoemulsion **B**. A previously reported method was adopted [35] and required validation parameters such as specificity, linearity, accuracy, intra-day and inter-day precision, limit of quantification and limit of detection were evaluated for celecoxib. Reverse phase chromatography was performed using C18 column (Hypersil Gold C18 150 mm×4.6 mm, 5 µm pore size) and 75:25 methanol-water combination. Analysis was performed at isocratic conditions with the flow rate of 1 mL/min at 25°C column temperature. The detection wavelength was 252 nm. Celecoxib showed a sharp peak at 3.8 min retention time. HPLC was calibrated in the concentration range of 0.15–20 µg/mL celecoxib (correlation coefficient R^2>0.999). To assess drug content, nanoemulsion **B** (250 µL) was dissolved in 10 mL methanol and vigorously vortexed. The mixture was centrifuged at 4000 rpm (Centrifuge 5804 R, 15 amp version) for 10 min. Supernatant was collected and analyzed for celecoxib. Analysis was carried out in triplicates. All the formulation ingredients were analyzed separately for possible interference using same chromatographic conditions.

Cell Culture

Cell viability. Cell viability was assessed using CellTiter-Glo® luminescence assay. Briefly, mouse macrophages (RAW 264.7) were plated in 96 well plate at 10,000 cells/well. After overnight incubation at 37°C and 5% CO_2, culture medium was removed and adhered cells were exposed to nanoemulsions **A** and **B** (prediluted in complete medium) at different PFPE concentrations and incubated overnight. 50 µL of the medium was carefully removed and 25 µL of CellTiter-Glo® analyte was added to each well. The plate was shaken for 20 min at room temperature to induce cell lysis. 60 µL of the cell lysate was transferred to a white opaque 96 well plate and luminescence was recorded on Perkin Elmer Victor 2 Microplate Reader.

Cell labeling. To assess the *in vitro* behavior of the nanoemulsions, cell labeling studies were conducted on mouse macrophages. Cells were cultured in 6 well plates at 0.3 million per well for 48 h. After aspirating the medium, cultured cells were washed with medium and phosphate-buffered saline (PBS). Cells were exposed to celecoxib and NIRF dye loaded nanoemulsion **B** (prediluted in medium) with concentration of PFPE ranging from 0.09 to 1.4 mg/mL. 2 mL of nanoemulsion **B** containing medium was added to each well. Cells were incubated for 24 h at 37°C and 5% CO_2. Cells were washed (2×) with complete medium to remove non-internalized nanoemulsion and detached using trypsin. Detached cells were collected and centrifuged at 1100 rpm for 5 min. The supernatant was removed and the cell pellet was resuspended in complete medium and counted using Neubauer hemocytometer. To count the cells, equal volume of cell suspension and 0.4% Trypan blue cell staining solution were mixed and 25 µL of this mixture was used for cell counting. Cells were centrifuged again at 2000 rpm for 10 min to ensure complete removal of non-internalized nanoemulsion. After removing the supernatant, 180 µL of deionized water and 200 µL of 0.02% v/v aqueous TFA solution was added to the cell pellet, vortexed and transferred to 5 mm borosilicate NMR tubes.

^{19}F NMR measurements of labeled cells. NMR tubes with the labeled cell lysate (~0.4 mL) prepared as described above were subjected to ^{19}F NMR analysis to quantify the total fluorine content in the cells. The number of ^{19}F per cell (Fc) was calculated using the following formula Fc = [(Ic/Ir)Nr]/Nc [36] , where (Ic/

Ir) is the ratio of the integrated values of the PFPE peak in the cell pellet around -91.5 ppm corresponding to 40 fluorine nuclei divided by the TFA reference peak at -76.0 ppm, Nr is the total number of ^{19}F in the TFA reference sample and Nc is the total cell number in the pellet.

NIRF measurements of labeled cells. NMR tubes containing labeled cells, TFA and water were directly imaged in Odyssey® Infrared Imaging system. Briefly, the NMR tubes were aligned and carefully taped to a paper, placed in the sample compartment and imaged. Images at 785 nm excitation wavelength and emission above 810 nm were collected. Imaging parameters include an intensity setting of 8 and 2.5 mm focus offset. Images were quantified using the instrument software and unexposed cells were used for background correction.

Fluorescence microscopy. Images of nanoemulsion labeled mouse macrophages were captured using confocal microscopy (Leica TCS SP2 spectral confocal microscope, Leica Microsystems) to assess the intracellular distribution of the nanoemulsion. Macrophages were cultured for 24 h on glass cover slips (Fisherfinest, 22×22-1) placed in a 6-well plate at a concentration of 10^5 cells per well. Cultured macrophages were exposed to nanoemulsion **C** (21 µL nanoemulsion/mL medium; 2 mL total) for 24 h. After removing 1 mL medium, cells were fixed in 1 mL of 4% paraformaldehyde for 30 min. The medium in the cultured confocal plates (with glass cover slips) was carefully removed and washed with PBS (supplemented with 1% FBS). A stock solution of FITC dye conjugated mouse antimouse CD45.1 antibody (CD45-FITC) in 1% FBS in PBS was prepared at 1 µg/mL concentration. Cells in each well were exposed to 1 mL of the stock solution and left undisturbed at room temperature. After 15 min, dye solution was removed and washed with 1% FBS in PBS twice. Each cover slip was transferred to a microscopy slide with antifade mounting medium (ProLong® Gold, Invitrogen). Images were captured on a spectral analyzer confocal microscope. For visualizing FITC, excitation was achieved with the blue Ar laser 488 nm and emission window of 500 nm to 590 nm. Visualizing the Cellvue® Burgundy dye was achieved with the red HeNe 633 nm laser excitation and emission window of 640 nm to 850 nm. A transmission DIC image is acquired simultaneous to each confocal scan.

PGE$_2$ assay. To investigate the *in vitro* therapeutic efficacy of the drug carrier, effect of nanoemulsions on PGE$_2$ production by macrophages was assessed. Efficacy of nanoemulsion as drug carrier was assessed by comparing the effect on PGE$_2$ production with free drug. Cells were plated in 6 well plates at 0.3 million cells/well and incubated overnight. Cells were exposed to nanoemulsion **B** at 1.4 mg/mL PFPE concentration (9.28 µM celecoxib), free drug dissolved in DMSO (9.28 µM) and DMSO. Fresh medium was added to unexposed cells. After overnight incubation, all wells were washed (2×) with medium and PBS. Bacterial toxin lipopolysaccharide (LPS) at 1 µg/mL diluted in medium (2 mL total) was added to each well with cells (exposed and unexposed) and incubated. Unexposed cells treated with LPS were designated as control and unexposed cells without LPS activation were designated as untreated. After 4 h incubation, supernatant was collected and analyzed using commercially available PGE$_2$ ELISA kit. Samples were analyzed at two different dilutions (1:4 and 1:9) and two replicates of each dilution were used. Assessment of PGE$_2$ production in the supernatant and data analysis was performed according to the manufacturer instructions.

Results and Discussion

A novel COX-2 inhibiting PFC theranostic nanoemulsion with dual imaging capabilities (NIRF and ^{19}F MR) was prepared. Design, formulation and *in vitro* evaluation are discussed in detail.

Theranostic PFPE nanoemulsion design

Presented here is a novel PFC nanoemulsion designed to label macrophages upon exposure and inhibit their COX-2 activity. In this study, the PFC nanoemulsion has three key components (a) the anti-inflammatory drug celecoxib (b) NIRF dye for fluorescence imaging and (c) PFPE for ^{19}F MRI. A proposed schematic of the nanoemulsion droplet is shown in Figure 1.

PFPE was chosen as the ^{19}F imaging tracer to facilitate future *in vivo* imaging of the theranostic with ^{19}F MRI and *ex vivo* ^{19}F NMR cell loading quantification [29]. PFPE is desirable for *in vivo* ^{19}F MRI due to the large number of magnetically equivalent fluorine nuclei. Further, this molecule shows high chemical and biological inertness. To date no metabolizing enzymes have been known to breakdown PFCs that can produce reactive intermediates [37]. PFPE shows a single main peak around -91.5 ppm [38] in the ^{19}F NMR spectrum corresponding to the monomer repeats CF_2CF_2O. The total number of magnetically equivalent fluorines around -91.5 ppm is 40. A small peak around -59 ppm in the PFPE spectrum is not MRI detectable and hence its presence does not affect the image analysis [36]. PFPE was previously used for *in vivo* ^{19}F MRI tracking of *ex vivo* labeled immune cells [29,36] and is currently tested in cancer patients as immunotherapy imaging agent [39]. Due to high biological inertness, PFPE elimination is slow and relies on the reticuloendothelial system followed by expiration through lungs [37]. This is the general clearance profile for most PFCs used in biomedical applications [27,37]. To enable intracellular fluorescence microscopy and future *in vivo* NIRF imaging of the theranostic, CellVue® NIR815 (excitation max = 786 nm, emission max = 814 nm) or Burgundy (excitation max = 683 nm, emission max = 707 nm) lipophilic dyes were selected. We incorporated two imaging agents to provide complimentary information about *in vivo* nanoemulsion accumulation by ^{19}F MR and NIRF imaging modalities. Nanoemulsions can be imaged quantitatively in deep tissues using ^{19}F MRI. NIRF imaging can enable visualization of nanoemulsion accumulation even at low amounts due to its sensitive nature.

Pure PFC as a major component of PFC nanoemulsions cannot incorporate lipophilic drugs due to its significant lipophobicity [40]. Previous reports showed the incorporation of therapeutic moieties in the surfactant layer surrounding PFC core of a nanoemulsion droplet [41,42]. Alternatively, coconut oil was used to solubilize lipophilic drug camptothecin in a PFC emulsion [43].

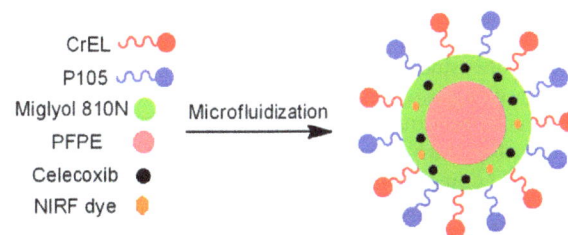

Figure 1. Proposed nanoemulsion droplet. Droplet carrying celecoxib, perfluoropoly (ethylene glycol) ether (PFPE) and near-infrared fluorescence (NIRF) dye. Cremophor EL® (CrEL) and Pluronic® P105 (P105) are the nonionic surfactants. Miglyol 810N is the hydrophobic oil phase.

A

Size Distribution by Intensity

B

C

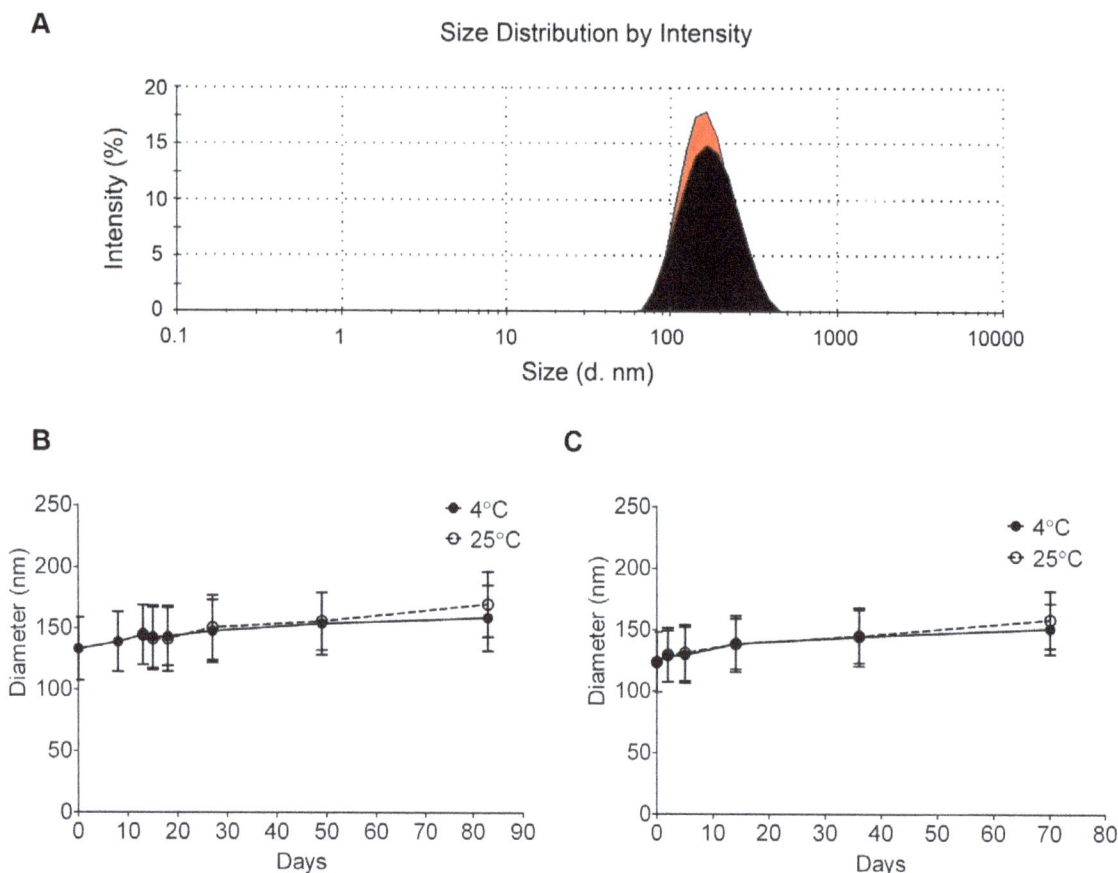

Figure 2. Shelf life of nanoemulsions with average droplet diameter (nm) at 4°C and 25°C. (A) Representative size distribution by intensity of nanoemulsions **A** (black) and **B** (red). (B) Stability of nanoemulsion **A**. (C) Stability of nanoemulsion **B**. Error bars represent half width of polydispersity index (PDIw/2).

Similarly, we used Miglyol 810N to solubilize celecoxib and a NIRF dye. Miglyol 810N, a medium-chain triglyceride of GRAS (generally regarded as safe) category, is widely used in parenteral nutrition emulsion formulations [44].

The challenging task of stabilizing immiscible hydrocarbon oil (Miglyol 810N) and PFPE was achieved by using a combination of nonionic surfactants CrEL and P105 under high shear liquid processing (microfluidization). P105 belongs to Pluronic® block copolymers (of ethylene oxide and propylene oxide subunits) which are commonly used for solubilization of hydrophobic drugs [45]. CrEL, produced by reacting castor oil and ethylene oxide in 1:35 molar [46] is used in pharmaceutical preparations as solubilizer for hydrophobic drugs and emulsifying agent [47,48]. It is important to rationalize the use of CrEL in this formulation, because of the studies showing associated toxicity. CrEL is associated with hypersensitivity reactions, hyperlipidemia, abnormal lipoprotein patterns, aggregation of erythrocytes and peripheral neuropathy which were observed with paclitaxel formulation, Taxol [47]. The amount of CrEL in Taxol is as high as 26 mL per administration, with each mL of formulation containing 527 mg of CrEL [47,49]. Paclitaxel formulations with reduced amount of CrEL showed significantly decreased allergic reactions suggesting that CrEL related toxicity is dose dependent [50,51]. The formulation reported in this work used only 13.8 mg of CrEL per mL emulsion, which is significantly lower (~38 fold compared to Taxol). Based on these calculations and prior reports [47], we suspect that allergic reactions are unlikely with the PFPE

formulations reported here. Nonetheless, the detailed toxicity studies in animal models is warranted and is part of future studies. Each component in this design has a unique role in achieving theranostic potential of the final nanoemulsion. The formulation ingredients were chosen to achieve a stable formulation of immiscible PFPE and hydrocarbon oil with dual imaging capabilities and drug delivery.

Nanoemulsion preparation and characterization

Nanoemulsions with and without drug/dye (**B**, **C** and **A** respectively; Table 1) were prepared using high pressure liquid processing on microfluidizer M110S (Microfluidics Corp. Newton, MA). Nanoemulsion **A** acts as a drug and dye free control for nanoemulsion **B**; nanoemulsion **C** (containing Cellvue® Burgundy) was formulated to obtain confocal images of labeled cells due to the unavailability of confocal excitation laser for Cellvue® NIR815. During processing, use of organic solvents and thin film emulsification method was avoided as residual solvents in the final formulation could lead to cell toxicity in test cultures. DLS measurements showed an average droplet size and polydispersity index (PDI) of less than 160 nm and 0.15 respectively. Shelf life was determined by following the droplet size and PDI upon storage at 4°C and 25°C (Figure 2). The inclusion of drug and dye in the nanoemulsion had no significant effect (p = 0.1275, Mann Whitney test, GraphPad Prism) on droplet size over time upon storage at 4°C (Figure 2). Nanoemulsions **A** and **B** were stable for at least 70 days. However, when stored at 25°C minimal average

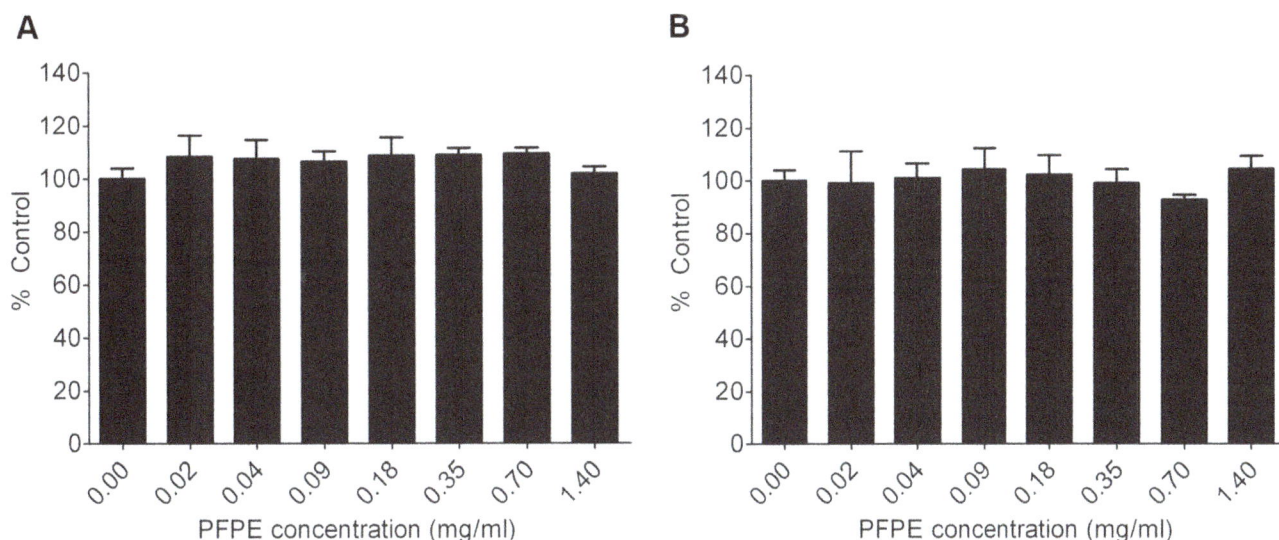

Figure 3. Macrophage cell viability post labeling. (A) Nanoemulsion **A** (B) Nanoemulsion **B**. Each data point represent mean of at least three replicates and the error bars are standard deviation of the mean. Values are reported as percent control (0 mg/mL PFPE).

size increase was observed. Therefore, the nanoemulsions are recommended to be stored at 4°C. A representative size distribution graph of nanoemulsions **A** and **B** is shown in Figure 2A. Small droplet size helps end-process sterilization by filtration [51] which is needed for future *in vivo* experiments. To further evaluate stability, zeta potential of the nanoemulsions was measured. Large values of zeta potential (> ±30 mV) ensure

greater repulsion between the nanodroplets leading to a stable nanoemulsion [52]. Both drug free and drug loaded nanoemulsions, sterically stabilized by nonionic surfactants, showed a moderate zeta potential value around -17 ± 6 mV (Figure S1).

Nanoemulsions were further characterized by ^{19}F NMR and NIRF imaging. In nanoemulsions **A** and **B**, ^{19}F NMR peak at -91.5 ppm has not shown any changes in chemical shift and line

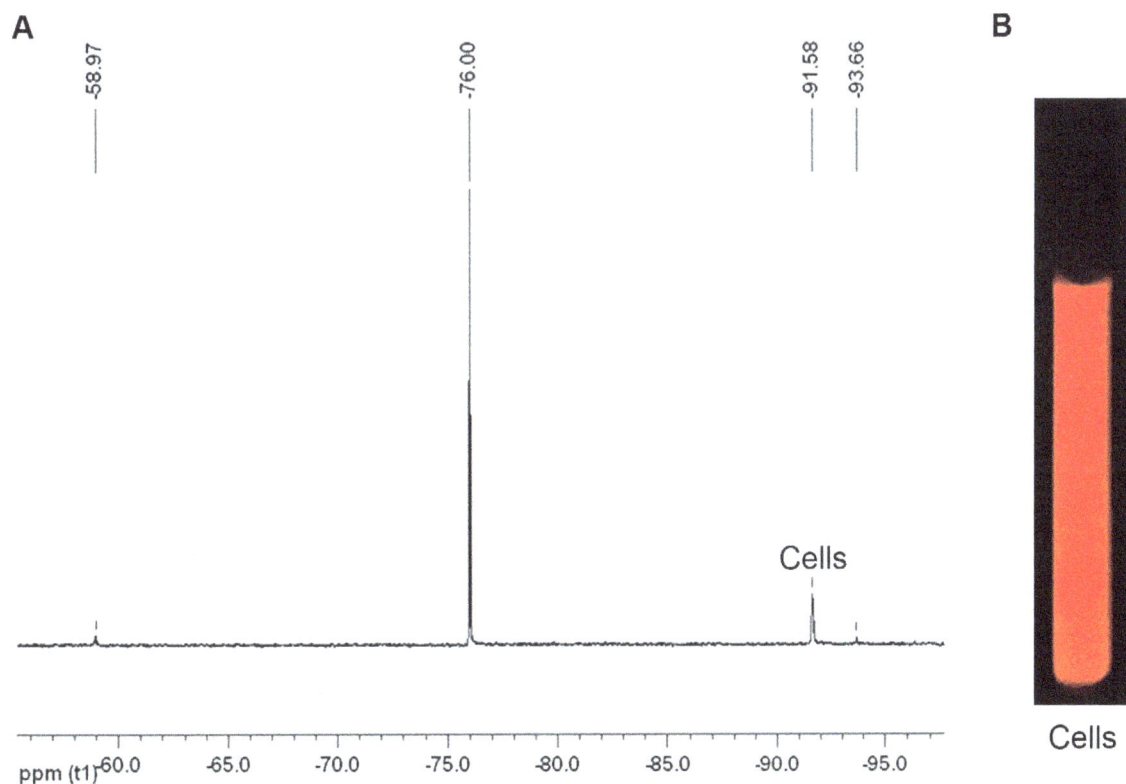

Figure 4. ^{19}F NMR and NIRF imaging of nanoemulsion B labeled macrophages. (A) ^{19}F NMR of cells labeled with nanoemulsion **B**. 0.02% v/v aqueous TFA set at -76.00 ppm was used as reference for ^{19}F NMR. (B) NIRF image (at 800 nm) of cells labeled with nanoemulsion **B** in NMR tube.

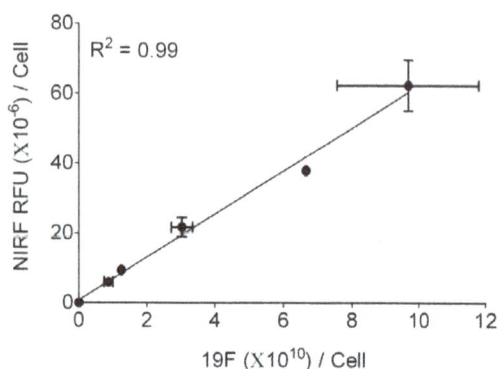

Figure 5. ^{19}F NMR-NIRF correlation of labeled macrophage cells. Data points represent cells labeled with different concentrations of nanoemulsion **B** (0–1.4 mg/mL PFPE).

shape ([38]; Figures S2 and S3). This result demonstrates the chemical stability of PFPE in the final product and during processing. ^{19}F NMR and NIRF images were recorded for nanoemulsion **B** dilutions in deionized water. NIRF images of nanoemulsion **B** are shown in Figure S4 and signal intensities in Table S1. A linear relationship was obtained for fluorine nuclei and NIRF signal for the dilution series (Figure S5). Based on this result, we believe that the estimates of *in vitro* cell labeling can be obtained by NIRF imaging alone without the need for ^{19}F NMR.

Reverse phase HPLC was utilized to evaluate drug loading in nanoemulsion **B**. All the formulation ingredients were individually run for any possible interference with the celecoxib peak. Excipients did not show UV absorbance around 252 nm (data not shown). Predicted celecoxib concentration based on calibration model was found to be 139.3±8.7 µg/mL nanoemulsion. To summarize, DLS results confirm the formation of nanoemulsion with stable droplet size. ^{19}F NMR and NIRF imaging clearly showed the incorporation of PFPE and NIRF dye in the nanoemulsion. HPLC analysis quantified the drug content in nanoemulsion **B**.

In vitro toxicity and uptake studies in macrophages

Before performing *in vitro* biological tests, colloidal stability of nanoemulsions in cell culture medium was evaluated by monitoring changes in droplet size. Nanoemulsions **A** and **B** were incubated in the complete cell culture medium for 24 h. No considerable change in droplet size and PDI was noted under cell culture relevant conditions (Figure S6 and Table S2). This is a crucial finding as any structural changes in the nanodroplets during incubation with cells could give misleading results on the nanodroplet cellular uptake and toxicity profile, which would further render the nanoemulsions unsuitable for *in vivo* testing. With this result, *in vitro* toxicity studies were conducted using Celltiter-Glo® Luminescence Cell Viability Assay to assess the suitability of the prepared nanoemulsions for biomedical applications. The assay makes use of the amount of ATP present in the culture to quantitate the number of metabolically active or viable cells. Mouse macrophages (RAW 264.7) were chosen as the model inflammatory cells. As shown in Figure 3, no considerable effect on cell viability was detected after 24 h exposure to nanoemulsions. Cell viability was between 92–104% of the control group (untreated cells).

To investigate the utility of nanoemulsion **B** for future *in vivo* imaging studies, *in vitro* cell uptake tests were performed on macrophages. Macrophages were incubated with nanoemulsion **B** at different PFPE concentrations for 24 h and exposed cells were analyzed by ^{19}F NMR and NIRF imaging to assess the intracellular uptake. Representative ^{19}F NMR and NIRF image of labeled cells is shown in Figure 4. As shown in Figure 4A, PFPE line shape and peak position at -91.5 ppm was unchanged upon uptake in cells when compared with PFPE in nanoemulsion **B** (Figure S3). This result suggests the chemical stability of PFPE in cells which is crucial for their use as an imaging tracer. ^{19}F NMR and NIRF measurements of labeled cells showed a dose-dependent uptake of the nanoemulsion (Figure S8). NIRF signal intensities and images of nanoemulsion **B** labeled cells at different dilutions is shown in Table S3 and Figure S7 respectively. Macrophages labeled with varying concentrations of nanoemulsion **B** showed a linear correlation ($R^2 = 0.99$) between ^{19}F signal and NIRF intensity per cell (Figure 5). Interestingly, linear correlation was obtained without chemically conjugating PFPE and fluorescent dye as reported earlier [29]. Based on these results, it can be

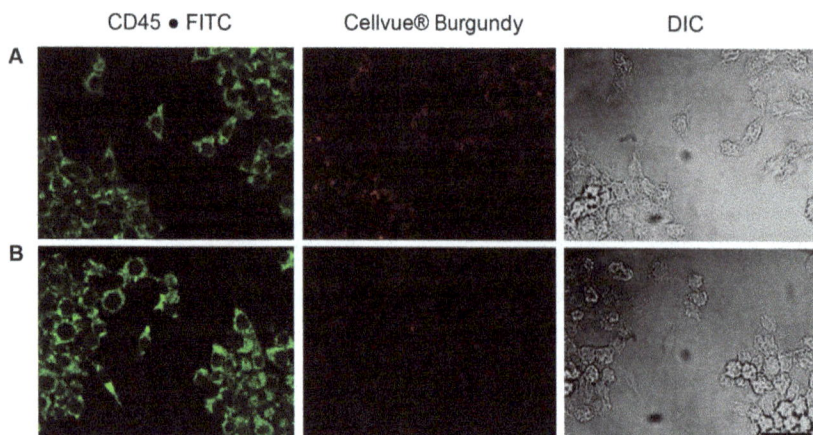

Figure 6. Fluorescence images of macrophages. (A) Cells labeled with anti-CD45 (FITC) green and incorporated nanoemulsion **C** containing celecoxib and Cellvue® Burgundy dye represented as red. (B) Cells not exposed to the nanoemulsion **C** exhibit CD45 labeling with FITC (green) but no red signal. Transmitted light DIC image acquired simultaneously shows field of view (Bar = 30 µm). The microscope image acquisition parameters were identical between the experimental and control, and the images were all acquired within 15 min of one another.

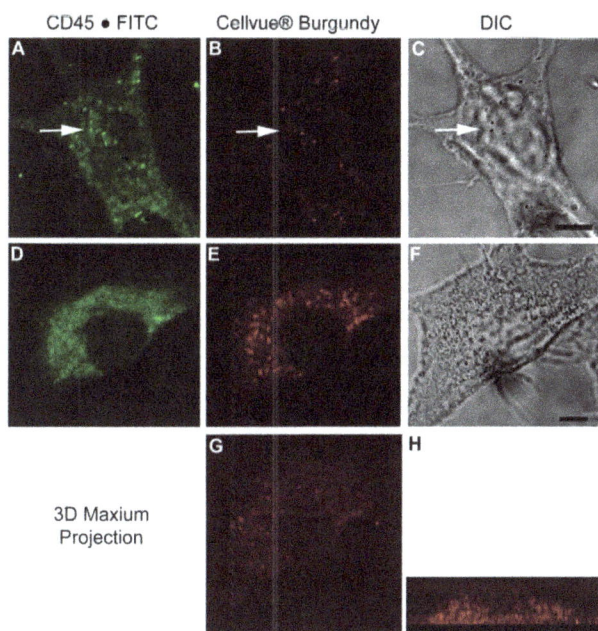

CD45 • FITC	Cellvue® Burgundy	DIC

3D Maxium
Projection

Figure 7. Magnified fluorescence images of individual macrophages exposed to nanoemulsion C. (A) Cells labeled with CD45-FITC (green) and incorporated nanoemulsion **C** containing celecoxib and Cellvue® Burgundy dye exhibit broad expression of CD45 as well as localized points of fluorescent signal indicating internalization of CD45 protein. (B) The same cell and focal plane as viewed in panel A reveals the internalized Cellvue® Burgundy labeled nanoemulsions as discrete particles. (C) The transmitted light DIC view of the cell reveals the black refractive droplets, coincident with the red and green fluorescent signals (Bar = 5 μm). (D) A different cell labeled with CD45-FITC (green), (E) internalized Cellvue® Burgundy (red) and (F) transmitted light DIC view reveals discrete droplets (Bar = 5 μm). (G) The cell shown in panel D was imaged in serial section and rendered by maximum-projection to represent all of the Cellvue® Burgundy labeled particles viewed from above and (H) in 90° cross-section, to reveal that the droplets are distributed throughout the cell cytoplasm.

proposed that the nanoemulsion was not destabilized before entering the cell in the labeling medium. Any instability of nanoemulsion would lead to poor or no correlation between ^{19}F NMR and NIRF signals due to the differences in uptake of imaging agents. A strong correlation between signals corresponding to two imaging agents is a requisite to utilize the nanoemulsion for in vitro and in vivo dual mode imaging studies.

In vitro ^{19}F NMR was used to evaluate the utility of the nanoemulsion for future in vivo ^{19}F MRI. Presented nanoemulsions have lower amount (7.2% w/v) of PFPE than our earlier reported cell tracking formulations [37,53]. However, we found that at a very low PFPE concentration of only 1.4 mg/mL, significant cell uptake (1.0×10^{11} fluorine atoms per cell) is achieved. With this labeling efficiency, approximately 7.5×10^5 cells per voxel are required to obtain in vivo ^{19}F MR images at 11.7 T [27,36]. These findings strongly suggest that we would be able to detect our nanoemulsion accumulation in vivo. Detailed dosing studies of the reported celecoxib loaded PFPE nanoemulsion in preclinical animal models are beyond the scope of this report and will be reported in the future.

Although ^{19}F NMR and NIRF imaging of the nanoemulsion labeled cells showed dose-dependent cell labeling, conclusions about membrane adsorbed versus phagocytosed nanodroplets cannot be made from this data alone. Therefore, fluorescence

confocal microscopy was performed on macrophages exposed to nanoemulsion. To more closely match the excitation and emission capabilities of the confocal microscope system, an alternate drug loaded nanoemulsion (nanoemulsion **C**, Table 1) was prepared with Cellvue® Burgundy dye (683 nm/707 nm). In vitro characterization of nanoemulsion **C** is shown in Figure S9. Cells exposed to nanoemulsion **C** were stained with anti-CD45.1 antibody conjugated with FITC dye (CD45-FITC). CD45 is a protein tyrosine phosphatase, receptor type C cell membrane associated protein. The nanoemulsion uptake was visualized by Cellvue® Burgundy dye. Figure 6A clearly shows the presence of the CD45 protein (green) and the Cellvue® Burgundy labeled nanodroplets (red) in the cytoplasm. As a control, cells not exposed to nanoemulsion were labeled with CD45-FITC (Figure 6B). No evidence of NIRF signal was observed in the control group. Figure 7 shows a higher magnification view of individual cells that reveal the particle nature of the Cellvue® Burgundy labeled nanodroplets, which are also evident in the transmitted DIC view of the cells as black refractive particles (Figure 7C, F). Three-dimensional rendering of a z series of 33 optical sections for the 10.4 μm thickness of the cell revealed that the nanodroplets are within the cytoplasm, specifically in the maximum projection cross-sectional view of the cell (Figure 7H). It is interesting to note that the endocytic engulfment of the nanodroplet has also internalized the CD45 protein and many of the nanodroplets co-present with green and red fluorescent signals. CD45 internalization has been previously reported [54]. ^{19}F NMR, NIRF imaging and confocal fluorescence microscopy clearly demonstrates the uptake of nanoemulsion droplets by exposed macrophages in vitro. In a separate experiment, presence of nanoemulsion droplets in the intracellular compartments was assessed by labeling lysosomes of macrophages with Lysotracker® Green (Protocol S1). It appears that nanoemulsion droplets are distributed in the entire volume of the cytoplasm and no preferential accumulation in the lysosomes was observed (Figure S11).

COX-2 inhibition in macrophages

Macrophages in the tumor environment express elevated levels of COX-2 which is involved in the biosynthesis of PGE_2. The potential anti-inflammatory effect of celecoxib loaded nanoemulsion **B** on the production of PGE_2 by LPS activated macrophages was studied. Macrophages were first exposed to nanoemulsion **B** for 24 h, washed with medium and then activated using LPS for 4 h. Amount of PGE_2 released into the medium was quantified using a commercially available ELISA assay. One-way ANOVA with Tukey's multiple comparison test was employed to evaluate the statistical significance between the treatments. Results are shown in Figure 8. LPS activated cells showed up to 10 fold increase in PGE_2 as compared to untreated and a statistically significant difference was observed (p<0.0001). A statistically significant difference (p<0.0001) between nanoemulsion **B** and LPS treated control was observed. Cells labeled with nanoemulsion **B** produced, on average, 50.6 ± 8.2 pg of PGE_2 per mL as compared to 504.5 ± 41.2 pg/mL by LPS activated control. Exposing LPS activated macrophages to DMSO (vehicle for free drug) has not shown any effect on PGE_2 production compared to control. Although, PGE_2 reduction by nanoemulsion **B** is not statistically different from free drug, nanoemulsion mediated celecoxib delivery may be advantageous in reducing systemic exposure to the drug and related side effects. Additionally, dual mode imaging capabilities allow for non-invasive imaging of nanoemulsion biodistribution. In a separate experiment, effect of nanoemulsions **A** and **B** on PGE_2 production was studied. Nanoemulsion **B** showed significant reduction in PGE_2 production

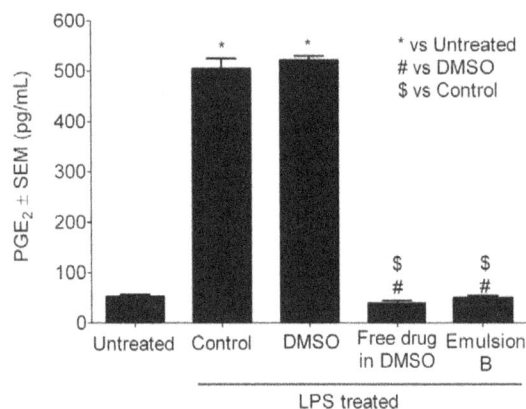

Figure 8. Production of PGE$_2$ in macrophages assessed after LPS treatment. LPS treatment was performed post cell labeling with nanoemulsion **B**, free drug dissolved in DMSO and DMSO. Cells not exposed to LPS were designated as untreated. * # $ represents statistical significance comparisons (p<0.0001) between treatments. Each data point represents the average of four independent measurements, where the error bars are the standard error of the mean (SEM).

compared to nanoemulsion **A** and control (Figure S10). Nanoemulsion **A** has not shown any significant contribution to changes in PGE$_2$ levels proving that the drug free vehicle is inert towards PGE$_2$ production. The presented theranostic PFPE nanoemulsion showed celecoxib delivery and COX-2 inhibition in macrophages. Introduction of celecoxib directly to macrophages by nanoemulsions loading may change their phenotype from tumor promoting M2 to tumor suppressing M1-like.

Conclusion

This paper presented novel drug carrying nanoemulsion formulation equipped with dual mode (^{19}F MR and NIRF) imaging capabilities. The prepared nanoemulsions showed good stability for at least 70 days. The utility of dual mode imaging was shown by a strong correlation between NIRF and ^{19}F NMR signals of labeled cells. Confocal imaging clearly demonstrated that the nanoemulsion droplets are incorporated into the cytoplasm of engulfing cells. Nanoemulsion delivery of celecoxib is demonstrated in macrophages by their inhibitory effect on PGE$_2$ production and release. The formulation platform developed here can be used to incorporate other lipophilic drugs and can act as a dual imaging tracer to label phagocytic cells such as macrophages. Drug release and *in vitro/in vivo* activity studies on breast cancer models are currently under investigation.

Supporting Information

Equation S1 PFPE amount (mg/mL nanoemulsion) calculation.

Figure S1 Zeta potential distribution as measured by Zetasizer Nano (Malvern, UK). Zeta potential of nanoemulsion **A** (red, -17 ± 6.6 mV) and nanoemulsion **B** (black, -17.7 ± 6.7 mV) in deionized water at 1:39 v/v dilution.

Figure S2 Representative ^{19}F NMR of nanoemulsion A. NMR was recorded on Bruker Instruments, Inc., Billerica, MA at 470 MHz in water with TFA reference at -76.00 ppm.

Figure S3 Representative ^{19}F NMR of nanoemulsion B. NMR was recorded on Bruker Instruments, Inc., Billerica, MA at 470 MHz in water with TFA reference at -76.00 ppm.

Figure S4 Representative NIRF imaging of nanoemulsion B dilutions. Decreasing concentration of the emulsion from left to right. Images at 785 nm excitation wavelength and emission above 810 nm were collected on Li-COR Odyssey® Infrared imaging system in 5 mm Borosilicate NMR tubes. For NIRF signal intensity, see Table S1.

Figure S5 Nanoemulsion B dilutions (1:1 v/v) in 0.02% v/v TFA (A) Plot of ^{19}F atoms (of PFPE around -91.5 ppm) with percent emulsion in NMR sample. (B) Plot of NIRF RFU with percent emulsion in the NMR sample (5 mm borosilicate NMR tubes, 0.4 mL total volume).

Figure S6 Stability of nanoemulsions in cell culture medium. Droplet size distribution before incubation is shown in red and after incubation in black. (A) Nanoemulsion **A** (B) Nanoemulsion **B**. No significant change in dropletsize was seen for both nanoemulsions after 24 h incubation with cell culture medium (Table S2). Analysis performed on Zetasizer Nano (Malvern, UK).

Figure S7 Representative NIRF imaging of cells labeled with nanoemulsion B. Images at 785 nm excitation wavelength and emission above 810 nm were collected on Li-COR Odyssey® Infrared Imaging system in 5 mm Borosilicate NMR tubes. For NIRF signal intensity, see Table S3.

Figure S8 Dose dependent uptake of nanoemulsion B. (A) ^{19}F atoms/cell at different concentrations of PFPE. (B) NIR fluorescence/cell at different concentrations of PFPE.

Figure S9 Characterization of nanoemulsion C. Nanoemulsion **C** was prepared to facilitate confocal microscopy. (A) Stability at 4°C and 25°C (B) Macrophage viability post 24 h exposure.

Figure S10 Production of PGE$_2$ in activated macrophages. Macrophages were exposed to either of the nanoemulsions **A** and **B** at 1.4 mg/mL PFPE concentration. LPS treatment was performed post cell labeling with nanoemulsions **A** or **B** for 3 h. Fresh medium was added to unexposed cells (untreated). Control represents LPS activated unexposed cells. PGE$_2$ production was quantified in the supernatant using PGE$_2$ ELISA kit (Cayman Chemicals). Each data point represents the average of at least nine independent measurements, where the error bars are the standard error of the mean (SEM). Statistically significant difference was obtained between nanoemulsion **B** and all other treatments. One-way ANOVA with Tukey's multiple comparison test was conducted to evaluate statistical significance.

Figure S11 Fluorescence microscopy of macrophages exposed to nanoemulsion C and lysosome specific fluorescent probe. (A) The transmitted light DIC view of the cells; (B) Fluorescent image of nucleus (blue) and lysosomes (green); (C) Fluorescent image of nucleus (blue) and nanoemulsion C (red) and (D) Fluorescent image of nucleus, lysosomes and nanoemulsion droplets. The scale bar represents 10 μm.

Table S1 NIR signal intensity (relative fluorescence units, RFU) of nanoemulsion B dilutions in 0.02% aq. TFA (1:1 v/v). Sample G represents nanoemulsion A (without NIRF dye) in aqueous TFA to correct for background.

Table S2 Average droplet diameter and PDI of nanoemulsions A and B before and after incubation in media.

Table S3 NIR signal intensity (relative fluorescence units, RFU) of cells labeled with nanoemulsion B (A–E) and unlabeled control cells (F). Unlabeled cells were used to correct for the background fluorescence signal from the cell suspension.

Protocol S1 Lysosomal labeling and confocal imaging procedure.

Acknowledgments

Authors like to thank Virgil Simpleanu (Carnegie Mellon University, Pittsburgh NMR Center for Biomedical Research) for his help with ^{19}F NMR measurements, Haibing Teng (Carnegie Mellon University, Molecular Biosensor and Imaging Center) for her assistance with confocal microscopy, Konjit Amede (Duquesne University) for assisting with DLS measurements and Haripriya Vemana (Duquesne University) for her assistance with cell culture.

Author Contributions

Conceived and designed the experiments: SKP YZ JAP JMJ. Performed the experiments: SKP YZ JAP JMJ. Analyzed the data: SKP YZ JAP. Contributed reagents/materials/analysis tools: JAP, JMJ. Wrote the paper: SKP JMJ.

References

1. Coussens LM, Werb Z (2002) Inflammation and cancer. Nature 420: 860–867.
2. Yu JL, Rak JW (2003) Host microenvironment in breast cancer development: Inflammatory and immune cells in tumour angiogenesis and arteriogenesis. Breast Cancer Res 5: 83–88.
3. Mehibel M, Singh S, Chinje EC, Cowen RL, Stratford IJ (2009) Effects of cytokine-induced macrophages on the response of tumor cells to banoxantrone (AQ4N). Mol Cancer Ther 8: 1261–1269.
4. Reddy BS, Hirose Y, Lubet R, Steele V, Kelloff G, et al. (2000) Chemoprevention of colon cancer by specific cyclooxygenase-2 inhibitor, celecoxib, administered during different stages of carcinogenesis. Cancer Res 60: 293–297.
5. Nakao S, Kuwano T, Tsutsumi-Miyahara C, Ueda S, Kimura YN, et al. (2005) Infiltration of COX-2-expressing macrophages is a prerequisite for IL-1 beta-induced neovascularization and tumor growth. J Clin Invest 115: 2979–2991.
6. Kalinski P (2012) Regulation of immune responses by prostaglandin E$_2$. J Immunol 188: 21–28.
7. Basu GD, Pathangey LB, Tinder TL, Gendler SJ, Mukherjee P (2005) Mechanisms underlying the growth inhibitory effects of the cyclo-oxygenase-2 inhibitor celecoxib in human breast cancer cells. Breast Cancer Res 7: R422–R435.
8. Harris RE, Alshafie GA, Abou-Issa H, Seibert K (2000) Chemoprevention of breast cancer in rats by celecoxib, a cyclooxygenase 2 inhibitor. Cancer Res 60: 2101–2103.
9. Thun MJ, Henley SJ, Patrono C (2002) Nonsteroidal anti-inflammatory drugs as anticancer agents: mechanistic, pharmacologic, and clinical issues. J Natl Cancer Inst 94: 252–266.
10. Jendrossek V (2011) Targeting apoptosis pathways by celecoxib in cancer. Cancer Lett In press.
11. Nakanishi Y, Nakatsuji M, Seno H, Ishizu S, Akitake-Kawano R, et al. (2011) COX-2 inhibition alters the phenotype of tumor-associated macrophages from M2 to M1 in ApcMin/+ mouse polyps. Carcinogenesis 32: 1333–1339.
12. Seedher N, Bhatia S (2003) Solubility enhancement of Cox-2 inhibitors using various solvent systems. AAPS PharmSciTech 4: 36–34.
13. Paulson SK, Vaughn MB, Jessen SM, Lawal Y, Gresk CJ, et al. (2001) Pharmacokinetics of celecoxib after oral administration in dogs and humans: effect of food and site of absorption. J Pharmacol Exp Ther 297: 638–645.
14. Paulson SK, Zhang JY, Breau AP, Hribar JD, Liu NWK, et al. (2000) Pharmacokinetics, tissue distribution, metabolism, and excretion of celecoxib in rats. Drug Metab Dispos 28: 514–521.
15. Venkatesan P, Puvvada N, Dash R, Prashanth Kumar BN, Sarkar D, et al. (2011) The potential of celecoxib-loaded hydroxyapatite-chitosan nanocomposite for the treatment of colon cancer. Biomaterials 32: 3794–3806.
16. Solomon SD, McMurray JJV, Pfeffer MA, Wittes J, Fowler R, et al. (2005) Cardiovascular risk associated with celecoxib in a clinical trial for colorectal adenoma prevention. N Engl J Med 352: 1071–1080.
17. Sumer B, Gao J (2008) Theranostic nanomedicines for cancer. Nanomedicine (Lond) 3: 137–140.
18. Bartlett DW, Su H, Hildebrandt IJ, Weber WA, Davis ME (2007) Impact of tumor-specific targeting on the biodistribution and efficacy of siRNA nanoparticles measured by multimodality in vivo imaging. Proc Natl Acad Sci USA 104: 15549–15554.
19. Gianella A, Jarzyna PA, Mani V, Ramachandran S, Calcagno C, et al. (2011) Multifunctional nanoemulsion platform for imaging guided therapy evaluated in experimental cancer. ACS Nano 5: 4422–4433.
20. Rapoport N, Nam KH, Gupta R, Gao Z, Mohan P, et al. (2011). Ultrasound-mediated tumor imaging and nanotherapy using drug loaded, block copolymer stabilized perfluorocarbon nanoemulsions. J Control Release 153: 4–15.
21. Shiraishi K, Endoh R, Furuhata H, Nishihara M, Suzuki R, et al. (2011). A facile preparation method of a PFC-containing nano-sized emulsion for theranostics of solid tumors. Int J Pharm 421: 379–387.
22. Akers WJ, Kim C, Berezin M, Guo K, Fuhrhop R, et al. (2011). Noninvasive photoacoustic and fluorescence sentinel lymph node identification using dye-loaded perfluorocarbon nanoparticles. ACS Nano 5: 173–182.
23. Frangioni JV (2003). In vivo near-infrared fluorescence imaging. Curr Opin Chem Biol 7: 626–634.
24. Lim YT, Noh YW, Kwon JN, Chung BH (2009) Multifunctional perfluorocarbon nanoemulsions for (19)F-based magnetic resonance and near-infrared optical imaging of dendritic cells. Chem Commun (Camb) 45: 6952–6954.
25. Pogue BW, Leblond F, Krishnaswamy V, Paulsen KD (2010) Radiologic and near-infrared/optical spectroscopic imaging: where is the synergy? AJR Am J Roentgenol 195: 321–332.
26. Morawski AM, Winter PM, Yu X, Fuhrhop RW, Scott MJ, et al. (2004) Quantitative "magnetic resonance immunohistochemistry" with ligand-targeted ^{19}F nanoparticles. Magn Reson Med 52: 1255–1262.
27. Srinivas M, Heerschap A, Ahrens ET, Figdor CG, de Vries IJM (2010) 19F MRI for quantitative in vivo cell tracking. Trends Biotechnol 28: 363–370.
28. Bible E, Dell'Acqua F, Solanky B, Balducci A, Crapo PM, et al. (2012) Non-invasive imaging of transplanted human neural stem cells and ECM scaffold remodeling in the stroke-damaged rat brain by ^{19}F- and diffusion-MRI. Biomaterials 33: 2858–2871.
29. Janjic JM, Srinivas M, Kadayakkara DKK, Ahrens ET (2008) Self-delivering nanoemulsions for dual fluorine-19 MRI and fluorescence detection. J Am Chem Soc 130: 2832–2841.
30. Lanza GM, Winter PM, Neubauer AM, Caruthers SD, Hockett FD, et al. (2005) ^{1}H/^{19}F magnetic resonance molecular imaging with perfluorocarbon nanoparticles. Curr Top Dev Biol 70: 57–76.
31. Winter PM, Cai K, Caruthers SD, Wickline SA, Lanza GM (2007) Emerging nanomedicine opportunities with perfluorocarbon nanoparticles. Expert Rev Med Devices 4: 137–145.
32. O'Hanlon CE, Amede KG, O'Hear MR, Janjic JM (2012) NIR-labeled perfluoropolyether nanoemulsions for drug delivery and imaging. J Fluor Chem 137: 27–33.
33. Hertlein T, Sturm V, Kircher S, Basse-Lüsebrink T, Haddad D, et al. (2011) Visualization of abscess formation in a murine thigh infection model of Staphylococcus aureus by ^{19}F-magnetic resonance imaging (MRI). PLoS One 6: e18246.
34. Hitchens TK, Ye Q, Eytan DF, Janjic JM, Ahrens ET, et al.(2011) ^{19}F MRI detection of acute allograft rejection with in vivo perfluorocarbon labeling of immune cells. Magn Reson Med 65: 1144–1153.
35. Baboota S, Faiyaz S, Ahuja A, Ali J, Shafiq S, et al. (2007) Development and validation of a stability-indicating HPLC method for analysis of celecoxib (CXB) in bulk drug and microemulsion formulations. Acta Chromatographica 18: 116–129.
36. Srinivas M, Morel PA, Ernst LA, Laidlaw DH, Ahrens ET (2007) Fluorine-19 MRI for visualization and quantification of cell migration in a diabetes model. Magn Reson Med 58: 725–734.
37. Janjic JM, Ahrens ET (2009) Fluorine-containing nanoemulsions for MRI cell tracking. Wiley Interdiscip Rev Nanomed Nanobiotechnol 1: 492–501.
38. Gerhardt GE, Lagow RJ (1978) Synthesis of the perfluoropoly(ethylene glycol) ethers by direct fluorination. J Org Chem 43: 4505–4509.

39. Celsense Inc. website. Celsense, Inc. Announces First Clinical Trial Authorization by the US Food and Drug Administration (FDA) for Use of Its Cell Tracking Imaging Product Cell Sense. Available: http://www.celsense.com/june27_first_clinical_trial.html. Accessed: 7 January 2013.

40. Krafft MP, Riess JG (2007) Perfluorocarbons: life sciences and biomedical uses Dedicated to the memory of Professor Guy Ourisson, a true RENAISSANCE man. J Polym Sci A Polym Chem 45: 1185–1198.

41. Kaneda MM, Sasaki Y, Lanza GM, Milbrandt J, Wickline SA (2010) Mechanisms of nucleotide trafficking during siRNA delivery to endothelial cells using perfluorocarbon nanoemulsions. Biomaterials 31: 3079–3086.

42. Lee SJ, Schlesinger PH, Wickline SA, Lanza GM, Baker NA (2011) Interaction of melittin peptides with perfluorocarbon nanoemulsion particles. J Phys Chem B 115: 15271–15279.

43. Fang JY, Hung CF, Hua SC, Hwang TL (2009) Acoustically active perfluorocarbon nanoemulsions as drug delivery carriers for camptothecin: drug release and cytotoxicity against cancer cells. Ultrasonics 49: 39–46.

44. Moss G (2009) Medium-chain triglycerides. In: Rowe RC, Sheskey PJ, Quinn ME, editors. Handbook of Pharmaceutical Excipients, 6th ed. Newyork: Pharmaceutical Press. pp. 429–431.

45. Wang Y, Yu L, Han L, Sha X, Fang X (2007) Difunctional Pluronic copolymer micelles for paclitaxel delivery: Synergistic effect of folate-mediated targeting and Pluronic-mediated overcoming multidrug resistance in tumor cell lines. Int J Pharm 337: 63–73.

46. Hoffman H (1984) Polyoxythylenglycerol triricinoleat 35 DAC 1979. Pharm Zeit 129: 1730–1733.

47. Gelderblom H, Verweij J, Nooter K, Sparreboom A (2001) Cremophor EL: the drawbacks and advantages of vehicle selection for drug formulation. Eur J Cancer 37: 1590–1598.

48. Singh KK (2009) Polyoxyethylene castor oil derivatives. In: Rowe RC, Sheskey PJ, Quinn ME, editors. Handbook of Pharmaceutical Excipients, 6th ed. Newyork: Pharmaceutical Press. pp. 572–579.

49. Taxol® (paclitaxel) INJECTION website. Available: http://packageinserts.bms.com/pi/pi_taxol.pdf. Accessed: 7 January 2013.

50. Chao TC, Chu Z, Tseng LM, Chiou TJ, Hsieh RK, et al. (2005) Paclitaxel in a novel formulation containing less Cremophor EL as first-line therapy for advanced breast cancer: a phase II trial. Invest New Drugs 23: 171–177.

51. Wang Y, Wu KC, Zhao BX, Zhao X, Wang X, et al. (2011) A novel paclitaxel microemulsion containing a reduced amount of Cremophor EL: pharmacokinetics, biodistribution, and in vivo antitumor efficacy and safety. J Biomed Biotechnol 2011: 1–10.

52. Uskoković V, Odsinada R, Djordjevic S, Habelitz S (2011) Dynamic light scattering and zeta potential of colloidal mixtures of amelogenin and hydroxyapatite in calcium and phosphate rich ionic milieus. Arch Oral Biol 56: 521–532.

53. Janjic JM, Ahrens ET (2012) Compositions and methods for producing cellular labels for nuclear magnetic resonance techniques. US 8,277,610 B2.

54. Rieger AM, Hall BE, Barreda DR (2010) Macrophage activation differentially modulates particle binding, phagocytosis and downstream antimicrobial mechanisms. Dev Comp Immunol 34: 1144–1159.

19

Automatic Extraction of Nanoparticle Properties Using Natural Language Processing: NanoSifter an Application to Acquire PAMAM Dendrimer Properties

David E. Jones[1]*, **Sean Igo**[1,2], **John Hurdle**[1], **Julio C. Facelli**[1,2]

1 Department of Biomedical Informatics, University of Utah, Salt Lake City, Utah, United States of America, **2** Center for High Performance Computing, University of Utah, Salt Lake City, Utah, United States of America

Abstract

In this study, we demonstrate the use of natural language processing methods to extract, from nanomedicine literature, numeric values of biomedical property terms of poly(amidoamine) dendrimers. We have developed a method for extracting these values for properties taken from the NanoParticle Ontology, using the General Architecture for Text Engineering and a Nearly-New Information Extraction System. We also created a method for associating the identified numeric values with their corresponding dendrimer properties, called NanoSifter. We demonstrate that our system can correctly extract numeric values of dendrimer properties reported in the cancer treatment literature with high recall, precision, and f-measure. The micro-averaged recall was 0.99, precision was 0.84, and f-measure was 0.91. Similarly, the macro-averaged recall was 0.99, precision was 0.87, and f-measure was 0.92. To our knowledge, these results are the first application of text mining to extract and associate dendrimer property terms and their corresponding numeric values.

Editor: Valentin Ceña, Universidad de Castilla-La Mancha, Spain

Funding: The project described was supported by Grant Numbers T15LM007124 and R01-LM010981 from the National Library of Medicine. The funders had no role in study design, data collection and analysis, decision to publish, or preparation of the manuscript. The funders had no role in study design, data collection and analysis, decision to publish, or preparation of the manuscript.

Competing Interests: The authors have declared that no competing interests exist.

* E-mail: davide.jones@utah.edu

Introduction

Nanomedicine is the field of study that considers the application of nanoparticles and nanoscience techniques to health care and medical research [1]. A main focus of nanomedicine includes the use of nanoparticles as delivery vectors for pharmaceutics, diagnostic devices, and tissue replacement materials [2]. This field is relatively new, however it is producing large numbers of publications and substantial new data each year [3]. Data being published contains valuable information regarding how the structure of these nanoparticles relates to their biochemical and biophysical properties, which include but are not limited to their diameter, molecular weight, surface charge, zeta potential, bioavailability, cytotoxicity, etc. [4].

We have chosen dendrimers for our initial application of natural language processing (NLP) to nanomedicine, because they are well-defined, highly branched polymeric nanoparticles that can easily be modified to differing specifications. There is also substantial literature reporting their biological, chemical, and physical properties. Dendrimers are composed of a central core that is surrounded by concentric shells [5,6]. The number of shells that extend out from the central core determines the particular generation of the dendrimer. Due to their structure, these molecules form very symmetric, three-dimensional particles that promise to be highly useful in the fields of pharmaceutics and medicine as delivery vectors [7]. The scaffold structure of dendrimers has been found to be a suitable carrier for a variety of drugs and siRNA, improving the solubility and bioavailability of poorly soluble agents. Currently there are several classes of dendrimers in use or under consideration for biomedical applications. This study focused on poly(amidoamine) (PAMAM) dendrimers that show promise for cancer treatment.

Databases and repositories containing information relevant to biomedical nanoparticles, especially their biochemical and biophysical properties, are critical for both primary research as well as secondary uses such as data mining and predictive modeling. The American National Standards Institute's Nanotechnology Standards Panel (ANSI-NSP) has created a Nanotechnology Standards database which is a free for individuals and groups seeking information about standards and other relevant documents related to nanomaterials and nanotechnology-related products and processes [8]. The database does not directly host standards and other similar documents, however it provides a place for standards developing organizations to add their relevant documents. This may someday be an important resource for the future development of standardized terminology in the field of nanotechnology and nanomedicine, but it does not contain an extensive collection of values of biological properties of medical nanomaterials.

nanoHUB.org is the premier site for computational nanotechnology research, education, and collaboration [9]. This resource provides an environment for collaboration and aggregation of tools used in simulating nanoscale phenomena. But with this resource, the researchers must provide their own nanomaterial-specific data to utilize the host of simulation tools provided. To our knowledge, there is no authoritative, up-to-date database where researchers consistently contribute results from new publications on biomedical nanoparticles and their properties. Some attempts have been reported in the literature, like caNanoLab, a database created by the National Cancer Institute for sharing nanoparticle information [10]. However, caNanoLab contains a limited number of nanoparticles, and for those it often has incomplete information regarding their biological, chemical, and physical properties. Also, there are only limited capabilities to query this system. No data model exists to support comparing the properties of a molecule to its biochemical and biophysical activity. These properties are necessary to advance research on nanoparticles, but the only way to retrieve this information currently is by manual extraction from the primary literature.

Though manual extraction is a very time consuming and resource intensive process, little research has been done to apply computational methods to obtain nanoparticle property data from the vast biomedical literature on nanoparticles. Information extraction (IE) efforts are widely acknowledged to be important in harnessing the rapid advance of biomedical knowledge, particularly in areas where important factual information is published in diverse literature [11]. In particular, NLP is a family of methods based on syntactic/semantic analysis that can extract information automatically from the literature [12].

NLP has been used effectively in other biomedical domains. For instance, Chaussabel utilized NLP algorithms to extract data from the literature on cell line profiling. He observed that this approach could be applied beyond genomic data analysis [13]. Garten et al. successfully applied NLP methods to the pharmacogenomics literature to create structured databases built on data from unstructured text [14]. Hunter et al. created a system called OpenDMAP that extracts protein transport, interaction, and gene expression assertions [11]. In the field of nanoinformatics there has been an attempt at harnessing the utility of NLP in the nanomedicine literature by Garcia-Remesal and colleagues. They developed a method utilizing named entity recognition to identify four different categories of information: nanoparticle names, routes of exposure, toxic effects, and particle targets [15]. The method that this group developed was moderately successful, but it was designed as a proof-of-concept with limited quantitative detail. Our goal is to gather detailed quantitative data associated with dendrimer properties.

In this study, we evaluate the use of NLP methods to extract numeric values for the properties of biomedical dendrimers reported in the cancer treatment literature. We use open source tools for extracting particle property values, using the NanoParticle Ontology (NPO) [4] as a starting point. In particular, the tools we use are a processing pipeline called the General Architecture for Text Engineering (GATE) and its IE module ANNIE (a Nearly-New Information Extraction System) [16]. In a real-world sentence, a nanoparticle property term can appear arbitrarily far from its associated value, so we also created a method of associating the two. We demonstrate that our system can correctly extract dendrimer property terms and their corresponding numeric values as evaluated by the typical NLP metrics of recall, precision, and f-measure score.

Materials and Methods

Literature Corpus

We collected from PubMedCentral relevant articles on dendrimer nanoparticles as reported in the cancer treatment literature. Articles were retrieved in pdf format. The search criteria used was "PAMAM dendrimers AND cancer treatment." This search yielded 420 journal articles on March 4, 2013. Articles were excluded from this study if they did not contain explicit numeric values of biological, chemical, and/or physical properties of dendrimers. From this pool of 420 articles, we randomly selected 200 journal articles. A subset of 100 articles was used as the training set for our system. The other subset of 100 articles was used for the creation of the test set for our system. Citations for both the training and test set of documents can be found in the supplementary information (Appendix S1 and Appendix S2). For similar applications in related fields, the selection of a test set of approximately 100 documents is a common target that represents a compromise of quality and cost of the manual review. For instance Zaremba et al. used a test set of 138 abstracts to analyze enteropathogenic bacteria, such as *Escherichia coli* and *Salmonella*, literature [17].

NLP Method Development

The NLP system reported here uses a two-step process to extract the desired property terms and numeric values. The first step involves the actual identification and annotation of the numeric values and dendrimer property terms. This corpus annotation pipeline was built using the Java Annotations Patterns Engine (JAPE) and integrating components from ANNIE within GATE. In order to search for the numeric values, we had to develop a regular expression model (Appendix S3). The specific dendrimer property terms were selected from the NPO and represent the properties of nanoparticles. The dendrimer property terms were selected from the NPO with the ultimate goal of linking the NPO with our tool to provide metadata for the data extractions from the nanomedicine literature. The initial nanoparticle property terms list was confirmed to be relevant for the nanomedicine community by expert review by the members of Dr. Hamidreza S. Ghandehari's research lab (http://nanoinstitute. utah.edu/research/ustar-clusters/ghandehari-lab/ghandehari-PI. php) at the University of Utah. The list of terms considered here includes hydrodynamic diameter (NPO_1915), particle diameter (NPO_1539), molecular weight (NPO_1171), zeta potential (NPO_1302), cytotoxicity (NPO_1340), IC50 (NPO_1195), cell viability (NPO_1343), encapsulation efficiency (NPO_1336), loading efficiency (NPO_1334), and transfection efficiency (NPO_1335). The property terms, their corresponding NPO identification code, and their definitions can be found in Table 1. To search for these property terms, the system utilizes a simple keyword identification scheme.

The training set of documents was manually annotated for numeric values and dendrimer property terms using GATE. Following the annotation, the numeric values associated with each property term were extracted manually and organized in a tabular format for ease of use and comparison. Once the pipeline was able to successfully annotate the numeric values and the dendrimer property terms, we developed an algorithm that would associate numeric values and dendrimer property terms that occurred within the same sentence using proximity metrics. We selected a proximity distance metric of 200 characters because our preliminary experiments have shown that the sensitivity and specificity of the system was best for this distance in the training set. For instance we observed that if we increased it, the number of false positives increased without any improvement in the observed recall of the system. Finally, we optimized performance iteratively before moving on to the test set of documents.

Reference Standard Creation

Two domain experts were selected from the nanotechnology program at the University of Utah. Before allowing them to review the test subset of 100 articles, they independently reviewed,

Table 1. Listing of the NPO Property Terms.

PROPERTY TERM	NPO CODE	DEFINITION
Hydrodynamic Diameter	NPO_1915	The hydrodynamic size which is the diameter of a particle or molecule (approximated as a sphere) in an aqueous solution.
Particle Diameter	NPO_1539	Diameter which inheres in a particle.
Molecular Weight	NPO_1171	The sum of the relative atomic masses of the constituent atoms of a molecule.
Zeta Potential	NPO_1302	The potential difference between the bulk dispersion medium (liquid) and the stationary layer of liquid near the surface of the dispersed particulate.
Cytotoxicity	NPO_1340	Toxicity that impairs or damages cells, and it is a desired property of the dispersed particulate.
IC50	NPO_1195	A measure of toxicity which is the concentration of a drug or inhibitor that is required to inhibit a biological process or a participant's activity in that process by half.
Cell Viability	NPO_1343	Viability of a cell to proliferate, grow, divide, or repair damaged cell components.
Encapsulation Efficiency	NPO_1336	The efficiency of inhering in a nanomaterial or supramolecular structure by virtue of its capacity to encapsulate an amount of molecular entity, isotope or nanomaterial.
Loading Efficiency	NPO_1334	A quality inhering in a material entity by virtue of it having the capacity to carry an amount of another material entity.
Transfection Efficiency	NPO_1335	The efficiency inhering in a bearer's ability to facilitate transfection.

annotated, and extracted information from the training set of articles using GATE. The annotations consisted of numeric values and dendrimer property terms selected from the NPO. Their annotations were compared and Cohen's kappa was calculated. Cohen's kappa is a statistical measure of inter-rater reliability, and for this study we required it to be ≥80%, which has been categorized as excellent by Fleiss at a value of 75% or higher [18].

Upon achieving an inter-rater reliability of 80%, the annotators independently reviewed, annotated, and extracted information from the test set of articles. Again, the numeric values and dendrimer property terms were taken from the NPO and were annotated using GATE. Following the annotation, the numeric values associated with each property term were extracted and organized in a tabular format.

NLP System Performance

The subset of 100 test articles was processed by our new NLP system. The output from the system was organized in a tabular format for ease of use and comparison.

Data Analysis

Our NLP and manual results were compared on a by-nanoparticle property term basis. The extracted numeric values associated to the dendrimer property terms were evaluated and determined to be true positive, false positive, or false negative. First, we calculated the recall, precision, and f-measure of each nanoparticle property term. We then calculated the micro-averaged and macro-averaged recall, precision, and f-measure. When using micro-averaged measurements, each "source" (e.g. document) is given the same weight, and calculations are made on a pooled contingency table [19]. Macro-averaged measurements are calculated by giving the same weight to each concept category or class (e.g., dendrimer property term) [19].

The recall, precision, and f-measure were calculated using the following equations:

$$Recall = TP/(TP + FN) \qquad (1)$$

$$Precision = TP/(TP + FP) \qquad (2)$$

$$F - measure =$$
$$((1 + \beta^2) * Precision * Recall)/((\beta^2 * Precision) + Recall) \qquad (3)$$

In these equations TP is true positive, FP is false positive, FN is false negative, and β is the weighting applied to the relationship between precision and recall. For our purposes we decided to weight the precision and recall evenly, so $\beta = 1$.

Results

Table 2 summarizes the results of the evaluation of the NLP system that we created. The results of the system are compared against the manually annotated reference standard. The table shows the recall, precision, and f-measure for each of the nanoparticle property terms and numeric value relationships. Table 3 displays both the micro-averaged and macro-averaged recall, precision, and f-measure values.

As can be seen in Table 2, our NLP system yields recall values ranging from 0.95 to 1 and precision values ranging from 0.59 to 1. The f-measure values range from 0.73 to 1. The micro-averaged values for recall was 0.99, precision was 0.84, and f-measure was 0.91. Similarly, the macro-averaged values for recall was 0.99, precision was 0.87, and f-measure was 0.92.

Discussion

The tables show an important difference between recall and precision. In this task, high recall is preferred to high precision, because we do not want our system to miss instances of property terms and their associated numeric values. The number of articles returned for any given search (e.g., our "PAMAM dendrimers AND cancer treatment" search) is too large for routine manual search, but reviewing NanoSifter *results* is quite traceable. The results can be manually reviewed post-processing without much additional effort. From the results, it can be seen that "encapsu-

Table 2. Results from the Evaluation of the Nanosifter NLP System.

Nanoparticle Property Term	TP	FP	FN	Recall	Precision	F-measure	Occurrences by Article
Hydrodynamic Diameter	8	0	0	1	1	1	6
Particle Diameter	211	39	1	0.995283	0.844	0.91341991	56
Molecular Weight	143	23	2	0.986207	0.86145	0.91961415	25
Zeta Potential	41	0	1	0.97619	1	0.98795181	16
Cytotoxicity	124	18	1	0.992	0.87324	0.92883895	29
IC50	47	8	1	0.979167	0.85455	0.91262136	15
Cell Viability	78	31	0	1	0.7156	0.8342246	25
Encapsulation Efficiency	1	0	0	1	1	1	1
Loading Efficiency	5	0	0	1	1	1	1
Transfection Efficiency	19	13	1	0.95	0.59375	0.73076923	9

lation efficiency" and "loading efficiency" were the best property terms extracted with recall, precision, and f-measure values of 1. These scores are likely due to the low prevalence of these properties appearing in our literature corpus. "Transfection efficiency" was the property term that was the least well extracted from nanomedicine literature. It had a recall value of 0.95, a precision value of 0.59, and an f-measure value of 0.73.

These results indicate that the NanoSifter NLP system can, generally, extract numeric values associated with particle property terms from dendrimers reported in the cancer treatment literature with high recall, precision, and f-measure scores. To the authors' knowledge, these results are the first application of text mining to extract numeric values associated to dendrimer property terms from nanomedicine literature. With regards to our application, the high recall values are more important than the moderate precision values. This is because the lack of precision is manageable and can be quickly corrected by manual post processing of the annotated text.

As can be seen from the results, there was a fair amount of fluctuation in the values for precision for each property term. There were a few property terms that yielded precisions of 1 including "hydrodynamic diameter," "zeta potential," "encapsulation efficiency," and "loading efficiency." This can be accounted for by the limited number of instances that these terms appeared in the literature. Of all of the property terms used in this study, these were the least common. The next tier of precision values of interest are those that were greater than 0.80, these include "particle diameter," "molecular weight," "cytotoxicity," and "IC50." These property terms yielded quite reasonable precision values, as we expected based upon their occurrences in the literature and the specificity of the syntax used when describing these property terms and their numeric values.

Table 3. Micro-averaged and Macro-averaged Recall, Precision, and F-measure

Type of Average	Recall	Precision	F-measure
Micro	0.989766	0.83684	0.90689886
Macro	0.987885	0.87426	0.922744

The lowest precision values could be seen for "cell viability" (0.72) and "transfection efficiency" (0.59). One reason for these lower precision values is that the numeric units for these properties are percentages. There was a significant number of false positives in the literature corpus because the number of occurrences of percentages for other, non-particle items within the 200-character proximity metric was large. With specific regard to "transfection efficiency," precision values for this term were the lowest because the terminology used to refer to this property is not standardized. There are many different ways in which the literature refers to this property, making it difficult not to overfit a method of retrieving the numeric values of this property.

Limitations

NanoSifter uses a method that appears to be generally reliable and accurate. However, there are imperfections that were observed while processing and analyzing the data from this study. First, the data extracted by our method is not always directly associated with a dendrimer nanoparticle. For instance, many times the system correctly finds, annotates, and extracts a "molecular weight measurement", but this measurement may be associated with a subunit utilized in the synthesis of a PAMAM dendrimer or another material used in one of the articles. A method to address this limitation could include post-analysis manual review of the system's performance. Another limitation of our system is that the NanoSifter algorithm can only pair a nanoparticle property term with a single numeric value annotation before and after itself. This causes a problem when a sentence is more complex and contains a property term, random text, numeric value, random text, or another numeric value. In NLP, this is a problem called co-reference resolution, and it could be addressed with a more sophisticated language model than the one used in this study.

Another limitation is that our system would only retrieve the first numeric value expressed following the property term. This situation accounts for some of the false negatives ("particle diameter," "cytotoxicity," "IC50," and "transfection efficiency") found in our analysis. This could also be addressed by using a more sophisticated language model than the one used in this study. Finally, the other false negatives, "molecular weight" and "zeta potential," account for another limitation of our system. Since we were processing pdf documents in this study, occasionally there would be an instance where a property term exceeded a single line of text, so a dash would be inserted in the word and it would continue on the next line. The method used in developing this system did not account for this artifact. Therefore the NanoSifter NLP system would not annotate this property term, and no association would be made to the corresponding numeric value. A method for addressing this would be to use XML documents instead of pdfs in future analyses. These limitations are not novel to our approach, as they are common throughout the field of NLP. Nonetheless they are counterbalanced by the ability to extract information from journal articles at a much lower cost than manual review.

Future Work

Since this is early work in an important but neglected area of nanoinformatics, there are many directions this research could be taken. The first priority will be to make corrections to our system to try to improve our recall, precision, and f-measure values. Another priority will be to attempt to use this system to annotate and extract information from another subclass of nanoparticles. This will help to validate the ability of this system to generalize across the field of nanoparticles. One of the most important next steps would be to expand the property terms and numeric values that the system targets. Some specific properties that we are considering include "exposure times" and "cell types" interacting with the nanoparticles. This would allow for greater databases to be created regarding PAMAM dendrimers and nanoparticles in general. Another goal would be to more seamlessly integrate the NPO into our system so that the annotations and extractions contain descriptive metadata. Finally, it is important that we attempt to implement some sort of negation analysis tool into our system. This would specifically help in the instances where an article states that the dendrimer nanoparticles were not toxic at a certain concentration.

Conclusion

In this paper, we have presented a nanoinformatics method based on NLP approaches for automatically extracting numeric values associated with dendrimer property terms from the nanomedicine literature. The results from our analysis demonstrate that the NanoSifter NLP system can be used to reliably and accurately extract information from dendrimers developed for cancer treatment literature and shows promise for the future of text mining in the field of nanoinformatics. This initial research in the field of applying NLP to nanomedicine literature could assist in significant advances for the nanomedicine community. This work could lead to the creation of databases containing valuable information regarding nanoparticles at a much lower cost than using manual review. The readily available data on nanomedical relevant particles could be further analyzed for many secondary uses of the data. In particular, the acquired data could be used for data mining to find correlations between properties, create predictive models like quantitative structure activity relationships, and eventually reach the point where potential candidate molecules can be created *in silico* and modeled to theoretically predict their biochemical activity before synthesis. This would reduce the search space for novel, effective nanoparticles for use in medicine and pharmaceutics.

Author Contributions

Conceived and designed the experiments: DEJ SI JH JCF. Performed the experiments: DEJ. Analyzed the data: DEJ. Contributed reagents/materials/analysis tools: DEJ SI. Wrote the paper: DEJ SI JH JCF.

References

1. Jain K (2008) The Handbook of Nanomedicine. Totowa, New Jersey: Humana.
2. Staggers N, McCasky T, Brazelton N, Kennedy R (2008) Nanotechnology: the coming revolution and its implications for consumers, clinicians, and informatics. Nurs Outlook 56: 268–274.
3. de la Iglesia D, Maojo V, Chiesa S, Martin-Sanchez F, Kern J, et al. (2011) International efforts in nanoinformatics research applied to nanomedicine. Methods Inf Med 50: 84–95.
4. Thomas DG, Pappu RV, Baker NA (2011) NanoParticle Ontology for cancer nanotechnology research. J Biomed Inform 44: 59–74.
5. Wood KC, Little SR, Langer R, Hammond PT (2005) A family of hierarchically self-assembling linear-dendritic hybrid polymers for highly efficient targeted gene delivery. Angew Chem Int Ed Engl 44: 6704–6708.
6. Kolhe P, Misra E, Kannan RM, Kannan S, Lieh-Lai M (2003) Drug complexation, in vitro release and cellular entry of dendrimers and hyperbranched polymers. Int J Pharm 259: 143–160.
7. du Toit LC, Pillay V, Choonara YE, Pillay S, Harilall SL (2007) Patenting of nanopharmaceuticals in drug delivery: no small issue. Recent Pat Drug Deliv Formul 1: 131–142.
8. Institute ANS (2013) ANSI-NSP Launches Nanotechnology Standards Database. New York: ANSI News and Publications.
9. nanoHUB.org (2013) nanoHUB.org Online Simulation and More for Nanotechnology About Us.
10. National Cancer Institute (2011) caNanoLab. pp. Welcome to the cancer Nanotechnology Laboratory (caNanoLab) portal. caNanoLab is a data sharing portal designed to facilitate information sharing in the biomedical nanotechnology research community to expedite and validate the use of nanotechnology in biomedicine. caNanoLab provides support for the annotation of nanomaterials with characterizations resulting from physico-chemical and in vitro assays and the sharing of these characterizations and associated nanotechnology protocols in a secure fashion.
11. Hunter L, Lu Z, Firby J, Baumgartner WA, Jr., Johnson HL, et al. (2008) OpenDMAP: an open source, ontology-driven concept analysis engine, with applications to capturing knowledge regarding protein transport, protein interactions and cell-type-specific gene expression. BMC Bioinformatics 9: 78.
12. Liu K, Hogan WR, Crowley RS (2011) Natural Language Processing methods and systems for biomedical ontology learning. Journal of Biomedical Informatics 44: 163–179.
13. Chaussabel D (2004) Biomedical literature mining: Challenges and solutions in the 'omics' era. American Journal of PharmacoGenomics 4: 383–393.
14. Garten Y, Coulet A, Altman RB (2010) Recent progress in automatically extracting information from the pharmacogenomic literature. Pharmacogenomics 11: 1467–1489.
15. Garcia-Remesal M, Garcia-Ruiz A, Perez-Rey D, de la Iglesia D, Maojo V (2013) Using nanoinformatics methods for automatically identifying relevant nanotoxicology entities from the literature. Biomed Res Int 2013: 410294.
16. Cunningham H, al e (2011) Text Processing with GATE. University of Sheffield Department of Computer Science.
17. Zaremba S, Ramos-Santacruz M, Hampton T, Shetty P, Fedorko J, et al. (2009) Text-mining of PubMed abstracts by natural language processing to create a public knowledge base on molecular mechanisms of bacterial enteropathogens. BMC Bioinformatics 10: 177.
18. Fleiss JL (1981) Statistical methods for rates and proportions. New York: John Wiley.
19. Yang Y (1999) An Evaluation of Statistical Approaches to Text Categorization. Information Retrieval 1: 69–90.

20

Intracellular Gold Nanoparticles Increase Neuronal Excitability and Aggravate Seizure Activity in the Mouse Brain

Seungmoon Jung[1,9]**, Minji Bang**[1,9]**, Byung Sun Kim**[1]**, Sungmun Lee**[2]**, Nicholas A. Kotov**[3]**, Bongsoo Kim**[4]**, Daejong Jeon**[1]*****

1 Department of Bio and Brain Engineering, Korea Advanced Institute of Science and Technology (KAIST), Daejeon, Republic of Korea, **2** Department of Biomedical Engineering, Khalifa University of Science, Technology, and Research, Abu Dhabi, United Arab Emirates, **3** Department of Chemical Engineering, University of Michigan, Ann Arbor, Michigan, United States of America, **4** Department of Chemistry, Korea Advanced Institute of Science and Technology (KAIST), Daejeon, Republic of Korea

Abstract

Due to their inert property, gold nanoparticles (AuNPs) have drawn considerable attention; their biological application has recently expanded to include nanomedicine and neuroscience. However, the effect of AuNPs on the bioelectrical properties of a single neuron remains unknown. Here we present the effect of AuNPs on a single neuron under physiological and pathological conditions *in vitro*. AuNPs were intracellularly applied to hippocampal CA1 neurons from the mouse brain. The electrophysiological property of CA1 neurons treated with 5- or 40-nm AuNPs was assessed using the whole-cell patch-clamp technique. Intracellular application of AuNPs increased both the number of action potentials (APs) and input resistance. The threshold and duration of APs and the after hyperpolarization (AHP) were decreased by the intracellular AuNPs. In addition, intracellular AuNPs elicited paroxysmal depolarizing shift-like firing patterns during sustained repetitive firings (SRF) induced by prolonged depolarization (10 sec). Furthermore, low Mg^{2+}-induced epileptiform activity was aggravated by the intracellular AuNPs. In this study, we demonstrated that intracellular AuNPs alter the intrinsic properties of neurons toward increasing their excitability, and may have deleterious effects on neurons under pathological conditions, such as seizure. These results provide some considerable direction on application of AuNPs into central nervous system (CNS).

Editor: Gennady Cymbalyuk, Georgia State University, United States of America

Funding: This work was supported by the Korea Health 21 R&D grant (A120051) funded by Ministry of Health and Welfare, and also supported by the KUSTAR-KAIST Institute, Korea, under the R&D program supervised by the KAIST. The funders had no role in study design, data collection and analysis, decision to publish, or preparation of the manuscript.

Competing Interests: The authors have declared that no competing interests exist.

* E-mail: clark@kaist.ac.kr

⑨ These authors contributed equally to this work.

Introduction

In the past decade, nanoparticles (NPs) have been used in biological and biomedical applications such as drug delivery, photothermal therapy, biosensing, and bioimaging [1,2]. NPs are typically transferred into the cells by endocytosis [3]. NPs can be used for delivery of genes or drugs into the cytosol and subcellular organelles, including the nucleus [4]. In particular, intracellular dynamics could be monitored by intracellular delivery of nano-sized contrast agents, and targeting nanomedicine into subcellular organelles could vastly improve the efficacy of therapeutic regimens such as proapoptotic drugs, lysosomal enzymes, gene therapy, and photodynamic therapy [4,5]. However, many toxic effects of NPs on various types of cells have been widely known [6,7].

Gold NPs (AuNPs) are of particular interest due to their excellent stability and various biocompatibility properties, including nontoxicity, non-immunogenicity, and high tissue permeability without hampering cell functionality [8,9]. The distinct properties of AuNPs suggest their potential for the delivery of therapeutic substances—such as drugs or small nucleotides—into the brain for treatment of various neurological diseases or disorders [10]. Although novel AuNPs have been developed as carriers for delivery of therapeutic substances to neuronal cells across the blood-brain barrier (BBB) [11,12], the effect of AuNPs on the electrophysiological activity of neurons has not been investigated. Recently, several NPs, such as silver (Ag), copper oxide (CuO), zinc oxide (ZnO), and tungsten carbide (WC), were reported to alter some properties of ion channels and neuronal excitability [13–19]. Although it remained unclear that these NPs affected inside or outside of neuronal cells, these works suggested some deleterious or toxic effects of NPs on bioelectrical properties of neurons in the brain.

The neuron is a cellular unit of the nervous system. Neuronal activity is represented by changes in membrane potential, such as action potentials (APs), and the related synaptic transmission by means of neurotransmitters, which is associated with receiving, integrating, and transmitting information in the brain [20]. The intrinsic properties of ion channels in neurons determine or regulate neuronal activity [21], and alterations in neuronal activity could impact both physiological and pathophysiological conditions. For instance, abnormal neuronal activity or imbalanced excitation/inhibition is associated with several neurological diseases, such as epilepsy [22,23]. Therefore, for biological and biomedical applications of AuNPs as intracellular carriers of specific molecules (e.g., a drug, antibody, or oligonucleotide) to the brain, a thorough understanding of the interplay between intracellular AuNPs and neuronal intrinsic properties is necessary [24,25].

Patch-clamp recording is an approach to measure bioelectrical properties of the cells, and is especially useful in the study of excitable cells such as neurons [26–28]. One of advantages in whole-cell patch-clamp recordings is to deliver substances directly inside of a living cell through a glass micropipette. Thus, it is possible to investigate the intracellular effect of substances on the bioelectrical properties of a single neuron by whole-cell patch-clamp recordings. In this study, we measured the electrophysiological properties of neurons using the whole-cell patch-clamp technique after delivering AuNPs intracellularly into hippocampal CA1 neurons from the mouse brain. We also investigated the effect of intracellular AuNPs on two *in vitro* seizure models (prolonged sustained repetitive firings and low Mg^{2+}-induced epileptiform burst discharges). Herein is the first report of the effects of AuNPs on the bioelectrical properties of a single neuron under physiological and pathological conditions.

Materials and Methods

Ethics Statement

Animal care and handling were conducted according to the guidelines approved by the Institutional Animal Care and Use Committee (Approval Number: KA2013-33) of the Korea Advanced Institute of Science and Technology (KAIST). All efforts were made to minimize suffering.

Animals

Young male *C57BL/6* mice (4–5 weeks old) were used in the present study. Four mice were housed as a group under a 12-hr light/dark cycle with free access to food and water.

Nanoparticles and application to brain slice

AuNPs (OD 1, stabilized suspension in citrate buffer, negatively charged) of 5- and 40-nm diameter were purchased from Sigma (St. Louis, MO, USA). Fluorophore-labeled (maximum absorbance: 600 nm) spherical AuNPs (40 nm, methyl conjugated) were purchased from Nanopartz Inc. (Loveland, CO, USA). AuNPs were stored at 4°C before use. For intracellular application of AuNPs, the AuNPs were diluted with an intrapipette solution (approximately 1.1×10^{11} NPs/mL for 5-nm AuNPs, and 1.44×10^{8} NPs/mL for 40-nm AuNPs). Electrophysiological recordings with citrate-suspended, non-fluorescent AuNPs were performed 10 min after cell rupture.

Brain slice preparation and patch-clamp recordings

Preparation of hippocampal slices and the whole-cell patch-clamp recording method have been described previously [29,30]. Fully anesthetized mice were decapitated and the horizontal hippocampal slices (310 µm) were prepared in oxygenated (95% O_2, 5% CO_2), cold, ACSF (124 mM NaCl, 3.0 mM KCl, 1.23 mM NaH_2PO_4, 2.2 mM $CaCl_2$, 1.2 mM $MgCl_2$, 26 mM $NaHCO_3$, and 10 mM glucose, pH 7.4). After 1 hr recovery, brain slices were incubated in ACSF and whole-cell recordings were obtained from hippocampal CA1 neurons at 31 °C using glass pipette electrodes (3–6 MΩ). To measure APs and miniature

excitatory postsynaptic currents (mEPSCs), glass pipettes were filled with an internal solution (135 mM K-gluconate, 5 mM KCl, 2 mM $MgCl_2$, 5 mM EGTA, 10 mM HEPES, 0.5 mM $CaCl_2$, 5 mM Mg-ATP, and 0.3 mM Na-GTP) which was buffered to pH 7.4 with KOH. APs were triggered by a step-current injection (30 pA steps) from −150 pA to +150 pA in current-clamp mode for 1 sec. The numbers, threshold, and latency of APs evoked by the injected currents in AuNPs-treated and untreated neurons were analyzed. The duration of the first AP was measured at half amplitude above the threshold. The after hyperpolarization (AHP) amplitude was isolated from the first AP. Spontaneous firings in CA1 neurons were measured at −50 mV. After the neurons had been voltage-clamped at −60 mV, the mEPSC experiment was performed in the presence of 1 µM tetrodotoxin (TTX), 10 µM bicuculline (GABA_A receptor antagonist), and 5 µM CGP 55845 (GABA_B receptor antagonist). For the prolonged sustained repetitive firing (SRF) experiment, 80–90-pA currents were injected into the cell under current-clamp configuration for 10 sec [31]. The SRF experiment was conducted more than two times in each of cell. To precipitate epileptiform burst discharges, brain slices were incubated in low-Mg^{2+}/high-K^+ ACSF containing the following (in mM): 124 mM NaCl, 5.0 mM KCl, 1.23 mM NaH_2PO_4, 2.2 mM $CaCl_2$, 26 mM $NaHCO_3$, and 10 mM glucose [32]. After 1 hr incubation, epileptiform activity was measured with the same low-Mg^{2+}/high-K^+ ACSF. The low-Mg^{2+}/high-K^+ ACSF elicited bursts of spikes, and the number of bursts showing more than three spikes was analyzed for 3 min. In all the patch-clamp recordings, large cells in hippocampal CA1 region were visually chosen, and pyramidal neurons and interneurons were identified on the basis of their distinctive intrinsic membrane properties including firing patterns as described previously [33–36]. Cells showing intrinsic membrane properties of interneurons were removed from analysis. Patch-clamp recordings were performed using a MultiClamp 700 B amplifier and a Digidata1440 (Axon instruments), and the acquired data were analyzed using the pCLAMP version 10.2 (Axon Instruments) and Mini-Analysis Program (Synaptosoft).

Statistical analysis

All data are presented as means ± standard error of the mean (SEM). Statistical analyses were conducted using the SPSS software (SPSS, Chicago, IL, USA) and R (Software Foundation, Boston, MA, USA). Data were analyzed by analysis of variance (ANOVA) followed by *post hoc* comparisons. Student's *t*-test was used to identify main effects. A *p*-value <0.05 was considered to indicate statistical significance.

Results

Altered passive electrical properties by an intracellular application of AuNPs

To investigate the effect of AuNPs on the electrophysiological properties of a single neuron, AuNPs were added to the hippocampal CA1 neurons from a mouse brain slice using a glass micropipette after being mixed with an intrapipette solution. We first verified the intracellular distribution of AuNPs by using fluorophore-conjugated 40-nm AuNPs, which showed fluorescent signals inside of neurons (Figure 1A). The spherical cell shape was clearly displayed by the fluorescent signals, which indicates that the AuNPs were evenly and broadly distributed throughout the cell body and membrane (Figure 1A, right). Then, non-fluorescent AuNPs of 5- and 40-nm diameter were used for the intracellular application throughout the experiments. Passive membrane properties, such as input resistance, threshold potential for AP

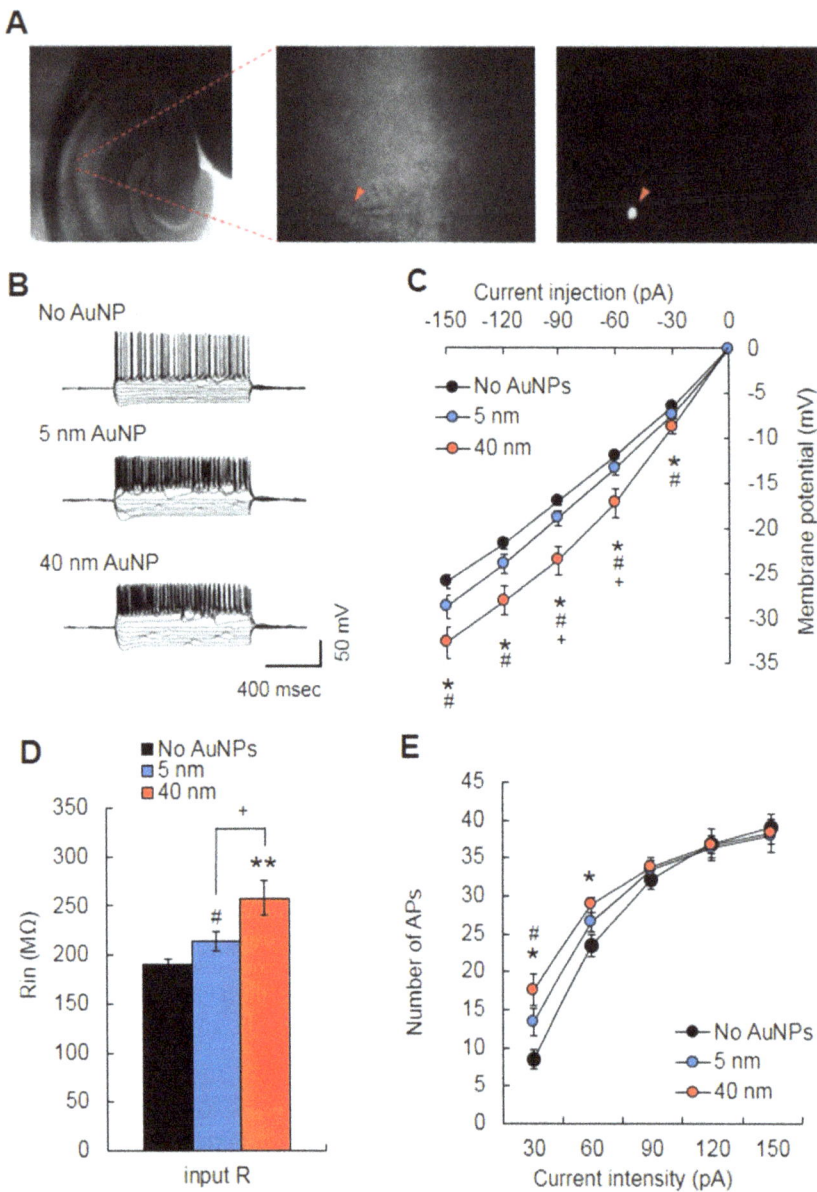

Figure 1. Effects of intracellular treatment with 5- or 40-nm AuNPs on the passive electrical properties of hippocampal CA1 neurons from a mouse hippocampal slice. (A) DIC images of a brain slice (left, 50×) and hippocampal CA1 layer (middle, 630×), and a fluorescence image of CA1 neurons (right, 630×) loaded with fluorophore-conjugated AuNPs through a patch pipette (middle) after breaking the gigaohm seal. The fluorescence signal indicates the infusion of AuNPs into the cell. (B) Representative traces of membrane potential changes and APs elicited by step-current injections for 1 sec from AuNP-treated and untreated (no AuNPs) hippocampal CA1 neurons. (C) AuNPs of both sizes considerably increased the changes in membrane potential. (D) Input resistance was significantly increased by AuNPs of both sizes. (E) The 5- or 40-nm AuNPs increased the number of APs substantially at low current intensity (at 30- or 60-pA depolarizing current injection). $*p<0.05$, $**p<0.01$, Student's t-test, No AuNPs $vs.$ 40-nm AuNPs; $\#p<0.05$, Student's t-test, No AuNPs $vs.$ 5-nm AuNPs; $+p<0.05$, Student's t-test, 5-nm AuNPs $vs.$ 40-nm AuNPs.

generation, and firing frequencies against the amplitude of injected currents, were measured using a current-clamp configuration. APs were generated by current injections with 30 pA steps (Figure 1B). Current-voltage relationships were obtained from values measured at the middle (500 msec) of hyperpolarizing pulses (Figure 1C). The plotted relationship curve shifted downward in both 5- ($n = 18$) and 40-nm ($n = 12$) AuNP-treated neurons compared to non-treated CA1 neurons ($n = 29$) ($F_{(2,56)} = 45.91$, $p<0.001$, two-way ANOVA), and changed more with the 40-nm AuNPs than with 5-nm AuNPs ($F_{(1,28)} = 22.32$, $p<0.001$, two-way ANOVA)

(Figure 1C). There was also a significant difference in the input resistance between AuNP-treated and non-treated neurons ($F_{(2,56)} = 11.25$, $p<0.05$, one-way ANOVA). The input resistance was increased in both 5- (214.22 ± 9.94 MΩ, $p<0.05$, Student's t-test) and 40-nm (258.1 ± 17.45 MΩ, $p<0.001$, Student's t-test) AuNP-treated neurons compared to non-treated CA1 neurons (190.87 ± 5.59 MΩ) (Figure 1D), and changed more with the 40-nm AuNPs than with 5-nm AuNPs ($p<0.05$, Student's t-test). As a measure of neuronal excitability, we plotted firing frequencies against the intensity of injected currents. Increased number of

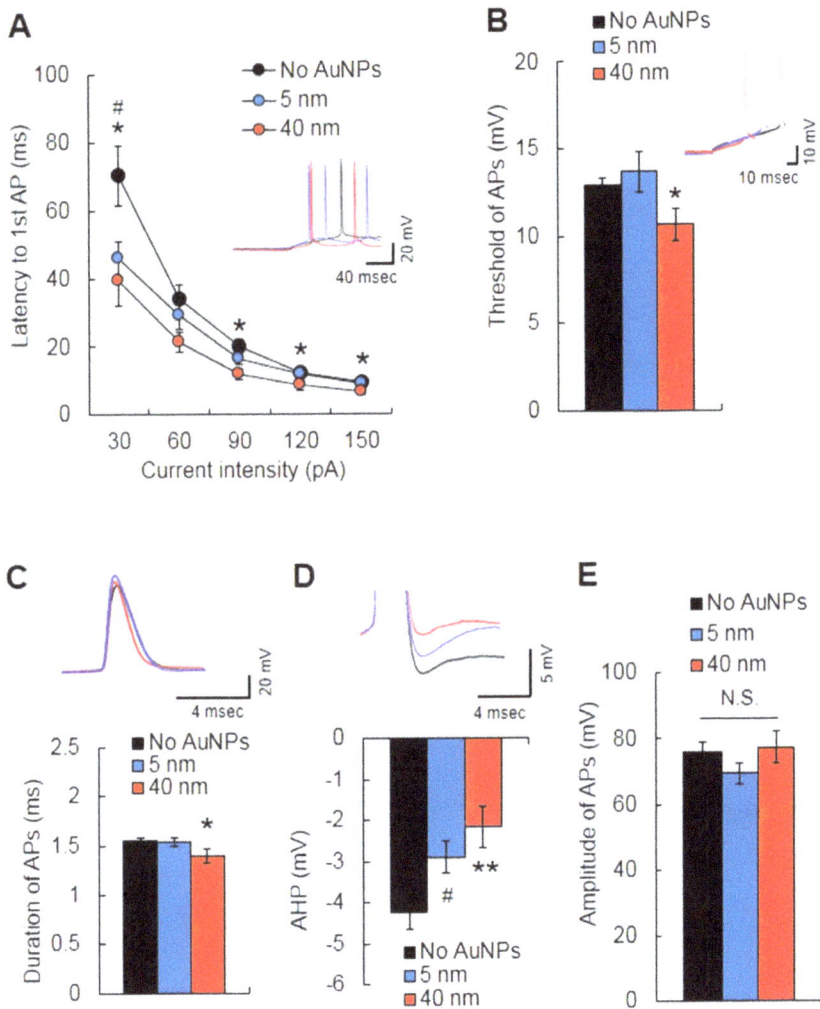

Figure 2. Effects of intracellular treatment with 5- or 40-nm AuNPs on AP properties in hippocampal CA1 neurons. (A) AuNPs of both sizes significantly decreased the latency to the first AP. The 40-nm AuNPs significantly reduced the AP threshold (B) and AP duration (C). (D) AuNPs of both sizes significantly increased AHP. (E) AuNPs did not affect the AP amplitude. *$p<0.05$, **$p<0.01$, Student's t-test, No AuNPs vs. 40-nm AuNPs; #$p<0.05$, Student's t-test, No AuNPs vs. 5-nm AuNPs; N.S., no significance.

spikes was observed at low intensities of current injection. Both 5- (13.33 ± 1.76) and 40-nm AuNP-treated neurons (17.58 ± 2.04) showed significantly more spikes at +30 pA injections than the non-treated neurons (8.51 ± 1.22) ($p<0.05$, Student's t-test) (Figure 1E). The 40-nm (28.83 ± 0.99), but not 5-nm, AuNP-treated neurons also showed significantly more spikes at +60 pA injections than the non-treated neurons (23.51 ± 1.43) ($p<0.05$, Student's t-test). Collectively, these results suggest that intracellular AuNPs alter the basic bioelectrical properties of hippocampal CA1 neurons.

The properties of AP are altered by an intracellular application of AuNPs

Regarding neuronal excitability, we further analyzed the properties of APs in CA1 neurons treated with AuNPs intracellularly. The latency to the first spikes at injection of a current of each intensity was noticeably decreased after the intracellular application of 40-nm AuNP-treated neurons ($F(1,41)=15.70$, $p<0.001$, two-way ANOVA) (Figure 2A). The 5-nm AuNP-treated neurons showed significant change at only 30 pA injection ($p<0.05$, Student's t-test). In addition, 40-nm AuNPs decreased the

threshold amplitude of the first AP generation (non-treatment, $n=23$, 12.89 ± 0.43 mV; 40 nm, $n=12$, 10.69 ± 0.91 mV, $p<0.05$, Student's t-test) (Figure 2B) and shortened the duration of the first AP (Figure 2C) compared with non-treatment (non-treatment, 1.54 ± 0.03 ms; 40 nm, 1.39 ± 0.07 ms, $p<0.05$, Student's t-test). However, 5-nm AuNPs did not affect the threshold amplitude (13.7 ± 1.18 mV) and duration (1.53 ± 0.04 ms) of the first AP generation. The AHP was significantly reduced with both AuNP sizes (5 nm, -2.9 ± 0.39 mV; 40 nm, -2.16 ± 0.51 mV) compared with non-treatment (-4.24 ± 0.39 mV) ($F(2,50)=6.10$, $p<0.005$) (Figure 2D), and there was no significant difference in the AHP amplitude between the 5-nm and 40-nm AuNP-treated neurons. No difference in the amplitude of APs was observed between the AuNP-treated and non-treated neurons (Figure 2E). These results demonstrate that intracellular AuNPs may lead CA1 neurons to become more excitable. Considering the alteration in basic bioelectrical properties, not only the ion channels active in the subthreshold range but also the channels activated during APs are subjected to alterations in CA1 neurons treated intracellularly with AuNP.

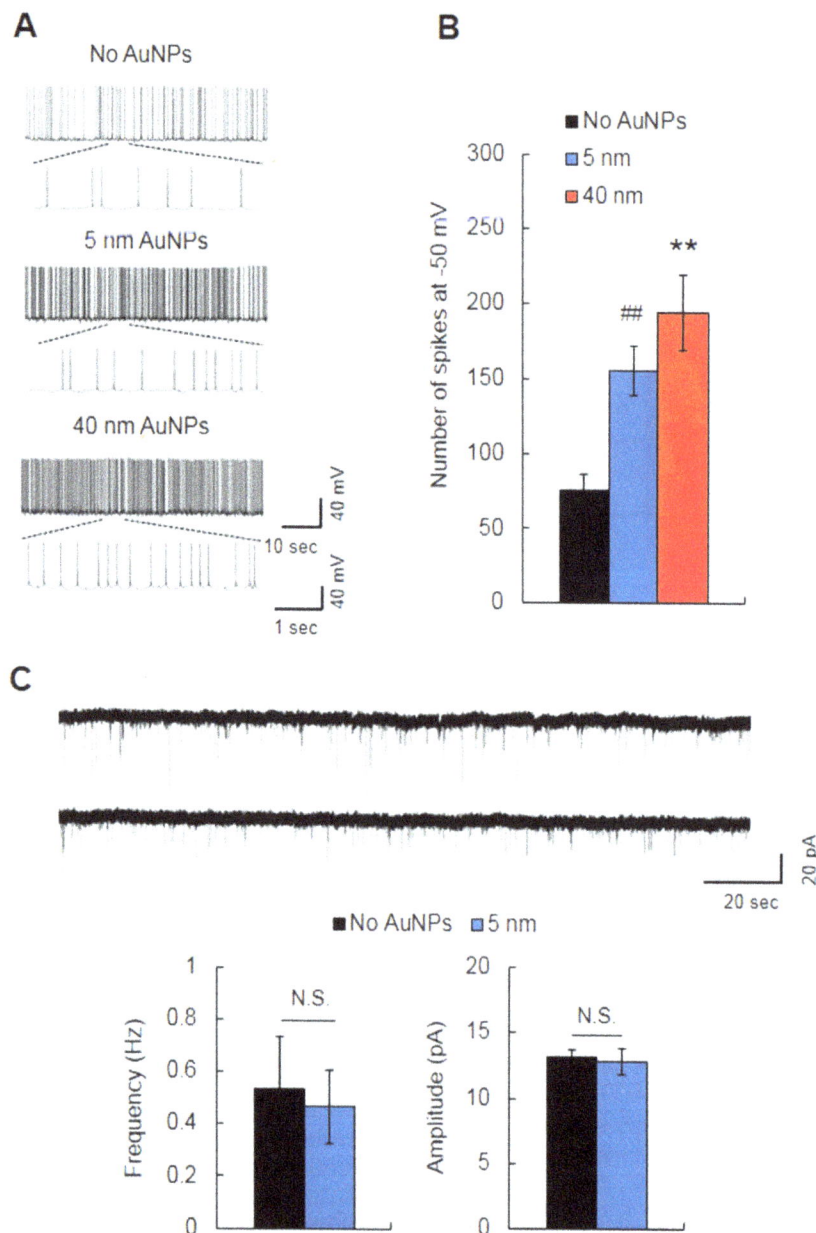

Figure 3. Effects of intracellular treatment with 5- or 40-nm AuNPs on spontaneous firing and excitatory synaptic transmission. (A) The representative traces of spontaneous firing from AuNP-treated and non-treated (No AuNPs) hippocampal CA1 neurons. (B) AuNPs of both sizes significantly increased the rate of spontaneous firing. (C) Neurons treated with 5-nm AuNP showed similar mEPSC frequency and amplitude to non-treated neurons. **$p<0.01$, Student's t-test, No AuNPs $vs.$ 40-nm AuNPs; ##$p<0.01$, Student's t-test, No AuNPs $vs.$ 5-nm AuNPs; N.S., no significance.

Effects on spontaneous firings and mEPSC of intracellular application of AuNPs

Next, we examined whether spontaneous firings in CA1 neurons are affected by intracellular treatment with AuNPs. Spontaneous APs were measured at -50 mV in current-clamp mode (Figure 3A). The 5- ($n=7$, 155.86 ± 16.52) and 40-nm ($n=11$, 194.55 ± 25.29) AuNP-treated neurons displayed more than twice the number of spikes than non-treated neurons ($n=13$, 75.46 ± 10.73) ($F(2,28)=12.36$, $p<0.001$, one-way ANOVA) (Figure 3B). To determine whether the AuNPs affected excitatory synaptic transmission—which can alter spontaneous firing—we measured mEPSC in 5-nm AuNP-treated neurons. There was little difference in the frequency (Figure 3C, Left) and amplitude

(Figure 3C, Right) of mEPSC between AuNP-treated ($n=6$) and non-treated neurons ($n=7$). Thus, the increased firing rate may be due to the altered intrinsic properties of CA1 neurons themselves rather than increased excitatory synaptic transmission. These results indicate that intracellular treatment with AuNPs enhanced the excitability of CA1 neurons.

Effects of AuNPs on prolonged depolarization and low Mg²⁺-induced epileptiform discharges

Increased excitability can lead to neurological diseases, such as epilepsy. Thus, we investigated the effects of intracellular treatment with AuNPs on two in $vitro$ seizure models (prolonged SRF and low Mg^{2+}-induced epileptiform burst discharges) [31,32].

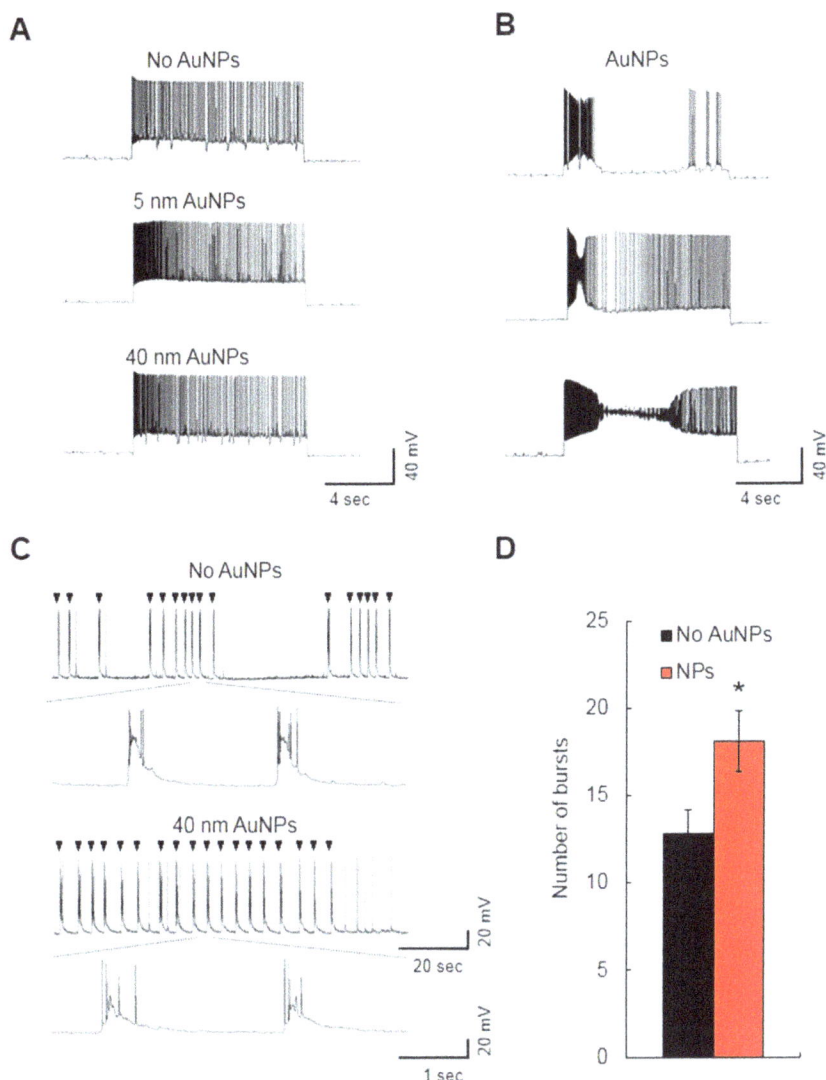

Figure 4. Effects of intracellular treatment with 5- or 40-nm AuNPs on seizure models. (A, B) Prolonged SRF experiment. (A) The representative traces of repetitive firings elicited by long (10 sec) depolarizing current pulses. (B) PDS-like spikes, an epiletiform activity, were observed from a portion of AuNP-treated hippocampal CA1 neurons. (C) The representative traces of low Mg^{2+}-induced bursts of spikes. Arrow head indicates a burst of spikes. (D) Intracellular 40-nm AuNPs increased the number of bursts. Average number of bursts per min were presented. *$p < 0.05$, Student's t-test; N.S., no significance.

In prolonged SRF experiment, about 70% of AuNP-treated CA1 neurons showed a similar firing pattern to non-treated neurons (5-nm AuNP-treated, $n = 8/11$; 40-nm AuNP-treated, $n = 10/14$; untreated CA1 neurons $n = 18$) (Figure 4A). However, approximately 30% neurons (7 of 25 neurons) repetitively displayed abnormal eccentric firing behaviors during the long depolarization (Figure 4B). Interestingly, the abnormal firings induced by intracellular AuNPs look very similar to a paroxysmal depolarizing shift (PDS), a cellular manifestation of epileptic seizure caused by excessive ionic currents, an imbalance in the ionic distributions, or dysfunctions of ion channels such as Na^+ or Ca^{2+} channels [37]. The abnormal firing patterns were never observed in non-treated CA1 neurons. With regard to low-Mg^{2+}-induced epileptiform activity, AuNP-treated (40 nm, $n = 12$, 18.08 ± 1.74) neurons showed significantly increased number of bursts compared to untreated CA1 neurons ($n = 8$, 12.82 ± 1.31) ($p < 0.05$, Student's t-test) (Figure 4C and 4D). Taken together, these results indicate that the increased excitability of AuNP-treated neurons is likely to

result in the hyperexcitability implicated in pathological conditions such as seizure.

Discussion

The unique physical properties of NPs, including their small size and ability to cross the BBB, are potential advantages for delivering drugs, genes and other small molecules into the brain [38,39]. AuNPs are widely used in biotechnology because of their excellent biocompatibility. However, their physiological influence on neuronal cells has to date not been investigated extensively. In this study, we examined the effects of intracellular application of AuNPs on the activity of a single neuron using patch-clamp recording. Intracellular treatment with AuNPs increased the input resistance and number of spikes, decreased the latency and threshold of AP generation, reduced the AHP, and enhanced spontaneous firing rate in hippocampal CA1 neurons. Furthermore, the AuNPs elicited PDS-like epileptiform activity in a portion of neurons during a long depolarization and increased the number of bursts in low Mg^{2+}-induced *in vitro* seizure model. These results demonstrate that intracellular AuNPs increase the excitability of neurons and aggravate the irritability of neurons in pathological conditions, such as seizure. In this regard, our data suggest that the delivery of AuNPs as vehicles to carry therapeutic agents into CNS should be carefully considered despite their well-known advantage and biocompatibility.

Many studies have reported that NPs including AuNPs have toxicity involving cell damage or death [20,40–42]. The kinetics of bioactive molecules such as a drug or protein, are likely to differ considerably intra- and extracellularly [43,44]. Furthermore, an increasing number of studies on drug delivery to the brain has focused on intracellular labeling of proteins to enhance the effectiveness of the cargo molecules in nanomedicine [4,5]. Recently, application of several NPs, such as Ag, CuO, ZnO, and WC, has been reported to alter the properties of ion channels and excitability in neurons of the brain [13–19]. Ag-NPs have been shown to enhance glutamatergic synaptic transmission and the neuronal firing rate in rat hippocampal slices [15,18]. CuO- and ZnO-NPs were suggested to affect sodium or potassium currents and enhance the excitability of acutely isolated rat hippocampal CA3 neurons [13,14]. In a study of WC-NPs, the NPs reduced the number of APs [19]. However, these studies did not provide the direct evidence on the pathophysiological effects of the NPs on the neuronal cells. The present study showed that intracellular AuNPs led to abnormal firing patterns and aggravated epileptic activity under pathological conditions. Thus, our study suggest a possibility that intracellular AuNPs can cause or worsen neuronal dysfunction or damage in the brain.

In neurons, ions diffuse along the electrochemical gradient, and the passive diffusion of ions through open channels create currents. These currents can then alter neuronal membrane potentials. Ion channels in the plasma membrane are the primary determinant of the bioelectrical properties of neurons in the brain. Various modes of the effects of AuNPs on ionic flow could be hypothesized. Although we could not identify the ion channel affected by AuNPs in this study, AuNPs might interact with Na^+ and K^+ channels. Voltage-gated Na^+ and K^+ channels play a critical role in the generation or shaping of APs [45], and Ca^{2+}-activated K^+ channels together with voltage-gated K^+ channels regulate AHP or spike frequency in neurons [46,47]. In our study, the amplitude of AP was not altered by the AuNPs; therefore, the AuNPs were unlikely to significantly affect the voltage-gated Na^+ channels.

However, modulation of the kinetics of voltage-gated Na^+ channels by the AuNPs cannot be ruled out [46]. K^+ channels including Ca^{2+}-activated K^+ channels appear to have the capacity to interact with AuNPs. For instance, pharmacological blocks or genetic mutations of $K_v12.2$, a slowly activating voltage-gated K^+ channel, led to increased input resistance and number of APs in compliance with small electric stimuli [48]. Pharmacological block of Ca^{2+}-activated K^+ channels also cause increased excitability by reducing AHP and increasing the spike frequency in neurons [49]. The intracellular interaction of AuNPs with specific modules of ion channels may be regarded as another aspect of the effects of AuNPs. Voltage-gated ion channels are generally composed of two main parts, the pore-forming transmembrane domains (including voltage-sensing modules), and cytoplasmic loops [50]. Recent studies showed that the pores of ion channels such as hERG and nicotinic acetylcholine receptor (nAChRs) can be directly clogged with ultra-small AuNPs (~1.4-nm diameter) and impede the movement of ions through the channel pore [51,52]. However, the estimated pore sizes of Na^+ and K^+ channels are either below or approximately 10 Å in diameter [53,54]; therefore, the AuNPs used in our study were too large to physically internalize within the channels' pore. Further studies should identify the ion channels affected by the AuNPs by measuring the ionic currents within channels and examine the mechanisms underlying the interaction of the AuNPs with the ion channels.

In conclusion, we demonstrated that intracellular AuNPs caused hippocampal CA1 neurons to be more excitable in terms of generating more APs, which might result from the reduced threshold and duration of AP, increased input resistance, and reduced AHP amplitude. We also examined the effects of AuNPs on *in vitro* pathological conditions. Occasionally, intracellular AuNPs elicited eccentric firing patterns during a long-depolarization, and aggravated epileptiform activity in a seizure model *in vitro*. Thus, the intracellular AuNPs might lead to disturbances in neuronal functions and hyperexcitability in pathological conditions such as seizure. Our results provide valuable information for the use of AuNPs in nanomedicine as an intracellular drug-delivery system.

Author Contributions

Conceived and designed the experiments: SL NAK BK DJ. Performed the experiments: SJ MB. Analyzed the data: BSK DJ. Wrote the paper: SJ BSK SL NAK BK DJ.

References

1. Cho K, Wang X, Nie S, Chen ZG, Shin DM (2008) Therapeutic nanoparticles for drug delivery in cancer. Clin Cancer Res 14: 1310–1316.

2. Selvan ST, Tan TT, Yi DK, Jana NR (2010) Functional and multifunctional nanoparticles for bioimaging and biosensing. Langmuir 26: 11631–11641.

3. Zhang Y, Kohler N, Zhang MQ (2002) Surface modification of superparamagnetic magnetite nanoparticles and their intracellular uptake. Biomaterials 23: 1553–1561.

4. Torchilin VP (2006) Recent approaches to intracellular delivery of drugs and DNA and organelle targeting. Annu Rev Biomed Eng 8: 343–375.

5. Chou LY, Ming K, Chan WC (2011) Strategies for the intracellular delivery of nanoparticles. Chem Soc Rev 40: 233–245.

6. Crosera M, Bovenzi M, Maina G, Adami G, Zanette C, et al. (2009) Nanoparticle dermal absorption and toxicity: a review of the literature. Int Arch Occup Environ Health 82: 1043–1055.

7. Marquis BJ, Love SA, Braun KL, Haynes CL (2009) Analytical methods to assess nanoparticle toxicity. Analyst 134: 425–439.

8. Shukla R, Bansal V, Chaudhary M, Basu A, Bhonde RR, et al. (2005) Biocompatibility of gold nanoparticles and their endocytotic fate inside the cellular compartment: A microscopic overview. Langmuir 21: 10644–10654.

9. Boisselier E, Astruc D (2009) Gold nanoparticles in nanomedicine: preparations, imaging, diagnostics, therapies and toxicity. Chem Soc Rev 38: 1759–1782.

10. Begley DJ (2004) Delivery of therapeutic agents to the central nervous system: the problems and the possibilities. Pharmacol Ther 104: 29–45.

11. Prades R, Guerrero S, Araya E, Molina C, Salas E, et al. (2012) Delivery of gold nanoparticles to the brain by conjugation with a peptide that recognizes the transferrin receptor. Biomaterials 33: 7194–7205.

12. Etame AB, Diaz RJ, O'Reilly MA, Smith CA, Mainprize TG, et al. (2012) Enhanced delivery of gold nanoparticles with therapeutic potential into the brain using MRI-guided focused ultrasound. Nanomedicine 8: 1133–1142.

13. Zhao J, Xu L, Zhang T, Ren G, Yang Z (2009) Influences of nanoparticle zinc oxide on acutely isolated rat hippocampal CA3 pyramidal neurons. Neurotoxicology 30: 220–230.

14. Xu LJ, Zhao JX, Zhang T, Ren GG, Yang Z (2009) In vitro study on influence of nano particles of CuO on CA1 pyramidal neurons of rat hippocampus potassium currents. Environ Toxicol 24: 211–217.

15. Liu Z, Zhang T, Ren G, Yang Z (2012) Nano-Ag inhibiting action potential independent glutamatergic synaptic transmission but increasing excitability in rat CA1 pyramidal neurons. Nanotoxicology 6: 414–423.

16. Shan D, Xie Y, Ren G, Yang Z (2012) Inhibitory effect of tungsten carbide nanoparticles on voltage-gated potassium currents of hippocampal CA1 neurons. Toxicol Lett 209: 129–135.

17. Xie Y, Wang Y, Zhang T, Ren G, Yang Z (2012) Effects of nanoparticle zinc oxide on spatial cognition and synaptic plasticity in mice with depressive-like behaviors. J Biomed Sci 19: 14.

18. Liu Z, Ren G, Zhang T, Yang Z (2009) Action potential changes associated with the inhibitory effects on voltage-gated sodium current of hippocampal CA1 neurons by silver nanoparticles. Toxicology 264: 179–184.

19. Shan D, Xie Y, Ren G, Yang Z (2013) Attenuated effect of tungsten carbide nanoparticles on voltage-gated sodium current of hippocampal CA1 pyramidal neurons. Toxicol In Vitro 27: 299–304.

20. Yang Z, Liu ZW, Allaker RP, Reip P, Oxford J, et al. (2010) A review of nanoparticle functionality and toxicity on the central nervous system. J R Soc Interface 7 Suppl 4: S411–422.

21. Schulz DJ, Baines RA, Hempel CM, Li L, Liss B, et al. (2006) Cellular excitability and the regulation of functional neuronal identity: from gene expression to neuromodulation. J Neurosci 26: 10362–10367.

22. Dube C, Richichi C, Bender RA, Chung G, Litt B, et al. (2006) Temporal lobe epilepsy after experimental prolonged febrile seizures: prospective analysis. Brain 129: 911–922.

23. Sachdev PS (2007) Alternating and postictal psychoses: review and a unifying hypothesis. Schizophr Bull 33: 1029–1037.

24. Paulo CS, Pires das Neves R, Ferreira LS (2011) Nanoparticles for intracellular-targeted drug delivery. Nanotechnology 22: 494002.

25. Wang TT, Bai J, Jiang X, Nienhaus GU (2012) Cellular Uptake of Nanoparticles by Membrane Penetration: A Study Combining Confocal Microscopy with FTIR Spectroelectrochemistry. Acs Nano 6: 1251–1259.

26. Neher E, Sakmann B (1976) Single-channel currents recorded from membrane of denervated frog muscle fibres. Nature 260: 799–802.

27. Hamill OP, Marty A, Neher E, Sakmann B, Sigworth FJ (1981) Improved patch-clamp techniques for high-resolution current recording from cells and cell-free membrane patches. Pflugers Arch 391: 85–100.

28. Hille B (2001) Ion channels of excitable membranes.Sinauer Associates, Inc. 3rd Edition edition.

29. Jeon D, Song I, Guido W, Kim K, Kim E, et al. (2008) Ablation of Ca2+ channel beta3 subunit leads to enhanced N-methyl-D-aspartate receptor-dependent long term potentiation and improved long term memory. J Biol Chem 283: 12093–12101.

30. Jung S, Yang H, Kim BS, Chu K, Lee SK, et al. (2012) The immunosuppressant cyclosporin A inhibits recurrent seizures in an experimental model of temporal lobe epilepsy. Neurosci Lett 529: 133–138.

31. Errington AC, Stohr T, Heers C, Lees G (2008) The investigational anticonvulsant lacosamide selectively enhances slow inactivation of voltage-gated sodium channels. Mol Pharmacol 73: 157–169.

32. Kajsa M, Igelström CHS, Heyward PM (2011) Low-magnesium medium induces epileptiform activity in mouse olfactory bulb slices. J Neurophysiol 106: 2593–2605.

33. Bean BP (2007) The action potential in mammalian central neurons. Nat Rev Neurosci 8: 451–465.

34. Morin F, Beaulieu C, Lacaille JC (1996) Membrane properties and synaptic currents evoked in CA1 interneuron subtypes in rat hippocampal slices. J Neurophysiol 76: 1–16.

35. Martina M, Schultz JH, Ehmke H, Monyer H, Jonas P (1998) Functional and molecular differences between voltage-gated K+ channels of fast-spiking interneurons and pyramidal neurons of rat hippocampus. Journal of Neuroscience 18: 8111–8125.

36. Taverna S, Tkatch T, Metz AE, Martina M (2005) Differential expression of TASK channels between horizontal interneurons and pyramidal cells of rat hippocampus. J Neurosci 25: 9162–9170.

37. Johnston D, Brown TH (1984) The Synaptic Nature of the Paroxysmal Depolarizing Shift in Hippocampal-Neurons. Annals of Neurology 16: S65–S71.

38. Garcia-Garcia E, Andrieux K, Gil S, Couvreur P (2005) Colloidal carriers and blood-brain barrier (BBB) translocation: a way to deliver drugs to the brain? Int J Pharm 298: 274–292.

39. Koziara JM, Lockman PR, Allen DD, Mumper RJ (2003) In situ blood-brain barrier transport of nanoparticles. Pharmaceutical Research 20: 1772–1778.

40. Khlebtsov N, Dykman L (2011) Biodistribution and toxicity of engineered gold nanoparticles: a review of in vitro and in vivo studies. Chem Soc Rev 40: 1647–1671.

41. Pan Y, Neuss S, Leifert A, Fischler M, Wen F, et al. (2007) Size-dependent cytotoxicity of gold nanoparticles. Small 3: 1941–1949.

42. Chen YS, Hung YC, Lin LW, Liau I, Hong MY, et al. (2010) Size-dependent impairment of cognition in mice caused by the injection of gold nanoparticles. Nanotechnology 21: 485102.

43. Hartkoorn RC, Chandler B, Owen A, Ward SA, Bertel Squire S, et al. (2007) Differential drug susceptibility of intracellular and extracellular tuberculosis, and the impact of P-glycoprotein. Tuberculosis (Edinb) 87: 248–255.

44. Schmitt E, Gehrmann M, Brunet M, Multhoff G, Garrido C (2007) Intracellular and extracellular functions of heat shock proteins: repercussions in cancer therapy. J Leukoc Biol 81: 15–27.

45. Hodgkin AL, Huxley AF (1952) a quantitative description of membrane current and its application to conduction and excitation in nerve. J Physiol 117: 500–544.

46. Kress GJ, Mennerick S (2009) Action potential initiation and propagation: upstream influences on neurotransmission. Neuroscience 158: 211–222.

47. Stocker M (2004) Ca(2+)-activated K+ channels: molecular determinants and function of the SK family. Nat Rev Neurosci 5: 758–770.

48. Zhang X, Bertaso F, Yoo JW, Baumgartel K, Clancy SM, et al. (2010) Deletion of the potassium channel Kv12.2 causes hippocampal hyperexcitability and epilepsy. Nat Neurosci 13: 1056–1058.

49. Berkefeld H, Fakler B, Schulte U (2010) Ca2+-Activated K+ Channels: From Protein Complexes to Function. Physiological Reviews 90: 1437–1459.

50. Lehmann-Horn F, Jurkat-Rott K (1999) Voltage-gated ion channels and hereditary disease. Physiological Reviews 79: 1317–1372.

51. Chin C, Kim IK, Lim DY, Kim KS, Lee HA, et al. (2010) Gold nanoparticle-choline complexes can block nicotinic acetylcholine receptors. International Journal of Nanomedicine 5: 315–321.

52. Leifert A, Pan Y, Kinkeldey A, Schiefer F, Setzler J, et al. (2013) Differential hERG ion channel activity of ultrasmall gold nanoparticles. Proc Natl Acad Sci U S A 110: 8004–8009.

53. Doyle DA (1998) The Structure of the Potassium Channel: Molecular Basis of K+ Conduction and Selectivity. Science 280: 69–77.

54. Payandeh J, Scheuer T, Zheng N, Catterall WA (2011) The crystal structure of a voltage-gated sodium channel. Nature 475: 353–358.

21

Nanog1 in NTERA-2 and Recombinant NanogP8 from Somatic Cancer Cells Adopt Multiple Protein Conformations and Migrate at Multiple M.W Species

Bigang Liu[1], Mark D. Badeaux[1], Grace Choy[1], Dhyan Chandra[1], Irvin Shen[1], Collene R. Jeter[1], Kiera Rycaj[1], Chia-Fang Lee[2], Maria D. Person[2], Can Liu[1], Yueping Chen[1], Jianjun Shen[1], Sung Yun Jung[3], Jun Qin[3], Dean G. Tang[1,4]*

1 Department of Molecular Carcinogenesis, University of Texas M.D Anderson Cancer Center, Science Park, Smithville, Texas, United States of America, **2** College of Pharmacy, University of Texas, Austin, Texas, United States of America, **3** Department of Biochemistry, Baylor College of Medicine, Houston, Texas, United States of America, **4** Cancer Stem Cell Institute, Research Center for Translational Medicine, East Hospital, Tongji University School of Medicine, Shanghai, China

Abstract

Human Nanog1 is a 305-amino acid (aa) homeodomain-containing transcription factor critical for the pluripotency of embryonic stem (ES) and embryonal carcinoma (EC) cells. Somatic cancer cells predominantly express a retrogene homolog of Nanog1 called NanogP8, which is ~99% similar to Nanog at the aa level. Although the predicted M.W of Nanog1/NanogP8 is ~35 kD, both have been reported to migrate, on Western blotting (WB), at apparent molecular masses of 29–80 kD. Whether all these reported protein bands represent authentic Nanog proteins is unclear. Furthermore, detailed biochemical studies on Nanog1/NanogpP8 have been lacking. By combining WB using 8 anti-Nanog1 antibodies, immunoprecipitation, mass spectrometry, and studies using recombinant proteins, here we provide *direct* evidence that the Nanog1 protein in NTERA-2 EC cells exists as multiple M.W species from ~22 kD to 100 kD with a major 42 kD band detectable on WB. We then demonstrate that recombinant NanogP8 (rNanogP8) proteins made in bacteria using cDNAs from multiple cancer cells also migrate, on denaturing SDS-PAGE, at ~28 kD to 180 kD. Interestingly, different anti-Nanog1 antibodies exhibit differential reactivity towards rNanogP8 proteins, which can spontaneously form high M.W protein species. Finally, we show that most long-term cultured cancer cell lines seem to express very low levels of or different endogenous NanogP8 protein that cannot be readily detected by immunoprecipitation. Altogether, the current study reveals unique biochemical properties of Nanog1 in EC cells and NanogP8 in somatic cancer cells.

Editor: Rajvir Dahiya, UCSF / VA Medical Center, United States of America

Funding: This work was supported, in part, by grants from NIH (R01-CA155693), Department of Defense (W81XWH-13-1-0352), CPRIT (RP120380), the MDACC Center for Cancer Epigenetics and RNA Center-Laura & John Arnold Foundation grant (D.G. Tang), and by two Center Grants (CCSG-5 P30 CA166672 and ES007784). C. Jeter, M. Person, and C. Liu were supported, in part, by CPRIT RP120394, CPRIT RP110782, and DOD post-doctoral fellowship (PC121553), respectively. The funders had no role in study design, data collection and analysis, decision to publish, or preparation of the manuscript.

* E-mail: dtang@mdanderson.org

Introduction

Nanog1 (commonly called Nanog) is encoded by the *nanog* gene located on Chr. 12p13.31 (Fig. S1A). The gene has 4 exons and encodes a homeodomain transcription factor that is crucial for the self-renewal of embryonic stem (ES) cells [1,2]. Nanog1 overexpression in mouse ES cells (mESCs) overcomes the requirement of leukemia inhibitory factor for maintaining the pluripotency [1,3] whereas disruption of *nanog* results in mESC differentiation to extraembryonic endoderm [4]. Down-regulation of Nanog1 in human ESCs (hESCs) also leads to the loss of pluripotency and differentiation to extraembryonic cell lineages [5]. Furthermore, in association with other reprogramming factors, Nanog1 overcomes reprogramming barriers and promotes somatic cell reprogramming [6,7]. Thus, Nanog1 is a core intrinsic element of the transcriptional network for sustaining the self-renewal of ESCs.

Human Nanog1 protein has 305 amino acids (aa) and 5 functional subdomains, i.e., N-terminal domain (ND), homeodo-main (HD), C1-terminal domain (CD1), tryptophan-rich domain (WR) and C2-terminal domain (CD2) [8–11] (Fig. 1A). The ND is involved in transcription interference and C-terminal region contains the transcription activator. The HD domain is required for Nanog nuclear localization and transactivation and the WR region mediates the dimerization of Nanog protein, which is required for pluripotency activity [12,13]. Of interest, human *Nanog* has 11 pseudogenes [14], among which *NanogP8*, located on Chr. 15q14, has a complete open reading frame [14,15] (Fig. S1B) that possesses an Alu element in the 3′-UTR homologous to the one in *Nanog1* gene, suggesting that *NanogP8* is a retrogene rather than a pseudogene. NanogP8 has at least 6 conserved nucleotide (nt) differences compared to Nanog1, which may result in some aa changes [15].

Interestingly, many somatic cancer cells have been reported to express Nanog mRNA and/or protein [15–46]. Several caveats are associated with many of these studies. ***First***, at the mRNA

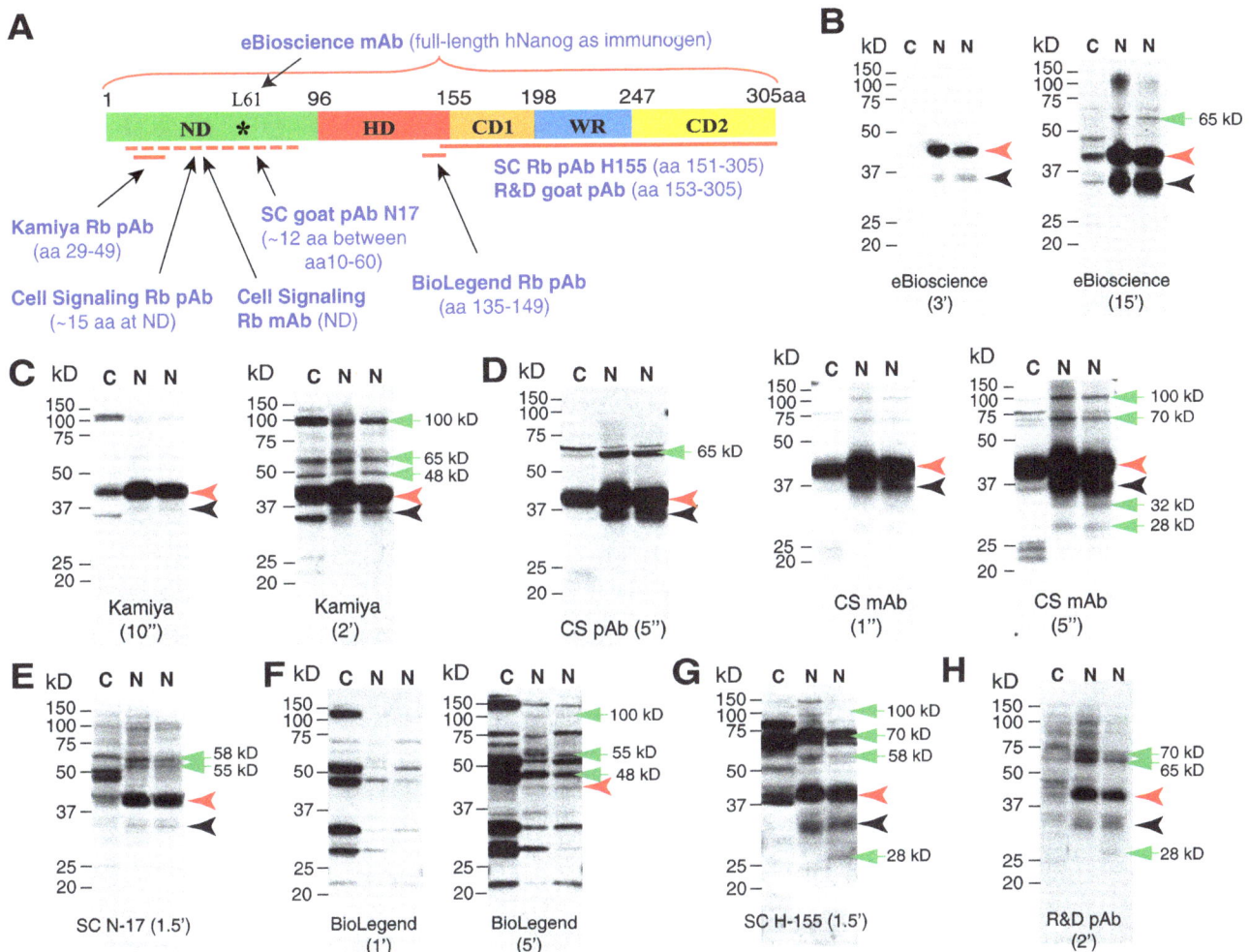

Figure 1. WB analysis of endogenous Nanog1 protein species in NTERA-2 cells. (A) Schematic of the human Nanog protein and 8 anti-Nanog Abs used in this study. Shown in parentheses are epitopes of individual Abs. ND, N-terminus domain; HD, homeodomain; CD1 and CD2, C-terminus domain 1 and 2; WR, tryptophan-rich domain. The asterisk in ND indicates the Leu61 residue recognized by the eBioscience mAb (arrow) mapped from our present studies. **(B–H)** WB analysis in NTERA-2 NE (N, two different batches) or cytosol (C) using 8 anti-Nanog Abs. Individual Ab is indicated at the bottom and M.W on the left. Black arrowhead, the predicted 35 kD Nanog protein; red arrowhead, the main 42 kD band; green arrows, additional bands (especially after longer exposures).

level, rigorous studies employing differential RT-PCR combined with sequencing or differential sensitivity to the restriction enzyme AlwN1 [15,17,31,32,34,46], have demonstrated that somatic cancer cells preferentially express the transcript of the *NanogP8* gene (Fig. S1B; see below) rather than *Nanog1*. Indeed, we have observed that the *nanog1* locus is silenced in some somatic cancer cells [15]. Our sequencing analyses reveal that embryonal carcinoma (EC) NTERA-2 cells express *Nanog1* but not *NanogP8* mRNA whereas all somatic cancer cells (6 different types including primary prostate tumor-derived cells) show 5 of the 6 nt differences specific to *NanogP8* [15]. Among the 5 conserved nt changes in *NanogP8*, one (nt759) should result in aa change (Q253H) although some cancer cells show other non-conserved nt changes (i.e., polymorphisms) that could also result in aa changes (e.g., L61P for HPCa6) [15]. Making the distinction between Nanog1 and NanogP8 is important because the two transcripts are derived from separate genomic loci and have differences at the nt sequence levels.

Second, many previous studies have been merely correlative without probing the functional importance of NanogP8 expression

in cancer cells. Using human prostate cancer (PCa) as a model, we have shown [15] that: 1) NanogP8 protein is expressed as a gradient in PCa cells with readily detectable nuclear NanogP8 in only a small fraction of PCa cells; 2) NanogP8 protein-expressing cells are increased in primary PCa compared to matching benign tissues; 3) *NanogP8* mRNA and NanogP8 protein are enriched in $CD44^+$ and $CD44^+CD133^+$ primary PCa cells; 4) shRNA-mediated knockdown of *NanogP8* inhibits tumor regeneration of prostate, breast, and colon cancer cells; and 5) The tumor-inhibitory effects of *NanogP8* knockdown are associated with inhibition of cell proliferation and clonal expansion of tumor cells and disruption of differentiation. Our recent studies demonstrate that inducible NanogP8 expression in bulk PCa cells is sufficient to confer cancer stem cell (CSC) properties and promote androgen-independent PCa growth [34] and that NanogP8 is enriched in undifferentiated ($PSA^{-/lo}$) PCa cells and its knockdown retards outgrowth of castration-resistant PCa [41]. Our studies [15,34,41] point to potential pro-oncogenic functions of NanogP8.

Third, the predicted Nanog1 and NanogP8 proteins are ~99% identical [15]. Consequently, the term 'Nanog' is often used to generically refer to either protein (that is also the case in this paper). Nevertheless, the specificity of the majority of commercially available anti-Nanog antibodies (which were all raised against Nanog1; Table 1) for Nanog1 and in particular, for NanogP8 remains uncharacterized. Therefore, it is unclear whether the putative Nanog protein bands shown on Western blotting (WB), in which frequently only a cropped strip is shown, or the Nanog protein staining shown in immunohistochemistry (IHC) truly represent the Nanog protein. *Finally*, intriguingly, although the predicated molecular mass of Nanog protein (for both Nanog1 and NanogP8) is ~35 kD, numerous studies have reported putative Nanog proteins migrating, on SDS-PAGE, at

Table 1. Examples of Nanog antibodies and their recognized protein bands.

Antibody*	Remarks	Protein band(s)	Reference(s)
eBioscience mAb (Cat# 14-5768)	Affinity-purified mAb using full-length hNanog as immunogen	~**43 kD** band in N-tera lysate; 2 **bands** of unspecified M.W	Product sheet, 24
Kamiya Rb pAb (PC-102)	Affinity-purified pAb using hNanog aa 29-49 as immunogen	~**42 kD band** in hESCs	Product sheet
SC goat pAb-N17 (SC-30331; N17)	Affinity-purified pAb against a hNanog N-terminus peptide	~**43 kD band** using rhNanog; ~**34 kD** in cancer cell lysates	Product sheet, 25
Cell Signaling pAb (3580)	Affinity-purified Rb pAb against N-terminus of hNanog	**42~45 kD** band in N-tera lysate	Product sheet
Cell Signaling mAb (5232)	Affinity-purified Rb mAb against N-terminus of hNanog	This antibody is used for ChIP	Product sheet
BioLegend Rb pAb (632002)	Affinity-purified pAb against hNanog aa 135-149	~**46 kD band** in N-tera cells	Product sheet
SC Rb pAb-H155 (SC-33759; H155)	Using hNanog aa151-305 peptide as immunogen	**40 kD band** in human ES cells; ~**40 kD** in adipose-derived stem cells; ~**40 kD** in lymphoma cell lines.	Product sheet, 47, 33
R & D goat pAb (AF1997)	Affinity-purified pAb against rhNanog aa 153-305 peptide	**40~42 kD** band using rhNanog; **35 kD** band in hESCs & EC cells; Several bands at ≥**34 kD** in HepG2 & OS732 cells; 36~**37 kD** in trophoblastic samples; ~**42 kD** in human colorectal cancer	Product sheet, 5, 16, 48, 26, 46
Abcam Rb pAb (21624)	Same as Kamiya pAb	~**40 kD** band in hESCs; A cluster of bands (unknown MW)	Product sheet, 49
Abcam Rb pAb (21603)	Using full-length mNanog as immunogen	~**39 kD** band in mESCs A cluster of bands (unknown MW) ~**38 kD** band in hESCs	Product sheet, 50, 51
Abnova mouse mAb (Clone 2C11)	Raised against C-terminal Nanog	**48 kD, 35 kD** and **29 kD** bands in N-tera cells	52
Bethyl Labs Rb pAb (BL1662)	Affinity purified pAb raised against mNanog	At least two bands between **37** and **49 kD** in mESCs	53
Chemicon Rb pAb (AB5731)	Affinity purified pAb raised against mNanog N-terminus peptide	~**35 and 55 kD** bands in mESCs; A cluster of bands in mESCs	Product sheet, 54
Chemicon Rb pAb	Raised against mNanog	**40 kD** and **80 kD** bands in mESCs; Dog testis and pig testis	55
Home-made Rb pAb	Raised against mNanog	2-3 bands at ~**40 kD** in mESCs	2
Home-made Rb pAb	Raised against mNanog	>3 bands with the largest at ~**44 kD** in mESCs, and EG cells	56
Home-made Rb pAb	Raised against hNanog (aa 168-183)	>3 bands of ~**36 kD** to ~**50 kD** in 293T cells transfected with hNanog	57
Home-made Rb pAb	FLAG-tagged Nanog stably expressed in mESCs	Two pulldown Nanog protein bands at ~**45 kD** and **37 kD**	58
Home-made Rb pAb	Raised against mNanog aa 1-95	Several bands at >**35 kD**	59

*This table lists the information for the 8 antibodies used in our current study, together with several others commercial and home-made antibodies. Shown in parenthesis are catalog numbers.
Abbreviations: EC, embryonal carcinoma; EG, embryonic germ cells; hESCs, human embryonic stem cells; hNanog, human Nanog; mESCs, mouse embryonic stem cells; mAb, monoclonal antibody; mNanog, mouse Nanog protein; pAb, polyclonal antibody; Rb, rabbit; rhNanog, recombinant human Nanog protein; SC, Santa Cruz.

apparent molecular mass of 29 to 80 kD in ES, EC, and somatic cancer cells (Table 1)[2,5,17,24–26,46–59]. Even more puzzlingly, the same antibody often seems to detect different 'Nanog' protein bands (Table 1). Whether all these reported putative Nanog protein species are true Nanog proteins has yet to be *directly* determined.

The present study was undertaken to address the last two questions. We first provide direct evidence that the Nanog1 protein in NTERA-2 EC cells migrates at <30 to ~100 kD. We then show that recombinant NanogP8 proteins can also migrate at ~28 kD to ~180 kD. These results suggest that the Nanog1/NanogP8 proteins likely adopt multiple conformations. We finally demonstrate that most long-term cultured somatic cancer cells seem to express very low levels of or biochemically divergent endogenous NanogP8 such that it cannot be readily immunoprecipitated down by the 8 commercial anti-NanogP8 antibodies.

Materials and Methods

Cells and Reagents

Various human cancer cell lines, including prostate cancer (PC3, Du145, LNCaP), breast cancer (MCF-7), colonic carcinoma (Colo320), and melanoma (WM-562) cells, were obtained from ATCC (American Type Culture Collection, Manassas, VA) and cultured in the recommended media containing 10% of heat-inactivated FBS (fetal bovine serum). Teratocarcinoma cells (NTERA-2, clone D; CRL-1973) were purchased from ATCC and cultured in mTeSR™1 medium (STEMCELL Technologies, Vancouver, Canada). Nuclear extraction kit was from Thermo Scientific (Rockford, IL). Eight anti-Nanog primary antibodies (Abs) were obtained from commercial companies (Table 1; Fig. 1A). Secondary and control Abs were purchased from Santa Cruz Biotechnology. ECL reagents were bought from PerkinElmer, Inc (MA, USA). The current research does not involve animal experiments or human subjects (i.e., living individuals or identifiable private information). All other studies presented herein were the investigator-initiated and did not require approval from other regulatory bodies.

siRNA- and shRNA-mediated Nanog knockdown experiments

For the siRNA knockdown experiments (Fig. 2), we used siGENOME SMARTpool siRNA against human Nanog1 and siCONTROL non-targeting siRNAs obtained from Dharmacon (Lafayette, CO). Cells plated 24 h earlier on 6-well dishes were transfected with siRNAs using Lipofectamine 2000. 48 h later, cells were harvested for WB experiments.

For the shRNA Nanog knockdown experiments, lentiviruses including pLL3.7, Nanog-shRNA and TRC-shRNA were produced in 293FT packaging cells (Clontech Laboratories, Mountain View, CA) using the modified protocols previously described [15,60]. In brief, geneticin-selected, early-passage 293FT cells (6×10^6/15 cm dish) were transfected with the RRE (6 µg), REV (4 µg) and VSVg (4 µg) packaging plasmids, along with a lentiviral vector (6 µg) using Lipofectamine 2000 to carry out the transfection. At 36–48 h, virus-containing media were collected and fresh media added. After an additional 12–24 h of culture, viral supernatants were again collected, pooled, and ultracentrifuged to produce concentrated viral stocks. Individual titers were determined for the GFP-tagged virus using HT1080 cells. The non-GFP tagged TRC-shRNA virus was prepared in parallel and assigned the control (pLL3.7) titer. Target cells were plated 24 h earlier and infected at approximately 50% cell density and harvested for *in vitro* or *in vivo* experiments 48–72 h post infection.

Bacterial expression and purification of recombinant Nanog (rNanog) proteins

NanogP8 or *Nanog1* cDNA in cultured NTERA-2, MCF-7, and LNCaP cells or in primary human PCa samples (HPCa1, HPCa5, and HPCa6) was amplified by RT-PCR using LDF1/LDR1 primers (LDF1 5'-TCTTCCTCTATACTAACATGAGT-3'; LDR1 5'-AGGATTCAGCCAGTGTCCA-3') and cloned into pCR2.1 [15]. The EcoRI/NotI or BamHI/SalI fragments containing the Nanog coding sequence were then subcloned into the same sites in either pET-28a (+) or pET-28b (+) bacterial expression vector to generate His-tagged fusion proteins. After insert verification by restriction enzyme digestion analysis and DNA sequencing, the plasmids were used to transform BL21(DE)₃ competent bacteria to express the fusion proteins. Nanog protein was induced by 0.5 mM IPTG for 3–6 h at 37°C and bacterial pellet was kept at −80°C. For purification, bacterial pellets containing His-tagged Nanog protein were lysed in lysis buffer (50 mM NaH_2PO_4, pH 8.0, 300 mM NaCl, 10 mM imidazole, protease inhibitor cocktail, and 1 mg/ml lysozyme), sonicated for 10 sec (one pulse). His-tagged Nanog in the supernatant was purified by nickel beads per manufacturer's instructions (Qiagen).

WB using recombinant proteins (rNanog), whole cell lysate (WCL), or nuclear extract (NE)

rNanog proteins were obtained and purified as described above. WCL and NE (from cultured cells) were prepared as previously described [15]. 50–80 µg proteins (rNanog and NTERA-2 NE) were analyzed by 12.5% SDS-PAGE and the gels were transferred onto an immobilon-P transfer membrane (Millipore, Bedford, MA). The membrane was blocked with 5% non-fat dried milk in TBST (10 mM Tris-HCl, 150 mM NaCl and 0.1% Tween-20) for 1 h at room temperature, and incubated overnight at 4°C with primary antibody. Membranes were washed three times with TBST buffer, then incubated for 1 h with 1:2000 secondary antibody, and developed with ECL Plus WB detection reagent (PerkinElmer).

Immunoprecipitation (IP)

Recombinant proteins (500 µg) or NE (800 µg or depending on experimental purposes) were incubated with the indicated anti-Nanog antibodies overnight at 4°C. Then the solution was incubated with protein A/G-agarose (Sigma, MO, USA) for another 1 h at 4°C. After incubation, the beads were washed three times with RIPA buffer. Proteins bound to protein A-agarose were eluted with SDS-PAGE loading buffer and boiled for 8 min and subjected to SDS-PAGE.

Dialysis refolding experiments

To isolate and purify a large amount of rNanog, bacterial pellets were sonicated 6 times with 10 seconds pulses. The sonication solution was centrifuged at 10,000 rpm for 15 min at 4°C. The pellets were washed with cold PBS for three times. The precipitated material was the inclusion body of rNanog protein. To dissolve the inclusion body, a buffer containing 7 M urea was used and then the solution was subjected to dialysis against the buffers of 1.5 M, 0.6 M, 0.3 M, and 0.1 M urea (in 1.5 M NaCl, 10 mM Tris HCl, pH 7.0). The dialysis solution was subjected to SDS-PAGE to determine the existence of soluble Nanog protein.

Nanog protein identification (ID) by mass spectrometry (MS) analyses

Three sets of MS-based protein ID experiments were performed on either rNanog proteins or NE prepared from NTERA-2 cells.

A

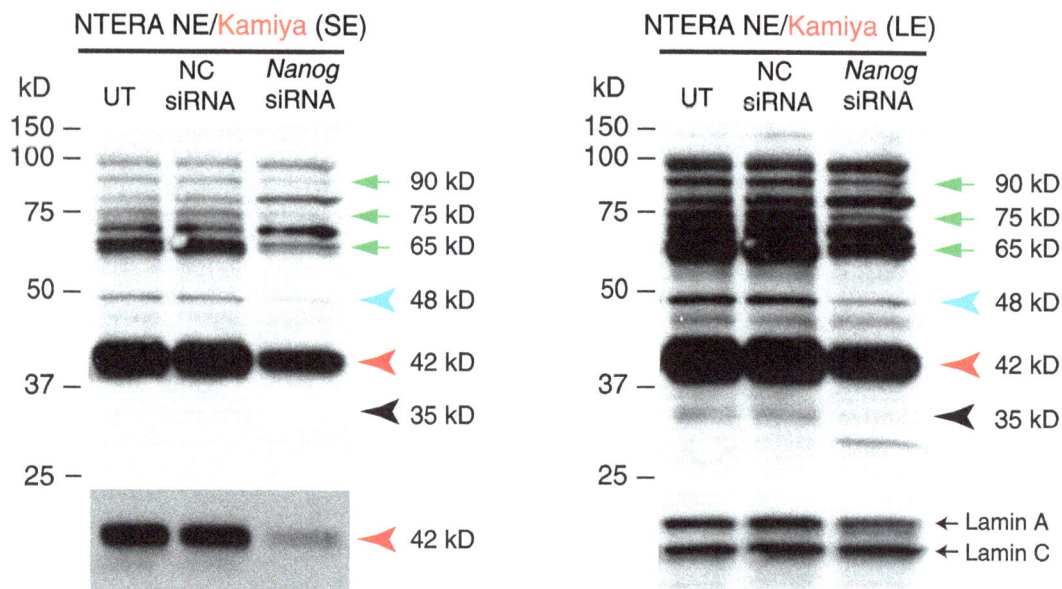

B

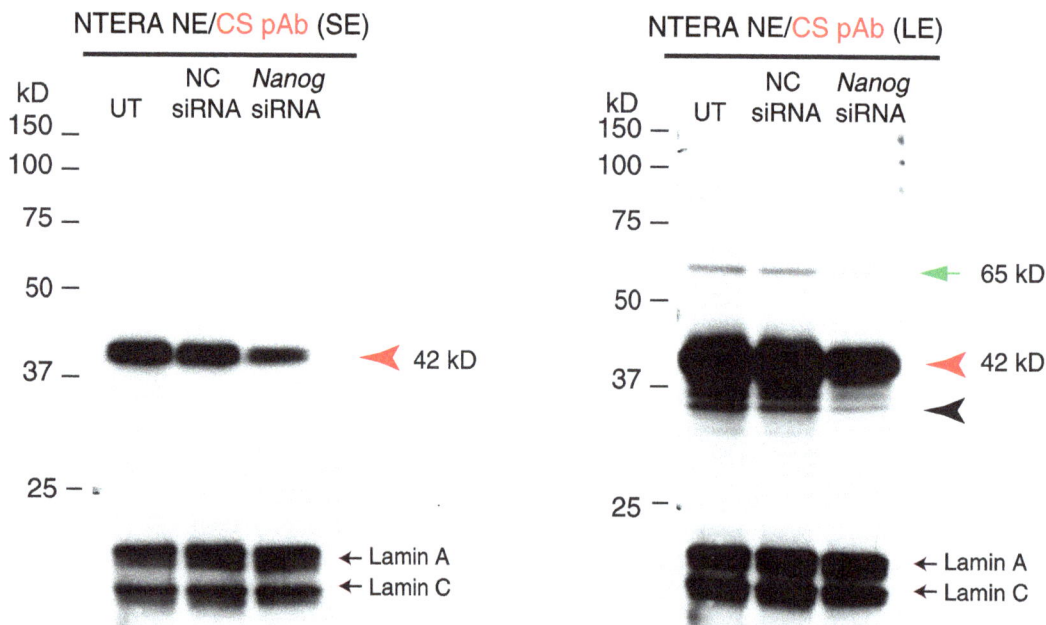

Figure 2. Characterization of Nanog proteins in NTERA-2 cells upon siRNA-mediated knockdown. (**A**) Nanog siRNA experiments in NTERA-2 cells. NTERA-2 cells were transfected with a pool of Nanog-specific siRNAs for 48 h, harvested, and used to isolate the NE for WB with the Kamiya Ab. The left and right panels represent the short and long exposure (SE and LE, respectively) films (note that the left bottom panel is the shortest exposed film to illustrate the significant knockdown of the 42 kD band). Lamin A/C was the loading control. UT, untransfected; NC siRNA, negative control (siCONTROL non-targeting) siRNAs; Nanog siRNA, SMARTpool siRNA against Nanog. Red, black, and blue arrowheads indicate the major 42 kD, minor 35 kD, and 48 kD protein bands, respectively, that were reduced upon Nanog knockdown. The green arrows refer to additional bands that also showed reduction. (**B**) The WB was performed with the CS anti-Nanog rabbit pAb.

In the *first* ID experiment, we immunoprecipitated a large amount of NTERA-2 cell NE (~2 mg total) with the R&D goat pAb. After washing (3x) with RIPA buffer, the proteins bound to the protein G-agarose were eluted with SDS-PAGE loading buffer and boiled for 8 min and subjected to SDS-PAGE. The gel was

then stained with SYPRO Ruby solution. Gel slices/areas containing bands of interest were cut out, proteins eluted, and subjected to trypsin digestion. The tryptic digests were analyzed by LC-MS/MS using the Dionex Ultimate 3000 RSLCnano LC coupled to the Thermo Orbitrap Elite. Prior to HPLC separation, the peptides were desalted using Millipore U-C18 ZipTip Pipette Tips following the manufacturer's protocol. A 2 cm long×100 μm I.D. C18 5 μm trap column (Proxeon EASY Column) was followed by a 75 μm I.D. ×15 cm long analytical column packed with C18 3 μm material (Dionex Acclaim PepMap 100). Buffer A was composed of 0.1% formic acid in water and Buffer B 0.1% formic acid in acetonitrile. Data was acquired for 35 min using an HPLC gradient of 5% B to 45% B over 30 min with a flow rate of 300 nl/min. The FT-MS resolution is set to 120,000, and top 20 MS/MS are acquired in CID ion trap mode. Raw data was processed using SEQUEST embedded in Proteome Discoverer v1.3 using the following parameters: full trypsin digest with maximum 2 missed cleavages, fixed modification carbamido-methylation of cysteine, variable modification oxidation of methionine and deaminidation of asparagine and glutamine, searching the human reference proteome from Uniprot from March 2012 (80990 entries). The mass accuracy for precursors was set to 10 ppm monoisotopic mass, for fragment ions was 0.8 Da monoisotopic mass. A decoy database was generated from the Uniprot human database and used by Peptide Validator and Scaffold for calculating false positive values. X!Tandem database searches (The GPM, thegpm.org, version CYCLONE (2010.12.01.1)) were performed embedded in Scaffold 3 Q+ (Proteome Software) using the same search parameters as SEQUEST. Scaffold was used for validation of peptide and protein identifications with confidence filtering of 95% confidence for two peptides, and a 99.9% protein confidence cutoff. False positive peptide and protein values were calculated as 0.02% and 0.1% respectively by Scaffold.

In the *second* set of experiments using the NTERA-2 cell NE, we performed the tandem IP using the Kamiya pAb followed by the R&D pAb and the immunoprecipitates were separated on a linear ion-trap mass spectrometry (LTQ, Oorbitrap Velos) as described [61]. Simply, the excised fragments were destained with destaining solution (40% methanol, 50 mM NaHCO₃ in water) and subjected to in-gel digestion using 100 ng trypsin in 50 mM NH₄HCO₃, pH 8.5, for 12 h. Peptides were then extracted with acetonitrile and dried in a Speed-Vac dryer (Thermo Savant). Each dried sample was dissolved in 20 μl of 5% methanol/95% water/0.01% formic acid solution and injected into the Surveyor HPLC system (Thermo Finnigan) using an autosampler. A 100 mm×75 μm C18 column (5 μm, 300 Å pore diameter, PicoFrit; New Objective) with mobile phases of A (0.1% formic acid in water) and B (0.1% formic acid in methanol) was used, with a gradient of 5–95% of phase B over 45 min followed by 95% of phase B for 5 min at 200 nl/min. Peptides were directly electrosprayed into the (LTQ Oorbitrap Velos (Thermo Scientific) using a nanospray source. The LTQ was operated in the data-dependent mode to acquire fragmentation spectra of the 20 strongest ions under direct control of Xcalibur software (Thermo Scientific). Obtained MS/MS spectra were searched against target-decoy Human refseq database in Proteome Discoverer 1.2 interface (Thermo Fisher) with Mascot algorithm (Mascot 2.1, Matrix Science). Variable modification of Acetylation (lysine), di-Glycine (lysine), Phosphorylation (Serine and Threonine) and Oxidation (Methionine) was allowed. Also, static modification of DeStreak (Cystein) was allowed. The precursor mass tolerance was confined within 10 ppm with fragment mass tolerance of 0.5 dalton and a maximum of two missed cleavages allowed. Assigned peptides were filtered with 5% false discover rate and subject to manual verifications.

In the *third* ID experiment, rNanog1/rNanogP8 proteins were purified and gel slices containing various protein bands were used in MALDI-TOF/TOF identification as described previously [62].

Tryptic digests were analyzed using a 4700 Proteomics Analyzer MALDI-TOF/TOF (AB Sciex, Foster City, CA). Samples were desalted with μC18 ZipTips (Millipore, Billerica, MA) with elution directly onto the MALDI target using the matrix α-cyano-4-hydroxycinnamic acid at 5 mg/ml in 67% acetonitrile/0.1% trifluoroacetic acid. MS and MS/MS spectra were acquired automatically using 4000 Series Explorer V 3.0 RC1. Up to 20 peaks with S/N 20 were selected for MS/MS fragmentation, excluding matrix, trypsin, and keratin peaks. Additional peak processing and database search were performed using GPS Explorer v3.5. MASCOT V2.0 or V2.2. Spectra were searched against the Swiss-Prot database including Human sequences (Sept. 1, 2009, 20495 entries). The search parameters chosen were cleavage by trypsin/P with up to 2 missed cleavages, fixed modification of carbamidomethylation of cysteine and variable modifications of protein N-terminal acetylation, methionine oxidation and pyroglutamic acid modification of peptide N-terminal glutamine residues. The database search used 50 ppm mass tolerance for monoisotopic MS masses and 0.2 Da for MS/MS, up to 100 peaks with minimum S/N 15 were selected for MS, and up to 65 fragment ions with minimum S/N 3. The search output combines the scores from MS search and the MS/MS search using a probabilistic MOWSE algorithm. The MASCOT score is defined as −10*logP, where P is the probability that the observed match is a random event. The MASCOT score 56 which corresponds to p<0.05 is chosen as the cutoff for a significant hit, and those proteins exceeding the cutoff value are reported.

Results

Seven of the eight anti-Nanog antibodies detect, on WB, the major 42 kD and minor 35 kD bands in NTERA-2 NE

Human *Nanog1* gene (gi 13376297), localized on Chr. 12p13.31 and primarily expressed in ES and EC cells, has a 915-bp open reading frame (Fig. S1A). It encodes a 305-aa protein that is commonly referred to as Nanog (Fig. 1A). Human Nanog1 protein is predicted to have a molecular mass of ~35 kD. Intriguingly, putative Nanog proteins migrating at apparent molecular masses of 29–80 kD on WB have been reported in ES, EC, and somatic cancer cells (Table 1). It is unclear, however, whether these reported Nanog protein species truly represent the Nanog proteins. To address this issue and to directly determine the molecular mass(es) of endogenous Nanog protein, we first performed comprehensive WB analyses in the EC NTERA-2 cells using 8 commercial anti-Nanog (i.e., anti-Nanog1) antibodies (Abs) directed against different regions of the human Nanog protein (Fig. 1A; Table 1, the first 8 Abs). Among the 8 anti-Nanog Abs, four [Kamiya Rb pAb, SC (Santa Cruz) goat pAb N17, Cell Signaling (CS) Rb pAb and Rb mAb] are directed to the ND, two (SC Rb pAb H-155 and R&D goat pAb) to the C-terminus, and one (BioLegend Rb pAb) to the HD (homeodomain) whereas another (eBioscience mAb) is raised against the full-length human Nanog1 protein (Fig. 1A).

The WB results revealed that different antibodies displayed distinct reactivity and each antibody detected multiple bands in NTERA-2 NE (we focused on nuclear expression as Nanog is a nuclear transcription factor) with M.W ranging from ~28 kD to ~100 kD (Fig. 1B-H). Specifically, the results demonstrated that: **1)** the eBioscience mAb showed the *cleanest* and most *specific* reactivity, detecting only a strong band of ~42 kD (Fig. 1B, red arrowhead) and a weak band of ~35 kD (Fig. 1B, black arrowhead) in the NE with no immunoreactive bands in the cytosol with 3 min exposure of the film (Fig. 1B). On the other hand, the eBioscience mAb exhibited the lowest sensitivity among the 8 antibodies. The 35 kD band and another ~65 kD band in the NE became apparent after extended exposure time (15 min; Fig. 1B). **2**) The Kamiya pAb showed *the second cleanest* reactivity (Fig. 1C). Upon short exposure (10 sec), it detected a strong band of 42 kD with a very faint band of 35 kD in the NE (Fig. 1C; red and black arrowheads, respectively). In the cytosol, it detected the 42 kD band and several other bands (Fig. 1C). Upon longer exposure (2 min), the Kamiya pAb also detected three upper bands at ~48 kD, ~65 kD and ~90 kD in both NE and cytosol (Fig. 1C, green arrows in right panel). **3**) Both CS Rb pAb and mAb were *the most sensitive* antibodies such that they detected the prominent 42 kD and 35 kD bands with only 1-5 sec exposure (Fig. 1D). Both antibodies also detected several additional bands (Fig. 1D, green arrows). **4**) The SC goat pAb (N17) detected an obvious 42 kD band and a faint 35 kD band, together with two upper bands of ~55 kD and ~58 kD (Fig. 1E). **5**) The Biolegend Rb pAb was the 'dirtiest' Ab and detected a series of strong bands in the cytosol and only faint 42 kD band in the NE (Fig. 1F, red arrow). **6**) The SC Rb pAb (H-155) detected the obvious 42 kD band and the weaker 35 kD band (Fig. 1G, red and black arrowheads, respectively), together with several other bands at 100 kD, 70 kD, 58 kD and ~28 kD (Fig. 1G, green arrows). **7**) Finally, the R&D goat pAb detected an apparent band of 42 kD and a minor band of 35 kD (Fig. 1H, red and black arrowheads, respectively), together with several additional bands (Fig. 1H, green arrowheads).

In summary, this comprehensive WB analysis using 8 anti-Nanog Abs and the NTERA-2 NE has revealed that: **1**) except for the BioLegend Ab, the other 7 anti-Nanog Abs all *commonly* recognize a major 42 kD and a minor 35 kD protein band; **2**) Three antibodies raised against the N-terminus, i.e., Kamiya Rb pAb, CS pAb and CS mAb, give clean WB results upon short exposure of films; **3**) with longer exposure, all antibodies detect additional protein bands ranging from ~28 kD to >100 kD. However, different Abs detect different patterns of reactive protein bands; **4**) With respect to sensitivity, the CS mAb and CS pAb are the most sensitive followed by the Kamiya Ab whereas the eBioscience mAb is the least sensitive; **5**) The BioLegend anti-Nanog Ab does not detect the 42 kD as the major protein band in the NTERA-2 NE; and **6**) Different antibodies may preferentially recognize different protein bands. For example, the Kamiya pAb, but not the CS pAb or mAb reacted with the 48 kD band.

siRNA-mediated knock-down reveals the 42-kD protein band as the predominant Nanog protein isoform in NTERA-2 cells

Since the predicted Nanog protein is ~35 kD, we can infer that the minor 35 kD protein band commonly detected on WB most likely is the Nanog protein. To confirm this inference and to determine whether the predominant 42 kD and any of the additional bands are also Nanog proteins, we performed siRNA knock-down experiments in NTERA-2 cells. As shown in Fig. 2A, the Kamiya Rb pAb consistently detected the strong 42 kD protein band, which decreased in response to Nanog siRNA. In addition, the minor 35 kD band and several other weaker bands migrating at 48 kD, 65 kD, 75 kD, and 90 kD all decreased in Nanog siRNA-treated cells (Fig. 2A). When we performed WB using the CS Rb pAb, the Nanog siRNAs also knocked down the 42 kD and 35 kD bands (Fig. 2B). In addition, an upper 65 kD band was also reduced (Fig. 2B). Note that the CS pAb did not detect 48 kD or other upper bands except the 65 kD band (Fig. 2B), consistent with the earlier WB results with the CS pAb and mAb (Fig. 1D). Collectively, the siRNA knock-down experiments suggest that *the Nanog protein in NTERA-2 cells seems to*

migrate, on denaturing SDS-PAGE, at multiple molecular masses including approximately 35, 42, 48, 65, 75, and 90 kD with the 42-kD band being the dominant "isoform".

Characterization of Nanog protein species in NTERA-2 cells by immunoprecipitation (IP) followed by mass spectrometry (MS) identification (ID)

To substantiate the siRNA knockdown results, we performed two sets of MS-based protein ID experiments. *In the first*, we performed IP experiments using **500 μg** of NE from NTERA-2 and somatic cancer cells and with 5 anti-Nanog Abs, i.e., eBioscience mAb, Kamiya pAb, CS Rb mAb, SC pAb H155, and R&D goat pAb (Fig. 3; Fig. S2A-B; data not shown). All 5 antibodies immunoprecipitated the 42 kD Nanog band in NTERA-2 NE. For example, when IP was done with the Kamiya pAb followed by WB using either eBioscience mAb (Fig. 3A, lane 9) or R&D goat pAb (Fig. S2A, lane 10), the 42 kD protein was immunoprecipitated. IP with the CS Rb mAb, also directed against the N-terminus of Nanog, pulled down the 42 kD band (WB using the R&D goat pAb; Fig. 3B, lane 9). When IP was done with the SC pAb H-155, directed towards the C-terminus of Nanog, followed by WB using either eBioscience mAb (Fig. 3C, lane 8) or R&D goat pAb (Fig. S2B), the 42 kD protein was immunoprecipitated in the NTERA-2 NE. Similarly, when IP was done with another C-terminus directed anti-Nanog Ab, i.e., the R&D goat pAb, followed by WB using eBioscience mAb, we again detected the 42 kD Nanog protein in NTERA-2 NE (Fig. 3D, lane 3). Finally, IP with the eBioscience mAb followed by WB with the R&D goat pAb also detected the 42 kD Nanog protein (not shown). In ALL these IP experiments, the IP products showed an enrichment compared to WB using NTERA-2 NE (Fig. 3A–D; Fig. S2A–B). Since in all these experiments, IP was performed with one anti-Nanog Ab whereas WB with another, *the results provide further evidence that the major Nanog protein species in NTERA-2 cells migrates at ~42 kD on WB analysis* (see below for discussion on IP results in somatic cancer cells). Note that other than the 42 kD protein band, the IP did not pull down appreciable amount of 35 kD or other high M.W bands (Fig. 3A–D), likely due to much lower amounts of these protein species.

Subsequently, we performed IP and MS ID experiments in which we prepared **2 mg** of NTERA-2 NE and immunoprecipitated the Nanog proteins using the R&D goat pAb, which pulled down three bands whose corresponding molecular masses were approximately 100 kD, 75 kD, and 55 kD, respectively (Fig. 3E). These three gel bands, together with the gel area corresponding to 42 kD (surprisingly, the 42 kD major band routinely detected on WB was not the major band pulled down by the R&D goat pAb), which were labeled as NTRD1 to NTRD4 (Fig. 3E), were cut out, and the proteins eluted and identified by LC-MS/MS (see Methods). We recovered Nanog peptides from all four gel slices, among which the NTRD4, corresponding to 42 kD region, showed the highest peptide spectral counts (Fig. 3F; Fig. S2C). Most recovered peptides were mapped to the N-terminus and homeodomain of the Nanog protein (Fig. S1C) although the IP Ab used was the R & D goat pAb directed against the C-terminus (Fig. 1A).

Tandem IP and MS identify Nanog protein species of 20–70 kD in NTERA-2 cells

In the second set of protein ID experiments, we employed tandem IP coupled with MS to identify the Nanog protein species in NTERA-2 cells (Fig. 4). We employed **6.5 mg** NTERA-2 cell NE in tandem IP with two anti-Nanog antibodies, i.e., the Kamiya

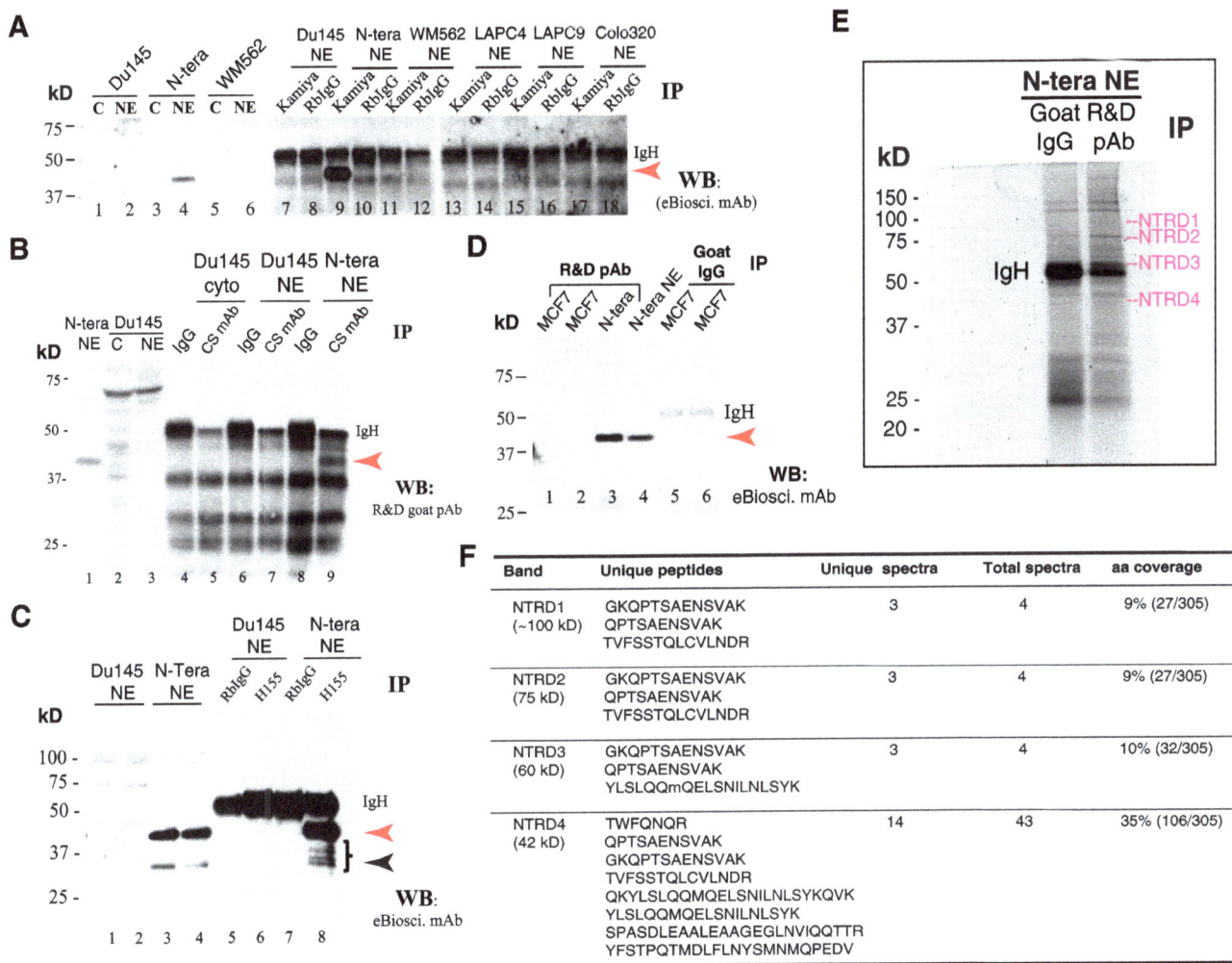

Figure 3. IP and ID studies with 5 anti-Nanog antibodies in NTERA-2 and cancer cells. (**A**) The NE of NTERA-2 and various cancer cells was used in IP with the Kamiya anti-Nanog Rb pAb followed by WB with the eBioscience mAb. Lanes 1–6 were regular WB with either cytosol (C) or NE. Red arrowhead, the 42 kD Nanog band; IgH, IgG heavy chain (~53 kD). Note that the prominent 42 kD protein band was detected on WB (lane 4) and immunoprecipitated down (lane 9) *only* in NTERA-2 NE. (**B**) The NTERA-2 NE or Du145 cytosol (cyto) or NE was used in IP with the CS anti-Nanog rabbit mAb followed by WB with the R&D goat pAb. Lanes 1–3 were regular WB. Red arrowhead, the 42 kD Nanog band; IgH, IgG heavy chain. Note that the 42 kD Nanog protein was detected on WB (lane 1) and immunoprecipitated down (lane 9) only in NTERA-2 NE. (**C**) The NE of NTERA-2 and Du145 cells was used in IP with the SC anti-Nanog rabbit pAb H155 followed by WB with eBioscience mAb. Lanes 1–4 were regular WB using two independent preparations of Du145 or NTERA-2 NE. Red arrowhead, the 42-kD band; black arrowhead, the 35-kD Nanog band; IgH, IgG heavy chain. Note that the 42-kD protein band was detected on WB (lane 3 and 4) and immunoprecipitated down (lane 8) *only* in NTERA-2 NE. The right-pointing bracket indicates the cluster of Nanog protein bands below the dominant 42 kD band. (**D**) The NE of NTERA-2 cells and MCF7 cells (two independent preparations) was used in IP with the R&D anti-Nanog goat pAb (goat IgG used as the control) followed by WB with the eBioscience mAb. Red arrowhead, the 42 kD Nanog band; IgH, IgG heavy chain. Note that the 42-kD Nanog protein band was detected on WB (lane 4; input) and immunoprecipitated down (lane 3) only in NTERA-2 NE. (**E–F**) Nanog protein ID by MALDI-TOF/TOF in NTERA-2 NE following IP using the R&D goat pAb. Shown are SYPRO Ruby gel image (E; NTRD1-4 refer to the 4 gel slices cut out for protein elution) and the Nanog peptides recovered from each gel slice (F).

Rb pAb followed by R&D goat pAb (Fig. 4A). Eleven gel slices covering from ~20 kD all the way to >250 kD were cut out and proteins eluted and subjected to LTQ MS analysis (Fig. 4B). We recovered Nanog peptides in gel slices 1, 2, 3, 4, 5, 6, and 8 whose M.W ranged from ~20 kD to 70 kD. Most recovered peptides were also mapped to the N-terminus and homeodomain (Fig. S1D). The highest spectral counts of Nanog peptides were recovered from gel slices 2 and 3, corresponding to 35–45 kD (Fig. 4B–C). These results indicate that *in NTERA-2 cells, Nanog protein migrates at apparent molecular masses of 20–70 kD.* Since the numbers of Nanog peptides recovered from gel slices 2 and 3 are very similar but on WB all anti-Nanog Abs preferentially detects

the ~42 kD band (in gel slice 3), these results further suggest that: 1) *the endogenous Nanog protein species migrate at various molecular masses probably by adopting distinct protein conformations and modifications (such as post-translational modifications or PTM, cleavage, etc), which are recognized by different anti-Nanog Abs*; and 2) *the conformation of the 42 kD Nanog is preferentially recognized by most anti-Nanog antibodies.*

Exogenous NanogP8 protein is recognized by the 8 anti-Nanog1 Abs and migrates at 42 kD on WB

The above WB, siRNA knockdown, and two sets of MS–based protein ID experiments indicate that the Nanog1 protein in NTERA-2 cells migrates, on reducing SDS-PAGE, at multiple

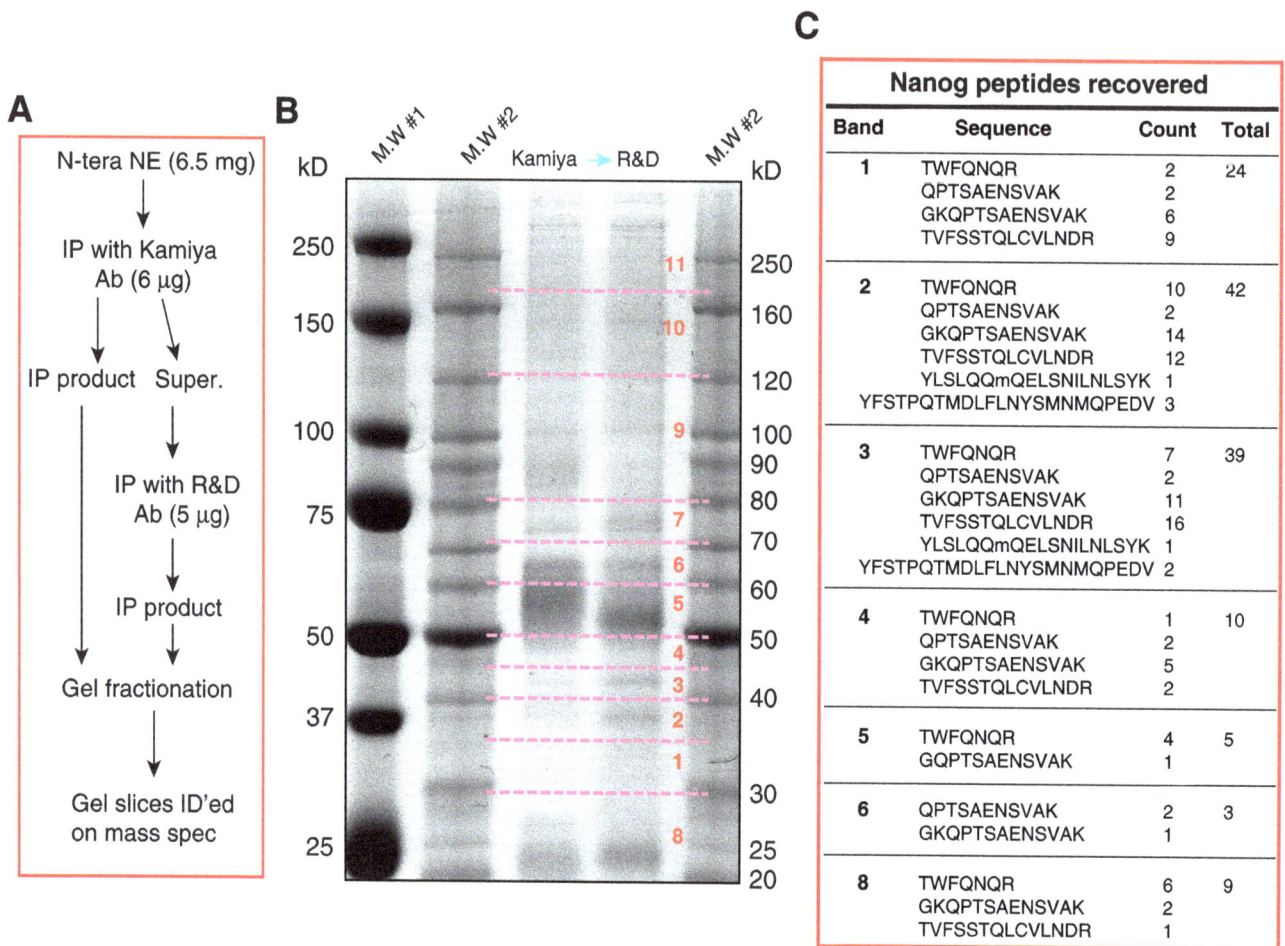

A Flow chart of tandem IP and sample preparation for LTQ mass spec analysis:
N-tera NE (6.5 mg) → IP with Kamiya Ab (6 μg) → IP product / Super. → IP with R&D Ab (5 μg) → IP product → Gel fractionation → Gel slices ID'ed on mass spec

B The gel image of SYPRO Ruby staining.

C Nanog peptides recovered

Band	Sequence	Count	Total
1	TWFQNQR	2	24
	QPTSAENSVAK	2	
	GKQPTSAENSVAK	6	
	TVFSSTQLCVLNDR	9	
2	TWFQNQR	10	42
	QPTSAENSVAK	2	
	GKQPTSAENSVAK	14	
	TVFSSTQLCVLNDR	12	
	YLSLQQmQELSNILNLSYK	1	
	YFSTPQTMDLFLNYSMNMQPEDV	3	
3	TWFQNQR	7	39
	QPTSAENSVAK	2	
	GKQPTSAENSVAK	11	
	TVFSSTQLCVLNDR	16	
	YLSLQQmQELSNILNLSYK	1	
	YFSTPQTMDLFLNYSMNMQPEDV	2	
4	TWFQNQR	1	10
	QPTSAENSVAK	2	
	GKQPTSAENSVAK	5	
	TVFSSTQLCVLNDR	2	
5	TWFQNQR	4	5
	GQPTSAENSVAK	1	
6	QPTSAENSVAK	2	3
	GKQPTSAENSVAK	1	
8	TWFQNQR	6	9
	GKQPTSAENSVAK	2	
	TVFSSTQLCVLNDR	1	

Figure 4. Mass spectrometry analysis of Nanog peptides recovered from NTERA-2 NE. (**A**) Flow chart of tandem IP and the sample preparation for LTQ mass spec analysis. (**B**) The gel image of SYPRO Ruby staining. The NE from NTERA-2 cells was used to perform the tandem IP with two anti-Nanog antibodies (i.e., Kamiya rabbit pAb and R&D goat pAb). The immunoprecipitates were separated by SDS-PAGE and the gel was stained with SYPRO Ruby. Gel slices as indicated (dashed lines and numbers) were cut out to elute proteins for LTQ mass spectrometry analysis. M.W#1 and M.W#2 were two protein markers. (**C**) Nanog peptides (sequences and total counts indicated) recovered for each gel slice.

apparent molecular masses (most abundant at 35–45 kD) with the 42 kD band being the predominant band on WB. Next, we turned our attention to Nanog protein(s) in somatic cancer cells. There exists strong experimental evidence [15,17,31,32,34,46,63] that somatic cancer cells preferentially express the retrogene *NanogP8* located on Chr. 15q14 (gi 47777342) (Fig. S1B). *NanogP8* is very similar to *Nanog1* at the mRNA level with only six reported nucleotide (nt) differences [14,15]. Indeed, our own sequencing [15] and differential qRT-PCR [34] analysis has demonstrated that the *Nanog* mRNA in multiple somatic cancer cell types (including primary prostate tumors) is derived from the *NanogP8* locus.

The *NanogP8* mRNA is predicted to encode a protein ~99% identical to the Nanog1 protein in ES and EC cells except for one aa difference, i.e., **Q253H** [15]. Our earlier sequencing studies also identified, in somatic cancer cells, 3 nt changes or polymorphisms that could potentially lead to aa changes, i.e., **L61P**, **D64Y**, and **N130S** [15]. *Very little is known about the biochemical properties of NanogP8 protein in cancer cells.* To address this important deficiency, we first asked whether NanogP8 protein could be recognized by the 7 anti-Nanog1 Abs (except the BioLegend Ab) by taking advantage of our recent K14-NanogP8

transgenic (Tg) animal model in which the NanogP8 cDNA derived from a primary PCa (i.e., HPCa5) was driven by a cytokeratin 14 promoter [64]. The IHC staining revealed that all 7 anti-Nanog1 Abs detected NanogP8 in the nuclei of Tg keratinocytes (Fig. S3). Of interest, the NanogP8 protein in the Tg tissues migrates, on denaturing SDS-PAGE and WB, at 42 kD [64]. Strikingly, both NTERA-2 *Nanog1* and HPCa5 *NanogP8* cDNAs, when overexpressed from in LNCaP PCa cells using a doxycycline inducible system [34], encoded proteins that also migrated at 42 kD on WB using the Kamiya pAb (Fig. 5A, red arrowheads), just like the endogenous Nanog1 protein in NTERA-2 cells (Fig. 5A). These results were consistent with our earlier findings [34]. On longer exposure of the films, the 35 kD band (Fig. 5A, bottom panel, black arrowhead) and a 28 kD band (Fig. 5A, bottom panel, green arrow) could also be detected. Note that the Kamiya pAb detected a prominent non-specific ~38 kD protein band (Fig. 5A, asterisks), which was identified by mass spec as the sorbital dehydrogenase (data not shown). Both 42 kD and 35 kD proteins could be immunoprecipitated down by the R&D goal pAb and detected on WB by the Kamiya pAb (Fig. 5B, lanes 5 and 6), which co-migrated with the endogenous Nanog1 proteins

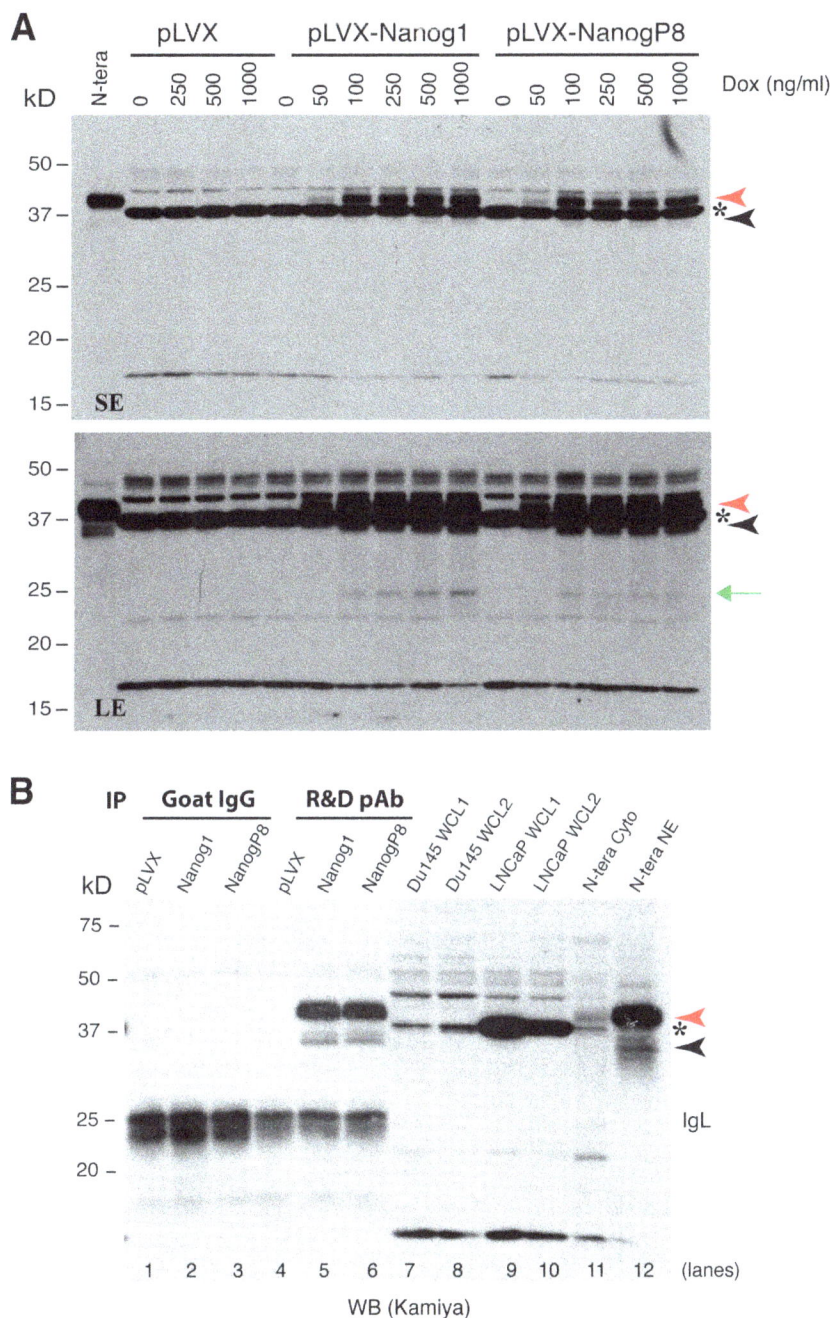

Figure 5. Exogenous NanogP8 expressed in LNCaP cells migrates mainly at 42 kD. (**A**) Exogenous NanogP8 migrates at 42 kD on WB. Whole cell lysate (80 μg/lane) prepared from control LNCaP (pLVX) or Nanog1/NanogP8 overexpressing LNCaP cells [34] in the presence of increasing amounts of doxycycline (Dox) was used in WB with the Kamiya anti-Nanog rabbit pAb. The red and black arrowheads indicate the main 42 kD and minor 35 kD Nanog bands, respectively. Green arrow, a ~28 kD band that also increased upon Dox induction. Asterisk, a non-specific band. N-tera, NE of NTERA-2 cell; SE, short exposure; LE, long exposure. (**B**) The exogenous 42 kD and 35 kD bands could be immunoprecipitated down by the R&D anti-Nanog pAb. Whole-cell lysate (WCL; 500 μg) derived from pLVX, pLVX-NANOG1 and pLVX-NANOGP8 LNCaP cells [34] were used in IP with the R&D anti-Nanog goat pAb (goat IgG used as the control) followed by WB with Kamiya anti-Nanog rabbit pAb. The red and black arrowheads indicate the main 42 kD and minor 35 kD Nanog bands, respectively. Asterisk, a non-specific band. NE, nuclear extract; cyto, cytosol protein. Note that the 42-kD and the 35-kD protein bands only from pLVX-NANOG1 and pLVX-NANOGP8 LNCaP cells (but not from LNCaP-pLVX cells) were IP'ed down and detected on WB (lanes 5,6). Goat IgG did not pull down any specific bands. Also, WCL from two batches (1 and 2) of Du145 s and LNCaP cells (80 μg/lane) did not reveal the 42 kD and 35 kD protein bands on WB (lanes 7-10). The 37 kD non-specific band (asterisk) was not immunoprecipitated down by the R&D goat pAb (lane 5–6).

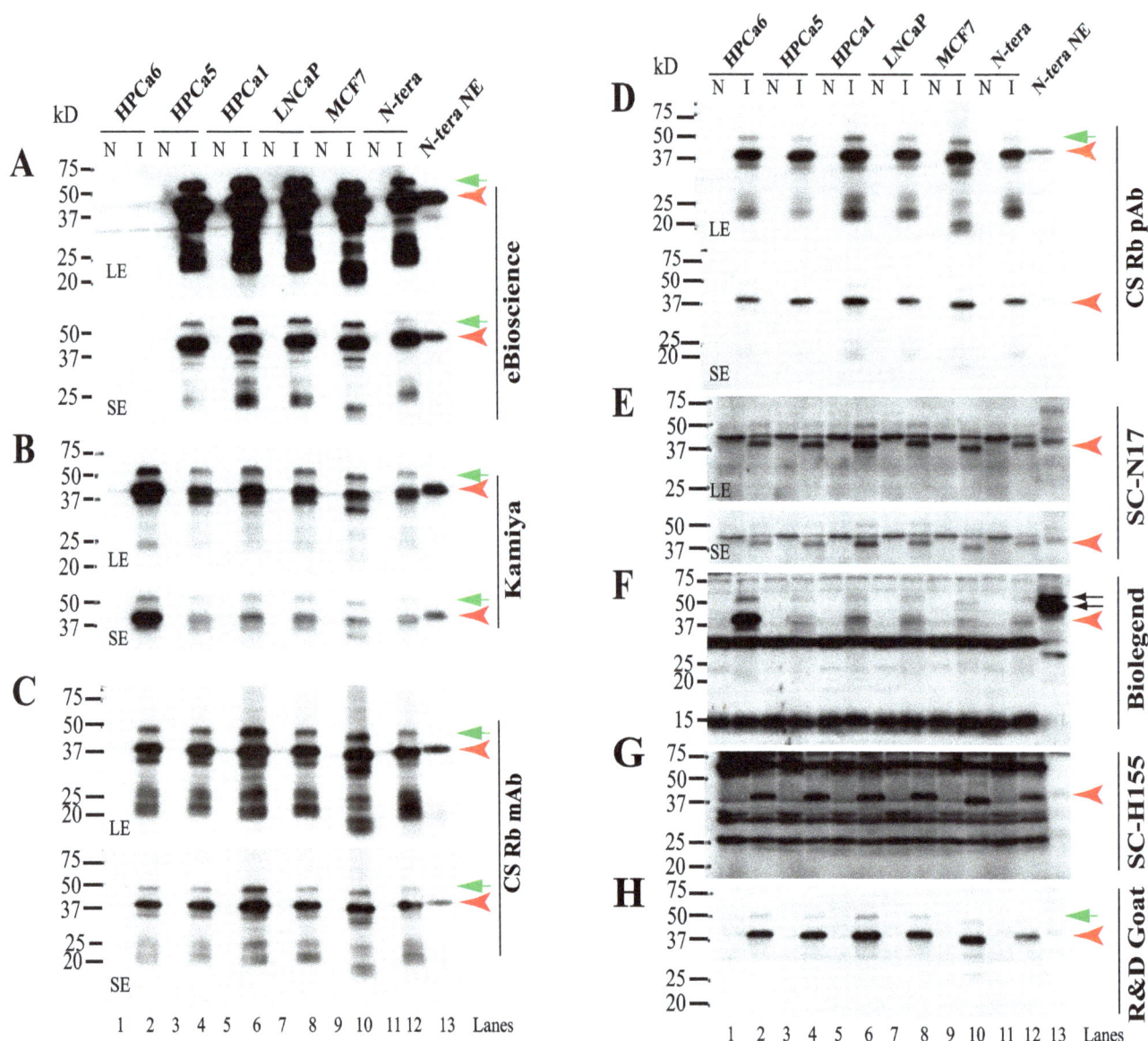

Figure 6. Reactivity of rNanogP8 and rNanog proteins towards 8 anti-Nanog Abs. WB analysis using 8 anti-Nanog Abs (A-H). Cell types from which the initial cDNAs were cloned are indicated on top. Individual Abs are indicated on the right and M.W on the left. For some Abs, both a long (LE) and short (SE) exposures were shown. N: non-induced; I: induced by IPTG (see Methods). The red arrowheads in each panel indicate the 42 kD major Nanog protein and green arrows point to minor upper bands. In panel F, the two arrows point to the ~48/54 kD doublets recognized by the BioLegend Rb pAb.

in NTERA2 NE (Fig. 5B, lane 12; see below for further discussions on NanogP8 in cancer cells).

Biochemical characterizations of recombinant Nanog1 (rNanog) and NanogP8 (rNanogP8) proteins using the 8 anti-Nanog Abs

Next, we freshly made rNanogP8 proteins from the cDNAs of a cultured PCa cell line (LNCaP), three primary prostate tumors (HPCa1, 5, and 6), and one breast cancer cell line (MCF7) (Fig. 6). For comparison, we also made rNanog1 from NTERA-2 cells (Fig. 6). The rNanogP8 and rNanog1 proteins were run on WB and probed with the 8 anti-Nanog Abs, which again exhibited distinct reactivity (Fig. 6). We observed that the majority of rNanogP8 proteins behaved overall similarly to the rNanog1 protein. Like in NTERA-2 NE, most Abs detected a major 42 kD

band (Fig. 6, red arrowheads) as well as a minor upper band of either ~48 kD (SC N17, R&D goat pAb, CS Rb pAb and Rb mAb) or ~55 kD (Kamiya, BioLegend and eBioscience) (Fig. 6, green arrows). Five Abs (eBioscience, Kamiya, CS pAb and mAb, and R&D pAb) showed clean reactions on WB whereas the other three antibodies (BioLegend pAb, SC H-155, and SC N17) detected many non-specific bands (i.e., bands detected under non-induced conditions) (Fig. 6).

Also, different rNanog proteins displayed differential reactivity to different anti-Nanog Abs. Five antibodies (CS Rb pAb and Rb mAb, SC N17, SC pAb H-155, and R&D goat pAb) detected all six rNanog proteins with similar efficiency but two antibodies (Kamiya pAb and BioLegend pAb) preferentially reacted with the HPCa6 rNanogP8 whereas the eBioscience mAb, remarkably, did not recognize the HPCa6 rNanogP8 at all (Fig. 6A). Very

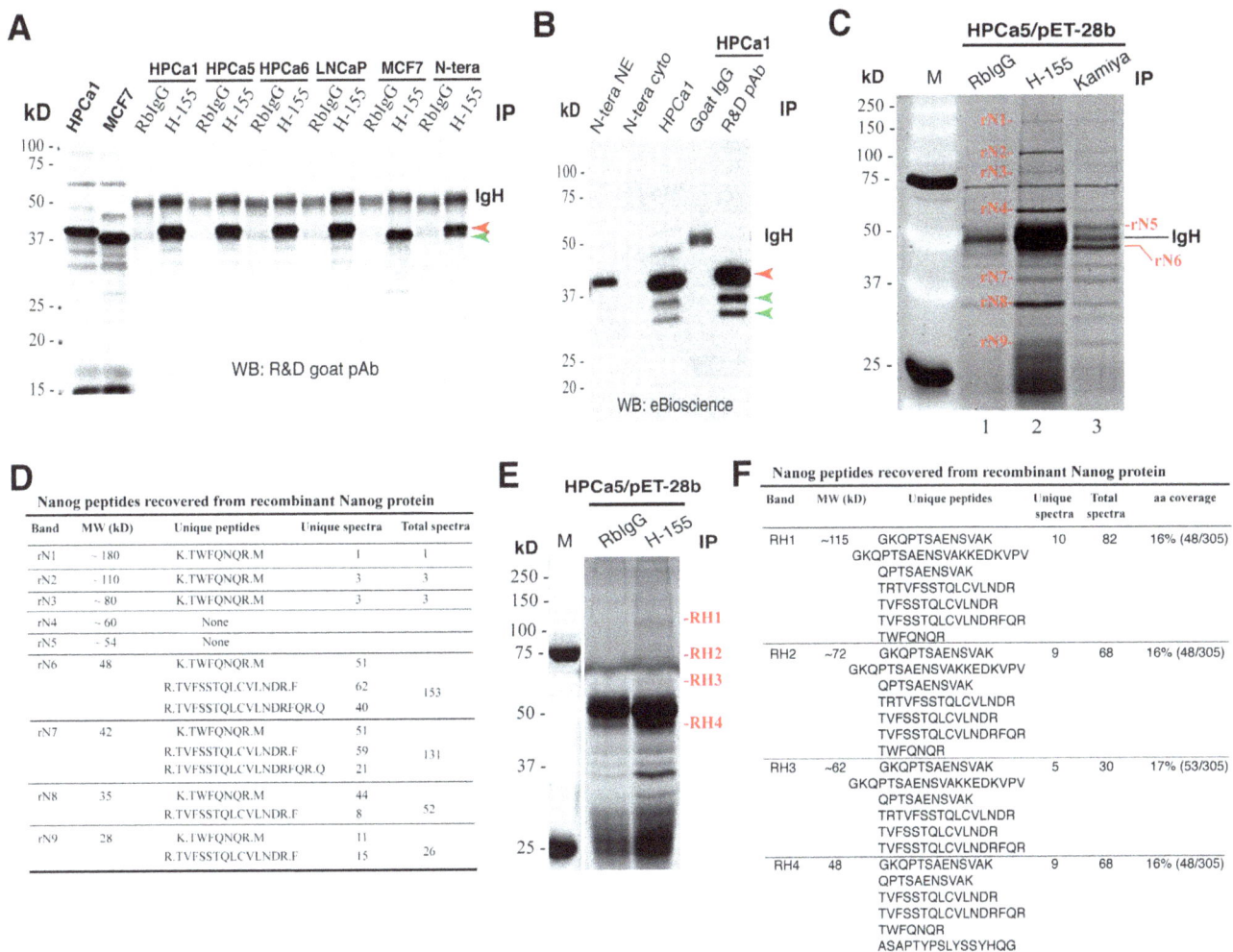

Figure 7. rNanogP8 protein ID using IP and mass spectrometry. (**A**) IP in rNanogP8 proteins using the SC pAb H-155 followed by WB using the R& D goat pAb. RbIgG was used as the control Ab. Red arrowhead, the 42 kD band; green arrowhead, the faster migrating major band in MCF7 rNanogP8; IgH, IgG heavy chain. (**B**) IP using the R&D goat pAb followed by WB using the eBioscience mAb. Goat IgG was used as the control Ab. Red arrowhead, the 42 kD band; green arrowheads, the two lower bands; IgH, IgG heavy chain. In this experiment, NTERA-2 NE and cytosol and HPCa rNanogP8 were also loaded in WB analysis. (**C–D**) Mass spectrometry ID of rNanogP8 proteins. The HPCa5 rNanogP8 protein made in pET-28b was used in IP with either H-155 or Kamiya Rb pAb (RbIgG as the Ab control). The immunoprecipitates were subjected to SDS-PAGE, stained with SYPRO Ruby, and 9 bands (rN1 - rN9) cut out for protein ID (C). M, protein marker. The identified peptides were presented in D. (**E–F**) Mass spectrometry ID of rNanogP8 proteins in a separate experiment using conditions as above. Four bands (RH1–RH4) were cut out for protein ID (E) and identified peptides presented in F.

interestingly, the MCF7 rNanogP8 consistently migrated slightly faster, at ~37 kD, than other rNanogP8 proteins (Fig. 6). Taken together, these results suggest that **1)** *the rNanogP8 proteins, like endogenous Nanog1 and rNanog1 in NTERA-2 cells, seem to be able to adopt different conformations that exhibit varying mobility on reducing SDS-PAGE;* **2)** *somatic cancer cells may encode different 'isoforms' of NanogP8 due to sequence polymorphisms (such as those in HPCa6 and MCF7 cells); and* **3)** *NanogP8 proteins of different conformations or polymorphisms may be differentially recognized by various anti-Nanog Abs with the 42 kD still being the major band on WB.*

rNanogP8 proteins also migrate at multiple M.W species as revealed by IP followed by MALDI-TOF/TOF

We performed IP experiments in rNanog1 protein from NTERA-2 and rNanogP8 proteins from HPCa1, HPCa5, HPCa6, LNCaP and MCF7 (Fig. 7A–C; data not shown). IP with the SC Rb pAb H-155 pulled down the major 42 kD (Fig. 8A, red

arrowhead; green arrowhead pointed to the faster migrating major band in MCF7 cells) and some minor bands (apparent upon longer exposure). IP with the R&D goat pAb also pulled down the major 42 kD (Fig. 7B, red arrowhead) and two lower bands (Fig. 7B; green arrowheads). Similarly, IP with the Kamiya pAb and eBioscience mAb pulled down rNanogP8 proteins (Fig. 7C; data not shown). These results indicate that like the endogenous Nanog1 protein in NTERA-2 NE, rNanogP8 proteins derived from cDNAs in somatic cancer cells can be readily immunoprecipitated down by various anti-Nanog Abs and that the 42 kD protein also represents the major protein band on WB.

Subsequently, we immunoprecipitated the HPCa5 rNanogP8 using two anti-Nanog antibodies, i.e., SC pAb H-155 and Kamiya pAb, and, after SDS-PAGE, we stained the gel with SYPRO Ruby (Fig. 7C). Compared with rabbit IgG control, at least 9 specific bands from ~28 kD to ~180 kD were identified by the two anti-Nanog Abs (Fig. 7C). Mass spectrometry ID experiments revealed

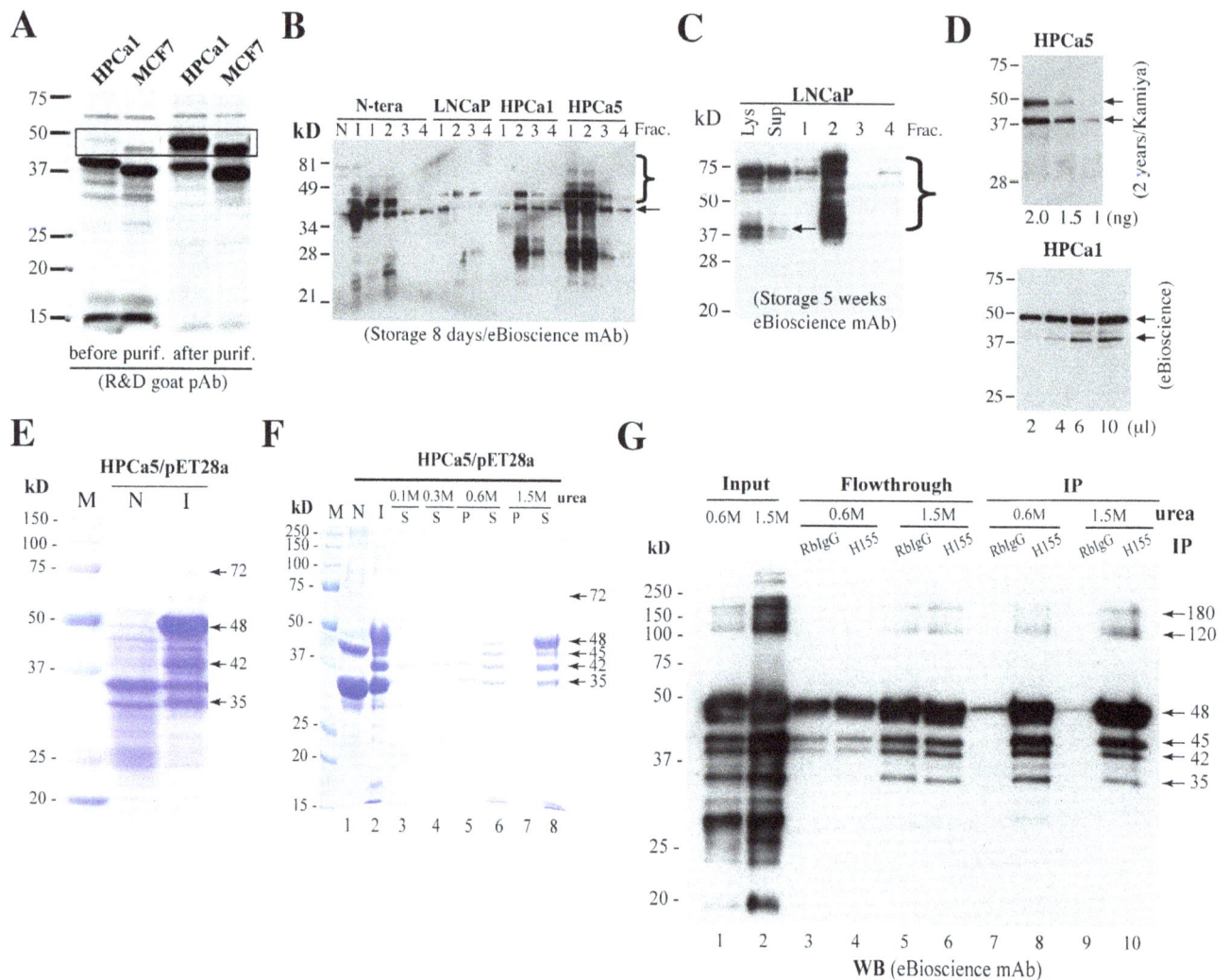

Figure 8. Evidence that rNanogP8 proteins can spontaneously form high M.W species. (**A**) rNanogP8 proteins from HPCa1 and MCF7, before and after purification, were used in WB using the R&D goat pAb. Note that prior to purification, the HPCa1 and MCF7 rNanogP8 proteins migrated at ~42 kD and ~37 kD, respectively, with a minor upper band detected for both proteins. After purification, the intensity of the upper bands (in the rectangle) became significantly stronger. (**B–C**) rNanogP8 proteins (from the indicated cell types) stored at −80°C for 8 days (B) or 5 weeks (C) were used in protein purification. Aliquots (20 μl) of 4 fractions (Frac.) for each sample, together with NTERA-2 non-induced (N) or induced (I) bacterial lysate (B) or LNCaP total bacterial lysate (Lys) or supernatant (Sup) (C), were used in WB with the eBioscience mAb. The arrows indicate the ~42 kD rNanogP8 proteins and right-hand brackets indicate high M.W ladders. (**D**) The HPCa5 and HPCa1 rNanogP8 proteins stored for 2 years was utilized in WB with the Kamiya pAb and eBioscience mAb, respectively. (**E**) Coomassie brilliant blue R-250 staining of rNanogP8 protein made from HPCa5 cDNA in pET-28a. The arrows indicate the high levels of rNanogP8 proteins under the induced conditions. M, protein marker; N, non-induced; I, induced by IPTG. (**F**) Coomassie brilliant blue R-250 staining of HPCa5 rNanogP8 proteins from refolding dialysis experiment. The inclusion bodies of rNanogP8 protein were first dissolved in 7 M urea and then subjected to dialysis against decreasing concentrations of urea. N, non-induced; I, induced by IPTG; S, soluble portion; P, pellet (precipitated portion). Arrows indicate rNanogP8 proteins of different molecular weights. (**G**) IP analysis confirms that the refolded bands are rNanogP8. The dialysis samples containing 1.5 M or 0.6 M urea were subjected to IP with the SC pAb (H-155) or RbIgG (control) followed by WB analysis with the eBioscience mAb. Arrows indicate rNanogP8 proteins of different molecular weights.

that 7 of the 9 bands (named rN1-rN9) with M.W of approximately 28 kD, 35 kD, 42 kD, 48 kD, 80 kD, 110 kD, and 180 kD, contained Nanog peptides (Fig. 7D). A repeat IP and MALDI-TOF/TOF experiment similarly uncovered Nanog peptides at approximately 48, 62, 72, and 115 kD (Fig. 7E-F).

Evidence that rNanogP8 proteins spontaneously form high M.W species

During multiple experiments with rNanogP8 proteins, we frequently observed <40 kD protein bands, which might have resulted from degradation. However, we also consistently observed >40-kD Nanog protein bands (Fig. 5; Fig. 6B; Fig. 7C–F),

particularly during protein purification with stored samples (Fig. 8A–D). For instance, the HPCa1 rNanogP8, prior to purification, migrated at ~42 kD with a minor band at 48 kD (Fig. 8A). Strikingly, after purification (i.e., with increased protein concentration), the 48 kD band became predominant (Fig. 8A). Likewise, the MCF7 rNanogP8 migrated at ~37 kD with a minor band at ~45 kD, the latter of which significantly increased upon purification (Fig. 8A). These spontaneously formed high M.W rNanogP8 species were even more apparent when the samples stored at −80°C for different time intervals were used in protein purification (Fig. 8B–C). For example, distinct tapering ladders of ≥40-kD were observed in the purified NTERA-2, LNCaP,

HPCa1, and HPCa5 rNanog(P8) proteins stored for 8 days (Fig. 8B). In LNCaP rNanogP8 stored for 5 weeks (prior to being used in purification), it appeared that most 42 kD and intermediate protein species had migrated to ~72 kD (Fig. 8C). When purified rNanogP8 stored long-term (e.g., 2 years) were used in WB, we routinely detected two bands at 42 kD and 48 kD with some minor bands in between (Fig. 8D).

When we subcloned the full-length HPCa5 NanogP8 cDNA into the pET-28a bacterial expression vector that allows high levels of recombinant protein expression, we detected at least 4 obvious recombinant rNanogP8 bands at approximately 75 kD, 48 kD, 42 kD and 35 kD upon IPTG induction, among which the 48 kD band had the highest yield (Fig. 8E). We found that the recombinant HPCa5 rNanogP8 protein was mostly present in the inclusion bodies. Consequently, we conducted a 'refolding' analysis in buffers containing various concentrations of urea. After isolating the inclusion bodies of the HPCa5 rNanogP8 protein from pET-28a E. coli, we first used 7 M urea solution to dissolve the inclusion bodies, and then carried out the refolding dialysis experiments, in decreasing concentrations of urea (i.e., 1.5 M, 0.6 M, 0.3 M and 0.1 M), to let the denatured rNanogP8 protein refold into its potentially native form (Fig. 8F). We next conducted SDS-PAGE with these dialysis samples and stained with Coomassie brilliant blue R-250. As shown in Fig. 8F, there were at least 5 soluble rNanogP8 bands with the M.W of approximately 72 kD, 48 kD, 45 kD, 42 kD and 35 kD in the 1.5 M urea dialysis buffer and the 48 kD band was the most prominent (Fig. 8F, lane 8). In 0.6 M urea, the abundance of the 48 kD band was dramatically reduced (Fig. 8F, lane 6). Strikingly, the 42 kD HPCa5 rNanogP8 protein remained soluble even in 0.3 M and 0.1 M urea dialysis buffers (Fig. 8F, lane 3 and 4), suggesting that the 42 kD band might represent the 'native' (i.e., most soluble) NanogP8 protein. Finally, we immunoprecipitated the 0.6 M and 1.5 M urea dialysis samples using the SC Rb pAb H-155 and analyzed them on WB. The H-155 Ab immunoprecipitated down 6 bands with M.W ranging from 35 kD to ~180 kD (Fig. 8G, lanes 8 and 10).

Anti-Nanog1 antibodies fail to IP down endogenous NanogP8 proteins in long-term cultured or xenograft-derived somatic cancer cells

Many studies have reported Nanog protein expression in somatic cancer cells (see Introduction; Table 1) [63], Given that most somatic cancer cells predominantly or exclusively express NanogP8 mRNA, it stands to reason that any putative Nanog protein they express should be NanogP8 protein. As frequently only a cropped strip of WB image or a 'representative' IHC panel is shown in many of these studies, it is unclear whether the Nanog protein reported on WB or IHC truly represents NanogP8. In preliminary studies, we carried out IP experiments using 4 anti-Nanog Abs, i.e., Kamiya Rb pAb, CS Rb mAb, SC H-155 Rb pAb, and R&D goat pAb using ~500 μg NE prepared from several cultured cancer cell lines including PCa (Du145), breast cancer (MCF7), colon cancer (Colo320), and melanoma (WM562) (Fig. 3A–D; Fig. S2A). Although these Abs readily immunoprecipitated down the 42 kD endogenous Nanog protein in NTERA-2 NE (Fig. 3A–D; Fig. S2), and the rNanogP8 proteins made from somatic cancer cells (Fig. 7), they did not pull down the 42 kD or other NanogP8 proteins in cultured cancer cell NE (Fig. 3A–D; Fig. S2A). Similarly, the Kamiya pAb did not immunoprecipitate down NanogP8 in the NE of LAPC4 and LAPC9 PCa xenografts (Fig. 3A). Finally, although the R&D pAb immunoprecipitated down both 42 kD and 35 kD Nanog1 and NanogP8 proteins in Dox-induced LNCaP cells, it did not pull down any of these proteins in un-induced cells (Fig. 5B). In fact, regular WB using cytosol and NE from these cancer cells did

not reveal the major 42 kD Nanog, which was always readily detected in NTERA-2 NE (Fig. 3A–D; Fig. S2A; Fig. 5). It should be noted that the Kamiya pAb recognized a non-specific ~38 kD band, which was not NanogP8 as it was not pulled down by the R&D goat pAb (Fig. 5B).

Discussion

The current project was undertaken to characterize the biochemical properties of Nanog protein in EC and, potentially, in somatic cancer cells. We have made the following novel findings. **FIRST**, Nanog1 in NTERA-2 cells is detected on WB primarily as a 42 kD protein band by 7 of the 8 Abs tested. **SECOND**, Nanog1 in NTERA-2 cells exists at multiple M.W species at up to ~100 kD (Table 2). Some of these Nanog1 species (e.g., the 35 kD) appear to be as abundant as the 42 kD protein (based on mass spec) although most of them are less abundant. **THIRD**, the 42 kD Nanog1 band in NTERA-2 cells can be reliably and readily immunoprecipitated down by most Abs and subsequently detected on WB. **FOURTH**, most anti-Nanog1 Abs tested recognize the EXOGENOUS NanogP8 protein in Tg mouse tissues and human cancer cells and the exogenous NanogP8 in these settings also migrates at 42 kD. **FIFTH**, both rNanog1 and rNanogP8 proteins are detected, on WB, mainly as the 42 kD band. **SIXTH**, like endogenous Nanog1 protein in NTERA-2 cells, rNanog1/rNanogP8 proteins can also be detected as multiple M.W species at up to ~180 kD. **SEVENTH**, rNanogP8 appears to be able to spontaneously form high M.W protein species. **EIGHTH**, the epitope of the eBioscience mAb raised against the full-length human Nanog1 protein is mapped to the L61 region in the N-terminus. **FINALLY**, anti-Nanog1 antibodies fail to IP down the endogenous 42 kD (or 35 kD or other) NanogP8 proteins in long-term cultured or xenograft-derived somatic cancer cells.

Nanog protein has 305 aa with a predicted M.W of ~35 kD. WB using NE, siRNA-mediated knockdown, and IP followed by MS-based peptide ID, combined, provide UNEQUIVOCAL evidence that the endogenous Nanog protein in NTERA-2 EC cells exists as multiple M.W species, migrating at <30 kD (e.g., 28 kD) all the way to ~100 kD (Fig. 1–4; Table 2). WB analysis of NTERA-2 NE demonstrates that 7 of the 8 anti-Nanog Abs, raised against the N- or C-terminus of human Nanog protein, preferentially reacts with the major 42 kD and minor 35 kD Nanog1 proteins. These results are consistent with the two MS-based ID experiments showing that the gel slices of 35–45 kD contain the highest Nanog peptide counts (Fig. 3F; Fig. 4B–C), suggesting that the endogenous Nanog1 proteins in NTERA-2 cells are most abundant at 35–45 kD range. On the other hand, the 42 kD Nanog seems to be the preferred protein species recognized by all anti-Nanog Abs in both WB and IP, which raises the possibility that different Nanog protein species may adopt different conformations, which dictate their gel mobility as well as Ab reactivity. In support, we find that in addition to the 42 kD and 35 kD bands, various anti-Nanog Abs also recognize additional protein bands but with slightly different patterns (Table 2). Surely, although not all protein bands detected on WB in NTERA-2 NE are Nanog1 proteins, knockdown experiments together with MS-based protein ID do suggest that many of these bands represent authentic Nanog1 protein.

How could a predicted 35 kD nuclear protein migrate at as large as ~100 kD on reducing and denaturing SDS-PAGE? First of all, the Nanog1 gene has been reported to undergo alternative splicing, which could potentially generate Nanog proteins of different molecular masses [52,57]. But alternative splicing mainly gives rises to smaller Nanog protein variants [57]. Furthermore,

Table 2. Nanog proteins in NTERA-2 human EC cells.*

Ab	IP/MALDI	Tandem IP/ LTQ	IP	WB	Knockdown
	(Fig. 3E–F)	(Fig. 4)	(Fig. 3A–D; Fig. S2)	(Fig. 1)	(Fig.2)
eBioscience mAb				42 kD>>35 kD>>65 kD	
Kamiya Rb pAb		~25–70 kD	42 kD	42 kD>>35 kD; 48, 65, 100 kD	42 kD>>35 kD; 48, 65, 75, 90 kD
CS Rb pAb				42 kD>>35 kD>>65 kD	42 kD>>35 kD; 65 kD
CS Rb mAb			42 kD	42 kD>>35 kD; 28,32, 70, 100 kD	
SC N-17 Rb pAb				42 kD>>35 kD>55/58 kD	
Biolegend Rb pAb				48 kD>35 kD>55 kD	
SC H-155			42 kD	42 kD>>35 kD; 55 kD, 58 kD	
R&D goat pAb	42, 60, 75, 100 kD	~25–70 kD	42 kD	42 kD>>35 kD; 28, 65, 70 kD	

*Presented are the results of various experiments using the 8 anti-Nanog Abs. The Biolegend Ab is the only Ab that does not recognize the 42 kD band as the major protein band and does not recognize the 35 kD band. Presented molecular masses are all estimated based on their migrations on SDS-PAGE.

cDNA-derived Nanog1 and rNanog1 proteins also migrate at 42 kD or above, arguing against alternative splicing being a mechanism generating high M.W Nanog protein species. *Secondly,* Nanog protein in ES cells has been reported to function as homodimers formed through the interactions between the Nanog's WR domain, which could weigh ~70–80 kD [12,13]. Moreover, Nanog protein in ES cells exists in gigantic protein complexes [53]. These observations, however, do not seem to be able to explain our present results because Nanog homodimers and protein complexes would have been disrupted by the reducing agents and denaturant SDS. *Thirdly,* Nanog protein can undergo two types of post-translational modifications (PTM) including phosphorylation [39,65] and ubiquitylation [66]. Indeed, in some of our analysis of endogenous Nanog1 in NTERA-2 cells, we frequently observed one or several bands right above the minor 35 kD Nanog protein (e.g., Fig. 1E and H; Fig. 3C; Fig. 5B), which likely represent the phosphorylated Nanog1 proteins. However, we believe that phosphorylation will unlikely explain dramatic M.W shifts of the Nanog1 protein (all the way up to ~100 kD). More important, the rNanog protein made from the NTERA-2 cDNA in bacteria, which lack the above-mentioned PTM machinery, also primarily migrates at 42 kD and higher, thus arguing against PTMs as the major mechanisms.

Comprehensive studies using rNanog1 from NTERA-2 and rNanogP8 proteins from multiple somatic cancer cells further suggest that Nanog1 and NanogP8, which are >99% identical at the aa levels, seem to possess unique INTRINSIC biochemical properties that can allow them to adopt multiple protein conformations and consequently migrate at multiple apparent M.W. The rNanogP8 proteins made from 4 PCa (LNCaP, and HPCa1, 5, and 6) cell types all migrate at 42 kD as the major band on both WB (Fig. 6) and IP (Fig. 7A–B) analyses. Strikingly, IP of the rNanogP8 proteins using both an N-terminus directed Ab (i.e., Kamiya) and a C-terminus directed Ab (i.e., SC H-155) followed by MS have uncovered Nanog peptides from 28 kD all the way to ~180 kD (Fig. 7C–F). Surprisingly, although the 42 kD protein represents the major band on WB, the highest number of Nanog peptides is recovered for the 48 kD band, which is recognized preferentially by the Kamiya Ab (Fig. 7C–D). In contrast, the SC H-155 Ab preferentially recognizes the 42 kD (the second highest peptide count) and 35 kD (the third highest peptide count) rNanogP8 proteins (Fig. 7C–D).

How could rNanogP8 proteins made in bacteria, just like endogenous Nanog proteins in NTERA-2 cells, migrate at multiple M.W species? Since the above-discussed experiment (Fig. 7C–F) is performed using two different Abs on the rNanogP8 proteins made from a single cancer cell type, i.e., HPCa5, the results strongly argue that the rNanogP8 protein, by itself, can adopt different conformations that are differentially recognized by different anti-Nanog Abs. In support, we have observed that the rNanog1 in NTERA-2 cells as well as the rNanogP8 proteins from somatic cancer cells can spontaneously form high M.W protein species during protein purification when the (local) concentrations of recombinant proteins dramatically increase (Fig. 8A–C). In further support, when rNanogP8 proteins are made in a bacteria strain that allows high levels of protein production, an apparent ladder of rNanogP8 protein species ranging from ~35 kD to 180 kD is observed (Fig. 8E–G). In such experiments, the 48 kD protein represents the major NanogP8 species (Fig. 8E–F), which can be efficiently pulled down by the SC H-155 Ab (Fig. 8G). However, in the urea-assisted 'denaturation and refolding' analysis, the 42 kD rNanogP8 is the only protein species observed in the lowest urea-containing buffers whereas all other rNanogP8 species can only be seen in buffers containing >0.3 M urea (Fig. 8F), suggesting that *the 42 kD rNanogP8 represents the 'native' and most stable (and most soluble) Nanog protein conformer*, explaining its preferential detection on WB in both NTERA-2 NE (Fig. 1) and rNanog1/rNanogP8 proteins (Fig. 6). This suggestion is consistent with the fact that even when Nanog1/NanogP8 cDNAs are overexpressed in cancer cells, the major protein detected on WB is 42 kD (Fig. 5). The intrinsic properties of Nanog1/NanogP8 proteins, to a certain degree, resemble c-Myc oncoprotein, which can be detected on WB at M.W from ~40 kD to >60 kD [67,68]. Precisely how the high M.W Nanog proteins are formed is currently being investigated in the lab.

The Nanog protein (i.e., NanogP8) has been reported to be overexpressed in tumors [18–27,54–57,63]. However, our preliminary IP experiments using NE extracts prepared from long-term cultured cancer cell lines or xenograft tumors fail to pull down the dominant 42 kD (or other) NanogP8 protein although all 5 anti-Nanog Abs can readily immunoprecipitate the 42 kD Nanog protein in the same amount (500 µg) of NTERA-2 NE (Fig. 3A–D; Fig. S2A; Fig. 5B). In fact, WB using NE or cytosol from these somatic cancer cells did not identify the 42 kD NanogP8 protein (e.g., Fig. 3A–D; Fig. 5). Several potential explanations may underlie this negative result. ***First***, long-term cultured or xenograft-derived somatic cancer cells express too low levels of NanogP8 to be detected by conventional IP with relatively small amounts of NE. ***Second***, long-term cancer cell cultures and perhaps most long-term xenografts contain too few NanogP8-expressing cells. This possibility is supported by our earlier studies showing NanogP8 expression in only a very small percentage of cultured cancer cells [15] and primary prostate tumors and early-generation xenograft prostate tumors expressing NanogP8 mRNA at levels several orders of magnitude higher than in long-term cultured cancer cells [34]. ***Third***, endogenous NanogP8 protein in somatic cancer cells is different from the Nanog1 protein in NTERA-2 cells such that the NanogP8 is not recognized well on IP and WB by anti-Nanog1 Abs. ***Finally***, the endogenous NanogP8 protein in somatic cancer cells may undergo unique PTMs and consequently possess a very short half-life. All these (and other) possibilities are currently been pursued in our lab.

Supporting Information

Figure S1   Genomic organization of *Nanog* and *NanogP8* genes and Nanog peptides recovered from mass spectrometry analysis in N-tera NE. (**A**) Schematic of *Nanog1* gene. Chr, chromosome; E, exon; TSS, transcription start site; UTR, untranslated region. The 22-bp region unique to the 3′-UTR of *Nanog1* (vertical black bar) was indicated, which was used in our earlier differential qRT-PCR analysis [34]. (**B**) Schematic of the *NanogP8* retrogene, which lacks introns and is located on Chr. 15q14. The red asterisk indicates the AlwN1 restriction site located on nt144. (**C–D**) Nanog protein structure. The 305 aa Nanog protein contains 5 protein domains, i.e., ND (N-terminus domain), HD (homeodomain), WR (tryptophan-rich domain), and CD1 and CD2 (C-terminus domain 1 and 2). Most Nanog peptides recovered in N-tera NE upon LC-MS/MS (Fig. 3E–F) and tandem IP/mass spectrometry (Fig. 4) are mapped to the ND and HD (indicated below; the numbers in the parentheses refer to the number of peptides).

Figure S2   IP studies and Nanog protein ID in NTERA2 NE. (**A**) N-tera NE was used in IP with the Kamiya pAb followed by WB with the R&D goat pAb. Lanes 1–7 were regular WB using cytosol (C) or NE from the cells indicated or using whole cell lysate (WCL) from NTERA2. The top and bottom panels are long exposure (LE) and short exposure (SE), respectively. Red arrowhead indicates the 42 kD Nanog and black arrowhead the 35 kD band (both circled) whereas green arrows indicate additional bands detected on WB only in NE. IgH, IgG heavy chain (~53 kD). (**B**) NTERA2 NE was used in IP with the SC pAb (H-155) followed by WB with the eBioscience mAb. Red arrowhead, the 42 kD Nanog band; IgH, IgG heavy chain. (**C**) Representative mass spectra of peptides detected in the 4 gel slices labeled as NTRD1 – NTRD4.

Figure S3   HPCa5-derived NanogP8 expressed in transgenic mouse epidermis is recognized by all 7 anti-Nanog Abs tested. Immunohistochemistry of skin sections stained with 7 anti-Nanog antibodies. WT, wild-type; TG, K14-NanogP8 transgenic mouse [64]. Boxes areas were enlarged and shown in insets. Dark brown color indicates the positive cells; blue color indicates nuclear counterstaining.

Acknowledgments

We thank S. Bratton and M. Bedford for valuable suggestions and critically reading the manuscript. We also thank other members of the Tang lab for discussions and support.

Author Contributions

Conceived and designed the experiments: BL MDB GC DC IS CRJ KR CFL MDP CL YC JS SYJ JQ DGT. Performed the experiments: BL MDB GC DC IS CRJ KR CFL MDP CL YC JS SYJ JQ DGT. Analyzed the data: BL MDB GC DC IS CRJ KR CFL MDP CL YC JS SYJ JQ DGT. Wrote the paper: BL MDP JS SYJ DGT.

References

1. Chambers I, Colby D, Robertson M, Nichols J, Lee S, et al. (2003) Functional expression cloning of Nanog, a pluripotency sustaining factor in embryonic stem cells. Cell 113: 643–655.
2. Mitsui K, Tokuzawa Y, Itoh H, Segawa K, Murakami M, et al. (2003) The homeoprotein Nanog is required for maintenance of pluripotency in mouse epiblast and ES cells. Cell 113: 631–642.
3. Pei D (2009) Regulation of pluripotency and reprogramming by transcription factors. J Biol Chem 284: 3365–3369.
4. Darr H, Mayshar Y, Benvenisty N (2006) Overexpression of NANOG in human ES cells enables feeder-free growth while inducing primitive ectoderm features. Development 133: 1193–1201.
5. Hyslop L, Stojkovic M, Armstrong L, Walter T, Stojkovic P, et al. (2005) Downregulation of NANOG induces differentiation of human embryonic stem cells to extraembryonic lineages. Stem Cells 23: 1035–1043.
6. Yu J, Vodyanik MA, Smuga-Otto K, Antosiewicz-Bourget J, Frane JL, et al. (2007) Induced pluripotent stem cell lines derived from human somatic cells. Science 318: 1917–1920.
7. Theunissen TW, van Oosten AL, Castelo-Branco G, Hall J, Smith A, et al. (2011) Nanog overcomes reprogramming barriers and induces pluripotency in minimal conditions. Curr Biol 21: 65–71.
8. Do HJ, Yang HM, Moon SY, Cha KY, Chung HM (2005) Identification of a putative transactivation domain in human Nanog. Exp Mol Med 37: 250–254.
9. Pan G, Pei D (2005) The stem cell pluripotency factor NANOG activates transcription with two unusually potent subdomains at its C terminus. J Biol Chem 280: 1401–1407.
10. Do HJ, Lim HY, Kim JH, Song H, Chung HM (2007) An intact homeobox domain is required for complete nuclear localization of human Nanog. Biochem Biophys Res Commun 353: 770–775.
11. Chang DF, Tsai SC, Wang XC, Xia P, Senadheera D, et al. (2009) Molecular characterization of the human NANOG protein. Stem Cells 27: 812–821.

12. Mullin NP, Yates A, Rowe AJ, Nijmeijer B, Colby D, et al. (2008) The pluripotency rheostat Nanog functions as a dimer. Biochem J 411: 227–231.
13. Wang J, Levasseur DN, Orkin SH (2008) Requirement of Nanog dimerization for stem cell self-renewal and pluripotency. Proc Natl Acad Sci USA 105: 6326–6331.
14. Booth HA, Holland PW (2004) Eleven daughters of NANOG. Genomics 84: 229–238.
15. Jeter CR, Badeaux M, Choy G, Chandra D, Patrawala L, et al. (2009) Functional evidence that the self-renewal gene NANOG regulates human tumor development. Stem Cells 27: 993–1005.
16. Ezeh UI, Turek PJ, Reijo RA, Clark AT (2005) Human embryonic stem cell genes OCT4, NANOG, STELLAR, and GDF3 are expressed in both seminoma and breast carcinoma. Cancer 104: 2255–2265.
17. Zhang J, Wang X, Li M, Han J, Chen B, et al. (2006) NANOGP8 is a retrogene expressed in cancers. FEBS J 273: 1723–1730.
18. Gu G, Yuan J, Wills M, Kasper S (2007) Prostate cancer cells with stem cell characteristics reconstitute the original human tumor in vivo. Cancer Res 67: 4807–4815.
19. Chiou SH, Yu CC, Huang CY, Lin SC, Liu CJ, et al. (2008) Positive correlations of Oct-4 and Nanog in oral cancer stem-like cells and high-grade oral squamous cell carcinoma. Clin Cancer Res 14: 4085–4095.
20. Zhang S, Balch C, Chan MW, Lai HC, Matei D, et al. (2008) Identification and characterization of ovarian cancer-initiating cells from primary human tumors. Cancer Res 68: 4311–4320.
21. Alldridge L, Metodieva G, Greenwood C, Al-Janabi K, Thwaites L, et al. (2008) Proteome profiling of breast tumors by gel electrophoresis and nanoscale electrospray ionization mass spectrometry. J Proteome Res 7: 1458–1469.
22. Ye F, Zhou C, Cheng Q, Shen J, Chen H (2008) Stem-cell-abundant proteins Nanog, Nucleostemin and Musashi1 are highly expressed in malignant cervical epithelial cells. BMC Cancer 8:108.

23. Bussolati B, Bruno S, Grange C, Ferrando U, Camussi G (2008) Identification of a tumor-initiating stem cell population in human renal carcinomas. FASEB J 22: 3696–3705.

24. Kochupurakkal BS, Sarig R, Fuchs O, Piestun D, Rechavi G, et al. (2008) Nanog inhibits the switch of myogenic cells towards the osteogenic lineage. Biochem Biophys Res Commun 365: 846–850.

25. Bourguignon LY, Peyrollier K, Xia W, Gilad E (2008) Hyaluronan-CD44 interaction activates stem cell marker Nanog, Stat-3-mediated MDR1 gene expression, and ankyrin-regulated multidrug efflux in breast and ovarian tumor cells. J Biol Chem 283: 17635–17651.

26. Siu MK, Wong ES, Chan HY, Ngan IIY, Chan KY, et al. (2008) Overexpression of NANOG in gestational trophoblastic diseases: effect on apoptosis, cell invasion, and clinical outcome. Am J Pathol 173: 1165–1172.

27. Meng HM, Zheng P, Wang XY, Liu C, Sui HM, et al. (2010) Overexpression of nanog predicts tumor progression and poor prognosis in colorectal cancer. Cancer Biol Ther 9.

28. Meyer MJ, Fleming JM, Lin AF, Hussnain SA, Ginsburg E, et al. (2010) CD44posCD49fhiCD133/2hi defines xenograft-initiating cells in estrogen receptor-negative breast cancer. Cancer Res 70: 4624–4633.

29. Liu M, Sakamaki T, Casimiro MC, Willmarth NE, Quong AA, et al. (2010) The canonical NF-kappaB pathway governs mammary tumorigenesis in transgenic mice and tumor stem cell expansion. Cancer Res 70: 10464–10473.

30. Chiou SH, Wang ML, Chou YT, Chen CJ, Hong CF, et al. (2010) Coexpression of Oct4 and Nanog enhances malignancy in lung adenocarcinoma by inducing cancer stem cell-like properties and epithelial-mesenchymal transdifferentiation. Cancer Res 70: 10433–10444.

31. Po A, Ferretti E, Miele E, De Smaele E, Paganelli A, et al. (2010). Hedgehog controls neural stem cells through p53-independent regulation of Nanog. EMBO J 29: 2646–2658.

32. Zbinden M, Duquet A, Lorente-Trigos A, Ngwabyt SN, Borges I, et al. (2010) NANOG regulates glioma stem cells and is essential in vivo acting in a cross-functional network with GLI1 and p53. EMBO J 29: 2659–2674.

33. Salmina K, Jankevics E, Huna A, Perminov D, Radovica I, et al. (2010) Up-regulation of the embryonic self-renewal network through reversible polyploidy in irradiated p53-mutant tumour cells. Exp Cell Res 316: 2099–2112.

34. Jeter CR, Liu B, Liu X, Chen X, Liu C, et al. (2011) NANOG promotes cancer stem cell characteristics and prostate cancer resistance to androgen deprivation. Oncogene 30: 3833–3845.

35. Mathieu J, Zhang Z, Zhou W, Wang AJ, Heddleston JM, et al. (2011) HIF induces human embryonic stem cell markers in cancer cells. Cancer Res 71:4640–4652.

36. Lee TK, Castilho A, Cheung VC, Tang KH, Ma S, et al. (2011) CD24(+) liver tumor-initiating cells drive self-renewal and tumor initiation through STAT3-mediated NANOG regulation. Cell Stem Cell 9: 50–63.

37. Bourguignon LY, Earle C, Wong G, Spevak CC, Krueger K (2012) Stem cell marker (Nanog) and Stat-3 signaling promote MicroRNA-21 expression and chemoresistance in hyaluronan/CD44-activated head and neck squamous cell carcinoma cells. Oncogene 31: 149–160.

38. Noh KH, Lee YH, Jeon JH, Kang TH, Mao CP, et al. (2012) Cancer vaccination drives Nanog-dependent evolution of tumor cells toward an immune-resistant and stem-like phenotype. Cancer Res 72: 1717–1727.

39. Ho B, Olson G, Figel S, Gelman I, Cance WG, et al. (2012) Nanog increases focal adhesion kinase (FAK) promoter activity and expression and directly binds to FAK protein to be phosphorylated. J Biol Chem 287: 18656–18673.

40. Shan J, Shen J, Liu L, Xia F, Xu C, et al. (2012) Nanog regulates self-renewal of cancer stem cells through the insulin-like growth factor pathway in human hepatocellular carcinoma. Hepatol 56: 1004–1014.

41. Qin J, Liu X, Laffin B, Chen X, Choy G, et al. (2012) The PSA(-/lo) prostate cancer cell population harbors self-renewing long-term tumor-propagating cells that resist castration. Cell Stem Cell 10: 556–569.

42. Santini R, Vinci MC, Pandolfi S, Penachioni JY, Montagnani V, et al. (2012) Hedgehog-GLI signaling drives self-renewal and tumorigenicity of human melanoma-initiating cells. Stem Cells 30: 1808–1818.

43. Ibrahim EE, Babaei-Jadidi R, Saadeddin A, Spencer-Dene B, Hossaini S, et al. (2012) Embryonic NANOG activity defines colorectal cancer stem cells and modulates through AP1- and TCF-dependent mechanisms. Stem Cells 30: 2076–2087.

44. Noh KH, Kim BW, Song KH, Cho H, Lee YH, et al. (2012) Nanog signaling in cancer promotes stem-like phenotype and immune evasion. J Clin Invest 122: 4077–4093.

45. Siu MK, Wong ES, Kong DS, Chan HY, Jiang L, et al. (2013) Stem cell transcription factor NANOG controls cell migration and invasion via dysregulation of E-cadherin and FoxJ1 and contributes to adverse clinical outcome in ovarian cancers. Oncogene 32: 3500–3509.

46. Zhang J, Espinoza LA, Kinders RJ, Lawrence SM, Pfister TD, et al. (2013) NANOG modulates stemness in human colorectal cancer. Oncogene 32: 4397–4405.

47. Kalbermatten DF, Schaakxs D, Kingham PJ, Wiberg M (2011) Neurotrophic activity of human adipose stem cells isolated from deep and superficial layers of abdominal fat. Cell Tissue Res 344: 251–260.

48. Zhang X, Neganova I, Przyborski S, Yang C, Cooke M, et al (2009) A role for NANOG in G1 to S transition in human embryonic stem cells through direct binding of CDK6 and CDC25A. J Cell Biol 184: 67–82.

49. Pereira L, Yi F, Merrill BJ (2006) Repression of Nanog gene transcription by Tcf3 limits embryonic stem cell self-renewal. Mol Cell Biol 26: 7479–7491.

50. Storm MP, Bone HK, Beck CG, Bourillot PY, Schreiber V, et al. (2007) Regulation of Nanog expression by phosphoinositide 3-kinase-dependent signaling in murine embryonic stem cells. J Biol Chem 282: 6265–6273.

51. Chan KK, Zhang J, Chia NY, Chan YS, Sim HS, et al. (2009) KLF4 and PBX1 directly regulate NANOG expression in human embryonic stem cells. Stem Cells 27: 2114–2125.

52. Eberle I, Pless B, Braun M, Dingermann T, Marschalek R (2010) Transcriptional properties of human NANOG1 and NANOG2 in acute leukemic cells. Nucleic Acids Res 38: 5384–5395.

53. Liang J, Wan M, Zhang Y, Gu P, Xin H, et al. (2008) Nanog and Oct4 associate with unique transcriptional repression complexes in embryonic stem cells. Nat Cell Biol 10: 731–739.

54. Hamazaki T, Kehoe SM, Nakano T, Terada N (2006) The Grb2/Mek pathway represses Nanog in murine embryonic stem cells. Mol Cell Biol 26: 7539–7549.

55. Kuijk EW, de Gier J, Lopes SM, Chambers I, van Pelt AM, et al. (2010) A distinct expression pattern in mammalian testes indicates a conserved role for NANOG in spermatogenesis. PLoS One 5: e10987.

56. Hatano SY, Tada M, Kimura H, Yamaguchi S, Kono T, et al. (2005) Pluripotential competence of cells associated with Nanog activity. Mech Dev 122: 67–79.

57. Kim JS, Kim J, Kim BS, Chung HY, Lee YY, et al. (2005) Identification and functional characterization of an alternative splice variant within the fourth exon of human nanog. Exp Mol Med 37: 601–607.

58. Wu Q, Chen X, Zhang J, Loh YH, Low TY, et al. (2006) Sall4 interacts with Nanog and co-occupies Nanog genomic sites in embryonic stem cells. J Biol Chem 281: 24090–24094.

59. Torres J, Watt FM (2008) Nanog maintains pluripotency of mouse embryonic stem cells by inhibiting NFkappaB and cooperating with Stat3. Nat Cell Biol 10: 194–201.

60. Zaehres H, Lensch MW, Daheron L, Stewart SA, Itskovitz-Eldor J, et al. (2005) High-efficiency RNA interference in human embryonic stem cells. Stem Cells 23: 299–305.

61. Jung SY, Malovannaya A, Wei J, O'Malley BW, Qin J (2005) Proteomic analysis of steady-state nuclear hormone receptor coactivator complexes. Mol Endocrinol 19: 2451–2465.

62. Gorini G, Ponomareva O, Shore KS, Person MD, Harris RA, et al. (2010) Dynamin-1 co-associates with native mouse brain BKCα channels: proteomics analysis of synaptic protein complexes. FEBS Lett. 584: 845–851.

63. Palla AR, Piazzolla D, Abad M, Li H, Dominguez O, et al. (2013) Reprogramming activity of NANOGP8, a NANOG family member widely expressed in cancer. Oncogene Jun 10. doi: 10.1038/onc.2013.196. [Epub ahead of print].

64. Badeaux MA, Jeter CR, Gong S, Liu B, Suraneni MV, et al. (2013) In vivo functional studies of tumor-specific retrogene NanogP8 in transgenic animals. Cell Cycle 12:2395–2408.

65. Moretto-Zita M, Jin H, Shen Z, Zhao T, Briggs SP, et al. (2010) Phosphorylation stabilizes Nanog by promoting its interaction with Pin1. Proc Natl Acad USA 107: 13312–13317.

66. Ramakrishna S, Suresh B, Lim KH, Cha BH, Lee SH, et al. (2011) PEST motif sequence regulating human NANOG for proteasomal degradation. Stem Cells Dev 20: 1511–1519.

67. Persson H, Hennighausen L, Taub R, DeGrado W, Leder P (1984) Antibodies to human c-myc oncogene product: evidence of an evolutionarily conserved protein induced during cell proliferation. Science 225: 687–693.

68. Ramsay G, Evan GI, Bishop JM (1984) The protein encoded by the human proto-oncogene c-myc. Proc Natl Acad Sci USA 81: 7742–7746.

22

Molecular Mechanisms of Nanosized Titanium Dioxide–Induced Pulmonary Injury in Mice

Bing Li[1♥], **Yuguan Ze**[1♥], **Qingqing Sun**[1♥], **Ting Zhang**[2,3♥], **Xuezi Sang**[1], **Yaling Cui**[1], **Xiaochun Wang**[1], **Suxin Gui**[1], **Danlin Tan**[1], **Min Zhu**[1], **Xiaoyang Zhao**[1], **Lei Sheng**[1], **Ling Wang**[1], **Fashui Hong**[1]*, **Meng Tang**[2,3]*

1 Medical College of Soochow University, Suzhou, China, **2** Key Laboratory of Environmental Medicine and Engineering, Ministry of Education, School of Public Health, Southeast University, Nanjing, China, **3** Jiangsu key Laboratory for Biomaterials and Devices, Southeast University, Nanjing, China

Abstract

The pulmonary damage induced by nanosized titanium dioxide (nano-TiO$_2$) is of great concern, but the mechanism of how this damage may be incurred has yet to be elucidated. Here, we examined how multiple genes may be affected by nano-TiO$_2$ exposure to contribute to the observed damage. The results suggest that long-term exposure to nano-TiO$_2$ led to significant increases in inflammatory cells, and levels of lactate dehydrogenase, alkaline phosphate, and total protein, and promoted production of reactive oxygen species and peroxidation of lipid, protein and DNA in mouse lung tissue. We also observed nano-TiO$_2$ deposition in lung tissue via light and confocal Raman microscopy, which in turn led to severe pulmonary inflammation and pneumonocytic apoptosis in mice. Specifically, microarray analysis showed significant alterations in the expression of 847 genes in the nano-TiO$_2$-exposed lung tissues. Of 521 genes with known functions, 361 were up-regulated and 160 down-regulated, which were associated with the immune/inflammatory responses, apoptosis, oxidative stress, the cell cycle, stress responses, cell proliferation, the cytoskeleton, signal transduction, and metabolic processes. Therefore, the application of nano-TiO$_2$ should be carried out cautiously, especially in humans.

Editor: Min Wu, University of North Dakota, United States of America

Funding: This work was supported by the National Natural Science Foundation of China (grant numbers 81273036, 30901218, 81172697), a project funded by the Priority Academic Program Development of Jiangsu Higher Education Institutions, the Major State Basic Research Development Program of China (973 Program) (grant number 2006CB705602), National Important Project on Scientific Research of China (grant number 2011CB933404), National Natural Science Foundation of China (grant numbers 30671782, 30972504) and the National Ideas Foundation of Student of Soochow University (grant number 111028534). The funders had no role in study design, data collection and analysis, decision to publish, or preparation of the manuscript.

Competing Interests: The authors have declared that no competing interests exist.

* E-mail: Hongfsh_cn@sina.com (FH); tm@seu.ecu.cn (MT)

♥ These authors contributed equally to this work.

Introduction

Nanosized titanium dioxide (nano-TiO$_2$) particles, due to their high surface area to particle mass ratio, have been increasingly used as catalysts and are now being commercially manufactured for use in medical, diagnostic, energy, component, and cosmetic applications as opposed to bulk TiO$_2$ (micrometer-sized) [1,2]. However, concerns have been raised over the safety of nano-TiO$_2$ particles, as the toxicological effects of nano-TiO$_2$ have been demonstrated through several exposure routes, including dermal, oral, and pulmonary. Especially, following inhalation, nano-TiO$_2$ particles are internalized by clathrin-mediated endocytosis, caveolin-mediated endocytosis, and macropinocytosis by both phagocytic and non-phagocytic cells [3]. Reportedly, industrial nano-TiO$_2$ production, which includes a process that produces heavy nano-TiO$_2$ dust, increased the risk of pneumoconiosis to workers. Several reports have shown that human exposure to nano-TiO$_2$ occurs through different pathways, including inhalation and exposure of the integumentary system. The pulmonary responses induced by inhaled nanoparticles (NPs) are considered to be greater than those produced by micron-sized particles because of the increased surface area to particle mass ratio [4,5].

In vitro studies have demonstrated that both rutile and anatase nano-TiO$_2$ impaired cellular function of human dermal fibroblasts and decreased cellular area, proliferation, mobility, and ability to contract collagen, with the latter being more potent in inducing damage [6]. Animal experiments arrived at the same results regarding the relationship between nano-TiO$_2$ exposure and lung inflammation. Moreover, inhaled NPs, after deposition in the lungs, largely escaped the alveolar macrophage surveillance system and gained greater access to the pulmonary interstitium by translocation from alveolar spaces through the epithelium [7]. Liu et al. [8] reported that intratracheal administration of nano-TiO$_2$ (5 nm) led to significant increases in lactate dehydrogenase (LDH) and alkaline phosphatase (ALP) activities, infiltration of inflammatory cells, and interstitial thickening in the rat lung.

Our previous in vivo studies demonstrated that exposure to nano-TiO$_2$ induced pulmonary inflammation and apoptosis in mice, which were associated with expression levels of nuclear factor–κB, tumor necrosis factor-α, cyclooxygenase-2, nuclear factor erythroid 2-related factor 2, heme oxygenase 1, glutamate-cysteine ligase catalytic subunit, interleukin (IL)-2, -4, -6, -8, -10, -18, and -1β, cytochrome P450 1A1, NF-κB-inhibiting factor, and heat shock protein 70 in the mouse lung [9,10]. Although the above-mentioned studies clarified the toxicological effects of nano-TiO$_2$, further studies are needed to elucidate the synergistic molecular mechanisms of multiple genes activated by nano-TiO$_2$-induced pulmonary inflammation and apoptosis in animals and humans.

DNA microarrays have been used to identify gene clusters involved in the progression of pulmonary fibrosis and lung injury [11–14]. Furthermore, gene expression profiling has been performed to elucidate the toxicological effects of single-walled carbon nanotubes, nano-TiO$_2$, and C$_{60}$ fullerene particles [15–17]. In the present study, we investigated gene expression profiles of the murine lung to explore mechanisms of immune/inflammation responses, apoptosis, and oxidative stress induced by exposure to nano-TiO$_2$ for 90 consecutive days to serve as a reference for future mechanistic studies on the effects of nano-TiO$_2$ and other NPs in pulmonary toxicity to animals and humans.

Materials and Methods

Preparation and Characterization of TiO$_2$ NPs

Nanoparticulated anatase TiO$_2$ was prepared via controlled hydrolysis of titanium tetrabutoxide. The details of the synthesis and characterization of nano-TiO$_2$ have been previously described by our group [18,19]. TiO$_2$ powder was dispersed on the surface of 0.5% (w/v) hydroxypropyl methylcellulose (HPMC) K4M solution, treated ultrasonically for 15–20 min, and then mechanically vibrated for 2 or 3 min. X-ray-diffraction (XRD) patterns of TiO$_2$ NPs were obtained at room temperature with a charge-coupled device (CCD) diffractometer (Mercury 3 Versatile CCD Detector; Rigaku Corporation, Tokyo, Japan) using Ni-filtered Cu Kα radiation. The particle sizes of both the powder and the NPs suspended in 0.5% (w/v) HPMC solution after incubation for 24 h (5 mg/mL) were determined using transmission electron microscopy (TEM) (Tecnai G220; FEI Co., Hillsboro, OR, USA) operating at 100 kV. The mean particle size was determined by measuring >100 randomly sampled individual particles. XRD measurements showed that TiO$_2$ NPs exhibited an anatase structure with an average grain size of ~ 6 nm, as calculated from the broadening of the (101) XRD peak of anatase using the Scherrer's equation. TEM demonstrated that the average size of the particles suspended in HPMC solvent for 24 h was 5–6 nm. The surface area of the sample was 174.8 m^2/g. The average aggregate or agglomerate size of the nano-TiO$_2$ after incubation in 0.5% w/v HPMC solution for 24 h (5 mg/mL) was measured by dynamic light scattering using a Zeta PALS+BI-90 Plus zeta potential analyzer for nanoparticles (Brookhaven Instruments Corp., Holtsville, NY, USA) at a wavelength of 659 nm. The mean hydrodynamic diameter of nano-TiO$_2$ in HPMC solvent was 294 nm (range, 208–330 nm) and the zeta potential after 12 and 24 h of incubation was 7.57 and 9.28 mV, respectively [19].

Ethics Statement

All experiments were approved by the Animal Experimental Committee of the Soochow University (grant no.: 2111270) and performed in accordance with the National Institutes of Health Guidelines for the Care and Use of Laboratory Animals.

Animals and Treatments

One hundred and twenty female CD-1 (Imprinting Control Region) mice aged 5 weeks with an average body weight (BW) of 23±2 g were purchased from the Animal Center of Soochow University (Jiangsu, China). All mice were housed in stainless steel cages in a ventilated animal facility with a temperature maintained at 24±2°C and relative humidity of 60±10% under a 12-h light/dark cycle. Distilled water and sterilized food were available *ad libitum*. Prior to dosing, the mice were acclimated to the environment for 5 days.

Nano-TiO$_2$ powder was dispersed onto the surface of 0.5% w/v HPMC, treated ultrasonically for 30 min, and mechanically vibrated for 5 min. For the experiment, the mice were randomly divided into four groups (n = 30 each), including a control group (treated with 0.5% w/v HPMC) and three experimental groups (treated with 2.5, 5, and 10 mg/kg BW TiO$_2$ NPs, respectively). The mice were weighed and then the nano-TiO$_2$ suspensions were administered by nasal instillation every day for 90 days. All symptoms and deaths were carefully recorded daily. After the 90-day period, all mice were weighed, anesthetized with ether, and then sacrificed. Blood samples were collected from the eye vein by rapidly removing the eyeball and serum was collected by centrifuging the blood samples at 1,200×g for 10 min. The lungs were quickly removed and placed on ice and then dissected and frozen at −80°C.

Coefficients of Lung

After weighing the body and lungs, the coefficients of lung mass to BW were calculated as the ratio of lung (wet weight, mg) to BW (g).

Bronchoalveolar Lavage (BAL) Analysis

After blood collection, the mice were euthanized and the lungs from the control and treated groups were immediately lavaged twice with phosphate buffer saline (PBS). An average of >90% of the total instilled PBS volume was retrieved both times and the amounts did not differ among the groups. The resulting fluid was centrifuged at 400×g for 10 min at 4°C to separate the cells from the supernatant containing various surfactants and enzymes. The cell pellet was used for enumeration of total and differential cell counts as described by AshaRani et al. [20]. Macrophages, lymphocytes, neutrophils, and eosinophils recovered from the BALF were counted using dark field microscopy to assess the

Table 1. Body weight, relative weight of lung and titanium accumulation in the mouse lung after nasal administration with nano-TiO$_2$ for 90 consecutive days.

Index	Nano-TiO$_2$ (mg/kg BW)			
	0	2.5	5	10
Net increase of body weight (g)	20±1a	16±0.8b	11±0.55c	5±0.25d
Relative weight of lung (mg/g)	9.27±0.47a	9.67±0.48a	11.31±0.57b	14.28±0.71c
Ti content (ng/g tissue)	Not detected	65±3.25a	113±5.65b	207±10.35c

Letters indicate significant differences between groups (p<0.05). Values represent means ± SE(N = 10).

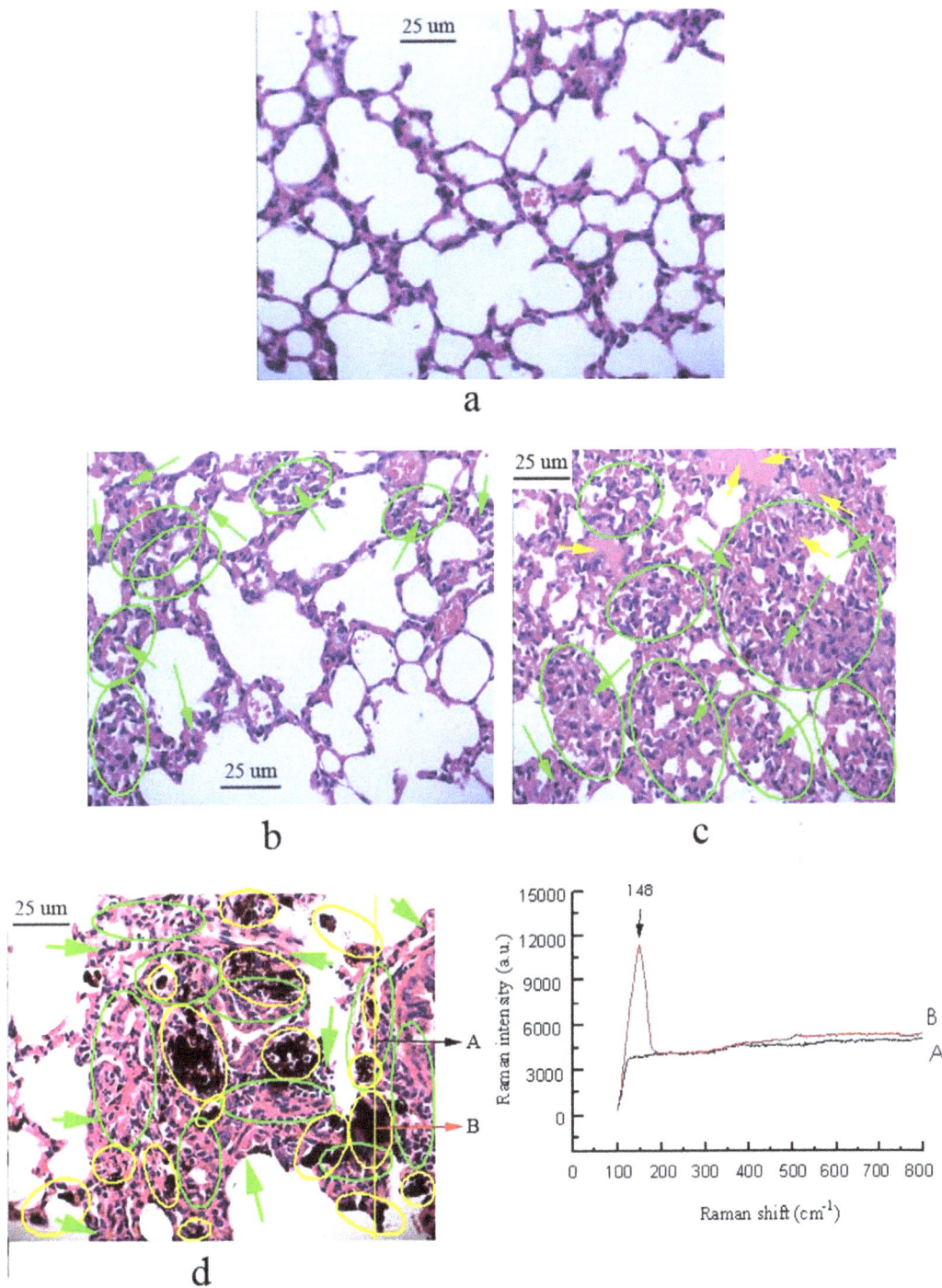

Figure 1. Histopathology of the lung tissue in ICR mice caused by nasal administration of nano-TiO₂ for 90 consecutive days. (a) Control group; (b) 2.5 mg/kg BW nano-TiO₂ group indicates inflammatory cell infiltration (green cycles) and thickening of pulmonary interstitium (green arrows); (c) 5 mg/kg BW nano-TiO₂ group indicates severs inflammatory cell infiltration (green circles), and great thickening of pulmonary interstitium (green arrows) and pulmonary edema (yellow arrows); (d) 10 mg/kg BW nano-TiO₂ group indicates severe inflammatory cell infiltration (green arrows) and great thickening of pulmonary interstitium (green arrows), yellow circles show black deposition in the lung. Arrow A spot is a representative cell that not engulfed the nano-TiO₂, while arrow B spot denotes a representative cell that loaded with nano-TiO₂. The right panels show the corresponding Raman spectra identifying the specific peak at about 148 cm-1.

extent of phagocytosis. LDH, ALP, and total protein (TP) in the cell-free lavage fluid were analyzed using an automated clinical chemical analyzer (Type 7170A; Hitachi, Ltd., Tokyo, Japan).

Lung Titanium Content Analysis

The frozen lung tissues were thawed and ~ 0.1 g samples were weighed, digested, and analyzed for titanium content. Briefly, prior to elemental analysis, the lung tissues were digested overnight with nitric acid (ultrapure grade). After adding 0.5 mL of H_2O_2,

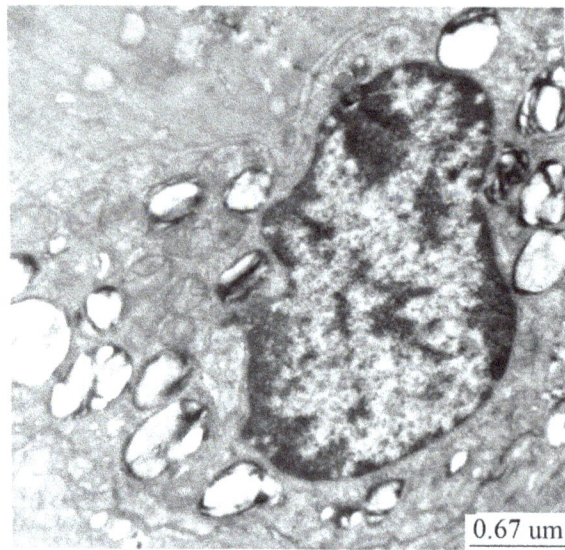

a

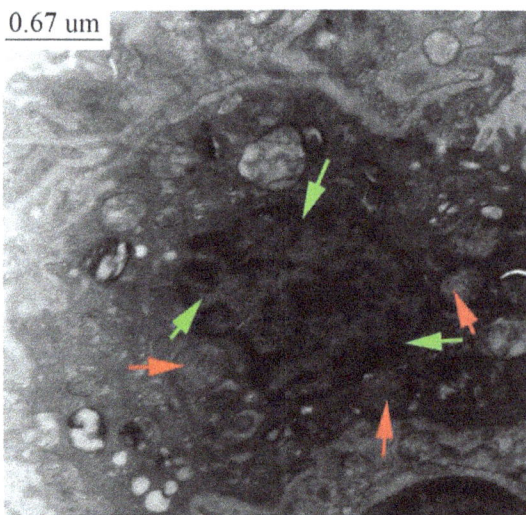

b

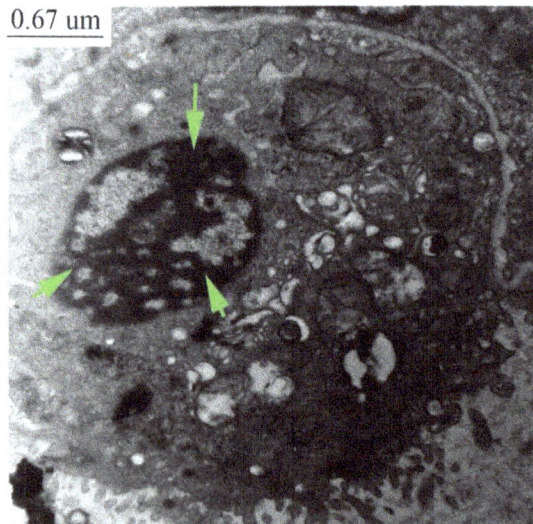

c

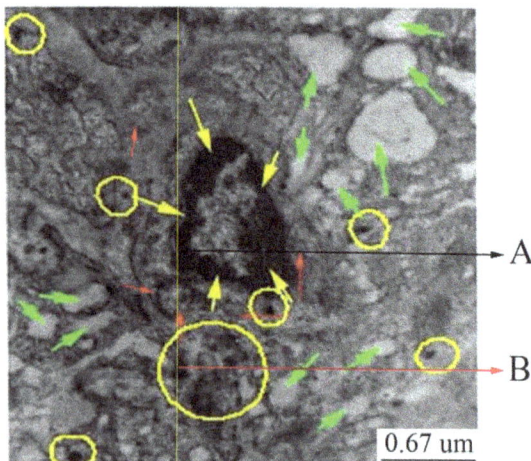

d

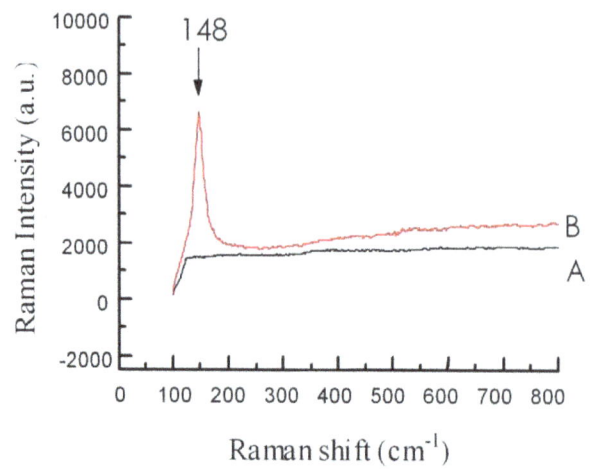

Figure 2. Ultrastructure of pneumonocyte in female mice lung caused by nasal administration of nano-TiO2 for 90 consecutive days. (a) Control: chromatin is well distributed, normal lamellar bodies; (b) 2.5 mg/kg BW nano-TiO$_2$ indicates a significant shrinkage and chromatin marginalization of the nucleus (yellow arrows), mitochondria swelling(red arrows); (c) 5 mg/kg BW nano-TiO$_2$ indicates a significant nucleus pyknosis (green arrows); (d) 10 mg/kg BW nano-TiO$_2$ indicate a significant nucleus pyknosis (yellow arrows), mitochondria swelling(red arrows) as well as evacuation of lamellar bodies (green arrows), circles show black deposition. Arrow A spot is a representative cell that not engulfed the nano-TiO$_2$, while arrow B spot denotes a representative cell that loaded with nano-TiO$_2$. (c) The right panels show the corresponding Raman spectra identifying the specific peak at about 148 cm^{-1}.

the mixed solutions were incubated at 160°C in high-pressure reaction containers in an oven until the samples were completely digested. Then, the solutions were incubated at 120°C to remove any remaining nitric acid until the solutions were colorless and clear. Finally, the remaining solutions were diluted to 3 mL with 2% nitric acid. Inductively coupled plasma-mass spectrometry (Thermo Elemental X7; Thermo Electron Co., Waltham, MA, USA) was used to determine the titanium concentration in the samples. Indium (20 ng/mL) was chosen as an internal standard element. The detection limitation of titanium was 0.074 ng/mL and data are expressed as ng/g of fresh tissue.

Histopathological Analysis

For pathological studies, all histopathological examinations were performed using standard laboratory procedures. The lungs were embedded in paraffin blocks, then sliced (5-μm thickness), and placed on glass slides. After hematoxylin–eosin staining, the stained sections were evaluated by a histopathologist unaware of the treatments using light microscopy (U-III Multi-point Sensor System; Nikon, Tokyo, Japan).

Observation of Pulmonary Ultrastructure

Lungs were fixed in fresh 0.1 M sodium cacodylate buffer containing 2.5% glutaraldehyde and 2% formaldehyde followed by a 2 h fixation period at 4°C with 1% osmium tetroxide in 50 mM sodium cacodylate (pH 7.2–7.4). Staining was performed overnight with 0.5% aqueous uranyl acetate, then the specimens were dehydrated in a graded series of ethanol (75, 85, 95, and 100%) and embedded in Epon 812 resin. Ultrathin sections were made, contrasted with uranyl acetate and lead citrate, and observed by TEM (model H600; Hitachi, Ltd., Tokyo, Japan). Lung apoptosis was determined based on the changes in nuclear morphology (e.g., chromatin condensation and fragmentation).

Confocal Raman Microscopy of Lung Sections

Raman analysis of pulmonary glass or TEM slides was performed using backscattering geometry in a confocal configuration at room temperature with an HR-800 Raman microscope system equipped with a 632.817 nm He-Ne laser (JY Co., Fort De, France). Laser power and resolution were approximately 20 mW and 0.3 cm^{-1}, respectively, while the integration time was adjusted to 1 s.

Oxidative Stress Assay

Reactive oxygen species (ROS) (O$_2^-$ and H$_2$O$_2$) production and levels of malondialdehyde (MDA), protein carbonyl (PC), and 8-hydroxy deoxyguanosine (8-OHdG) in the lung tissues were assayed using commercial enzyme-linked immunosorbent assay kits (Nanjing Jiancheng Bioengineering Institute, Jiangsu, China) according to the manufacturer's instructions.

Microarray and Data Analysis

Gene expression profiles of the lung tissues isolated from control and nano-TiO$_2$-treated mice were compared by microarray analysis using Illumina BeadChip technology (Affymetrix, Santa Clara, CA, USA). Total RNA was isolated using the Ambion Illumina RNA Amplification Kit (cat no.1755; Ambion, Inc., Austin, TX, USA) according to the manufacturer's protocol and stored at −80°C. RNA amplification has become the standard method for preparing RNA samples for array analysis [21]. Total RNA was then submitted to Biostar Genechip, Inc. (Shanghai, China) to analyze RNA quality using a bioanalyzer and complementary RNA (cRNA) was generated and labeled using the one-cycle target labeling method. cRNA from each mouse was hybridized to a single array according to standard Illumina RNA Amplification Kit protocols for all arrays.

Illumina BeadStudio data analysis software (Illumina, Inc., San Diego, CA, USA) was used to analyze the data generated in this study. This program identifies differentially expressed genes and establishes the biological significance based on the Gene Ontology

Table 2. Numbers of inflammatory cells and biochemical changes in BALF of mice after nasal administration with nano-TiO$_2$ for 90 consecutive days.

Parameter	Nano-TiO$_2$ (mg/kg BW)			
	0	2.5	5	10
Macrophages (10^4/per mouse)	11±0.55a	20±1.0b	36±1.80c	50±2.95d
Lymphocytes (10^4/per mouse)	3±0.15a	6±0.30b	11±0.55c	19±0.95d
Neutrophils (10^4/per mouse)	6±0.31a	14±0.70b	22±1.10c	36±1.80d
Eosinophils (10^4/per mouse)	5±0.25a	10±0.50b	17±0.85c	28±1.40d
LDH (unit/L)	582±29a	689±34b	778±39c	986±49d
ALP (unit/L)	100±5a	136±7b	188±9c	225±11d
TP (g/L)	26.68±1.34a	33.49±1.67b	41.96±2.10c	48.21±2.41d

Letters indicate significant differences between groups (p<0.05). Values represent means ± SE(N = 10).

Table 3. Oxidative stress in the mouse lung after nasal administration with nano-TiO₂ for 90 consecutive days.

Oxidative stress	TiO₂ NPs (mg/kg BW)			
	0	2.5	5	10
O₂⁻ (nmol/mg prot. min)	23±1.15a	30.27±1.51b	39.18±1.96c	50±2.50d
H₂O₂ (nmol/mg prot. min)	43±2.15a	61.22±3.06b	78.96±3.95c	110±5.50d
MDA (μmol/mg prot)	1.08±0.05a	1.59±0.08b	2.89±0.15c	5.15±0.26d
Carbonyl (μmol/mg prot)	0.54±0.03a	0.98±0.05b	1.85±0.09c	3.04±0.15d
8-OhdG (mg/g tissue)	0.42±0.02a	2.26±0.11b	4.25±0.21c	7.12±0.36d

Letters indicate significant differences between groups ($p<0.05$). Values represent means ± SE (N = 5).

Consortium database (http://www.geneontology.org/GO.doc.html). Differentially expressed genes were identified using the Student's t-test (two-tailed, unpaired) with a threshold of 13.0 to limit the data set to genes up-regulated or down-regulated (DiffScore >13).

Quantitative Real-time PCR

Expression levels of coagulation factor VII, hydroxymethylglutaryl CoA synthase 2, plasminogen activator - urokinase receptor, tubulin folding cofactor B, and adenosine deaminase *(Ada)* mRNA in the mouse lung tissues were determined using real-time quantitative reverse transcriptase polymerase chain reaction (qRT-PCR) [22–24]. Synthesized complimentary DNA was generated by qRT-PCR with primers designed with Primer Express Software (Applied Biosystems, Foster City, CA, USA) according to the software guidelines. PCR primer sequences are available upon request.

Statistical Analysis

All results are expressed as means ± standard error. Significant differences were examined using the Dunnett's pair-wise multiple comparison t-test using SPSS version 19 software (SPSS, Inc., Chicago, IL, USA). A p-value <0.05 was considered statistically significant.

Results

BW, Relative Lung Weight, and Titanium Accumulation

Titanium accumulation, BW, and relative lung weight of the mice are listed in Table 1. As shown, an increased nano-TiO₂ dose led to a gradual decrease in BW, whereas the relative lung weight and titanium content were significantly increased ($p<0.05$), indicating growth inhibition and lung damage in the mice. These findings were confirmed by subsequent pulmonary histological and ultrastructural observations and oxidative stress assays.

Histopathological Lung Evaluation

The histological changes in the lung specimens are shown in Fig. 1. Unexposed lung samples did not exhibit any histological changes (Fig. 1a), while those exposed to increasing nano-TiO₂ concentrations exhibited severe pathological changes, including infiltration of inflammatory cells, thickening of the pulmonary interstitium, and edema (Fig. 1b–d). In addition, we also observed significant black agglomerates in the lung samples exposed to 10 mg/kg of nano-TiO₂ (Fig. 1d). Confocal Raman microscopy further showed a characteristic nano-TiO₂ peak in the black agglomerate (148 cm⁻¹), which further confirmed the deposition of nano-TiO₂ in the lungs (see spectrum B in the Raman insets in Fig. 1d).

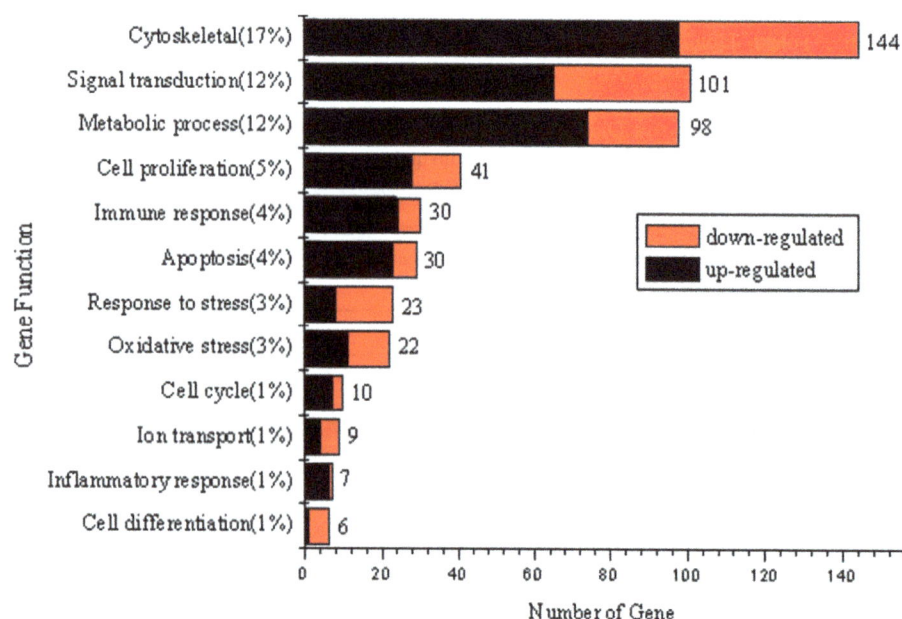

Figure 3. Functional categorization of 521 genes. Genes were functionally classied based on the ontology-driven clustering approach of PANTHER.

Table 4. Significant alteration of representative genes after nasal administration of 10 mg/kg BW TiO$_2$ NPs for 90 consecutive days.

Symbol		Gene ID	Ontology	DiffScore	Pval
Immune response	Defb4	NM_019728	defense response to bacterium	121.33	0
	H2-Oa	NM_008206	regulation of T cell differentiation	−26.34	0
Inflammatory response	Chi3l3	NM_009892	inflammatory response	93.12	0
	Alox5ap	NM_009663	leukotriene production involved in inflammatory response	20.37	0
	Il1b	NM_008361	inflammatory response	14.56	0
Apoptosis	Pdia2	NM_001081070	apoptosis	73.16	0
	Niacr1	NM_030701	apoptosis	67.22	0
	Ada	NM_007398	negative regulation of thymocyte apoptosis	28.04	0.18
	Sphk2	NM_203280	anti-apoptosis	−13.20	0
	Erbb2	NM_001003817	negative regulation of apoptosis	−14.43	0
Cell cycle	Cdkn1a	NM_001111099	cell cycle arrest	15.26	0
	Cdkn1c	NM_001161624	Cell cycle	−15.89	0
Oxidative stress	Cryab	NM_009964	oxygen and reactive oxygen species metabolic process	25.36	0
	Alkbh7	NM_025538	oxidoreductase activity	−19.72	0

Ultrastructural Changes of the Lung

Changes to the pneumonocytic ultrastructure in the mouse lung samples are presented in Fig. 2. As shown, the untreated mouse pneumonocytes (control) had no abnormal changes (Fig. 2a), whereas the pneumonocytic ultrastructure from the nano-TiO$_2$-treated groups indicated a classical morphology characteristic of apoptosis, including mitochondrial swelling, nuclear shrinkage, chromatin condensation, and evacuation of the pneumonocytic lamellar bodies (Fig. 2b–d). In addition, black deposits were observed in the pneumonocytes exposed to 10 mg/kg of nano-TiO$_2$ via TEM (Fig. 2d) and Raman signals of nano-TiO$_2$ were also exhibited via confocal Raman microscopy (Fig. 2d).

Inflammatory Cells and Biochemical Assessments in BALF

To further determine whether long-term nano-TiO$_2$ exposure induces lung inflammation, we analyzed inflammatory cell content and biochemical changes in BALF. As shown in Table 2, the numbers of macrophages, lymphocytes, neutrophils, and eosinophils, and LDH, ALP, and TP contents in the nano-TiO$_2$-exposed mice showed obvious increases with increasing nano-TiO$_2$ dose ($p<0.05$), indicating that nano-TiO$_2$ exposure caused severe inflammation and biochemical dysfunction in mice.

Table 5. Comparison of fold-difference between the control and 90 day 10 mg/kg BW dosage.

Gene	△△Ct	Fold	Microarray
F7	−1.786201	3.4490546778	2.458522
Hmgcs2	1.163294	0.44649192843	0.523989
Plaur	−1.536868	2.9016389168	1.98002
Tbcb	−0.004397	1.0030524173	1.562935
Ada	−2.280629	4.8588975045	6.867184

Oxidative Stress Analysis

The effects of nano-TiO$_2$ on the production of O$_2^-$ and H$_2$O$_2$ in mouse lung tissues are shown in Table 3. With increasing nano-TiO$_2$ dose, the rate of ROS generation in the nano-TiO$_2$-exposed groups was significantly elevated ($p<0.05$), suggesting that exposure to nano-TiO$_2$ accelerated ROS production in lung tissues.

To further demonstrate the effects of nano-TiO$_2$ on ROS generation in mouse lung tissue, the levels of lipid peroxidation (MDA), protein peroxidation (PC), and DNA damage (8-OHdG) were examined. As shown in Table 3, levels of MDA, PC, and 8-OHdG in tissues from the nano-TiO$_2$-exposed groups were markedly elevated ($p<0.05$), suggesting that nano-TiO$_2$–induced ROS accumulation led to lipid, protein, and DNA peroxidation in the lung.

Change in Gene Expression Profiles

Treatment with 10 mg/kg BW of nano-TiO$_2$ resulted in the most severe pulmonary damage and these tissues were used to detect gene expression profiles to further explore the mechanisms of pulmonary damage induced by nano-TiO$_2$. Whole-genome expression profiling using mRNAs from pulmonary tissues of vehicle control groups and those treated with 10 mg/kg BW of nano-TiO$_2$ for 90 consecutive days were analyzed with the Illumina Bead Chip. The nano-TiO$_2$-treated group was compared with the vehicle control under these criteria: DiffScore ≥13 or ≤ −13 and $p≤0.05$. The results showed that ∼ 1.16% of the total genes (521/45,000 genes with known functions) were significantly changed following nano-TiO$_2$ exposure. Of these 521 genes, 361 were up-regulated and 160 were down-regulated. The gene expression profile of the lung tissues from the TiO$_2$ NPs-treated mice was classified using the ontology-driven clustering algorithm included with the PANTHER Gene Expression Analysis Software (www.pantherdb.org/). The 521 genes were closely involved in immune responses, inflammatory responses, apoptosis, oxidative stress, metabolic processes, stress responses, signal transduction, cell proliferation, the cytoskeleton, cell differentiation, cell cycling, and so on (Fig. 3), whereas the functions of another 327 genes were unknown. Genes related to immune responses, inflammatory

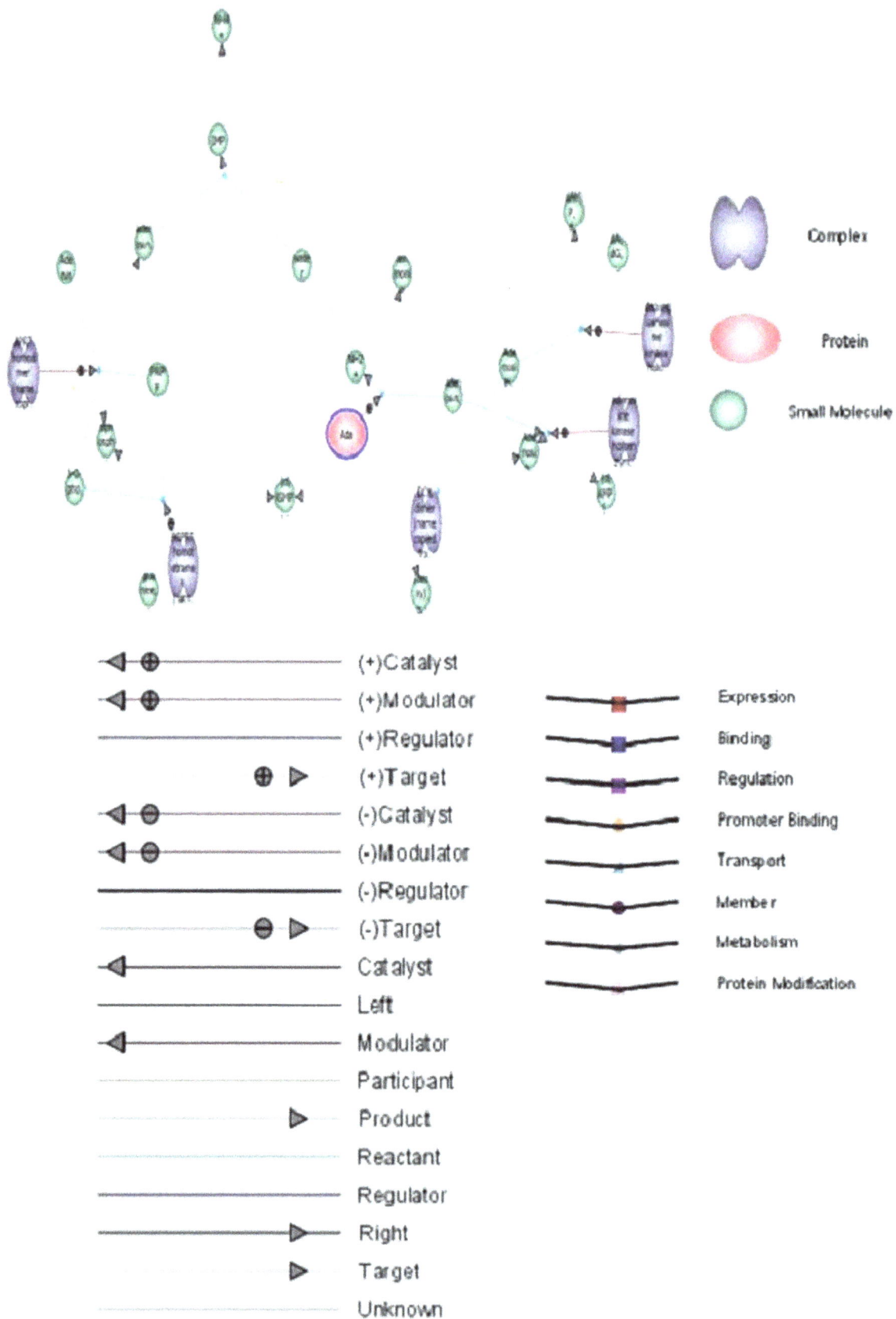

Figure 4. Ada network pathway obtained from network analysis of differentially expressed genes. Gene Spring software was used to construct and visualize molecular interaction networks.

responses, apoptosis, oxidative stress, and the cell cycle are listed in Table 4 (representative genes) and Table S1 (all data).

qRT-PCR

To verify the accuracy of the microarray analysis, five genes that demonstrated significantly different expression patterns were further evaluated by qRT-PCR due to their association with apoptosis, cell differentiation, blood coagulation, and the cytoskeleton. The qRT-PCR analysis of all five genes displayed expression patterns comparable with the microarray data (i.e., either up- or down-regulation; Table 5).

Discussion

The results of the present study indicated that nasal administration of 2.5, 5, and 10 mg/kg of nano-TiO_2 for 90 consecutive days induced BW reduction, increased relative lung mass, nano-TiO_2 deposition, (Table 1, Figs 1d, 2d), pulmonary inflammation, thickening of pulmonary interstitium, edema (Fig. 1), and pneumonocytic apoptosis (Fig. 2) in mouse lung tissues coupled with biochemical dysfunction, marked by increased LDH, ALP, and TP levels in the BALF (Table 2), and severe oxidative stress, marked by significant production of $O_2^{\cdot-}$ and H_2O_2, and peroxidation of lipids, proteins, and DNA (Table 3). Furthermore, nano-TiO_2 exposure significantly increased the influx of inflammatory cells, including macrophages, lymphocytes, neutrophils, and eosinophils, in the BALF (Table 2), further supporting the assertion that nano-TiO_2 exposure induced pulmonary inflammation. The pulmonary injuries and oxidative stress caused by nano-TiO_2 exposure may be involved in impaired immune function and antioxidant capacity in mice and, thus, may be associated with altered gene expression in lung tissue. To elucidate the molecular mechanisms of lung damage and identify specific biomarkers induced by nano-TiO_2 exposure, RNA microarray analysis of mouse lung tissue was performed to establish a global gene expression profile and identify toxicity-response genes in mice following exposure to 10 mg/kg BW of nano-TiO_2 for 90 consecutive days. Our analysis indicated that the expression levels of 847 genes were significantly changed and 521 of these genes were involved in immune responses, inflammatory responses, apoptosis, oxidative stress, metabolic processes, stress responses, signal transduction, cell proliferation, the cytoskeleton, cell differentiation, and the cell cycle.

In the present study, severe inflammatory responses in the lung tissue occurred due to nano-TiO_2-induced toxicity (Table 2, Fig. 2). Some studies have demonstrated that ultrafine particle exposure to the respiratory tract can induce pulmonary inflammation [9,10,25–27]. Latex nanomaterials instilled intratracheally enhanced neutrophilic lung inflammation with pulmonary vascular permeability related to LPS resulting from the activation of innate immune responses [28]. The present study was performed to assess pulmonary immune responses and toxicity in response to nasal administration of nano-TiO_2 and found that 38 genes (4.49% of 847 genes) involved in immune and inflammatory responses were significantly changed as shown by the microarray data (Table S1). Of these 38 genes, 31 were up-regulated and seven down-regulated. Beta-defensins contribute to the innate and adaptive immune responses in a role as chemoattractants, of which, beta-defensin 4 (Defb4), an antibiotic peptide which is locally regulated by inflammation [29], and the presence of H2-Oa (histocompatibility 2, O region alpha locus) in B cells may serve to focus presentation of antigens internalized by membrane immunoglobulins to increase the specificity of the immune response and avoid reactivity to self antigens [30]. Our data

showed that *Defb4* expression was increased by 121.13-fold and *H2-Oa* expression was decreased by 2.69-fold in the nano-TiO_2-exposed group (Table 4), suggesting that nano-TiO_2 induced *Defb4* expression and suppressed *H2-Oa* expression, which are both closely related to immune system impairment and inflammation generation (marked by significantly increased levels of macrophages, lymphocytes, neutrophils and eosinophils) in the mouse lung following nano-TiO_2-induced toxicity. Therefore, we suggest that *Defb4* and *H2-Oa* may be potential biomarkers of nano-TiO_2 exposure in the lung. In addition, the gene for chitinase 3-like 3 (*Chi3l3*) is characteristically expressed by alternatively activated macrophages. Previous studies have demonstrated innate *Chi3l3* expression in the lungs of infected severe combined immunodeficiency (SCID) mice [31] and eosinophils [32]. Increased Chi3l3 protein expression has been associated with inflammatory diseases, in particular with eosinophilic chemotaxis and promotion of cytokine production [33–35]. The arachidonate 5-lipoxygenase-activating protein (ALOX5AP) is involved in inflammation by mediating the activity of 5-lipoxygenase, which is a regulator of leukotriene biosynthesis, which are pro-inflammatory lipid mediators secreted by inflammatory cells [36,37]. IL-1 induces pro- and anti-inflammatory response of macrophages. The *IL-1* gene cluster contains three related genes *(IL-1A, IL-1B,* and *IL1-RN)*, which encode the proinflammatory cytokines IL-1α, IL-1β, as well as their endogenous receptor antagonist IL-1ra, respectively [38]. In the current study, nano-TiO_2 exposure resulted in significantly increased expression of the *Chi3l3, Alox5ap,* and *IL1b* genes with the DiffScores of 93.12, 20.37, and 14.56 (Table 4), respectively, indicating a pulmonary inflammatory response, which was closely related to excessive increases of inflammatory cells in the lung (Table 2). Taken together, *Chi3l3, Alox5ap,* and *Il1b* may be potential biomarkers of nano-TiO_2-induced pulmonary toxicity.

In the present study, classic morphological characteristics of apoptosis, such as mitochondrial swelling and nuclear chromatin condensation in the pneumonocytes was observed following exposure to 10 mg/kg BW of nano-TiO_2 (Fig. 1d). To further clarify the apoptotic molecular mechanisms, we analyzed microarray data and found that 31 genes (25 up-regulated and six down-regulated) were altered significantly by exposure to nano-TiO_2 (Table S1). The expression of several apoptotic mRNAs, including protein disulfide isomerase associated 2, niacin receptor 1, and *Ada* were significantly up-regulated, of which DiffScores were 73.16, 67.22, and 28.04, respectively; whereas sphingosine kinase 2 and v-erb-b2 erythroblastic leukemia viral oncogene homolog 2 were down-regulated with DiffScores of −13.20 and −14.43, respectively (Table 4). As shown in Fig. 4, specifically, the apoptotic pathway analysis showed that nano-TiO_2 regulated toxicological pathways by increasing the expression of a key factor, Ada, which is an essential enzyme of purine catabolism that is responsible for the hydrolytic deamination of adenosine and 2'-deoxyadenosine to inosine and 2'-deoxyinosine, respectively. These biochemical pathways are essential for maintaining homeostasis, as both Ada substrates have substantial signaling properties. Adenosine engages G protein–coupled receptors on the surface of target cells to evoke a variety of cellular responses, whereas 2'-deoxyadenosine is cytotoxic via mechanisms that interfere with cellular growth and differentiation or the promotion of apoptosis and inflammation [39]. Ada deficiency is a fatal autosomal recessive form of SCID, of which failure to thrive, impaired immune responses, and recurrent infections are characteristics [40,41]. Adenosine is generated in response to lung hypoxia and injury, and several studies have suggested that this signaling pathway might play an important role in chronic lung diseases, such as asthma and chronic obstructive pulmonary disease [42–44]. Therefore, increased *Ada* expression due to nano-TiO$_2$ exposure may reduce the accumulation of adenosine and 2'-deoxyadenosine in lung tissue, which in turn can cause cytoprotective or anti-inflammatory responses. *Ada* may be a potential biomarker of lung toxicity caused by nano-TiO$_2$ exposure. Since apoptosis is accompanied by altered cell cycle progression, our data suggest that 10 genes involved in the cell cycle were also significantly altered (Fig. 3 and Table S1). Of these 10 genes, seven were up-regulated and three down-regulated. For instance, cyclin-dependent kinase inhibitor (*Cdkn)1a* was increased with a DiffScore of 15.26, whereas *Cdkn1c* was reduced with a DiffScore of −15.89 (Table 4). Among the cell cycle regulatory proteins that are activated following DNA damage, CDKN1A plays essential roles in the DNA damage response by inducing cell cycle arrest, direct inhibition of DNA replication, as well as regulation of fundamental processes, such as apoptosis and transcription [45]. Excessive *Cdkn1a* expression following nano-TiO$_2$ exposure may affect DNA damage repair and promote apoptosis in the mouse lung. Since *Cdkn1c* is a cell cycle inhibitor, its role has been largely implicated as a tumor suppressor gene whose loss of function promotes tumor growth and progression [46]. Thus, inhibition of nano-TiO$_2$-induced *Cdkn1c* expression is speculated to contribute to apoptotic progression in lung tissue.

The present study suggested that nano-TiO$_2$ exposure promoted ROS production (such as O$_2^-$ and H$_2$O$_2$) and led to peroxidation of lipids, proteins, and DNA in mouse lung tissue, indicating oxidative stress, which may be associated with alterations of oxidative stress-related gene expression. Our microarray analysis showed that approximately 22 genes involved in oxidative stress were significantly changed in the nano-TiO$_2$-exposed lung (Fig. 3 and Table S1). Of these 22 genes, 11 were up-regulated and 11 down-regulated (Fig. 3 and Table S1). In this study, crystallin-alpha B (*Cryab*) was highly expressed following nano-TiO$_2$ exposure, with a DiffScore of 25.36, whereas alkylation repair homolog 7 (*Alkbh7*) was significantly suppressed, with a DiffScore of −19.72 (Table 4). Reportedly, *Cryab* expression in the retina is increased in response to oxidative stress and it has been postulated that this represents a protective mechanism against oxidative stress-induced apoptosis [47]. Elevated *Cryab* expression may increase in response to oxidative stress following nano-TiO$_2$-induced pulmonary damage. Alkbh7 is an oxidoreductase, which plays an important role in cardioprotection during ischemia/reperfusion by reducing oxidative stress [48]. In the current study, reduced *Alkbh7* expression induced by nano-TiO$_2$ exposure may cause pulmonary peroxisomal disorders and decrease antioxidative

capacity or detoxification. Therefore, *Cryab* and *Alkbh7* may be potential biomarkers of nano-TiO$_2$-induced pulmonary toxicity.

In regard to the dose selection in this study, we consulted a 1969 study from the World Health Organization, which reported a median lethal dose of TiO$_2$ of >12,000 mg/kg BW orally administered to rats. In the present study, we selected 2.5, 5, and 10 mg/kg BW of nano-TiO$_2$ and exposed mice to these concentrations every day for 90 days, which was equal to approximately 0.15–0.7 g of nano-TiO$_2$ in a human weighing 60–70 kg following such exposure. Although these doses were relatively safe, we recommend using caution for the long-term application of products containing nano-TiO$_2$ in humans.

Conclusion

After exposing mice to nano-TiO$_2$ for 90 consecutive days, depositions of nano-TiO$_2$ in pulmonary tissues and even in pneumonocytes were observed, which in turn resulted in significant infiltration of inflammatory cells, biochemical dysfunction, oxidative stress, and pneumonocytic apoptosis in mouse lung tissue. The pulmonary injuries following long-term nano-TiO$_2$ exposure may be closely associated with significant changes in the expression of genes involved in immune responses, inflammatory responses, apoptosis, oxidative stress, metabolic process, stress responses, signal transduction, cell proliferation, the cytoskeleton, cell differentiation, and cell cycle, specifically, with an increase in *Ada* expression. The obvious elevation in *Ada* expression following nano-TiO$_2$ exposure may trigger signaling cascades associated with inflammatory or apoptotic pathways. Therefore, the application of nano-TiO$_2$ should be carried out cautiously, especially in humans.

Supporting Information

Table S1 Genes of known function altered significantly after nasal administration of 10 mg/kg BW TiO$_2$ NPs for 90 consecutive days.

Author Contributions

Conceived and designed the experiments: FH BL YZ QS. Performed the experiments: FH BL YZ QS YZ TZ. Analyzed the data: FH BL YZ QS TZ XS YC XW SG DT MZ XZ LS LW MT. Contributed reagents/materials/analysis tools: YZ QS TZ XS YC XW SG. Wrote the paper: FH BL YZ QS.

References

1. Roco MS, Bainbridge WS, editors (2001) Societal implications of nanoscience and nanotechnology. National Science Foundation, NSET Workshop Report Kluwer Academic Publishers: Norwell, MA.
2. Jortner J, Rao CNR (2002) Nanostructured advanced materials: perspectives and directions. Pure Appl Chem 74: 1491–1506.
3. Thurn KT, Arora H, Paunesku T, Wu A, Brown EM, et al. (2011) Endocytosis of titanium dioxide nanoparticles in prostate cancer PC-3M cells. Nanomedicine 7: 123–130.
4. Donaldson K, Stone V, Clouter A, Renwick L, MacNee W (2001) Ultrafine particles. Occup Environ Med 58: 211–216.
5. Hoyt VW, Mason E (2008) Nanotechnology: emerging health issues. Chem Health Saf 15: 10–15.
6. Pan Z, Lee W, Slutsky L, Clark RAF, Pernodet N, et al. (2009) Adverse affects of titanium dioxide nanoparticles on human dermal fibroblasts and how to protect cells. Small 5: 511–520.
7. Oberdörster G, Oberdörster E, Oberdörster J (2005) Nanotechnology: an emerging discipline evolving from studied of ultrafine particles. Environ Health Perspect 113: 823–839.
8. Liu R, Yin LH, Pu YP, Liang GY, Zhang J, et al. (2009) Pulmonary toxicity induced by three forms of titanium dioxide nanoparticles via intra-tracheal instillation in rats. Prog Nat Sci 19, 573–579.
9. Sun QQ, Tan DL, Ze YG, Liu XR, Zhou QP, et al. (2012) Oxidative damage of lung and its protective mechanism in mice caused by long-term exposure to titanium dioxide nanoparticles. J Biomed Mater Res Part A 100(10): 2554–2562.
10. Sun QQ, Tan DL, Ze YG, Sang XZ, Liu XR, et al. (2012) Pulmotoxicological effects caused by long term titanium dioxide nanoparticules exposure in mice. J Hazard Mater 235–236: 47–53.
11. Katsuma S, Nishi K, Tanigawara K, Ikawa H, Shiojima S, et al. (2001) Molecular monitoring of bleomycin-induced pulmonary fibrosis by cDNA microarray-based gene expression profiling. Biochem Biophys Res Commun 288: 747–751.
12. McDowell SA, Gammon K, Zingarelli B, Bachurski CJ, Aronow BJ, et al. (2003) Inhibition of nitric oxide restores surfactant gene expression following nickel-induced acute lung injury. Am J Respir Cell Mol Biol 28: 188–198.
13. Kaminski N, Rosas IO (2006) Gene expression profiling as a window into idiopathic pulmonary fibrosis pathogenesis: can we identify the right target genes? Proc Am Thorac Soc 3: 339–344.
14. Studer SM, Kaminski N (2007) Towards systems biology of human pulmonary fibrosis. Proc Am Thorac Soc 4: 85–91.
15. Chen HW, Su SF, Chien CT, Lin WH, Yu SL, et al. (2006) Titanium dioxide nanoparticles induce emphysema-like lung injury in mice. FASEB J 20: 2393–2395.

16. Chou CC, Hsiao HY, Hong QS, Chen CH, Peng YW, et al. (2008) Single-walled carbon nanotubes can induce pulmonary injury in mouse model. Nano Lett 8: 437–445.

17. Fujita K, Morimoto Y, Ogami A, Myojy T, Tanaka I, et al. (2009) Gene expression profiles in rat lung after inhalation exposure to C_{60} fullerene particles. Toxicol 258: 47–55.

18. Yang P, Lu C, Hua N, Du Y (2002) Titanium dioxide nanoparticles co-doped with Fe^{3+} and Eu^{3+} ions for photocatalysis. Mater Lett 57: 794–801.

19. Hu RP, Zheng L, Zhang T, Cui YL, Gao GD, et al. (2011) Molecular mechanism of hippocampal apoptosis of mice following exposure to titanium dioxide nanoparticles. J Hazard Mater 191: 32–40.

20. AshaRani PV, Mun GLK, Hande MP, Valiyaveettil S (2009) Cytotoxicity and Ggenotoxicity of silver nanoparticles in human cells. Acs Nano 3: 279–290.

21. Kacharmina JE, Crino PB, Eberwine J (1999) Preparation of cDNA from single cells and subcellular regions. Method Enzymol 303: 13–18.

22. Ke LD, Chen Z (2000) A reliability test of standard-based quantitative PCR: exogenous vs endogenous standards. Mol. Cell Probes 14: 127–135.

23. Livak KJ, Schmittgen TD (2001) Analysis of relative gene expression data using real-time quantitative PCR and the 2(-Delta Delta C(T)) method. Methods 25: 402–408.

24. Liu WH, Saint DA (2002) Validation of a quantitative method for real time PCR kinetics. Biochem. Biophys Res Commun 294: 347–353.

25. Grassian VH, Oshaughnessy PT, Adamcakova-Dodd A, Pettibone JM, Thorne PS (2007) Inhalation exposure study of titanium dioxide nanoparticles with a primary particle size of 2 to 5 nm. Environ Health Perspect 115: 397–402.

26. Li J, Li Q, Xu J, Li J, Cai X, et al. (2007) Comparative study on the acute pulmonary toxicity induced by 3 and 20 nm TiO_2 primary particles in mice. Environ. Toxicol Pharmacol 24: 239–244.

27. Monteiller C, Tran L, MacNee W, Jones A, Miller B, et al. (2007) The pro-inflammatory effects of low-toxicity low-solubility particles, nanoparticles and fine particles, on epithelial cells in vitro: the role of surface area. Occup Environ Med 64: 609–615.

28. Inoue K, Takano H, Yanagisawa R, Koike E, Shimada A (2009) Size effects of latex nanomaterials on lung inflammation in mice. Toxicol Appl Pharmacol 234: 68–76.

29. Röhrl J, Yang D, Oppenheim JJ, Hehlgans T (2010) Human beta-defensin 2 and 3 and their mouse orthologs induce chemotaxis through interaction with CCR2. J Immunol 184: 6688–94.

30. Liljedahl M, Winqvist O, Surh CD, Wong P, Ngo K, et al. (1998) Altered antigen presentation in mice lacking H2-O. Immunity 8: 233–243.

31. Reece JJ, Siracusa MC, Scott AL (2006) Innate immune responses to lung-stage helminth infection induce alternatively activated alveolar macrophages. Infect Immun 74: 4970–4981.

32. Loke P, Gallagher I, Nair MG, Zang XX, Brombacher F, et al. (2007) Alternative activation is an innate response T Cells to be+to injury that requires CD4 sustained during chronic infection. J Immunol 179: 3926–3936.

33. Lee CG, Da Silva CA, Lee JY, Hartl D, Elias JA (2008) Chitin regulation of immune responses: an old molecule with new roles. Curr Opin Immunol 20: 684–689.

34. Owhashi M, Arita H, Hayai N (2000) Identification of a novel eosinophil chemotactic cytokine (ECF-L) as a chitinase family protein. J Biol Chem 275: 1279–1286.

35. Cai Y, Kumar RK, Zhou J, Foster PS, Webb DC (2009) Ym1/2 promotes Th2 cytokine expression by inhibiting 12/15(S)-lipoxygenase: identification of a novel pathway for regulating allergic inflammation. J Immunol 182: 5393–5399.

36. Peters-Golden M, Henderson Jr WR (2007) Leukotrienes. N Engl J Med 357: 1841–1854.

37. Kajimoto K, Shioji K, Ishida C, Iwanaga Y, Kokubo Y, et al. (2005) Validation of the association between the gene encoding 5-lipoxygenase-activating protein and myocardial infarction in a Japanese population. Circ J 69: 1029–1034.

38. El-Omar EM, Carrington M, Chow WH, McColl KE, Bream JH, et al. (2000) Interleukin-1 polymorphisms associated with increased risk of gastric cancer. Nature 404: 398–402.

39. Mohsenin A, Mi T, Xia Y, Kellems RE, Chen JF, et al. (2007) Genetic removal of the A2A adenosine receptor enhances pulmonary inflammation, mucin production, and angiogenesis in adenosine deaminase-deficient mice. Am J Physiol Lung Cell Mol Physiol 293: L753–L761.

40. Hirschorn R, Candotti F (2006) Immunodeficiency due to defects of purine metabolism. In: Ochs H, Smith C, Puck J, eds. Primary immunodeficiency diseases. Oxford, England: Oxford University Press, 169–190.

41. Albuquerque W, Gaspar HB (2004) Bilateral sensorineural deafness in adenosine deaminase-deficient severe combined immunodeficiency. J Pediatr 144: 278–80.

42. Jacobson MA, Bai TR (1997) The role of adenosine in asthma. In Purinergic Approaches in Experimental Therapeutics (Jacobson, K.A. and Jarvis, M.F., eds) 315–331, Wiley-Liss.

43. Fozard JR, Hannon JP (1999) Adenosine receptor ligands: potential as therapeutic agents in asthma and COPD. Pulm PharmacolTher 12: 111–114.

44. Blackburn MR (2003) Too much of a good thing: adenosine overload in adenosinedeaminase-deficient mice. TRENDS in Pharmacol Sci 24: 66–70.

45. Cazzalini O, Scovassi AI, Savio M, Stivala LA, Prosperi E (2010) Multiple roles of the cell cycle inhibitor p21(CDKN1A) in the DNA damage response. Mutat Res 704: 12–20.

46. Ito Y, Yoshida H, Nakano K, Kobayashi K, Yokozawa T, et al. (2002). Expression of p57/Kip2 protein in normal and neoplastic thyroid tissues. Int J Mol Med 9: 373–376.

47. Whiston EA, Sugi N, Kamradt MC, Sack C, Heimer SR, et al. (2008) αB-crystallin protects retinal during Staphylococcus aureus-induced endophthalmitis. Infect Immun 76: 1781–1790.

48. Koga K, Kenessey A, Powell SR, Sison CP, Miller EJ, et al. (2011) Macrophage migration inhibitory factor provides cardioprotection during ischemia/reperfusion by reducing oxidative stress. Antioxid Redox Signal 14: 1191–1202.

Permissions

All chapters in this book were first published in PLOS ONE, by The Public Library of Science; hereby published with permission under the Creative Commons Attribution License or equivalent. Every chapter published in this book has been scrutinized by our experts. Their significance has been extensively debated. The topics covered herein carry significant findings which will fuel the growth of the discipline. They may even be implemented as practical applications or may be referred to as a beginning point for another development.

The contributors of this book come from diverse backgrounds, making this book a truly international effort. This book will bring forth new frontiers with its revolutionizing research information and detailed analysis of the nascent developments around the world.

We would like to thank all the contributing authors for lending their expertise to make the book truly unique. They have played a crucial role in the development of this book. Without their invaluable contributions this book wouldn't have been possible. They have made vital efforts to compile up to date information on the varied aspects of this subject to make this book a valuable addition to the collection of many professionals and students.

This book was conceptualized with the vision of imparting up-to-date information and advanced data in this field. To ensure the same, a matchless editorial board was set up. Every individual on the board went through rigorous rounds of assessment to prove their worth. After which they invested a large part of their time researching and compiling the most relevant data for our readers.

The editorial board has been involved in producing this book since its inception. They have spent rigorous hours researching and exploring the diverse topics which have resulted in the successful publishing of this book. They have passed on their knowledge of decades through this book. To expedite this challenging task, the publisher supported the team at every step. A small team of assistant editors was also appointed to further simplify the editing procedure and attain best results for the readers.

Apart from the editorial board, the designing team has also invested a significant amount of their time in understanding the subject and creating the most relevant covers. They scrutinized every image to scout for the most suitable representation of the subject and create an appropriate cover for the book.

The publishing team has been an ardent support to the editorial, designing and production team. Their endless efforts to recruit the best for this project, has resulted in the accomplishment of this book. They are a veteran in the field of academics and their pool of knowledge is as vast as their experience in printing. Their expertise and guidance has proved useful at every step. Their uncompromising quality standards have made this book an exceptional effort. Their encouragement from time to time has been an inspiration for everyone.

The publisher and the editorial board hope that this book will prove to be a valuable piece of knowledge for researchers, students, practitioners and scholars across the globe.

List of Contributors

Zhengxia Liu, Yucheng Wu, Zhirui Guo, Ying Liu, Yujie Shen, Ping Zhou and Xiang Lu
Department of Geriatrics, the Second Affiliated Hospital, Nanjing Medical University, Jiangsu, China

Harold Kwok, Kyle Briggs and Vincent Tabard-Cossa
Department of Physics, University of Ottawa, Ottawa, Ontario, Canada

Rachel A. Kudgus, Annamaria Szabolcs, Jameel Ahmad Khan and Resham Bhattacharya
Department of Biochemistry and Molecular Biology, College of Medicine, Mayo Clinic, Rochester, Minnesota, United States of America

Chad A. Walden and Joel M. Reid
Department of Physiology and Biomedical Engineering, College of Medicine, Mayo Clinic, Rochester, Minnesota, United States of America

J. David Robertson
Department of Chemistry and University of Missouri Research Reactor, University of Missouri, Columbia, Missouri, United States of America

Priyabrata Mukherjee
Department of Biochemistry and Molecular Biology, College of Medicine, Mayo Clinic, Rochester, Minnesota, United States of America
Department of Physiology and Biomedical Engineering, College of Medicine, Mayo Clinic, Rochester, Minnesota, United States of America
Mayo Clinic Cancer Center, College of Medicine, Mayo Clinic, Rochester, Minnesota, United States of America

Hua Jin, Qian Liang and Xiaoping Wang
Department of Pain Management, The First Affiliated Hospital of Jinan University, Guangzhou, China

Tongsheng Chen
MOE Key Laboratory of Laser Life Science & Institute of Laser Life Science, South China Normal University, Guangzhou, China

Beatriz Veleirinho
QOPNA Research Unit, Department of Chemistry, University of Aveiro, Aveiro, Portugal
Biotechnology and Biosciences Post-Graduation Program, Federal University of Santa Catarina, Florianópolis, Brazil

Daniela S. Coelho and Paulo F. Dias
Department of Cell Biology, Embryology, and Genetics, Federal University of Santa Catarina, Florianópolis, Brazil

Marcelo Maraschin
Plant Morphogenesis and Biochemistry Laboratory, Federal University of Santa Catarina, Florianópolis, Brazil

Rúbia Pinto
Epagri, Florianopolis, Brazil

Eduardo Cargnin-Ferreira
Federal Institute of Education, Science, and Technology of Santa Catarina, Garopaba, Brazil

Ana Peixoto and José A. Souza
Department of Pediatrics, Federal University of Santa Catarina, Florianópolis, Brazil

Rosa M. Ribeiro-do-Valle
Biotechnology and Biosciences Post-Graduation Program, Federal University of Santa Catarina, Florianópolis, Brazil

José A. Lopes-da-Silva
QOPNA Research Unit, Department of Chemistry, University of Aveiro, Aveiro, Portugal

Athanasios A. Koutinas, Vasilios Sypsas, Panagiotis Kandylis, Andreas Michelis and Argyro Bekatorou
Food Biotechnology Group, Department of Chemistry, University of Patras, Patras, Greece

Yiannis Kourkoutas
Department of Molecular Biology, Democritus University of Thrace, Alexandroupolis, Greece

Christos Kordulis
Group of Catalysis and Interfacial Chemistry for Environmental Applications, University of Patras, Patras, Greece
Institute of Chemical Engineering and High Temperature Chemical Processes (FORTH/ICE-HT), Patras, Greece

Alexis Lycourghiotis
Group of Catalysis and Interfacial Chemistry for Environmental Applications, University of Patras, Patras, Greece

Ibrahim M. Banat5, Poonam Nigam and Roger Marchant
School of Biomedical Sciences, University of Ulster, Coleraine, United Kingdom

Myrsini Giannouli and Panagiotis Yianoulis
Department of Physics, University of Patras, Patras, Greece

Lily L. Wong and Quentin N. Pye
Department of Ophthalmology, University of Oklahoma Health Sciences Center, College of Medicine, and Dean McGee Eye Institute, Oklahoma City, Oklahoma, United States of America

Suzanne M. Hirst and Christopher M. Reilly
Biomedical Sciences and Pathobiology, Virginia Polytechnic Institute and State University, and Via College of Osteopathic Medicine, Blacksburg, Virginia, United States of America

Sudipta Seal
Advanced Materials Processing Analysis Center, Mechanical Materials Aerospace Engineering, Nanoscience and Technology Center, University of Central Florida, Orlando, Florida, United States of America

James F. McGinnis
Department of Ophthalmology, University of Oklahoma Health Sciences Center, College of Medicine, and Dean McGee Eye Institute, Oklahoma City, Oklahoma, United States of America
Department of Cell Biology and Oklahoma Center for Neuroscience, University of Oklahoma Health Sciences Center, Graduate College, Oklahoma City, Oklahoma, United States of America

Brian J. Roxworthy
Department of Electrical and Computer Engineering, University of Illinois at Urbana-Champaign, Urbana, Illinois, United States of America

Michael T. Johnston, Randy H. Ewoldt and Kimani C. Toussaint Jr.
Department of Mechanical Science and Engineering, University of Illinois at Urbana-Champaign, Urbana, Illinois, United States of America

Felipe T. Lee-Montiel and Princess I. Imoukhuede
Department of Bioengineering, University of Illinois at Urbana-Champaign, Urbana, Illinois, United States of America

Haider Sami and Ashok Kumar
Department of Biological Sciences and Bioengineering, Indian Institute of Technology Kanpur, Kanpur, Uttar Pradesh, India

Auhin K. Maparu and Sri Sivakumar
Unit of Excellence on Soft Nanofabrication, Department of Chemical Engineering, Indian Institute of Technology Kanpur, Kanpur, Uttar Pradesh, India

Jia-Yuan Shi and Gen-Tao Zhou
Key Laboratory of Crust-Mantle Materials and Environments, Chinese Academy of Sciences, School of Earth and Space Sciences, University of Science and Technology of China, Hefei, People's Republic of China

Qi-Zhi Yao and Xi-Ming Li
School of Chemistry and Materials, University of Science and Technology of China, Hefei, People's Republic of China

Sheng-Quan Fu
Hefei National Laboratory for Physical Sciences at Microscale, University of Science and Technology of China, Hefei, People's Republic of China

Shayesteh Sepehr, Azadeh Rahmani-Badi, Hamta Babaie-Naiej and Mohammad Reza Soudi
Department of Biology, Alzahra University, Tehran, Iran

Andrew S. Mikhail
Leslie Dan Faculty of Pharmacy, University of Toronto, Toronto, Ontario, Canada
Institute of Biomaterials and Biomedical Engineering, University of Toronto, Toronto, Ontario, Canada

Sina Eetezadi
Leslie Dan Faculty of Pharmacy, University of Toronto, Toronto, Ontario, Canada

Christine Allen
Leslie Dan Faculty of Pharmacy, University of Toronto, Toronto, Ontario, Canada
Institute of Biomaterials and Biomedical Engineering, University of Toronto, Toronto, Ontario, Canada
Spatio-Temporal Targeting and Amplification Radiation Response (STTARR) Innovation Centre, Toronto, Ontario, Canada

Reid A. Roberts
Department of Microbiology and Immunology, University of North Carolina, Chapel Hill, North Carolina, United States of America

Tammy Shen
Eshelman School of Pharmacy, University of North Carolina, Chapel Hill, North Carolina, United States of America

Irving C. Allen
Department of Biomedical Sciences and Pathobiology, Virginia-Maryland Regional
College of Veterinary Medicine, Virginia Polytechnic Institute and State University, Blacksburg, Virginia, United States of America

Warefta Hasan
Department of Chemistry, University of North Carolina, Chapel Hill, North Carolina, United States of America

Joseph M. DeSimone
Eshelman School of Pharmacy, University of North Carolina, Chapel Hill, North Carolina, United States of America
Department of Chemistry, University of North Carolina, Chapel Hill, North Carolina, United States of America
Lineberger Comprehensive Cancer Center, University of North Carolina, Chapel Hill, North Carolina, United States of America
Department of Biochemistry and Biophysics, University of North Carolina, Chapel Hill, North Carolina, United States of America
Carolina Center of Cancer Nanotechnology Excellence, University of North Carolina, Chapel Hill, North Carolina, United States of America
Department of Chemical and Biomolecular Engineering, North Carolina State University, Raleigh, North Carolina, United States of America
Department of Pharmacology, University of North Carolina, Chapel Hill, North Carolina, United States of America
Institute for Advanced Materials, University of North Carolina, Chapel Hill, North Carolina, United States of America
Institute for Nanomedicine, University of North Carolina, Chapel Hill, North Carolina, United States of America
Sloan-Kettering Institute for Cancer Research, Memorial Sloan-Kettering Cancer Center, New York, New York, United States of America

Jenny P. Y. Ting
Department of Microbiology and Immunology, University of North Carolina, Chapel Hill, North Carolina, United States of America
Lineberger Comprehensive Cancer Center, University of North Carolina, Chapel Hill, North Carolina, United States of America

Christos Tzitzilonis, Senyon Choe and Roland Riek
Laboratory of Physical Chemistry, Swiss Federal Institute of Technology, ETH-Hönggerberg, Zürich, Switzerland
Structural Biology Laboratory, The Salk Institute, La Jolla, California, United States of America

Cédric Eichmann
Laboratory of Physical Chemistry, Swiss Federal Institute of Technology, ETH-Hönggerberg, Zürich, Switzerland

Innokentiy Maslennikov
Structural Biology Laboratory, The Salk Institute, La Jolla, California, United States of America

Christine Anna Muth
Department of New Materials and Biosystems, Max Planck Institute for Intelligent Systems, Stuttgart, Germany
Department of Biophysical Chemistry, University of Heidelberg, Heidelberg, Germany

Carolin Steinl and Gerd Klein
Section for Transplantation Immunology and Immunohematology, Center for Medical Research, University of Tübingen, Tübingen, Germany

Cornelia Lee-Thedieck
Department of New Materials and Biosystems, Max Planck Institute for Intelligent Systems, Stuttgart, Germany
Department of Biophysical Chemistry, University of Heidelberg, Heidelberg, Germany
Institute of Functional Interfaces, Karlsruhe Institute of Technology (KIT), Eggenstein-Leopoldshafen, Germany

Sourabh Dwivedi and Abdulaziz A. AlKhedhairy
Department of Zoology, College of Science, King Saud University, Riyadh, Saudi Arabia

Maqusood Ahamed
King Abdullah Institute for Nanotechnology, King Saud University, Riyadh, Saudi Arabia

Javed Musarrat
Department of Agricultural Microbiology, Faculty of Agricultural Sciences, Aligarh Muslim University, Aligarh, India

Nahid N. Jetha and Andre Marziali
Department of Physics and Astronomy, University of British Columbia, Vancouver, British Columbia, Canada

Valentyna Semenchenko
National Institute for Nanotechnology, Edmonton, Alberta, Canada

David S. Wishart
National Institute for Nanotechnology, Edmonton, Alberta, Canada
Departments of Computing Science and Biological Sciences, University of Alberta, Edmonton, Alberta, Canada

Neil R. Cashman
Brain Research Centre, University of British Columbia, Vancouver, British Columbia, Canada

Sravan Kumar Patel, Yang Zhang and Jelena M. Janjic
Graduate School of Pharmaceutical Sciences, Mylan School of Pharmacy, Duquesne University, Pittsburgh, Pennsylvania, United States of America

John A. Pollock
Department of Biological Sciences, Bayer School of Natural and Environmental Sciences, Duquesne University, Pittsburgh, Pennsylvania, United States of America

David E. Jones and John Hurdle
Department of Biomedical Informatics, University of Utah, Salt Lake City, Utah, United States of America

Julio C. Facelli and Sean Igo
Department of Biomedical Informatics, University of Utah, Salt Lake City, Utah, United States of America
Center for High Performance Computing, University of Utah, Salt Lake City, Utah, United States of America

Seungmoon Jung, Minji Bang, Byung Sun Kim and Daejong Jeon
Department of Bio and Brain Engineering, Korea Advanced Institute of Science and Technology (KAIST), Daejeon, Republic of Korea

Sungmun Lee
Department of Biomedical Engineering, Khalifa University of Science, Technology, and Research, Abu Dhabi, United Arab Emirates

Nicholas A. Kotov
Department of Chemical Engineering, University of Michigan, Ann Arbor, Michigan, United States of America

Bongsoo Kim
Department of Chemistry, Korea Advanced Institute of Science and Technology (KAIST), Daejeon, Republic of Korea

Bigang Liu, Mark D. Badeaux, Grace Choy, Dhyan Chandra, Irvin Shen, Collene R. Jeter, Kiera Rycaj, Can Liu, Yueping Chen and Jianjun Shen
Department of Molecular Carcinogenesis, University of Texas M.D Anderson Cancer Center, Science Park, Smithville, Texas, United States of America

Chia-Fang Lee and Maria D. Person
College of Pharmacy, University of Texas, Austin, Texas, United States of America

Sung Yun Jung and Jun Qin
Department of Biochemistry, Baylor College of Medicine, Houston, Texas, United States of America

Dean G. Tang
Department of Molecular Carcinogenesis, University of Texas M.D Anderson Cancer Center, Science Park, Smithville, Texas, United States of America
Cancer Stem Cell Institute, Research Center for Translational Medicine, East Hospital, Tongji University School of Medicine, Shanghai, China

Bing Li, Yuguan Ze, Qingqing Sun, Xuezi Sang, Yaling Cui, Xiaochun Wang, Suxin Gui, Danlin Tan1, Min Zhu, Xiaoyang Zhao, Lei Sheng, Ling Wang and Fashui Hong
Medical College of Soochow University, Suzhou, China

Ting Zhang and Meng Tang
Key Laboratory of Environmental Medicine and Engineering, Ministry of Education, School of Public Health, Southeast University, Nanjing, China
Jiangsu key Laboratory for Biomaterials and Devices, Southeast University, Nanjing, China

Index